Using robots in hazardous environments

Using robots in hazardous environments

Landmine detection, de-mining and other applications

Edited by
Y. Baudoin and Maki K. Habib

Published in affiliation with the CLAWAR Association

WOODHEAD PUBLISHING

Oxford   Cambridge   Philadelphia   New Delhi

Published by Woodhead Publishing Limited, 80 High Street, Sawston,
Cambridge CB22 3HJ, UK
www.woodheadpublishing.com

Woodhead Publishing, 1518 Walnut Street, Suite 1100, Philadelphia,
PA 19102-3406, USA

Woodhead Publishing India Private Limited, G-2, Vardaan House, 7/28 Ansari Road,
Daryaganj, New Delhi – 110002, India
www.woodheadpublishingindia.com

First published 2011, Woodhead Publishing Limited
© Woodhead Publishing Limited, 2011
The authors have asserted their moral rights.

British Library Cataloguing in Publication Data
A catalogue record for this book is available from the British Library.

Library of Congress Cataloging in Publication Data
A catalog record for this book is available from the Library of Congress.

ISBN 978-0-08-101503-2 (print)
ISBN 978-0-85709-020-1 (online)

The publishers' policy is to use permanent paper from mills that operate a sustainable
forestry policy, and which has been manufactured from pulp which is processed using
acid-free and elemental chlorine-free practices. Furthermore, the publishers ensure that
the text paper and cover board used have met acceptable environmental accreditation
standards.

Typeset by RefineCatch Limited, Bungay, Suffolk, UK
Printed by TJI Digital, Padstow, Cornwall, UK

Contents

v

Contributor contact details

(* = main contact)

Chapter 1

Y. Baudoin*
Royal Military Academy
30 Avenue de la Renaissance
1000 Brussels
Belgium
Email: Yvan.baudoin@rma.ac.be

M. K. Habib
School of Sciences and Engineering
The American University in Cairo
113 Kasr El Aini Street
P.O. Box 2511
Cairo, 11511
Egypt
Email: maki@aucegypt.edu

I. Doroftei*
Faculty of Mechanical Engineering
Gh. Asachi Technical University of
    Lasi
B-dul Dimitri Mangeron no. 61–63
Lasi, cod 700050
Romania
Email: idorofte@mail.tuiasi.ro; ioan_
    doroftei@yahoo.com

Chapter 2

C. Parra*, C. Otalora and A. Forero
Pontificia Universidad Javeriana
Carrera 7ª No. 40–62
Bogotá
Colombia
Email: carlos.parra@javeriana.edu.co

M. Devy
Centre National de la Recherche
    Scientifique
Laboratory for Analysis and
    Architecture of Systems
7 Avenue du Colonel Roche
F-31077 Toulouse
France
Email: michel@laas.fr

Chapter 3

P. Santana* and L. Correia
Laboratory of Agent Modelling
    (LabMAg)
University of Lisbon
1749-016 Lisboa
Portugal
Email: pedro.santana@di.fc.ul.pt

J. Barata
UNINOVA
New University of Lisbon
2829-516 Monte de Caparica
Portugal

Chapter 4

E. E. Cepolina*
Snail Aid – Technology for
    Development
Via C. Cabella 10/12
16122 Genova
Italy
Email: emacepo@dimec.unige.it

M. Zoppi
Faculty of Engineering
Area scientifico-disciplinare 09
University of Genova
Settore ING-IND/13
Italy

Chapter 5

V. G. Gradetsky
Institute for Problems in Mechanics
Prospect Vernadskogo, 101-1
119526 Moscow
Russia
Email: gradet@mail.ru

Chapter 6

L. Nomdedeu*, J. Sales, R. Marín
    and E. Cervera
Computer Engineering and Science
University Jaume I (UJI)
1207 Castellón
Spain
E-mail: leo.nomdedeu@gmal.com

J. Saez
Materials and Engineering Research
    Institute
Sheffield Hallam University
City Campus
Sheffield
S1 1WB
UK

Chapter 7

S. Larionova
PromAutomation Ltd.
d. 17, c. 5 lit. A
Pirogovskaya emb.
194044 Saint-Petersburg
Russia
Email: ls@pa.ru

A. T. de Almeida and L. Marques
Institute of Systems and Robotics
Deparment of Electrical Engineering
University of Coimbra – Polo II
3030-290 Coimbra
Portugal
Email: adealmeida@isr.uc.pt

Chapter 8

P. Druyts, Y. Yvinec and M. Acheroy
Department CISS/ELEC
Royal Military Academy
30 Avenue de la Renaissance
1000 Brussels
Belgium
Email: pascal.druyts@rma.ac.be;
    yvinec@elec.rma.ac.be; marc.
    acheroy@rma.ac.be

Chapter 9

G. El-Qady*, A. Mohamed, M.
    Metwaly and M. Atya
National Research Institute of
    Astronomy and Geophysics
11722 Helwan
Egypt
Email: gadosan@nriag.sci.eg

Chapter 10

I. Y.-H. Gu*
Department of Signals and Systems
Chalmers University of Technology
Gothenburg
Sweden
Email: irenegu@chalmers.se

T. Tjahjadi
School of Engineering
University of Warwick
Coventry
UK

Chapter 11

S. A. Berrabah
Mechanical Department
Royal Military Academy, Belgium
Avenue de la Renaissance 30
1000 Brussels
Belgium
Email: sidahmed.berrabah@rma.ac.uk

Chapter 12

T. Fukuda
Department of Micro-Nano Systems
    Engineering
Nagoya University
Furo-cho
Chikusa-ku
Nagoya, 464-8603
Japan
Email: fukuda@mein.nagoya-u.ac.jp

Y. Hasegawa
Department of Intelligent
    Information Technology
University of Tsukuba
Tennoudia 1-1-1
Tsukuba 305-8573
Japan
Email: hase@esys.tsukuba.ac.jp

K. Kosuge
System Robotics Laboratory
Department of Bioengineering and
    Robotics
Tohoku University
Aoba-yama 6-6-01
Sendai 980-8579
Japan

K. Komoriya
Intelligent Systems Research
    Institute
National Institute of Advanced
    Industrial Science and
    Technology (AIST)
Tsukuba Central 2
1-1-1 Umezono
Tsukuba
Ibaraki 305-8568
Japan

F. Kitagawa
Mitsui Engineering and Shipbuilding
    Co. Ltd. (MES)
16 Raffles Quay
41-02 Hong Leong Building
Singapore 048581-6220 4065
Japan

T. Ikegami
TADANO Ltd.
Ko-34
Shindon-cho
Takamatsu
Kagawa 761-0185
Japan

Chapter 13

M. K. Habib*
School of Sciences and Engineering
The American University in Cairo
113 Kasr El Aini Street
P.O. Box 2511
Cairo, 11511
Egypt
Email: maki@aucegypt.edu

Y. Baudoin
Royal Military Academy
30 Avenue de la Renaissance
1000 Brussels
Belgium
Email: Yvan.baudoin@rma.ac.be

Chapter 14

Š. Havlík
Institute of Informatics
Slovak Academy of Sciences
Ďumbierska I
97411 Banská Bystrica
Slovakia
Email: havlik@savbb.sk

Chapter 15

C. Armbrust*, T. Braun, T. Föhst,
   M. Proetzsch, A. Renner,
   B. H. Schäfer and K. Berns
Robotics Research Lab
Department of Computer Sciences
University of Kaiserslautern
P. O. Box 3049
67653 Kaiserslautern
Germany
Email: armbrust@cs.uni-kl.de; braun@
   cs.uni-kl.de; foehst@cs.uni-kl.de;
   proetzsch@cs.uni-kl.de; renner@
   cs.uni-kl.de; b_schaef@cs.uni-kl.de;
   berns@cs.uni-kl.de

Chapter 16

G. Kowalski*, J. Bedkowski,
   P. Kowalski and A. Maslowski
Research Institute for Automation
   and Measurements PIAP
Al. Jerozolimskie 202
02-486 Warsaw
Poland
Email: piotek.kowalski@gmail.com

Chapter 17

A. Pajaziti*
Str. Zagrebi No. 13
10000 Prishtina
Kosovo
Email: apajazit@uni-pr.edu

I. Gojani, Sh. Buza and A. Shala
Mechanical Engineering Faculty
University of Prishtina
Kosova
Email: apajazit@uni-pr.edu;
   ismajlgojani@hotmail.com;
   ahmetshala_2000@yahoo.com;
   shabanbuza@fim.uni-pr.edu

Chapter 18

A. Pajaziti*, I. Gojani, Sh. Buza and
   A. Shala
Mechanical Engineering Faculty
University of Prishtina
Kosova
Email: apajazit@uni-pr.edu;
   ismajlgojani@hotmail.com;
   ahmetshala_2000@yahoo.com;
   shabanbuza@fim.uni-pr.edu

G. Capi
Faculty of Information Engineering
Fukuoka Institute of Technology
Fukuoka
Japan
E-mail: capi@fit.ac.jp

Chapter 19

A. Abbas
Department of Mechanical
    Engineering
British University in Egypt
Cairo-Suez Road
P. O. Box 43
El Sherouk City 11837
Egypt
Email: aabbas@bue.edu.eg

Chapter 20

G. De Cubber* and D. Doroftei
Royal Military Academy
Department of Mechanics (MECA)
30 Avenue de la Renaissance
1000 Brussels
Belgium
Email: geert.de.cubber@rma.ac.be;
    daniela.doroftei@rma.ac.be

Chapter 21

V. G. Gradetsky*, V. B. Veshnikov
    and V. G. Chashchukin
Ishlinksky's Institute for Problems in
    Mechanics of the Russian
    Academy of Sciences (IPMech
    RAS)
Prospect Vernadskogo, 101-1
119526 Moscow
Russia
Email: gradet@mail.ru

Chapter 22

L. Nomdedeu*, J. Sales, R. Marín
    and E. Cervera
Computer Engineering and
    Science
University Jaume I (UJI)
1207 Castellón
Spain
Email: leo.nomdedeu@gmail.com

Chapter 23

O. Çayirpunar*, V. Gazi and
    B. Tavli
Department of Computer
    Engineering
TOBB University of Economics and
    Technology
Sögütözü Cad. No:43
Sögütözü
Ankara, 06560
Turkey
Email: ocayirpunar@etu.edu.tr

E. Cervera
Department of Computer Science
    and Engineering
Jaume-I University
Campus Riu Sec
12071 Castelló
Spain
E-mail: ecervera@occ.uji.es

U. Witkowski
Heinz Nixdorf Institute
University of Paderborn
Fuerstenallee 11
33102 Paderborn
Germany
Email: witkowski@hni.upb.de

J. Penders
Materials and Engineering Research
    Institute
Sheffield Hallam University
Sheffield S1 1WB
UK
E-mail: j.penders@shu.ac.uk

Chapter 24

U. Witkowski*, S. Herbrechtsmeier
    and M. El-Habbal
Heinz Nixdorf Institute
University of Paderborn
Fuerstenallee 11
33102 Paderborn
Germany
Email: witkowski@hni.upb.de

Chapter 25

M. Defoort*
LAMIH, FRE CNRS 3304
Université de Valenciennes et du
    Hainaut-Cambresis
59313 Valenciennes
France
Email: Michael.defoort@univ-
    valenciennes.fr

T. Floquet, A. Kokosy and
    W. Perruquetti
LAGIS, FRE CNRS 3303
Ecole Centrale de Lille
BP 48
Cité Scientifique
59651 Villeneuve-d'Ascq
France
Email: thierry.floquet@ec-lille.fr;
    annemarie.kokosy@isen.fr; wilfrid.
    perruquetti@ec-lille.fr

J. Palos
ISEN 41 bvd Vauban
59046 Lille Cedex
France
Email: Jorge.palos@isen.fr

Chapter 26

J. Bedkowski* and A. Maslowski
Faculty of Materials Science and
    Engineering
Warsaw University of Technology
ul. Wołoska 141
02-507 Warszawa
Poland
Email: janusbedkowski@gmail.com

Chapter 27

Y. Atas*, O. Cayirpunar, S. Burak
    Akat and V. Gazi
Department of Electrical and
    Electronics Engineering
TOBB University of Economics and
    Technology
Sögütözü Cad. No:43
Sögütözü
Ankara, 06560
Turkey
Email: yunusatas@yahoo.com

L. Alboul
Materials and Engineering Research
    Institute
Sheffield Hallam University
Sheffield S1 1WB
UK

Chapter 28

L. Alboul, J. Penders and J. Saez
Materials and Engineering Research
  Institute
Sheffield Hallam University
Sheffield S1 1WB
UK
Email: l.alboul@shu.ac.uk; j.penders@
  shu.ac.uk; j.saez-pons@shu.ac.uk

L. Nomdedeu
Computer Engineering and Science
University Jaume I (UJI)
1207 Castellón, Spain
E-mail: leo.nomdedeu@gmail.com

Chapter 29

U. Delprato, M. Cristaldi and G.
  Tusa
Intelligence for Environment and
  Security – IES Solutions s.r.l
Via Monte Senario, 98
00141 – Roma
Italy
Email: m.cristaldi@i4es.it; u.delprato@
  i4es.it; g.tusa@i4es.it

Mobile robotics systems have begun to emerge in applications related to security and environmental surveillance, including the prevention of disasters and intervention during disasters with all possible kinds of missions to ensure the safety of human beings. Let us mention the well-known explosive ordnance disposal (EOD) and improvised explosive device disposal (IEDD) missions already entrusted to military services in charge of the localisation, neutralisation and/or removal of explosive devices. Humanitarian de-mining campaigns and intervention/inspection by terrorist threats are typical examples of missions that may be conducted with the support of mobile robots.

The general objective of the International Advanced Robotics Programme (IARP) (www.iarp-robotics.org) is to encourage the development of advanced robotic systems that can dispense with human work on difficult activities in harsh, demanding or dangerous environments, and to contribute to the revitalization and growth of the world economy.

Through this book, the IARP working group HUDEM (Robotics Assistance to Mine-clearing) intends to summarize some important results of research and development (R&D) activities focusing on robotics and associated technologies applied to the solution of humanitarian de-mining, while the working group RISE (Risky Intervention and Surveillance of the Environment) extends the same R&D activities to the use of robotics systems, including remote sensor systems for assisting human operators in similar dangerous operations, related to future requests of the Kyoto process as well as to several natural disasters, such as earthquakes, inundations, forest fires, etc.

Robotics solutions that are properly sized with suitable modularized mechanized structures and well adapted to local conditions of unstructured, sometimes unknown, fields can greatly improve the safety and the security of personnel as well as their work efficiency, productivity and flexibility. Solving this problem presents challenges in robotic mechanics and mobility, sensors and sensor fusion, autonomous or semi-autonomous navigation and machine intelligence.

This book reviews and discusses the available robotics-related technologies along with their limitations and highlights development efforts to automate crucial tasks wherever possible through robotics. Furthermore, mapping tools and, consequently, (remote) sensor systems are crucial in the use of mobile robots in order to control the optimal execution of missions and enable safe and efficient navigation. The chapters include the following topics:

- remote controlled explosive devices detection systems
- mobile robotics systems (design, control, command) for unstructured environments (unmanned ground vehicles (UGV), unmanned aerial vehicles (UAV), multi-robotics cooperation)
- applications (humanitarian de-mining, fire-fighting)
- sensors and sensor fusion for environmental surveillance.

Great achievements in distant operation of systems have been reported in recent years, but intelligent operation of complex systems is still some years ahead. Since some of these autonomous missions are very challenging it has been suggested that semi-autonomous human supervised tele-robotics systems should be developed rather than concentrating full efforts on autonomous solutions to complex tasks. Tele-operation carried out in recent years reported that operators had suffered a loss of spatial awareness (disorientation, loss of context), cognitive (inadequate mental model about what is really out there) and perceptual errors (distance judgment, display interpretation), poor performance (imprecise control, obstacle detection), as well as simulator sickness and fatigue.

Communications between the robot and the tele-operation unit suffered from latency, bandwidth and reliability. As a consequence it is not possible to send the human operator as much information as the system may gather, so some data will always be kept at the remote site. Therefore, some intelligence has to be implemented on board the robot and an intelligent machine will use controlling agents to perform autonomous tasks.

This book also introduces results focusing on multi-agent systems (MAS) and multi-robotics systems (MRS and swarm of robots), promising tools that take into account the modular design of a mobile robot and the use of several robots by multi-task missions.

An agent is a computer system capable of autonomous action in some environment. A general way in which the term agent is used is to denote a hardware- or software-based computer system that enjoys the following properties: autonomy, social ability, reactivity and pro-activeness (taking the initiative). Agents being autonomous, reactive and pro-active differ from objects, which encapsulate some state, and they are more than expert systems as they take action on issues that are located in their environment instead of just advising what to do.

When several robots need to work together, it is necessary to manage a given number of supplementary tasks that are not directly productive but serve to improve the way in which those activities are carried out. The coordination of actions is one of the main methods of ensuring cooperation between autonomous robots. Actions have to be coordinated for four main reasons:

- The robots need information and results produced by other robots.
- Resources are limited.
- Costs have to be optimized
- To allow the robots to have separate but interdependent objectives and meet them while profiting from this interdependence.

This book has received the support of IARP and more particularly the contribution of members of the above-mentioned working groups, HUDEM and RISE.

Both working groups were born a few years ago under the European Network Climbing and Walking Robots and Associated Technologies (CLAWAR) (www.clawar.org) through two task groups focusing on humanitarian de-mining and similar outdoor applications. Since 2006, the end of European funding of the network, the CLAWAR association continues to pursue the promotion of these kinds of robotics systems through yearly well-known international symposiums.

The yearly-organized IARP workshops, HUDEM, are also supported by the ITEP (International Test and Evaluation Programme www.itep.ws), through Task 3.1.4, focusing on robotics systems for detecting mines.

We wish the reader a fruitful exploitation of the chapters of this book and will be happy to welcome him or her to our working groups.

Yvan Baudoin
Chairman of the IARP WG HUDEM and RISE

Osman Tokhi
Executive Chairman of the CLAWAR Association

Maki Habib
Co-editor

Part I

Humanitarian de-mining: The evolution of robots and the challenges

Introduction: Mobile robotics systems for humanitarian de-mining and risky interventions

Y. BAUDOIN, Royal Military Academy, Belgium, M. K. HABIB, The American University in Cairo and I. DOROFTEI, 'Gh. Asachi' Technical University of Lasi, Romania

Abstract: Dirty, dangerous and dull tasks, all of which are found in landmine detection, can be greatly aided by tele-operation. It is very desirable to remove the operator from the vicinity of the landmine and from the repetitive, boring operations that lead to loss of attention and potential injury. Tele-operated platforms naturally support multiple sensors and data fusion, which are necessary for reliable detection. Tele-operation of handheld sensors or multi-sensor-heads can enhance the detection process by allowing more precise scanning, which is useful for optimization of the signal processing algorithms.

This chapter summarizes the technologies and experiences presented during seven IARP workshops HUDEM and three IARP workshops RISE, based on general considerations and illustrated by some contributions of our own laboratory, located at the Royal Military Academy of Brussels, focusing on the detection of unexploded devices and the implementation of mobile robotics systems on minefields.

Key words: mobile robotics, sensor systems, human–machine-interface, autonomous navigation, de-mining techniques.

1.1 Objective of the book

This is the first book devoted to robotics (in the general acceptance of this discipline) dedicated to risky interventions – humanitarian de-mining in particular and it is intended to:

* disseminate the results of research activities related to the development of mobile roboticized carriers of detection sensors.
* assist actual and future test and evaluation activities of the International Test and Evaluation Programme (ITEP) and similar programmes.
* help research centres focusing on these kinds of solutions to refine and continuously adapt and update generic modules that may be transferred to useful robotics systems.

Dirty, dangerous and dull tasks, all of which are found in landmine detection, fire-fighting manoeuvres, rescue operations after earthquakes, and others can be greatly aided by tele-operation. It is very desirable to remove the operator from the vicinity of the damaged site and from the repetitive, boring/fighting/detection operations that

3

lead to potential risk and injury. Tele-operated platforms naturally support multiple sensors and data fusion, which are necessary for reliable actions. Tele-operation of handheld sensors or multi-sensor-heads can enhance the mine/chemical/biological detection process by allowing more precise scanning, which is useful for the optimization of the signal processing algorithms. Surprisingly, although a number of prototype multi-sensor systems have been built, until recently there has not been much research on tele-operated ones. More attention has been paid, in humanitarian de-mining, for instance, to (often heavy) tele-operated mine clearers or vegetation cutters, which do not reach the quality assurance imposed by the UN standards (99.6 per cent), even though they considerably speed up the de-mining of large areas.

We hope that this first book will encourage more investment in a promising domain: the ability to automatically detect mines over large areas would make a significant contribution to humanitarian de-mining and such a capability could be used to delimit suspected mined areas, conduct mine clearance, or assist in quality assurance operations. Similarly, the ability to prevent accidents by improving the tele-detection of dangerous chemical sources during the spread of critical fires (forest fires, industrial Seveso sites, complex urban areas, etc) would make a significant contribution to the safety of fire fighters and the security of our societies.

1.2   Humanitarian de-mining: historical steps

So long as there is war and conflict in the world, there will be humanitarian emergencies. So long as there are landmines in the ground, people will be deprived of their basic right to a decent life; communities will be denied the opportunity to prosper; nations will be depleted of resources needed to rebuild and develop. Yet with the continued support of Member States, we have the means to end this suffering. To that end, the United Nations Mine Action Service (UNMAS) is one of our most precious resources.

Kofi Annan, Secretary-General of the United Nations, April 1999

The year 1994 may be considered as the first decisive year in the long struggle with the use and the removal of anti-personnel (AP) mines throughout the world. The next steps summarize the historical evolution that followed.

1994: Foundation of the United Nations Mine Action Service (UNMAS) with objectives including mine awareness and risk reduction education; a minefield survey; mapping, marking and clearance; assistance to victims; and the advocacy to support a total ban on AP mines.
1994: Creation of the Voluntary Trust Fund for Mine Action, financing the UN mine actions.
1999: Entry into force of the OTTAWA Treaty adopted in September 1997, the Convention on the Prohibition of the use, stockpiling, production and transfer of AP-mines and their destruction.

1999: UNMAS and the Geneva International Centre for Humanitarian De-mining are developing and maintaining an Information Management System for Mine Action.

1998–1999: the Joint Research Centre of ISPRA (JRC) coordinates the European research activities focusing on new technologies for de-mining tasks. A first European Network of Excellence ARIS was started, along with a new International Network for Tests and Evaluations of Proposed Technologies, ITEP.

1999–2007 Under the European Long-term FW 4, 5 and 6, European projects were funded, leading to significant technological progresses

2009 TARANTO ASEM Seminar on New Technologies for De-mining and Human Security.

The European Commission confirms its efforts towards a mine-free world: from 2002 to 2008, the EC has committed more than €300 million for mine action worldwide (more than €1.5 billion for the EU as a whole). The EC approach will continue to consider landmines and explosive remnants of war (ERW) within a broader context of humanitarian assistance as well as long-term and sustainable socio-economic development programmes.

1.3   Humanitarian de-mining: the problem

Mines were used for the first time during the American Civil War in the United States (1861–1865). Anti-tank (ATK) mines were later improvised and laid on the battlefields during the First World War. The mine-clearing operations did not pose major problems as the ATK mines were visible and easy-to-detect. This is the reason why anti-personnel mines were conceived and systematically used on the ATK minefields during the Second World War. Such mines prevented the enemy from easily de-mining the defence system.

Anti-personnel mines today are used more and more as offensive weapons and for sowing terror among the civilian population of a country affected by guerrilla war. Marking of the minefield no longer occurs and anti-personnel mines, often buried in the ground, remain active after the war. Today, about 60 million AP mines infest 70 countries all over the world, two-thirds of them in Africa and South-East Asia. AP mines from the Second World War still exist in all the countries of Europe.

At this time, 45 countries are assisted by the above-mentioned European Assistance, namely Afghanistan, Albania, Angola, Armenia, Azerbaijan, Belarus, Burundi, Cambodia, Chile, Colombia, Croatia, Cyprus, DRC, Ecuador, Eritrea, Ethiopia, Georgia, Guinea-Bissau, Honduras, Indonesia, Iraq, Jordan, Kosovo, Kyrgyzstan, Laos, Lebanon, Mozambique, Myanmar, Nepal, Nicaragua, Peru, Russia, Senegal, Serbia, Somalia, Sri Lanka, Sudan, Tajikistan, Tunisia, Uganda, Ukraine, Venezuela, Vietnam and Yemen. This list obviously evolves with new signatories.

1.3.1 Anti-personnel mines

Two definitions coexist: according to the military standards, an AP mine is a pyrotechnic instrument that has been developed to be activated by an involuntary action of the enemy, in order to stop him fighting. According to civil rights, an AP mine is an object placed on or under the ground or any surface, designed to explode by the presence, proximity or contact of a person or a vehicle. Generally, two models of mines exist: blasting mines and fragmenting ones. More than 700 known types of mines have been produced in about 55 countries, varying from each other by their explosive load, activation means, action range, the effects they have on the human body, etc. Figure 1.1 shows a sample of AP mines.

Table 1.1 illustrates the differences between selected types of AP mines.

Two values that draw our attention to the difficulty of mine detection and the longevity of mines are that their diameter is often less than 10 cm, and they have a life span that often exceeds 75 years. AP mines can be bought for a price varying from 2.5 to 25 euros.

1.3.2 Economic and human impact

Since 1975, mines have killed more than one million people, essentially civilians and children. This is about 70 victims per day or 26,000 victims per year and they have caused about 300,000 children to be severely disabled. The people affected by this plague are those who are less able to overcome the medical, social and economic consequences. The International Committee of the Red Cross (ICRC) estimates that two-thirds of the surviving victims will get into debt for life, their life expectancy will be reduced to between 40 and 50 years, and they will need about 20 prostheses and will pay about 120 euros each. In countries where the individual mean income varies from 10 to 25 euros per month, crutches are often the only tools they can afford for walking.

The removal of a mine costs between 250 and 750 euros, depending on the accessibility to the minefield, the professional qualification of the de-miner and the type of mines. The mean daily productivity of a de-miner can be between 20 and 50m^2, taking into account false alarms generated by metallic inert objects or the high iron content of the soil.

Table 1.1 The differences between selected types of AP mines

Identification	Type	Load	Activation	Range
PMN	Blasting	100/150 g	5–20 kg	local
Home-made Pelota Chimica	Blasting	Variable	Variable	
POMZ-2	Fragmenting	300 g	Wire	50/200 m
V69	Fragmenting	700 g	Wire	50/200 m

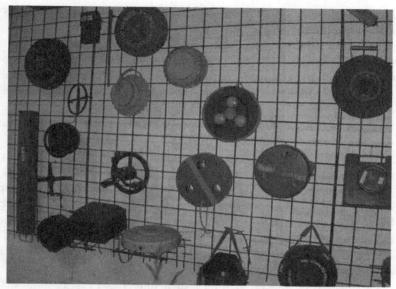

1.1 Samples of anti-personnel mines.

Starting from a pessimistic estimation of the infested surface, the de-mining cost can be evaluated at 100 billion (milliard) euros. The actual budget devoted to mine-clearing operations only allows the removal of 100,000 mines per year. In the same period, more than one million new AP mines were placed in countries affected by wars involving civilians. Médecins sans frontières (MSF), the non-governmental organization (NGO), recently estimated the mine-clearing time of the infested Afghan territory at 4,000 years.

Even if the actual statistics seem to be overestimates, it is a matter of urgency to develop detection, identification and removal techniques, and obviously to condemn and forbid the production and the use of such despicable weapons. In the most infested countries, 20 to 30 per cent of the economically exploitable surface is unusable. Added to the land that is of no economic interest, one easily understands that the effects on the local economy and the consequences on a worldwide scale are far from negligible.

Mine-clearing operations are very dangerous: for every 5,000 mines cleared, one de-miner is killed. In order to clear 60,000,000 mines we should expect 12,000 de-miners to be killed! This fact also explains why only a few experts from developed countries are working on the minefields. Everyone has to be covered by an annual mean insurance premium of 12,500 euros. As a consequence, the manual mine-clearing operations are entrusted to local civilians whose inexperience and fear of injuring themselves slow the progress of the de-mining work.

Experienced de-miners often limit their (important) role to the learning and training of local teams. In most affected countries, however, outstanding de-mining structures have been developed and entrusted to local managers, as in Cambodia where the Cambodian Mine Action Center (CMAC) coordinates the de-mining campaigns in a structured way, according to defined priorities.

Another problem involves the difference that exists between military and civilian mine clearance. Military de-mining operations accept low rates of clearance efficiency (CE) and it is often sufficient to punch a path through a minefield. For humanitarian de-mining purposes, on the other hand, a high CE is required (a CE of 99.6 per cent is required by UN). This can only be achieved through a 'keen carding of the terrain, an accurate scanning of the infested areas', which requires the use of sensitive sensors and their slow systematic displacements, according to well-defined procedures or drill rules, on the minefields. At present, handheld detectors seem to be the only and most efficient tools for identifying all unexploded ammunitions and mines but this first step doesn't solve the problem. The removal task and/or the neutralization and/or destruction task must follow and these last two tasks are time-consuming actions.

Let us end this introduction to the problem by concluding that the above-mentioned difficulties lead to the definition of global requirements that could accelerate the de-mining processes. The scientific community needs to focus on pursuing the research activities highlighted in Table 1.2.

Beside those technological priorities, the five pillars that have been defined and highlighted by the European Commission have to be taken into account. These are the clearance of mined areas according to priorities; mine risk education; the destruction of the anti-personnel landmines (APL) stockpiles; mine awareness; and victim assistance.

Table 1.2 List of key research activities to be pursued by the scientific community

Priority 1	The development of reliable sensors allowing the detection of minefields and, on those minefields, the detection of the mines (or similar explosive items).
Priority 2	The development of data processing algorithms confirming the detection and leading to the identification of the parameters needed for the next actions.
Priority 3	The development of fast removal techniques or neutralisation techniques.
Priority 4	The development of (tele-operated or not) mine-clearers and low-cost (tele-operated, semi-autonomous or not) mobile scanners.

1.4   Technological challenges of mobile robotics systems for humanitarian de-mining

Several workshops, among others organized by the European Network Climbing and Walking Robots and Associated Technologies, now the CLAWAR Association Ltd (CLAWAR) and the International Advanced Robotics Programme (IARP) (to which most chapters of this book have been major contributions), allowed discussions on the possible research and development activities for solving the problem(s). Robotic systems are not (yet) thought to be the most promising solutions, due to their high cost, use and maintenance difficulties, varying (daily changing) terrain conditions, etc. However, specific tasks could be entrusted to mechanical mine disposal systems (or, if efficient, roboticized sensor-carriers):

- The cutting of vegetation is a mechanical (tele-operated or not) process that does not need high-level research activities, but requires the adaptation of existing mobile cutters. See catalogues of the Geneva International Centre for Humanitarian Demining (GICHD).
- The mine-clearance of large agricultural areas. A detailed and recommended study has been managed by Håvard Bach, Head of the Operational Methods Section (GICHD, 2004).
- Detection tasks in very dense and dangerous areas (woody areas, mountains, etc. that constitute about 53 per cent of the infested areas). Such tasks could imply the realization of specific robots, for instance multi-legged robots, a difficult long-term challenge.
- The delineation of the borders of a suspected area. This is a task that could be entrusted to aerial tools and the EU encouraged several research and development projects focused on this task.
- The systematic scanning of a zone is a priority for manual de-mining with enhanced multi-sensor-heads, but also for assistance (safety) by mobile robotics systems on request.
- The inspection of an area after manual de-mining or mechanical clearance.

1.2 Mechanized flail de-mining track.

1.4.1   Sensors and multi-sensors systems

A huge number of anti-personnel (AP) mines are polluting the environment in about 60 countries. Thanks to the Ottawa Convention, mine clearing operations have been organized in a more controlled and effective way. Nevertheless, mine clearance remains a very slow process. It is estimated that, on average, a de-miner clears an area of 10 m² every working day if using conventional tools, i.e. metal detectors and prodders. To give an idea of the extent of the problem, in Cambodia, only approximately 260 km² have been cleared during the last 10 years. Encouraging results, however, emerge: as another example, 889,835 m² area has been cleared in Laos and 4,819 UXO (unexploded devices) removed.

However, humanitarian mine clearance operations must be understood and designed correctly, keeping in mind that their main goal is to provide an efficient aid to innocent people, who may be severely injured by these dreadful devices. Furthermore, the analysis of actual de-mining campaigns not only reveals that it takes a long time to clear polluted terrain, but also brings to the fore a far too large false alarm rate, the threat of plastic mines (which are difficult to detect by classical means i.e. by metal detectors), and the large variety of mine clearance scenarios, depending on the country, region, climate and the location of mines in villages, roads, cultivated fields, etc. The important parameters that characterize the mine detection problem are mine occurrence probability, the detection probability of a given material and the false alarm probability of a given material:

- *The mine occurrence probability* in a given position of a minefield expresses the local mine density of that minefield as well. Obviously, it is impossible to control this parameter because it depends on the actual terrain. Nevertheless, this parameter is very important for assessing the probability of an alarm in a given location on the minefield.
- *The detection probability* is the probability of having an alarm in a given position of a minefield for a given detection material, if there is a mine in that position. This probability indirectly gives a measure of the non-detection probability of that material as well.
- *The probability of false alarm* is the probability of having an alarm, for a given material, in a given location if there is no mine in that location. The latter two definitions are extremely important for understanding the humanitarian de-mining problem and for designing de-mining systems. Indeed, it is particularly important that the detection probability should be as close as possible to one. It is easy to show that evaluating the detection probability also amounts to evaluating the risk of the occurrence of a mine that has not been detected. This risk is concerned with human life and is therefore of the utmost importance. No risk is acceptable here and it is therefore an absolute requirement that a de-mining system should decrease the probability of such a risk to the lowest upper bound possible. Besides, although human lives are indirectly saved by decreasing the false alarm risk thanks to the acceleration of the de-mining operations, the false alarm risk is also a question of cost. Indeed, a de-mining method that minimizes the false alarm rate results in an acceleration of the de-mining operations and therefore in spending less money.

As a result, any de-mining operation enhancement must result in the highest possible detection probability (close to one) and in the smallest possible false alarm rate and that at the lowest price. Generally, it is accepted that the most efficient way for increasing the detection probability while minimizing the false alarm rate consists of using several complementary sensors in parallel and fusing the information collected by these sensors. As a matter of fact, it is imperative to evaluate the detection probability when optimizing the performances of a system. However, the detection probability, as defined before, assumes that a mine is present in the considered position. During organized trials, the positions of mines are well known, so the condition of the occurrence of a mine in the given position, when the performances of a system are being evaluated, is always realized. This latter remark is of particular importance because it justifies the organization of trials and the construction of models, to be validated by trials, in order to evaluate the detection probabilities.

Furthermore, assuming in the following as the first approximation that the sensors are independent, the detection probability can be maximized by separately optimizing the design of each sensor and of the associated signal processing. Next, it can easily be shown that the detection probability increases if the number of different sensors increases, and that maximizing the overall detection probability of a set of independent sensors is clearly the same as maximizing the detection capabilities of each individual sensor. This justifies the use of several complementary sensors and of data fusion techniques to increase the detection probability.

Among the most often cited sensors one finds metal detectors, radars and infrared sensors. Finally, the false alarm risk, i.e. the probability of having an alarm if there is no mine, cannot be as easily evaluated as the detection probability because of the use of data fusion methods which favour the manual or automatic cancellation of false alarms. Furthermore, it is very difficult to evaluate the risk of a false alarm because it is very complicated to define in a general way what is not a mine. In this context, it should be particularly inappropriate that a de-mining system, whatever it may be, makes a decision instead of the final user whose own physical security is involved. Therefore, a well-designed system should help the user in the decision making, not by replacing him, but by implementing efficient data fusion methods. For this purpose, methods that are able to deal with uncertainty by making proposals including doubt to the user seem to be promising (Acheroy, 2005).

1.4.2 Robotics systems

The robots will need to carry work packages of various sorts. The most important of these will be sensor packages. The following are typical detection sensor methods that are currently being used or are being researched at this present time:

- various metal detectors, ground penetrating radar (GPR)
- ultra-wide band GPR (UWB GPR), synthetically focused GPR, signal processing, data fusion, microwave imaging
- thermal imaging, infrared imaging, dynamic thermography, hyperspectral infrared, infrared polarimetric cameras
- nuclear quadrupole resonance, thermal neutron activation, lateral migration radiography, x-ray back-scatter (some of this group may be unsuitable due to complexity, weight, time scales to availability)
- prodding, seismo-acoustic, shock impulse and micro-electromechanical systems (MEMS)
- combination technologies, i.e. metal detection and GPR
- others.

It will be necessary to identify which are low key technologies and which are the higher ones, in order to take into account the development time scales required to bring to maturity the newer technologies. Some of these may be up to 10 years away from maturity.

The reliability of the equipment to detect and neutralize mines will need to be specified in quality assurance (QA) terms e.g. 99.9 per cent clearance guarantee. The European Joint Research Centre in Italy has published some quantified standards against which application trials should be measured.

Measurement and pinpointing accuracy will need to be defined. It is suggested that the limit on this as a benchmark could be a quarter of the diameter of the smallest AP mine, say 10–15 mm resolution. As a result the robot will need to be able to carry a marking work package and a means of communicating its position in the minefield. Accuracy of +/– 1 cm would be ideal.

The environment will play a large part in determining the attributes or characteristics of the robotics equipment. In fact it is unlikely that a multi-purpose, single machine will be developed that will cope with all forms of environment. The environmental issues fall into various categories and equipment operation will need to be defined as desirable or mandatory under these conditions. The main environmental characteristics that need to be taken into account are:

- weather: temperature, snow, ice, frost, rain, wind, humidity
- terrain: urban – street, inside buildings; rural – desert, rocky, dense vegetation, possible water scenarios.

Payload assessment must take into account two factors, expectations based on current engineering capability and those desired for the future, although the latter may be quite impractical at the present time. However, from knowledge of the various sensor systems being considered, a payload in the range 5–15kg is likely.

Various mechanism types are being considered. There is a need to concentrate on scenarios where tracked and wheeled vehicles will be unable to carry out the de-mining task. The likely configuration will therefore be a light-weight, articulated legged walker able to clamber over rough terrain, cross ditches, walk through dense vegetation without disturbing it, walk over hidden trip wires, climb steep slopes, etc. It will also have sufficient degrees of freedom of its body with respect to its legs to deliver sensor work packages and marking devices, probably on a boom or manipulator, to difficult to access positions accurately.

The actuation method is unclear. Electric motor driven joints seem the most likely, although pneumatics should not be ruled out. Either way, the power requirements of the vehicle are likely to significantly exceed those of the work package. Use of an umbilical, although possible, will seriously degrade the operational scope of the vehicle so suitable on-board power devices may be required. The weight of battery packs or motors for producing compressed air is significant and a balance between functionality and mission length may be hard to achieve. Soft pneumatic muscle actuators may provide some solution since, weight-for-weight, they are able to provide much higher power for lower pressure than pneumatic cylinders. In the short term, the use of an umbilical may be necessary whilst suitable on-board power technology is found.

Motion control will need to be highly sophisticated. General motion in difficult terrain will need advanced adaptive gait control such as is being developed at present in various research centres. Closely controlled motion will be required to deliver sensor packages to accurate positions when detection is in progress. The motion of the vehicle will demand by far the highest power requirements. Whilst some scenarios will allow the use of an umbilical, many will need more autonomy so an on-board power supply will be needed. Thus efficiency of motion will be most important, requiring advanced control algorithms. On the other hand, speed is unlikely to be paramount since detection will take time and will probably limit forward motion.

The modes of operation need to be specified. Most requirements will have a man-in-the-loop operation and there will be a direct line of sight operation at a safe distance. This safe distance will have to be specified, as will the method of ensuring that the safety restraints are carried out correctly. Typically, current methods for remote control from close in, up to 1–2 km distance, use tele-operation.

Examples of the advantages of tele-operation are that the task can be carried out by a single operator and that camera positions are easily selectable using a microwave link or fibre optic for a line of sight video transmission from the machine to the remote command station. To carry out complex tasks, the numbers of cameras needed and their positions will have to be considered. It is likely that at least two fixed or one rotational camera will need to be fitted to the vehicle to give all round viewing during operation. Operator control units can be fitted to display single or multi-image options. The communication link might be a 1.4 GHz video link as used in the systems for Grand Prix F1 racing. Fibre optic links that offer high bandwidth can be used but the trailing of cables can be a problem over long distances. A communications link to carry control and sensor feedback signals will also be required, probably using a fibre optic link.

In summary, machines for carrying out de-mining activities in place of human de-miners are generally likely to be wheeled or tracked. However, there is a possibility that in certain terrain, walkers will add value. There is little likelihood that pure climbers will be required. Assuming that the friction of its feet to the ground is sufficient to provide the traction required, provided the vehicle has the motive power to operate on steep inclines by modification of its gait, then a walker is sufficient. Such machines are likely to be light in weight. The control and communications system is likely to be of a nature which will facilitate the addition of higher order functionality such as sensor fusion, human–machine interface (HMI), navigation, etc.

The walker will need to carry several kilograms of work packages; a selection from vision cameras, infrared cameras, GPR, UWBR, metal detectors, chemical sensors and other more advanced detectors. Some may have to be held on a boom arm or manipulator. The machine will need to be able to traverse rough ground without operator intervention so a high degree of gait 'intelligence' is required. Since ground conditions will vary considerably during the mission, it will need to be able to sense ground condition and adjust its gait. It will also need to be able to hold detectors in a pre-determined relationship to the ground contour and to control delicate prodding movements.

The complete system will need to integrate the vehicle control and navigation systems with a data fusion system that will discriminate, to a high degree of confidence, between mine and 'no-mine' conditions.

Some such machines may need to be specified so that they can operate fully submerged in shallow seawater.

This book will detail some specific solutions fulfilling the above-mentioned requirements, while the next paragraphs summarize results of research and development presented during seven IARP workshops.

1.5   International advanced robotics programme (IARP) workshops: a major contribution to the exchange of information among the research and development (R&D) community

Seven IARP workshops have been organized, in Toulouse, Harare (Zimbabwe, 2000) then successively in Vienna, Prishtina, Brussels, Tokyo and Cairo. About 26 selected papers focused on the sensor technologies.

1.5.1   Remote sensing

In order to avoid a considerable waste of time, a first essential objective lies in the delimitation of the mine polluted areas: local information on observed explosions, craters, injured animals and/or on hospital casualty reports already involves the sending of local technical teams in charge of minefield-marking. This marking may never be precise as the suspected area may be much larger or even smaller than expected, even if performance ground sensors are used. Research efforts have been funded by the European Commission encouraging airborne surveys with colour, colour infrared and thermal cameras, multi-spectral sensors and other promising sensors.

Different projects have been initiated in this context (ARC, MINESEEKER, etc). We give only a very short description of one of the most promising projects on this matter, the SMART project. The goal of this project is to provide the human analyst with a GIS-based system – the SMART system – augmented with dedicated tools and methods designed to use multi-spectral and radar data in order to assist the analyst with an interpretation of the mined scene during the area reduction process. The usefulness of such image processing tools to help photo-interpretation has already been studied: the possibility of automatically processing a large amount of data and help with visual analysis is among their advantages.

The use of SMART includes a field survey and an archive analysis in order to collect knowledge about the site, satellite data collection, a flight campaign to record the data – multi-spectral with the Daedalus sensor and polarimetric SAR with the ESAR from DLR – and the exploitation of SMART tools by an operator to detect indicators of the presence or absence of mine-suspected areas. With the help of a data fusion module based on belief functions and fuzzy sets, the operator prepares thematic maps that synthesize all the knowledge gathered with these indicators. These maps of indicators can be transformed into danger maps showing how dangerous an area may be according to the location of known indicators and into priority maps indicating which areas to clear first, taking account of socio-economic impact and political priorities. These maps are designed to help the area reduction process. Preliminary results obtained with SMART showed a global substantial area reduction rate of 20 per cent and a misclassification rate of 0.1 per cent for what SMART considers as not mined and is actually mined.

The approach also has its limitations. The general knowledge used in SMART is strongly context-dependent. It has been currently derived from the study of three different test sites in Croatia chosen to be representative of south-east Europe. In the case of another context a new field campaign is needed in order to derive and implement new general rules. Before using SMART the list of indicators must be re-evaluated and adapted. For instance it has been noted that the assumption that a cultivated field is not mined, although quite valid in Croatia, may not apply in other countries such as Africa or Colombia. It must also be checked if the indicators can be identified on the data and if the new list is sufficient to reduce the suspected areas.

1.5.2   Close-in detection

Assuming the borders of a minefield have been defined, a systematic scanning of the field must follow. In order to assure the desired CE, the use of the following will be necessary: the best-known or proposed multi-sensor platform under investigation combines a metal detector, ground penetrating radar (GPR) or an ultra-wide-band (UWB), radar and an infrared camera. But, added to the possible use of other combinations, the optimization of existing sensors and/or the development of new sensors (nuclear quadrupole resonance (NQR) detection of nitrogen bonding in explosives), the simultaneous use of several sensors induces a certain number of problems that have to be solved. These include the fusion of the quite-different data provided by the sensors, the mutual interaction or inter-compatibility of the sensors, and the control of the positioning of the sensors above the inspected ground. Table 1.3 summarizes actual state-of-the-art detectors and their relevant characteristics if de-mining automatic technologies are envisaged (i.e. mounted on robots or vehicles).

For each type of sensor, specific signal processing techniques are used in order to extract useful information. The techniques used mainly include signal conditioning or pre-processing (e.g. signal detection, signal transformation, noise reduction, signal restoration and enhancement which are very important steps before further processing) and pattern recognition techniques aimed at increasing the expertise of each sensor separately. Nevertheless, it has been shown that no sensor is perfect for all scenarios and all conditions (moisture, depth, cost, etc). The analysis of the principles of operation of different sensors, their complementary information, and the factors that affect their operability, has led to the conclusion that their fusion should result in improved detectability and a reduced number of false alarms in various situations (different types of mines, soil, vegetation, moisture, etc.).

Table 1.3 Summary of the actual state-of-the-art of the detectors and their relevant characteristics if de-mining technologies are envisaged

Sensor family	Sensor	Maturity	Cost	Speed	Effectiveness
Prodder and acoustic	Prodder	In use	Low	Very low	High
	Smart prodder	In use In development	Low to medium	Very low	High
	Seismic and acoustic	R&D	High	Medium	High (in wet soil)
Electro magnetic	EMI devices	In use	Low to medium	Low to medium	High
	Magnetometer	In use	Low to medium	Low to medium	High
	Gradiometer	In use	Low to medium	Low to medium	High
	GPR	In use	Medium to high	Low to medium	High (in dry soil)
	MWR	In development	Medium to high	Low to medium	Medium
	Electrical Imp. Tom.	R&D	Low to medium	Low to medium	Unknown
	Electrography	R&D	Low to medium	Low to medium	Unknown
Electro-optic	Visible	In use	Low to medium	Medium	Low
	Infrared	In use	High	Medium	Medium
	Infrared polar.	R&D Prototype	High	Medium	Medium
	Multi and hyperspectal	R&D	High	Medium	Medium
	LIDAR	R&D	Very high	Medium	Low
	Terahertz	R&D	Very high	Medium	Low
	SLDV	R&D	Very high	Medium	Medium to high
Biosensors	Dog	In use	Medium to high	Medium to high	Medium to high
	Rodents	In development	Medium	Medium to high	Medium to high
	Artificial nose	R&D	Medium to high	Medium	Medium
Nuclear and chemical	NQR	R&D Prototype	Medium to high	Medium	Medium
	TNA	R&D Prototype	High	Medium	Medium
	FNA	R&D	Very high	Medium	Very high
	X-ray backscattering	R&D Prototype	High	Medium	Low
	X-ray fluorescence	R&D Prototype	High	Medium to high	Medium
	Chemical detectors	R&D	High	Medium	Unknown

The Japan Science and Technology Agency recently organized a test and evaluation for anti-personnel landmine detection systems using ground penetrating radar and a metal detector mounted on robotic vehicles. The test results showed that combining GPR with a metal detector can improve the probability of detection (PD) around a depth of 20cm, where it is difficult to detect targets by using only a metal detector, and that there is room for further improvement in the PD, for instance by feeding back the test results to testers to learn typical target images, where targets were not able to be detected in the blind tests (no pre-knowledge of the locations of the buried mines). It has also been learned that positioning control must be improved in scanning the ground with a sensor head, which is a key to making the best use of MDs mounted on vehicles (argument pro-robotics) (Ishikawa, 2005).

1.5.3   Mechanical mine clearance

Mechanical assistance consists of the use of motorized mine-clearers: adapted military vehicles or armoured vehicles may be used on large areas (agricultural areas, for instance) for as long as their access is granted and some mine-clearers now combine clearance and detection tools: the HITACHI landmine disposing machine illustrates this kind of system, combining a rake-grapple to cut the vegetation, a mounted metal detector for avoiding the rotary-cutter and a magnet system to remove the metal fragments (see Fig. 1.3). Intensive tests on mechanical mine clearance were carried out in Cambodia and Afghanistan and proved its efficiency: about 15,000 m^2 (300 times more than a human operator) may be cleared per day. One hundred per cent of ATK mines were removed and destroyed but only 90 per cent of AP mines. Although other clearance techniques (e.g. heavy tooled road rollers) already lead to higher efficiency (98 per cent), some AP mines may be pushed on size or buried deeper or partly damaged (and thus more dangerous).

1.5.4   Vehicle-mounted mine detector

Conventional vehicle-mounted mine detector systems employ an array of sensor elements to achieve a detection swath typically 2 to 4 m wide. Some systems employ more than one type of sensor technology. These systems, while being very useful are often expensive, unsafe, complex and inflexible. The IARP workshops have shown

1.3 Yamanashi Hitachi Construction Machinery Co., Ltd., Japan (6th IARP WS).

1.3 Continued

that the use of robotics systems (remote controlled vehicles) could improve the safety and the clearance efficiency and that they may be considered promising tools. However, the development of a robotics system (RS) involves the design, reliability and the cost-effectiveness of its modular components (see Fig. 1.4 and 1.5) (Nonami and Aoyama, 2005; Fukushima et al., 2005; Fukuda et al., 2005; Dorofteï et al., 2007).

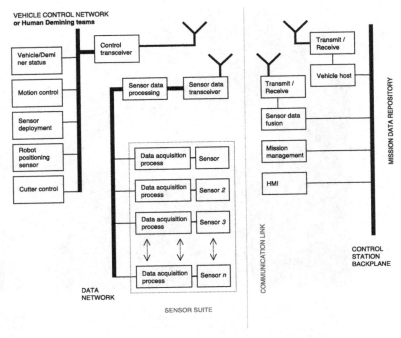

1.4 Modular description of a robotics system for the detection of explosive devices.

(a)

1.5 Some tele-operated sensor-carriers: (a) Gryphon-IV remote maneuvering experiment, Tokyo Institute of Technology, JP (b) Mine detection robot COMET-II – Chiba University, JP (c) Mine detected robot, Hunter Military Academy, BE.

(b)

(c)

1.5 Continued

1.5.5 Robotics control

The adaptive control usually implies the use of tele-operation, tele-presence and distributed intelligence. Tele-operation is the extension of a person's sensing and manipulating capability to a remote location creating communication channels to and from the human operator. The tele-presence defines the techniques that allow the human operator to feel himself physically present at the remote site. The intelligence combines sensory processing, world modelling, behaviour generation and value judgement to perform a variety of tasks under a-priori unknown conditions. The combination of tele-operation, tele-presence and human–machine distributed intelligence often defines the supervisory control. Through the introduction of AI techniques and the use of virtual or pseudo-virtual reality, the robotics system's teams today try to develop the concept of adaptive autonomy and virtual symbiosis.

As deduced from the previous scheme of a robotics system (top-down approach from the right side to the left side), an optimal approach in humanitarian de-mining should consist of providing a supervised autonomous unmanned ground vehicle (UGV) that can remove excess vegetation and then deploy a multi-sensing detector with sufficient precision to provide a reliable mapping system of detected mines. This will involve a combination of several different sensors, including:

- a sensor to determine the location of the robot vehicle within the area to be cleared
- an explosives proximity sensor to enable safe navigation
- a multi-sensing mine detector, incorporating, for instance, a 3D metal detector and a GPR
- a sensor to determine the position and orientation of the multi-sensing detector
- sensors to control the attitude of the vehicle
- a vision sensory to allow supervised autonomy (human operator (HO) in the loop).

Control station (TOP)

The application related to control and command has to implement three primary activities, defined in the blocks 'mission management, data fusion and HMI' concerning the mission management. It is important to make a clear distinction between high-level mission management (HLMM) or mission planning and the robotics system mission management (RSMM).

HLMM: Several problems are inherent to the process of supporting humanitarian de-mining campaigns with useful data. Within the framework of a de-mining campaign, contaminated areas are often very large and represent a lot of information. This huge amount of data has to be compiled and safely stored in a central repository to avoid loss or corruption of data. Also, this data needs to be represented in an explicit manner, in order for the user to work with it as effectively as possible.

Maps are key elements for campaign management and field work; however, as de-mining campaigns generally take place in developing countries, there is often a lack of accurate and recent maps for the zone of work. Information systems have been developed in order to solve those problems and are now used in countries affected by explosive remnants of war (ERW). IMSMA, the UN-standard information system for mine action, addresses this problem. It consists mainly of a database located at the Mine Action Centre (MAC) of a specific country or region, into which all mine action related data is centralized. It also contains management and planning tools that process the information in the database in order to help decision makers. PARADIS (Delhay et al., 2004) addresses the problem by compiling and organizing the data related to the campaign in a geographic database (GeoDb). As PARADIS is built on a GIS, it presents the data to the user on a map rather than on a form (x, y coordinates). The user is able to enter the data in the PARADIS system, then export them to IMSMA using the MAXML (Mine Action eXtended Markup Language – http://www.hdic.jmu.edu/conference/casualty/maxml_files/frame.htm). The manager interface is located in the MAC and is built on a fully featured GIS. As it needs to access the central geographic database and may run large processes while manipulating the data, it is based on a desktop (or laptop) computer (the desktop interface). The field operator or the RS controller will work on a lightweight GIS interface called the field interface. The latter runs on a personal digital assistant (PDA), the small size and light weight of which makes it the ideal tool for field work. It could also work on the RS laptop containing, in this last case, the field-manager interface as well as the human machine interface. A complete procedure of information exchange (manager to field, field to manager, MAC to IMSMA) is currently being developed, exploiting the advantages of internet facilities.

RSMM comprises:

- The planning process includes the evaluation of the environmental conditions compared to the available data (for instance, primary map of the minefield from aerial detection, with some information on natural relief, obstacles, PARADIS information, etc.) in order to generate a series of tasks (path-planning, cutting of the vegetation, inspection mode – choice of the sensor, sensor deployment, etc.).
- The directing process which defines with high precision and in an unambiguous way the operations (a set of operations or sub-tasks defines a task) that will be controlled by the field users, and the monitoring process involving the tele-operation aspects. The intelligence of the system lies essentially in the first two processes: an expert system (ES) will assist the local field manager (and, indirectly the field user during the monitoring process). Such an ES includes a comprehensive database (HLMM derived GIS, geological information, mines and UXO information/description, characteristics of the UGV, of the sensors, etc), a set of rules and a strategy allowing the best choice of rules according to the objective (minefield delineation, precise scanning of well-defined area, etc). The monitoring process requires the correct design of the HMI. As an example, the next figures describe the control-architecture of the Royal Military Academy hunter as well as the HMI (CORODE software) including the visualization of the signals delivered by the detection sensors (see Fig. 1.6) (Hong et al., 2001; Colon et al., 2001).

HLMM and RSMM or IMS: Both management tools may be combined to form an information management system (IMS) that carries out planning for the

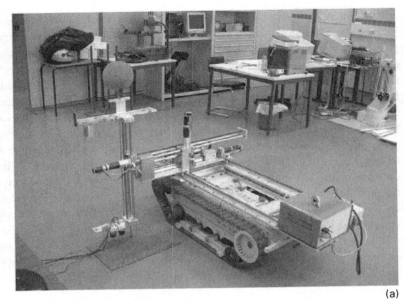

(a)

(b)

1.6 CORODE Project: (a) The global interface including both the tracking/positioning systems and the detection devices (b) HMI.

de-mining procedure and controls the sensing mobile robot(s), but also provides information on the current status or past de-mining results in order to share the information with operators and other de-mining organizations. The information management system is composed of three subsystems as follows: the controller of the sensing unit entrusted to the supervisor, the mine detection support system displaying the image processing results, and an integrated information interface for mine action (I3MA) based on the international standard Mine Action Extensible Mark-up Language: maXML.

The first IMS has been implemented by the inter-university team of Nagoya, Tsukuba and Tohoku under the sponsorship of the Japanese National Institute of Advanced Industrial Science and Technology (AIST) and with the support of the Mitsui Engineering and Shipbuilding Company.

Robot control (DOWN)

In the supervised mode, the safety and performance of the communication (particularly non-line-of-sight), as well as the computing speed capabilities of the informatics systems play a major role. A considerable amount of literature (among others for military ground missions) describes the constraints relating to those factors: for example, the standard 19.2 Kbit/s of compression needed for the high-bandwidth video signals, typically 20 MHz for one vehicle in frequency modulation mode. The vehicle control network and the data network impose the development of high level–low level control software. Here also a large amount of literature suggests solutions: as an example, the ANCEAEUS control system adopted on the JINGOSS mine-detection system developed by the Canadian Forces (DRES Defence Research Establishment Suffield) and mounted on an 8 × 8 wheeled vehicle ARGO (used in Somalia). The vehicle supervisor includes its own navigation module (semi-autonomous navigation, vehicle status monitoring, DGPS positioning functions, etc.) and its own application module (detection/marking) (Faust et al., 2003).

The objective of the supervised control is clear: to free the human operator(s) to concentrate on a higher level of control and optimally achieve the planned mission. No supervised control may be successfully implemented without having satisfied the following requirements:

- The use of a UGV adapted for the mission (adapted mechanical structure, locomotion mode, actuation, sensory, etc.).
- The training of the human operators (all ranks, thus including the commando levels) through an appropriate series of courses on the emerging information and control technologies, during on-the-field simulations, and then on dummy minefield trials under varying environmental conditions including uncertainties or randomly occurring events.
- The pursuit (and funding) of R&D activities related to the following issues: optimal allocation of information processing (interactive planning and control at the mission level (the above described TOP level), timely reactions on observed deviations), optimal allocation of control functions (high level/low level motion control of the UGV and orientation/positioning control of its sensors), multi-vehicle control (integration of navigation, task, sensory modules under predictable structured conditions and under uncertainties).

The first R&D results related to some of those issues, in real-time outdoor conditions are still unreliable.

Navigation

Although the most recommended control of a robotics platform is the tele-operated mode rather than the supervised one, high mobility on uneven terrain is also demanded, particularly after heavy mechanical mine-clearers have already turned over the soil. Furthermore local obstacles (tree stumps, rocks, and obviously detected explosive devices, etc.) have to be (automatically) avoided, the detection multi-sensor-head has to be (automatically) kept as close to the ground as possible (this task is extremely difficult for a human operator acting from a safe distance of the robot with poor vision on the soil relief), multiple units or agents (robots, manipulators, sensors) have to be coordinated, etc. The robotics researchers today favour the development of behaviour-based navigation (Doroftei et al., 2009).

Uncertainty in sensing and action, and changes in the environment can require frequent (re)planning, the cost of which may be prohibitive for complex systems. Path-planning based approaches have been criticized for making real-time reaction to sudden world changes impossible. Various approaches for achieving realtime reaction performance in autonomous agents have been proposed. Purely reactive bottom up approaches are featured in various implemented systems. They embed the agent's control strategy into a collection of pre-programmed condition/action pairs with minimal internal state. Reactive systems do not maintain internal models. Typically, they apply simple functional mapping between stimuli and appropriate responses, usually in the form of a lookup table, a set of reactive rules, a simple circuit, a vector field, or a connectionist network. All of those implementations are variations on the same theme of constant-time run-time direct encodings of the appropriate action for each input state. These mappings rely on a direct coupling between sensing and action, and fast feedback from the environment. Purely reactive strategies have proven effective for a variety of problems that can be completely specified at the design time. However, such strategies are inflexible at run-time due to their inability to store information dynamically. Hybrid architectures attempt a compromise between purely reactive and deliberative approaches, usually by employing a reactive system for low-level control and a planner for higher-level decision making. Hybrid systems span a large and diverse body of research. This includes reactive planning or reactive execution used in reactive action packages (RAP), higher-level primitives for planning that hide and take care of the details of execution, procedural reasoning system (PRS), and others. Hybrid solutions tend to separate the control system into two or more communicating but independent parts. In most cases, the low-level reactive process takes care of the immediate safety of the agent, while the higher level uses the planner to select action sequences.

Behaviour-based approaches are an extension of the reactive architectures and also fall between purely reactive and planner-based extremes. Although often conflated in the literature, reactive and behaviour-based systems are fundamentally different. While behaviourbased systems embody some of the properties of reactive systems, and usually contain reactive components, their computation is not limited to lookup and execution of simple functional mappings. Behaviours can be employed to store various forms of state and implement various types of representations. The organizational methodology of behaviour-based systems differs from classical hierarchical systems in its approach to modularity. The philosophy mandates that behaviour execution not be simply serialized, thus reducing the system to one that could be implemented through more traditional centralized means. The organizational methodology concerns the coordination of a multitude of behaviours, thus making behaviour arbitration one of the central design challenges of such systems. For the sake of simplicity, in the majority of systems the solution is a built-in, fixed control hierarchy imposing a priority ordering on the behaviours, much as such hierarchies have been used to employ priority schemes over reactive rules, such as for example in the subsumption architecture.

Coordination in subsumption has two primary mechanisms:

- Inhibition: used to prevent a signal being transmitted along a behavioural module's wire from reaching the actuators.
- Suppression: prevents the current signal from being transmitted and replaces that signal with the suppressing message.

More flexible, although often less tractable, solutions have been suggested, commonly based on selecting an output behaviour by computing a multi-variable function implicitly encoded in behaviour activation levels, such as voting schemes and spreading of activation. In behaviour-based robot navigation systems, goals are achieved by subdividing the overall task into small independent behaviours that focus on execution of specific subtasks. For example, a behaviour can be constructed which focuses on traversing from a start to a goal location, while another behaviour focuses on obstacle avoidance. In summary, the general constraints on behaviour-based systems roughly mandate that behaviours be relatively simple, incrementally added to the system, that their execution not be serialized, that they be more time-extended than simple atomic actions of the particular agent, and that they interact with other behaviours through the world rather than internally through the system.

Sensor positioning

The signal of GPR (normally used in combination with a metal detector) is strongly affected by a ground surface. If it is not flat and even, a reaction from the ground surface varies much more strongly than that from landmines. In addition, this variation of reaction from a ground surface disturbs imaging of landmines, and occasionally cancels it out. It is consequently mandatory to design an adaptive scanning of the ground surface to reduce the effect of bad positioning on the useful reflection signal. Proximity sensors attached directly to the sensor head can be a very simple solution for a reflexive control scheme to automatically adjust the vertical distance of the sensor head to the terrain. However, although technically more complex and expensive, in order to make a more efficient mapping and scanning of wider areas possible in a minimal time period, cameras

Table 1.4 Summary of the types of topographical map acquisition systems

Sensor type ╲ Feature	Cost	Accuracy	Acquisition speed
Active stereo vision	expensive	very good	fast
Passive stereo vision	affordable	good	fast
ID Laser range finder	affordable	very good	slow*
2D/3D Laser scanner	expensive	very good	fast

and/or laser range finders have to be used. Table 1.4 summarizes the types of topographical map acquisition systems.

The passive stereo system has been selected for the GRYPHON-IV, working in two steps: first the generation of a regular grid that will be superimposed on the terrain image, then the computation of the commands to the actuators of the five-DOF manipulator carrying the multi-sensor-head.

Depth from defocus is another original method used to recover information distances from textured images to displace mine detection sensors above the ground at a given distance without collision. If a fixed focal length camera is used the image of the object placed at a point will produce a sharp picture of the object in the focal plane. The more the object moves from its position, the more the image is blurred. The depth from defocus method uses the direct relationship between depth, camera parameters and the amount of blurring in images to derive the depth. Because the blurring in an image can be caused by either the imaging or the scene itself, at least two images taken under different camera configurations are generally required to eliminate the ambiguity. The practical implementation of this principle gives promising results too.

Robot positioning: tracking

The ability to track the pose of a mobile robot, relative to its environment, while simultaneously building a map of the environment itself, is a critical factor for successful navigation in a partially or totally unknown environment. Simultaneous localization and map building (SLAM) has therefore been a highly active research topic during the last decade. While most existing approaches to SLAM utilize sonar or laser scanners, the use of vision sensors, both stereo and monocular, has also been studied, mainly because vision can yield much richer information about the environment when compared to other kinds of range sensing devices.

It must be noted, however, that SLAM is intrinsically an incremental process, and consequently almost all the published approaches use recursive statistical techniques (Kalman or Bayesian filtering) which, although successful in the short term, suffer from error accumulation over time. Cumani and Guiducci (2005) propose an approach to SLAM which uses a panning stereo head as sensor, and an occupancy grid to store the acquired map. At regular intervals along its trajectory (in our case, at each detection-scanning step), the robot stops and 'looks around', i.e. acquires a set of stereo pair images by panning its head. Point and line features from each image pair are matched and their 3D estimated positions are used to build a local occupancy grid map, which is then merged into a global map after registration to the global reference frame, using the current estimate of the robot pose. The relative robot pose for registration is estimated by correcting the a priori (dead-reckoning) position estimate by map cross-correlation in x and y, while the heading correction is obtained by applying standard ego-motion estimation techniques to the images acquired while the robot moves between two consecutive stops. Combining the visual heading estimate with the translation estimate from map correlation yields a good compromise between speed and accuracy, by avoiding the need to perform a costly correlation search also in the angular domain. Instead of placing the camera on the mobile platform, a fixed camera may be used, located at a safe proximity to the scammed area: this is the approach followed by our research group.

Hong et al., 2001 proposed the use of a colour camera and the choice of the HIS model. In the HIS colour model, the characteristics used to distinguish one colour from another are brightness (I), hue (H), and saturation (S). Brightness embodies the chromatic notion of intensity. Hue is an attribute associated with the dominant wavelength in a mixture of light waves. Saturation refers to the relative purity of the amount of white light mixed with a hue. A red target is put on top of the tracked robot and algorithms have been defined to track the target, parameterized by its position, size and apparent diameter, with a good resolution (0.2 per cent).

Finally, let us also mention that a good positioning accuracy can be obtained with commercial systems such as differential global positioning systems (DGPS) so far as the communications allow their use.

1.6   Conclusions

The development of a robotics system not only depends on the technical aspects and modular components allowing the correct design of the remote controlled platform(s), the application related constraints also have to be carefully analysed in order to achieve the success of the whole system.

The constraints related to humanitarian de-mining, and more generally to outdoor risky interventions, may be summarized as follows: a high level of protection against the environmental conditions (dust, humidity, temperature, etc.), protection and resistance against vibration and mechanical shocks, long and continuous operation time between battery charging/changing or refuelling, wireless communication range depending on the terrain and minefield location, low cost, affordable prices by the use of off-the-shelf components (typical constraint for HUDEM due to the lack of a real commercial market), high reliability, fail-safeness, easy maintenance, easy to use, and application of mature technology.

In this chapter, some of the most relevant aspects of both technical and environmental aspects have been underlined.

1.7   Acknowledgements

We want to mention that this paper includes the contribution of our colleagues, Marc Acheroy, Director of the Signal Processing Centre of the Royal Military Academy (RMA) and our searchers involved in the HUDEM (humanitarian de-mining) project. We also want to thank our partners from the former European Network CLAWAR (Climbing and Walking Robotics) and from the WG HUDEM of the IARP (International Advanced Robotics Programme), as well as all the partners of the European funded projects (FW5-HOPE, SMART, FW6-VIEW-FINDER, GUARDIANS) who also contributed to this summary.

1.8   References

Acheroy, M. (2005) 'Mine action: status of sensor technology for close-in and remote detection of antipersonnel mines'. *IARP WS HUDEM 2005 Proceedings*, Tokyo.

Baudoin, Y. and Acheroy, M. (2002) 'Robotics systems for humanitarian demining: modular and generic approach and cooperation under IARP/ITEP/ERA Networks'. *IARP Hudem'02 WS Proceedings*, Vienna.

Colon, E. et al. (2001) 'An integrated robotic system for antipersonnel mines detection', IFAC Conference TA2001, July 24–26, Weingarten, Germany, pp. 129–134.

Cumani, A. and Guiducci, A. (2005). 'Improving mobile robot localisation and map building by stereo vision'. *ISMCR 2005 Proceedings*, Brussels.

Delhay, S., Lacroix, V., and Idrissa, M. (2004). 'PARADIS: Focusing on GIS Field Tools for Humanitarian Demining'. *The Fifth IARP WS HUDEM 2004 Proceedings*, Brussels.

Doroftei, D., Colon, E., Baudoin, Y., and Sahli, H. (2009) 'Development of a behaviour-based control and software architecture for a visually guided mine detection', *European Journal of Automated Systems* (JESA), Volume 43/3–2009, pp. 295–314.

Faust, A. A., Chesnay, R. H., Das, Y., McFee, J. E. and Russell, K. L. (2005) 'Canadian teleoperated landmine detection systems. Part II: antipersonnel landmine detection', *International Journal of Systems Science*, 36(9) July, pp. 529–543.

Fukuda, T., Matsuno, T., Kawai, Y., Yokoe, K., Hasegawa, Y., Kosuge, K., Hirata, Y., Shibata, T., Sugita, K., Kitagawa, F., Jyomuta, C., Kenmizako, T., Oka, F., Sato, K., Sakai, S., Aomori, N., Sakamoto, Y., Yoshida, T. and Hara, K. (2005). 'Environment-Adaptive Antipersonnel Mine Detection System – Advanced Mine Sweeper'. *IARP WS HUDEM 2005 Proceedings*, Tokyo.

Fukushima, F., Debenest, P., Tojo, Y., Takita, K., Freese, M., Radrich, H. and Hirose, S. (2005). 'Teleoperated Buggy Vehicle and Weight Balanced Arm for Mechanization of Mine Detection and Clearance Tasks'. *IARP WS HUDEM 2005 Proceedings*, Tokyo.

Geneva International Centre for Humanitarian Demining (GICHD) (2004). *A Study of Mechanical Application in Demining*. ISBN 2-88487-023-7, p. 44.

Hong, P., Sahli, H., Colon, E. and Baudoin, Y. (2001). 'Visual Servoing for Robot Navigation'. Int Symp CLAWAR. Karlsruhe, Germany (www.clawar.org).

Ishikawa, J., Kiyota, M. and Furuta, K. (2005). 'Evaluation of Test Results of GPR-based Anti-personnel Landmine Detection Systems Mounted on Robotic Vehicles'. *IARP WS HUDEM 2005 Proceedings*, Tokyo.

Nonami, K. and Aoyama, H. (2005). 'Research and Development of Mine Hunter Vehicle for Humanitarian Demining'. *IARP WS HUDEM 2005 Proceedings*, Tokyo.

2

Robots for non-conventional de-mining processes: From remote control to autonomy

C. PARRA, C. OTALORA and A. FORERO, Pontificia
Universidad Javeriana, Colombia and M. DEVY, CNRS, France

Abstract: This chapter aims to show the results of the perceptual strategy
of a global project being developed at Bogota in Colombia by Pontificia
Universidad Javeriana and at Toulouse in France by LAAS/CNRS. This
cooperation project, funded by the Colombian-French ECOS-NORD program,
includes an investigation about mine sensing technologies, path planning and
robotic platforms among others.

The chapter gives the general context and what are the main challenges when
coping with humanitarian demining, presents a method based on the analysis of
multisensory fused data to improve the landmine detection and its
implementation on an embedded system and presents two robotics demining
platforms called URSULA and AMARANTA. These robots have been designed
and built as mine hunting platforms to be used in developing countries: so these
platforms could be cheap and easy-to-use solutions for humanitarian demining.
The robots contain the perceptual capabilities described earlier in this chapter.

Key words: robotics, de-mining, sensor, data fusion.

2.1   Background

In Colombia, a landmine is any kind of non-conventional weapon designed to be
detonated by the victim, it usually remains active for few days after it was planted
and before it explodes. The fabrication cost is about 10 dollars to create one mine
and nearly 1,000 dollars to deactivate it. This great threat and the difficulty of
finding and destroying landmines make them a problem of global proportions.[17]

According to statistics from the System of Information Management for Action
of Mines of The Observatory of Colombian Mines (IMSMA), in Colombia during
the year 2007, 884 people were victims of anti-personnel mines and unused
ammunition.[25] Unfortunately 193 people died and 691 were wounded in 30 of the
total of 32 regions (Fig. 2.1) that make up the country.[25] These facts show that
humanitarian de-mining is a necessity in Colombia. Although it is a costly activity,
recent studies have shown that besides the benefit of social recuperation of the
affected communities it can be justified from a strictly economic standpoint.[18,8]
Nowadays, the de-mining process is very slow, expensive and most importantly,
it puts a lot of human lives at risk. De-miners are usually injured during the
minesweeping process, and they are exposed to constant danger and the threat of
accidents. Even with the help of dogs, the de-mining process has not improved
much during recent years.[18]

2.1 Locations of accidents with anti-personnel mines in Colombia.

Besides that, there are two main elements to the Colombia landmine problem. The first deals with the great number of different types of landmines, for example, landmines are usually made out of different materials, activated by different methods, and they vary in shape, weight and color. Also, they have different kinds of explosive, so their chemical characteristics vary. All these factors make landmines very difficult targets to detect with 100 percent reliability.

The second feature deals with the different terrains where mines are planted in Colombian territory. Landmines can be placed in deserts, mountains, swamps, roads, forests, etc. In order to overcome this problem, we have conceived soft vision techniques and analysis that can be carried out on places that have been modified by humans and which represent great importance for a community. These places mainly include hills, roads and agricultural land.

This chapter aims to show the results of the perceptual strategy of a global project being developed at Bogota in Colombia by Pontificia Universidad Javeriana and at Toulouse in France by LAAS/CNRS. This cooperation project, funded by the Colombian–French ECOS-NORD program, includes an investigation of mine-sensing technologies, path planning and robotic platforms among others.

The rest of the chapter is organized as follows: Section 2.2 gives the general context and the main challenges when carrying out humanitarian de-mining. Section 2.3 presents a method based on the analysis of multi-sensory fused data to improve landmine detection and its implementation on an embedded system. Sections 2.4 and 2.5 present two robotic de-mining platforms called Ursula and Amaranta. These robots have been designed and built as mine hunting platforms to be used in developing countries. The platforms are cheap and easy-to-use solutions for humanitarian de-mining. The robots contain the perceptual capabilities as illustrated in this chapter. Finally, the conclusions and future work are described in Section 6.

2.2   Humanitarian de-mining: a challenge for perception and robotics

There are two main issues that must be studied to address the de-mining problem. The first one deals with the great number of different types of landmines; they are usually made out of different materials, activated by different methods, and they vary in shape, weight and color. Also, they have different kinds of explosive, so their chemical characteristics are different. This makes landmines very difficult targets to detect with 100 percent reliability.

However, most of the mines have at least a small amount of metal, so metal detectors are still the number one method used in de-mining operations. The system deals mainly with mines made of metal as shown in Fig. 2.2. It could use complementary sensors: two methods based on vision, will be described below. First, the far infrared (FIR) camera has been evaluated in order to predict landmine presence using thermal effects of the landmine on the nearby ground; then color charge-coupled device (CCD) cameras looking at the ground close to the robot in order to detect areas in which landmines could be recently buried were studied. Furthermore, with investigation on the complementary method of detection, it is hoped to achieve the detection of different kinds of mines and to reduce the false alarm rate.

The second issue that must be studied to address the de-mining problem deals with the different terrains where mines can be buried. Landmines can be placed in deserts, mountains, swamps, roads, forests, etc. This means that if trying to build a robot for de-mining operations, its workspace has to be previously defined and limited. To work on humanitarian de-mining, this project is limited to operating on

2.2 Typical metallic detonation landmine.

places that have been modified by humans and that represent great importance for a community. These places mainly include roads, footpaths, urban areas and agricultural lands. Additionally, the system is designed for the Colombian territory.[19] This means that the robot was conceived to navigate in a tropical landscape. This terrain is usually covered in grass, rocks and sand and may contain hills and trees.

A generic robotic de-mining system must cope with two issues: mine detection and robot navigation on a potentially uneven terrain. Moreover, several difficult specifications must be fulfilled:

- A 100 percent detection rate: once a terrain has been de-mined, the accident risk must be null, so that the local population could make this area profitable again, for transport (road, footpaths, etc) or agriculture (fields).
- No danger for human operators: a de-mining machine must be autonomous or remotely controlled.
- A low cost de-mining machine, so that it can be replaced easily if it is damaged during the de-mining process.

The robotics platforms described in Sections 2.4 and 2.5 are the first attempts to satisfy these constraints; a more global system must be studied in the framework of our French–Colombian cooperation dedicated to humanitarian de-mining. The global system will consist of two subsystems:

- An aerial robot equipped with a FIR camera or with ground-penetrating radar (GPR) will fly over the terrain at a low altitude (typically 10 m) in order to build two discrete maps: a digital elevation map (DEM) and a probabilistic mine map (PMM).
- A terrestrial robot will be moved using the DEM, only on the more probable PMM area where mines could be buried, and local confirmation of the mine presence will be given by a metal detector and/or by vision. This global architecture has two main advantages: on the one hand, the more expensive sensors (FIR cameras and GPR) are kept far from the danger, and on the other hand, using the exhaustive exploration of the terrain by the aerial robot, difficult navigation problems will have to be solved only to reach places where a mine could be. In this way, the process will be made less expensive.

Our current robotics developments concern only the terrestrial component of this general system: the design, the implementation and the evaluation of low cost platforms equipped with a metal detector and with vision. It is important to mention that, although the de-mining problem is very difficult and much extra effort is needed, this project is a partial solution designed specifically for the Colombian problem, and can be easily extended to other areas and countries.

Even if the system is aimed at limiting the risk for human operators, considering the degree of difficulty of the task, an operator is required to supervise the de-mining process as a whole. The system is made up of two main components. The first component, called 'control station', is the center of supervising operations. With a PC equipped with video and radio communications, the control station is where all data is collected and shown to the operator, who can teleoperate the robot and configure and control its capabilities. The second component is the 'remote platform'. This is the robot, which sweeps the minefield, looks for mines, and gets all data from the environment.

The next section addresses the sensing challenge, i.e. how to detect landmines, and the following sub-sections will describe two terrestrial platforms designed in order to navigate on uneven terrains, moving sensors above areas potentially contaminated with landmines.

2.3   A complementary multi-sensory method for landmine detection

Nowadays, landmines are one of the greatest scourges for civil society during and after unconventional war. The humanitarian de-mining process is a very complex problem due to the large number of buried landmines, the conditions of the terrain where they are planted, and the large number of variables involved in the problem, making this kind of labor both intensive and risky.

Many people have suggested that technology can fix this problem and almost all research efforts are being directed towards an improved mine detector but due to landmines being made from different kinds of material (resin, plastic, etc), it is a complementary sensor for a metal detector that is really needed.[1]

Humanitarian de-mining robots equipped with a landmine detection system relying only on the data provided by a metal detector face two main problems: first, high false alarm rates, and second, their inability to detect non-metallic landmines.

Robotics researchers have proposed different solutions for the problem of humanitarian de-mining,[1,2,3,4] and nowadays are working on the development of new sensing methods, complementary to the metal detectors that are currently being used. Some of the complementary methods in use are ground penetrating radars (GPR),[5] electromagnetic induction (EMI), quadruple resonance (QR), and infrared vision (IR).[6]

Given that the main goal is to confirm that the ground being tested does not contain an explosive device, the mobile platforms oriented toward humanitarian de-mining applications, developed by Pontificia Universidad Javeriana and LAAS-CNRS (Laboratoire d'Analyse et d'Architectures des Systèmes-Centre National de la Recherche Scientifique), are equipped with a multisensory dataset composed of a metal detector, which has the ability to detect landmine presence if the mine has any metallic components, and an onboard camera looking at the ground close to the robot, used in order to detect areas where landmines could have been recently buried. Moreover, analysis of infrared images has been evaluated at LAAS-CNRS in order to predict landmine presence.

Finally real time image processing with an embedded digital signal processor (DSP) has been proposed in order to implement a navigation algorithm to find and follow semi-structured roads, and to analyze images of the road by taking into account color and texture characteristics.

The successive parts of this section will overview these complementary perceptual subsystems to be integrated in a robotic system dedicated to humanitarian de-mining.

2.3.1   Metal detector

Due to the fact that our project is focused first on the specific Colombian problem, a metal detector was needed because most of the landmines planted in Colombia have a metal component[8] and for this reason metal detectors are a key component of de-mining robots. The Intelligent Systems, Robotics and Perception Group (SIRP) of the Pontificia Universidad Javeriana designed a metal detector presented in Fig. 2.3. After a detailed study of the electromagnetic behavior of different coils and, based on information found in the Coinshooter metal detector by William Lahr,[9] it was concluded that with the correct positioning of the receiving coil over the transmitting coil, it was possible to eliminate the induced electromagnetic field at the receiver inductor.

Once both coils have been coupled to eliminate the induced field (as shown in Fig. 2.4), it is possible to observe oscillating waves at the receiver coil due to changes in the induced field. These changes are produced by the presence of metal objects close to the detector.

The metallic objects can be perceived at distances up to 20 centimeters under the ground (see Fig. 2.5). This distance is good for the objectives of the project

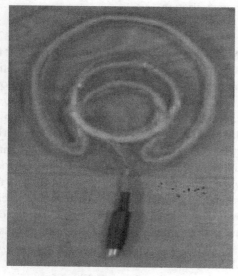

2.3 Metal detector module.

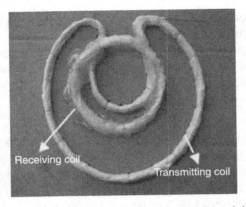

2.4 Coupling of the transmitting and receiving coils.

2.5 Metal detector characterization.

because landmines are usually located close to the surface. The detector has the capacity to perceive metallic objects of 5 grams at distances of up to 3 centimeters under ground. This is a great advantage for the robot because it allows the detection of mines with small metal components.

Another characterization was made in order to determine the detection pattern of the detector. Using a grid that was located over the coupling of the two coils, a metal part of 20 grams was put in each of the squares of the grid and the variations on the level of detection were measured. From this characterization the conclusion was made that the area in which the detector is the most sensitive is the one between the transmitter and receiver coils. It could also be seen that, because of the phase shifting between the emitted signal and the received one, the detector has the ability to discriminate between ferrous and non-ferrous metals. Even though this special characteristic of the detector was not used in this application, the successful development of this metal detector sets the base for future studies on this topic. The detection pattern obtained is shown in Fig. 2.6.

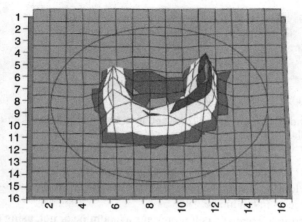

2.6 Detection pattern.

2.3.2   Texture analysis as a landmine detecting method

Sometimes the metal detector could not recognize landmines in an appropriate way, because they had been corroded, waterlogged and impregnated with mud or dirt. This fact makes it necessary to integrate another complementary sensory method, given that the main goal is to confirm that the ground being tested does not contain an explosive device.

For the reason mentioned above, we propose a method where an onboard camera along with an embedded DSP analyzes images of the road by taking account of color and texture characteristics. This extra information will be helpful for fusing data to improve landmine detection.

Landmine detection from color and texture analysis

For the landmine detection method proposed, 800 images were taken. The images processed are 512×512 pixels, and were shot 10 cm from the ground, covering a 25cm × 25cm area. The area of the surface covered by taking the picture at this height is equal to 25cm × 25cm and is smaller than the average area occupied by typical handmade landmines used in Colombia, which are the size of a tuna can (approximately 10 cm in diameter).

This research was limited to two types of terrain, one where the presence of mines is evident (grass), and one in which it is not (small rocks). Four hundred color images were taken for each type of terrain, and 200 of them correspond to images in which tuna cans were buried.

The image processing algorithm begins by dividing every image into 64 equally-sized cells (see Fig. 2.7). Thereby, the main goal is to determine which of the cells within the image represent the presence of a landmine. By dividing the

(a)   (b)

2.7 (a) and (b) Example of processed images divided into 64 equally sized cells.

image into cells, it is not necessary to implement a segmentation strategy – which is harder to develop for the application – given that there are no well-defined borders for the objects in the image.[10,11]

In order to determine whether a cell represents a landmine or not, using texture, a sum and difference histogram was calculated for each cell. Based on Unser (1986),[12] considering two picture elements in a relative position fixed by (d_1, d_2), as in Equation 2.1:

$$\begin{cases} y_1 = y_{k,l} \\ y_2 = y_{k+d_1, l+d_2} \end{cases} \qquad [2.1]$$

The sum and difference associated with the relative displacement $(d_1 = 1, d_2 = 1)$, were defined as in Equation 2.2:

$$\begin{cases} s_{k,l} = y_{k,l} + y_{k+d_1;\, l+d_2} = y_{k,l} + y_{k+1;l+1} \\ d_{k,l} = y_{k,l} - y_{k+d_1;\, l+d_2} = y_{k,l} - y_{k+1;l+1} \end{cases} \qquad [2.2]$$

Compared to a previously developed algorithm, in which co-occurrence matrixes were used in order to calculate the texture features for each cell,[13] sum and difference histograms proved to be a faster and less complicated method to develop the texture analysis proposed in this chapter. The results obtained, either by calculating co-occurrence matrixes or the sum and difference histograms, were exactly the same, proving that both are able to detect texture variations in the image due to the presence of buried objects.

Sum and difference histograms are a less complicated method and provide a faster way to make the texture analysis, and this method also resulted in an important reduction in the robot's memory requirements for developing its landmine detection and localization tasks. For the proposed method, color was chosen as a complementary feature to texture because it is often possible to see the devices that activate the landmines on the surface of the explored terrain. However, for the study developed it was considered that the devices that activate the landmines are not visible to the robot or whoever develops the landmine detection task. Color may be of great help if it becomes possible to characterize the activating device using a color space (see Fig. 2.8).

(a)   (b)

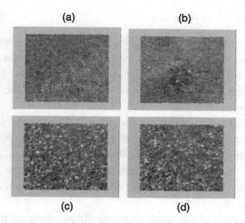

(c)   (d)

2.8 (a) and (b) types of terrain where the landmine presence is evident; (c) and (d) types of terrain where the landmine presence is not so evident.

Eight logic functions were implemented in order to fuse the data supplied by the binary matrices of the texture features that showed their capacity to detect the presence of landmines. The logic functions that showed better detection skills are shown next as texture logic functions 1, 2, and 3 (see Fig. 2.9).

The terms *M + feature X + b* must be interpreted as *Binary Matrix associated to feature X.* The fact that the previous logic functions proved to be good landmine detection algorithms becomes obvious because they are the result of operating outstanding detecting features (homogeneity and contrast) and ensembles of good landmine detecting features (energy, entropy, and cluster) with an AND logic operator. The result of each logic function is an 8 × 8 binary matrix.

Afterward, a threshold in the metal detector is selected for applying an OR logic operator with one of the three texture logic functions selected above. This kind of multiple detection function is proposed as an alternative to using texture analysis as a landmine detecting method, because of its ease of implementation, instead of using, for example, a fuzzy logic algorithm, and because of our major goal of decreasing the computational complexity so that the proposed method can be implemented on a dedicated hardware architecture.

$$(Mclusterb \cup Menergyb \cup Mentrophyb) \cap (Mcontrastb \cup Mhomogeneityb)$$
Texture logic function 1.

$$(Mclusterb \cup Menergyb \cup Mentrophyb \cup Mcontrastb) \cap Mhomogeneityb$$
Texture logic function 2.

$$(Mclusterb \cup Menergyb \cup Mentrophyb \cup Mhomogeneityb) \cap Mcontrastb$$
Texture logic function 3.

2.9 Texture logic functions for a texture analysis.

Evaluation of the landmine detection from vision

The results for the study were divided into two cases. For the first one, landmine detection and false alarm rates were calculated for each of the eight texture logic functions, using the images for the terrain composed mainly of grass. For the second case, the same rates were calculated, this time for the images of the terrain made up mainly of small rocks.

The detection algorithms that use texture logic functions 1, 2, and 3 showed the highest landmine detection rates, and the lowest false alarm rates. The landmine detection rates for these detection algorithms were 98 percent, 96 percent, and 70 percent, respectively, for the terrain composed mainly of grass. For the terrain made up mainly of small rocks, the landmine detection rates were 78 percent, 97 percent, and 97 percent, respectively.

The false alarm rates associated to the same three algorithms were 18 percent, 17 percent, and 7 percent, for the terrain composed mainly of grass, while for the terrain of small rocks, they were 39 percent, 39 percent, and 10 percent, respectively.

It is important to say that without using the data supplied by the binary matrix associated to the color analysis developed, the landmine detection rates for the same three algorithms would have been 77 percent, 68 percent, and 45 percent, for the terrain composed mainly of grass, and 78 percent, 79 percent, and 58 percent, for the terrain composed mainly of small rocks.

Evaluation of a far infrared camera

De-mining involves several stages, the most important being mine detection, which can be done in many different ways. Each detection system involves a different sensor. As mentioned earlier, the most popular sensor is the metal detector, used to detect metal that mines contain. Although the metal detector system is successful, it remains a problem that metal is not just part of mines but can also be found in a lot of objects or even in natural form in the ground.

From this, the resulting problem is that metal detectors give too many false alarms due to the wide spectrum of objects that they are able to identify. Another drawback of the method concerns the object–sensor distance, which must be as small as possible: the sensor must be positioned just above the buried landmine, making the presence of a terrestrial robot or a human operator in the minefield mandatory, and increasing the danger.

One way of reducing false alarms and danger when using metal detectors is to fuse the information given by this sensor with information from other sensors, such as ground penetrating radar (GPR), or far infrared cameras, etc.[21]

When the infrared camera is used as a sensor, the information that is obtained is an infrared image, not that different from a normal image; in fact, it can be processed in the same way as current images are processed. The idea is to identify the pattern of an underground mine in the infrared image to use it as a blueprint and compare it with the images of a possible minefield.

2.10 Infrared images.

This task is not easy to accomplish due to the fact that it is difficult to obtain the blueprint of a real mine in a real minefield.[22] This topic raises another issue that needs to be solved: processing the image to obtain the information required. Therefore we see that image segmentation is a necessary tool in the mine detection issue.

Our first approach to the problem has been made with some tests using an infrared camera, developed by the Micro Systems team at LAAS-CNRS. With this camera, images with different metal elements buried in the ground were acquired. Then these images were analyzed in order to identify what the elements looked like in the infrared image. Finally the images were processed to extract the information concerning the metal objects, such as their localization.

Figure 2.10 presents some infrared images on which this detection method was evaluated. In all of them, it is possible to identify two metal objects with a rectangular shape that are darker than the rest of the image. Our main objective in this test was to extract the two dark rectangles.

We used different segmentation techniques in order to extract these rectangular objects from the images, but all of them consist of finding edges in an intensity image.

The Sobel method returns edges at those points where the gradient of image is maximum. The Canny method finds edges by looking for local maxima of the gradient of the image. The gradient is calculated using the derivative of a Gaussian filter. The method uses two thresholds, to detect strong and weak edges, and includes the weak edges in the output only if they are connected to strong edges. Figure 2.11 shows one result with the Sobel gradient (top) and the Canny one (bottom).

These early results show the difficulty that may arise when trying to process the image without any pre-processing. The image has too much information, so it is not possible to find the objects we are looking for. One way to solve the problem is to reduce the amount of information in the image by creating a new binary thresholded image; with the right threshold the information remaining will be the one we need. Figure 2.12 shows the new binary image, once this procedure has been applied to one of the images.

At this point, it is recommended that the image is filtered to reduce the noise. After that, it is possible to apply one of the segmentation techniques again. Figure 2.13 shows the result of the segmentation, and two morphological operations to complete the mask: dilation and fill.

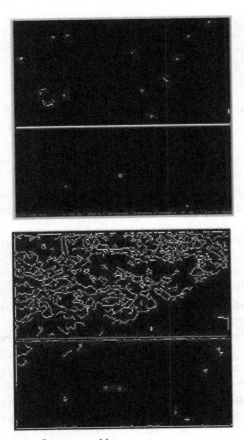

2.11 Segmented images.

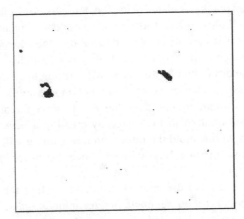

2.12 Binary images.

2.13 Final results.

2.3.3 Visual navigation

The approach used to solve the problem of navigation presented here is very similar to the one used with structured roads: find a path and then follow it.[10,15] Once the road or path has been identified, following it supposes a known problem in robotics. For this reason, this exposition is centered on the identification of the area that represents the path the robot will follow, and the extraction of some parameters needed for the control stage.

Detection of the path to be followed

As the camera is upward and horizontal to the ground, the representation of an image is composed of the sky, above the horizon; and the segment below the horizon contains a maximum of two areas: path and not path regions. Figure 2.14 shows the typical target image and the principal semantic characteristic in it: sky, upwards from base of barn; horizon, the base of barn; and road, from centre of base of barn towards the foreground.

Along with these ideas, other facts are implicit: the picture is taken horizontal to the ground, the sky is in the upper portion of the picture, there is sufficient light to distinguish the road or path, and there are no objects obstructing the view of the road.

2.14 Semantic characteristic in the image.

All these assumptions are easily fulfilled in real conditions. These characteristics are used as semantic information and help to delimit the scope of the problem. The processing of the images is done to exploit all the semantic characteristics mentioned before. At the end, semantic rules are applied to extract the essential information in the image: the route over the navigable terrain.

Real time implementation from a hardware architecture

The method implemented for this work was designed to be executed in an embedded system and has the capability to capture the image in the YCbCr colour space; it also takes the three channels of the space YCbCr, and uses each one as 256 level images. The dimension in all channels is reduced, so the portion of the sky is taken out. The image is reduced by geometric calculation. Loss of information, if some pixels from the ground near the horizon are lost, is not a problem because images in the future it will be corrected, before the robot reaches that point.

Afterward, a threshold is applied to each channel, and all the information in the three components is put together with an AND operator. This way, the information is merged by adding the coincident pixels and extracting those that only represent a hit in one or two channels.

Finally, the resulting image is filtered to reduce the noise and obtain a single path. A median filter with a 5×5 window is used for this purpose. See Fig. 2.15.[14]

After one region is selected as the path, the centroid and the direction of the path are extracted. This enables the control system to plan the path and overtake autonomous navigation.

To accomplish real time processing, the DSP EZKIT-LITE BF561 Blackfin® was used. Three special functions were implemented to bring about the operation in the desired hardware initialization. The Blackfin BF 561 processor and the development board ADSP EZKIT – LITE BF561 are prepared to capture video and audio and also to generate the interrupt and synchronism signals with the three peripherals: a video decoder ADV7183, AD1836 multichannel 96 kHz

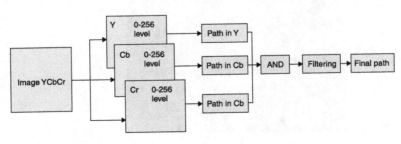

2.15 Flow diagram for YCbCr implementation.

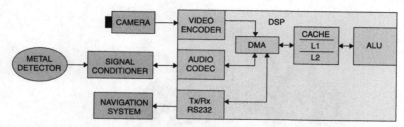

2.16 Implementation diagram.

audio codec and an embedded DMA. The video decoder transforms the analog video NTSC into digital video ITU-656, the audio codec takes the signal from the metal detector and the processor's DMA transfers the video and audio information into the RAM memory in the developed board.

The processing stage starts extracting each channel YCbCr from the image in memory. Then, it calls all the procedures that execute the algorithm: contrast estimation, histogram calculation, threshold calculation, and adds the three channels. Finally, it calculates the moments of the region of interest to extract the centroid and the orientation of the path. During transmission, which is the last step, the information concerning the path (centroid and the orientation) is transmitted by an RS-232 serial interface to a navigation module.

2.4   Ursula: our first approach toward a mobile platform for humanitarian de-mining purposes

Humanitarian de-mining is a difficult and dangerous activity for human beings. For this reason, investigators have made great efforts to develop robots that may lend assistance in the performance of this task.[26]

Mine hunting robots need to have specific characteristics in order to ensure their own safety to achieve the task. Mobility, modularity, portability, and endurance are important factors that robot designers have to consider to ensure the robot's performance.

Pontificia Universidad Javeriana has been working on the design of a robot capable of navigating on high-risk terrains, especially on semi-structured roads that are very common in the Colombian rural areas where mines are usually planted. Research and development of this type of robotic platform was initiated in 2003. The main objective was to design a modular robot composed of independent systems for mine detection. The product of this research has been robot Ursula, presented in Fig. 2.17.[7] It is a six-wheel mobile robot capable of detecting landmines using a metal detector[6] developed at Javeriana University. Ursula could be controlled with a laptop using a graphical user interface or it could be set to an autonomous mode where the robot decided where to go based on a navigation algorithm that followed non-structured roads.[16]

2.17 The Ursula platform.

2.4.1   Proposed architecture

The Ursula project is only concerned with the landmine detection problem; at this stage the mine removal task is not considered.

We will not review all possible robot architectures but rather focus on a specific architecture defined according to the requirements presented before.[24] Based on the background of the de-mining problem and considering all existing variables and possibilities, Ursula was created under a hierarchical structure. It was defined to give the maximum performance and to allow the robot to be an entirely modular system, completely repeatable and easy to modify and actualize. The proposed hierarchical architecture is shown in Fig. 2.18.

The first level looks to have independent micro-controlled systems to operate sensors and actuators mounted on the robot. Also, each module communicates with a central processor on the decisional level.

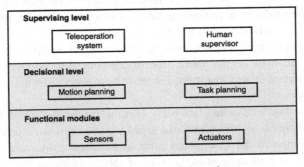

2.18 Hierarchical architecture of the robot.

2.4.2   Sensing modules on the Ursula robot

Positioning and heading modules

The position module measures the position of the robot and provides information on the distance traveled by the mobile. From the optical encoders used in the application, it is possible to obtain the relative position of the robot based on the odometer principles. The positioning module was implemented in a programmable logic device (PLD).

Due to the necessity of correcting the incremental error proper to the odometers, it was necessary to use a geomagnetic sensor (electronic compass). The heading provided by the compass allows the system to determine the deviation of the robot while it is moving. The compass and the encoders work as complementary sensors to give the robot an accurate positioning system.

Proximity sensors

To avoid the problem of possible collision, a proximity-sensing module based on the principles of the time of flight (TOF) theory was designed. Using two ultrasonic sensors (transmitter and receiver), the designed module is implemented with a microcontroller to process the information provided by the receiving sensor. This module allows the operator not only to determine the presence of an obstacle, but also to estimate its distance to the robot.

Acceleration measuring module

To estimate the inclination of the robot while making a sweep of the minefield and to help in the construction of a digital elevation map, a tilt measuring module based on the principles of acceleration measurement with an accelerometer was developed. Taking advantage of the sensitivity of the accelerometer to variations on the acceleration of gravity, it was possible to estimate the tilt of the platform by carrying out some mathematical operations using the signals provided by the sensor. These mathematical operations were implemented in a microcontroller in order to create an independent module open to future changes and improvements. This module gives the possibility of inertial navigation to the robot.

Other sensors for landmine detection and for navigation

Ursula is equipped with the metal detector described in Section 2.3.1, and with a color camera used both in order to evaluate methods proposed in Section 2.3.2 about the detection of areas on which landmines could be buried, and in Section 2.3.4, for visual navigation on uneven terrains.

2.4.3 Mechanical platform

The platform developed for the application is a sealed, rectangular, metallic box made of steel, and with two groups of three wheels. Its construction provides it with the necessary strength to protect it from damage due to collisions and explosions. Inside the platform two motors that provide differential drive to the robot are located, and also, the necessary circuitry to safely execute motions away from mines and obstacles detected in the explored field.

Based on the modular concept conceived for the robot, a subsystem was created that executes the directives for the automotive navigation of this. The module developed for the application controls the movement of the platform and allows simple communication between the processor of the system and the modules that control the displacement of the motors. In addition to these two essential tasks, the module for route control is equipped with analog channels, which provide the whole system with measurements like temperatures, currents and the position of the mobile. In a very specific way, this module has the capacity to control and execute instructions, in an independent manner, for up to four motors. This includes the position and velocity control of each of them by using a digital proportional-integral-derivative (PID) strategy implemented in a microcontroller.[24]

2.4.4   Detector arm

The detector arm in which the metal detector and the proximity sensors are located is presented in Fig. 2.19. It has two fundamental objectives. The first one is to imitate the movement that a blind person with a cane makes while walking; in this way it is possible to emulate the movements made by humans when de-mining a minefield. Second, given the two movements that the arm is able to make, it is possible to reduce the time in which the robot has to sweep the field. The arm has the ability not only to make movements in a vertical axis but in a horizontal axis as well. The movement of the arm in the horizontal axis allows the robot to sweep twice the land that it would sweep if the detector was static in front of the vehicle. So, as the arm is sweeping double the vehicle width every time it makes a horizontal displacement, the time that it will take to sweep the minefield is reduced by half. The vertical movement allows the robot to keep a fixed detection distance to the ground. Both movements the arm is able to make are done by the motors located on the arm.

2.19 The detector arm.

The arm is also equipped with three pairs of ultrasonic sensors around the coils of the metal detector. They are put in the arm as it is located in the front of the mobile. The arm will be the first to interact with the explored field. In the upper part of Fig. 2.19, the spy camera used to give the operator a real view of the field in which the robot is carrying out its mission can be seen. This camera is able to make visual navigation.[20,23]

2.4.5 Decision level

With this first de-mining system, the main task of the robot is to explore the entire minefield, to identify the mine positions and to return safely to the starting point. At this step, the terrestrial robot is in charge of all the de-mining process, including the incremental construction of a digital elevation map and a probabilistic mine map.

In order to accomplish this task, the robot has to be able to execute some basic routines. First of all, path planning and keeping track of its position on the minefield is necessary. This means that the robot has to know where it is at every point and has to plan where to go next. The human supervisor only has to give the robot the dimensions of the minefield (maximum 100 square meters) and the robot will sweep the assigned terrain. To keep track of the position the robot uses its odometry sensors and heading compass. By making the right measures, the robot corrects its orientation. By merging both odometry and the compass it calculates its relative position.

Otherwise, path planning is based on pre-established routes. After analyzing the robot and field characteristics, it was decided to sweep the field sequentially in an S form. That is, the system divides the minefield in parallel lanes one meter wide and positions the robot over the extreme-left lane. Once there, the robot starts sweeping the lane until it gets to the end of it, then moves to the next lane and starts traveling in the opposite direction to the previous one, performing an S shape.

Additionally, to perform a better search of mines, each lane is also divided into cells 25 cm long. By making this division, the system ensures that a scan is made inside the valid detection range of the metal detector. In each of the cells the robot has to stop, scan the cell by rotating its arm one hundred degrees, and actualize all of its state variables. During the rotation of the arm, the robot is stopped and it searches for mines and obstacles, covering an area of 25 cm wide across 1 meter. This takes about two seconds, but because the rotation is repeated, each stop on any cell takes nearly five seconds. This scan procedure is inspired by the movement carried out by human operators in their manual search for mines.

At the end of each rotation and before the robot moves to the next cell, a couple of additional tasks are performed. First, heading is corrected by taking a measure from the electronic compass and after comparing it to the previous one, the robot decides whether to rotate or not. The measure of the robot's tilt is also taken. This measure is later used by the system to build the digital elevation map.

The robot checks if any obstacles were detected during the arm's rotation. If it did not detect anything, the robot moves to the next cell. If anything was detected

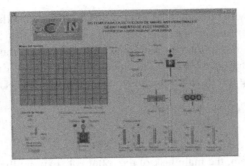

2.20 General supervision screen for the Ursula robot.

(mine or obstacle) the robot sends that information to the control station in order to get authorization to evade the obstacle or to wait for any other commands.

While the robot is sweeping the field, the human supervisor at the control station has all the robot's variables and the state of the task being performed. Speed, pitch and roll, position, temperatures, currents, heading, etc, are all presented to the user in the man–machine interface (MMI) screen, as well as the current status of the mine map and the images captured by a spy camera, only used for visual supervision by the operator.

At any time, the user can decide whether to stop the autonomous state of the robot or to switch it to teleoperation mode to take the robot to any desired position. The general MMI screen is shown in Fig. 2.20.

Finally, the results of the sweep are presented to the user. A digital elevation map and the number of mines detected with their location are reported and printed on the system's screen.

2.5   Amaranta: a legged robot platform

The Ursula robot had an acceptable performance, but presented the following operation issues: it was not able to overcome obstacles on rough terrain due to its short height and it was too heavy to be portable. These problems became the source of inspiration to design a new robot with better mobility, hence improving its functionality.

Legged robots[27,28,29] have been used for humanitarian de-mining because they have advantages over wheeled robots such as omnidirectionality, finite number of ground contact points, and robust navigation on uneven terrain.[30] Adding a wheel to the leg mechanism makes the robot faster than a robot with legs. Leg-wheel mechanisms give the robot a flexible traction, which is suitable for navigating on rough terrain. That is why institutions like NASA are currently working on leg-wheel rovers that serve as a mobile platform on the moon or other planets.[31]

The main objective of the research was to design a lightweight robot, equipped with an onboard metal detector to find buried metallic objects, a vision system that enables the robot to achieve navigation tasks and identify high-risk terrain, and a manipulator arm that facilitates interaction with the environment.

One of the main features of the Amaranta robot based on leg-wheel mechanisms is its modular architecture. Amaranta can be adapted to the type of terrain with three, four, five or six leg-wheel assemblies. The robot's sensory information is provided by two CCD cameras, a GPS, an electronic compass, a barometric altimeter, four force sensors, and 16 quadrature encoders. Amaranta is also equipped with a four DOF manipulator arm so the operator has basic manipulation capacities over the environment.

Amaranta has two modes of operation: the first one is a teleoperated mode where an operator can control the robot with a joystick or a laptop. The second type of operation is an autonomous mode where the robot's navigation is based on a vision algorithm that detects semi-structured roads.[12]

2.5.1 Amaranta's specifications

Figure 2.21 shows an overview of Amaranta. Amaranta has 12 DOF, it weighs 30 kg and is 40 cm high. It is composed of a main framework and four leg-wheel assemblies. Its main framework is a hexagonal shape and contains the power system, control system, communication system, and batteries. Up to six leg-wheel assemblies can be adapted on the sides of the robot, according to the type of terrain.

The mechanical design was made in Solidworks® (Fig. 2.22) and COSMOS® was used for stress analysis of each part. The mechanical structure and the legs are made of aluminum because of its good resistance and low weight.

Figure 2.23 shows a picture of one of Amaranta's leg-wheel assemblies. Each leg-wheel assembly has three DOF allowing the wheel to spin, yaw rotation in the wheel joint, and the flexion or extension of the leg. In addition, it has a four DOF manipulator that has two main purposes: a higher level of interaction with the environment (for example, grabbing small things that could be on the terrain) and carrying the metal detector. Table 2.1 shows the basic specifications of Amaranta.

2.21 Overview of Amaranta.

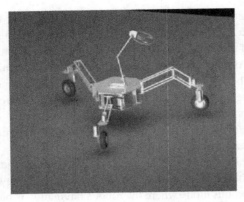

2.22 Mechanical design on Solidworks®.

2.23 Leg-wheel assembly.

Table 2.1 Specifications of Amaranta

CPU	16-bit digital signal controller (× 6)
	32-bit dual core DSP (× 1)
Vision	CCD camera (× 2)
Actuators	DC motor with reduction gear (× 16)
Sensors	Metal detector (× 1), encoder (× 16), force sensor (× 4), GPS (× 1), electronic compass (×1), and barometric altimeter (×1)
Battery	12 V @ 7A-h (× 2)
Weight	30 kg
Height	40 cm
Width	160 cm
Depth	160 cm

Amaranta has three different types of actuators. The joint in charge of leg extension is an 11 W motor with a 53:1 planetary gear head. This motor is connected to a nut-screw mechanism that changes rotational motion to translational motion. An 11 W motor with a 590:1 planetary gear head controls the wheel's yaw and a 4 W motor with a 131:1 planetary gear head drives the wheel's spin. This last motor is connected to the wheel with a 2:1 conical gear.

The proprioceptive sensors of the platform are 16 quadrature encoders and four force sensors. The encoders are used to determine the position and velocity of the motors while the force sensors measure the pressure that each leg is exerting on the ground.

The exteroceptive system has two onboard cameras, a GPS, and a metal detector. One camera is installed on Amaranta's front head and it is used for navigation while the other is attached to the arm manipulator. In this way, the operator controls the arm having a direct view of what he is manipulating. Amaranta also has a Garmin® Personal Navigator® that includes a GPS, an electronic compass and a barometric altimeter. The metal detector is used to find landmines that have metal components. It is installed on the tip of the manipulator enabling the robot to scan nearby terrain before advancing, hence ensuring its safety. The detector measures the variation of the induced magnetic field on a coil and compares it with a signal generated on the DSP.

2.5.2 System configuration

Figure 2.24 shows Amaranta's control system. The processing unit is composed of six 16-bit digital signal controllers (DSC) and one dual core DSP. The DSP executes a navigation algorithm[16] to find and follow non-structured roads and an algorithm to analyze images of the road in order to detect non-metallic landmines.[5] Non-metallic landmines can be detected analyzing color and texture variations over the image that may signal that the surface has been intentionally modified.

The DSP also receives information from the metal detector and, based on hard limiter's thresholds, determines if there is a possible landmine. The metallic objects can be detected at distances up to 20 centimeters underground.

The main DSC receives the control string that commands the robot's motion from the DSP. This string contains the velocity vector and the desired height of the robot's main framework. The velocity vector is obtained by the navigation algorithm or from the operator when it is in teleoperation mode. The main DSC processes this information to calculate wheel velocity and joint positions of each leg-wheel assembly so it can achieve the desired motion.

The main DSC communicates with the other five DSCs via a controller–area network (CAN) bus. Each leg-wheel assembly has a DSC that is in charge of performing the control algorithm for the three motors (two position controls and one velocity control for the wheel). The other DSC is in charge of the manipulator's joints position control and of processing the GPS data. These five

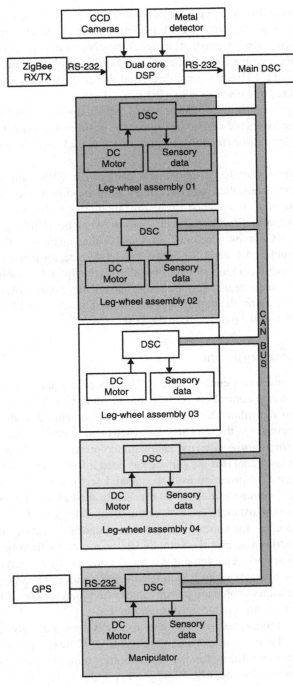

2.24 Control system block diagram.

2.25 GUI that controls Amaranta.

DSCs monitor motor current in order to detect excessive torques that could damage the motors.

Depending on the task to be performed by the robot, Amaranta can be controlled using a laptop or a joystick. A graphical user interface installed on the laptop is helpful because the operator can access all the data provided by the sensors. However, in some types of terrain carrying a laptop can be a problem. In such situations, the operator can control the robot using a remote controller. The remote controller has three modes of operation that enable the operator to control the motion of the robot, to drive the arm manipulator, and to move each leg independently. Figure 2.25 shows the graphical user interface developed in Visual C#.

The Zigbee® protocol enables wireless communication to the main station. The wireless control allows the operator to drive the robot at a distance of 300 m, minimizing the risks for the operator in case of an explosion.

2.5.3 Design highlights

Amaranta's design has three major advantages for navigation on high-risk terrain: flexibility, speed, and modularity.

Flexibility

Each leg-wheel assembly has three DOF that allow the extension and flexion of the mechanical structure (Fig. 2.26), hence controlling the robot's height. This

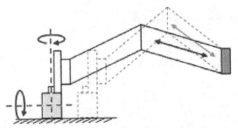

2.26 Three degrees of freedom leg-wheel assembly.

implies that Amaranta's center of gravity can be lowered in order to increase stability and it can be elevated to avoid obstacles that are present on the road. The robot's inclination can be controlled with the purpose of navigating on uneven terrain and to emulate the action of a shock absorber.

To achieve stability, the robot has to have a minimum of three leg-wheels; this way the polygon of support becomes a triangle. Therefore if the robot has more than three leg-wheels installed, each leg can be lifted up individually if a large obstacle is present on the road.

Velocity

Amaranta's configuration is characterized by having no specific front head, therefore it permits the robot to go in any direction in case it gets blocked. In addition, on a flat surface Amaranta can be driven at high speed lowering its center of gravity and applying full power to the motors of each wheel. A robot based only on leg assemblies will not reach high velocities mainly because it depends on its dynamical stability.

Due to its inherent omnidirectonality, if a tall obstacle is present on the road, Amaranta has the ability to achieve 90° turns in order to travel round the obstacle and continue with its commanded task.

Modularity

Mine hunting robots have to navigate on high-risk terrain and ensuring the robot's safety is a priority. This was the reason why the most important design criterion when developing Amaranta was modularity. There is a main DSC that executes main processing but each leg-wheel assembly has its own DSC in order to minimize damage in case of accident. If an explosion takes place, the damaged leg-wheel assembly can be replaced without affecting other parts of the robot. Since the robot may move using three of its four legs, this makes it capable of overcoming obstacles whereas a robot made up of only wheels would be trapped. This is also an advantage in case a leg-wheel assembly gets ruined by an accident.

Another advantage of modularity is portability because in order to transport the robot it can be disassembled, hence occupying less space. It is important to mention that Amaranta's main structure is made of aluminum so it is considerably less heavy than its predecessor Ursula.

2.5.4 Preliminary evaluation

To evaluate its performance, Amaranta was tested on flat and uneven surfaces. It was driven on simple paths like the ones shown in Fig. 2.27.

With a laptop, the operator sent speed, orientation and height commands to the robot. Speed can be adjusted from 0, which means stop to 100, which means full power (integer increments). The orientation command consists of an integer angle between 0 and 359. The height command ascends or descends the main framework.

The robot was capable of following straight lines with any startup position. Middle size objects (10 cm diameter) were placed on the road and the robot was capable of crossing over these.

At the present time the work on Amaranta includes the integration of new vision algorithms and also of inclination control algorithms on the platform.

Although the nut-screw mechanism in each leg-wheel assembly strongly connects the motor and the leg, in future work, a motor with higher torque could be installed directly on the joint in order to achieve a faster flexion or extension of the leg. Additionally, a model with one more DOF in each leg-wheel can be developed, which would allow the platform to walk by blocking the tires. Thus the robot would have two kinds of locomotion and it would be stronger for rough and irregular terrains.

2.6   Future trends

Despite the complexity of the de-mining task, the Ursula and the Amaranta platforms are initially low cost, modular, and easy to reproduce robotic solutions adapted to this task. A large step in functionality has been achieved by working with a global architecture for autonomous navigation.

Both robots have been developed for autonomous navigation on high-risk terrains. Ursula is a six-wheeled mobile robot, while the Amaranta mechanical model is based on a main framework with four leg-wheel assemblies. The

2.27 Examples of simple paths where Amaranta was tested.

electronic and mechanical characteristics of these robots and their advantages for navigation on uneven surfaces were discussed.

The results of our study on the fusion between different sensing modules have led us to the conclusion that the use of sensors in a complementary way provides a more efficient solution to the mine detection problem.[23]

Finally, Ursula and Amaranta will be the basis for future developments. Further studies in motion planning, exploring strategies and detection of non-metallic mines are part of the next stage in our cooperative French–Colombian project.

2.7   Acknowledgements

The project is partly supported by the PCP program (Colombia–COLCIENCIAS- and France), by the ECOS Nord project number COOM01 and by the Pontificia Universidad Javeriana.

Many master students from Pontificia Universidad Javeriana of Bogota (Colombia) have participated in developments presented in this chapter, specifically Henry Carrillo, Carlos Santacruz, Snaider Carrillo, Diego Botero, Alvaro Hilarión and Martha Manrique.

2.8   References

1 J. P. Trevelyan. Landmines: Problems and Solutions. *Asia-Pacific Magazine*. 1998.
2 Maki K. Habib. Humanitarian Demining: Reality and the Challenge of Technology – The State of the Art. *International Journal of Advanced Robotic Systems*. Volume 4, number 2. June 2007.
3 Svetlana, L. Marquez, L, Almeida, A. *Feature-level sensor fusion for a demining robot.* International IARP Workshop on Robotics and Mechanical Assistance in Humanitarian Demining and Similar Risky Interventions. Brussel – Leuven, Belgium. 2004.
4 Smith, C. White, T. *Development of intelligent autonomy for an unmanned Ground Vehicle.* International IARP Workshop on Robotics and Mechanical Assistance in Humanitarian Demining and Similar Risky Interventions. Brussel – Leuven, Belgium. 2004.
5 Lopera, Milisavljevic, Van den Bosch, Lambot, Gauthier. Analysis of segmentation techniques for landmine signature extraction from Ground Penetrating Radar 2D data. *Proceedings of the ANDESCON2004.* August 2004.
6 Campo, Coronado, Rizo, Otálora, Parra. Sistema sensorial para un robot aplicado a la detección y localización de minas antipersonales. *Proceedings of the IEEE Cas Tour.* Bogotá. November 2002.
7 Rizo, Coronado, Campo, Forero, Otálora, Devy, Parra. "Úrsula": Robotic Demining System. *Proceedings of the IEEE ICAR2003.*
8 Hap Hambric. Technology for mine clearance operations. "New approaches to mine detection". International conference on mine clearance technology. Copenhagen. Denmark.
9 Lahr William; COINSHOOTER METAL DETECTOR; Robo Deminer Contest 2002, International Demining Contest; Iran 19–21 de Agosto de 2002.
10 Aviña-Cervantes, Devy, Marín-Hernandez. *Lane Extraction and Tracking for Robot Navigation in Agricultural Applications.* LAAS-CNRS. Toulouse, Francia.

11 Geith Healey. Segmenting images using normalized color. *IEEE Transactions on Systems, Man and Cybernetics*. January/February 2002.

12 Unser, Michael. Sum and Difference Histograms for Texture Classification. *IEEE Transactions on Pattern Analysis and Machine Intelligence*. VOL. PAMI-8, No. 1. January 1986.

13 Coronado, Rizo, Parra. Análisis de textura para la detección de regiones sobre la superficie del suelo. *Proceedings of the ANDESCON 2004*. August 2004.

14 Berttozzi, M.; Broggi, A.; Cellario, M.; Fascioli, A.; Lombardi, P. & Porta, M. Artificial Vision on Roads Vehicles. *Proceedings of the IEEE*. Vol. 90. Issue 7. July 2002. ISSN: 0018-9219.

15 Forero A. & Parra, C. *Extraction of Roads from Out Door Images*. *Vision Systems Applications*. I-Tech Education and Publishing, Vienna, Austria. 2007.

16 Maldonado, A.; Forero A. & Parra, C. Real Time Navigation on Unstructured Roads. *Proceedings of Colombian Workshop on Robotics And Automation (CWRA/IEEE)*. Bogotá. Colombia. 2006.

17 http://www.landmines.org/glc/index-glc.asp. Website of Adopt-A-Minefield. United Nations Association of the USA Adopt-A-Minefield.

18 Organization of American States – OEA. *Reporte de Actividades del Programa de Acción Integral contra Minas Antipersonal de la OEA para la cuarta reunión de los Estados Americanos de la Convención de Ottawa*. September 2002.

19 Sembrando Minas, *Cosechando muerte*. CCCM y UNICEF. Bogotá D.C. 26 September 2000.

20 R. Murrieta-Cid, C. Parra and M. Devy. Visual Navigation in Natural Environments: From Range and Color Data to a Landmark-Based Model. In *Journal Autonomous Robot*. Vol 13, no. 2, pp. 143–168, September 2002.

21 C. Brushini and B. Gros. A Survey of Current Technology Research for the detection of Landmines. In *Proceedings of the International on Sustainable Demining*. Zagreb. Croatia. 1997.

22 O. Gonzalez, A. Katartziz, H. Salí and J. Cornelis. Pre-processing of polarimetric IR images for land mine detection. In *Proceedings of the 22nd Symposium on Information and Communication Theory in the Benelux*. 2001.

23 C. Parra, R. Murrieta-Cid, M. Devy and M. Briot. 3D modelling and robot localization from visual and range data in natural scenes. In *International Conference on Vision Systems (ICVS)*. *Lectures Notes in Computer Science*. Springer-Verlag, 1999.

24 R. Chatila. Deliberation and reactivity in autonomous mobile robots. *Robotics and Autonomous Systems*, 16(2–4), 1995.

25 Vicepresidencia de Colombia. (2008, February), Estadísticas Observatorio de Minas, [Available online]: http://www.derechoshumanos.gov.co/modules.php?name=informacion&file=article&sid=426.

26 J. P. Trevelyan, "Robots: a premature solution for the land mine problem," in *Proceedings of 8th International Symposium on Robotics Research*, Springer-Verlag London, 382–390, 1998.

27 C. Salinas, M. Armada, P. Gonzalez, "A New Approach for Terrain Description in Mobile Robots for Humanitarian Demining Missions," in IARTP/EURON Workshop on Robotics for Risky Interventions and Environmental Surveillance, 2008.

28 K. Nonami, Q. Huang, D. Comiso, N. Shimoi, and H. Uchida, "Humanitarian mine detection six-legged walking robot", *Proceedings of the 3rd International Conference on Climbing and Walking Robots*, pp. 861–868, Madrid, Spain, 2000.

29 S. Hirose, and K. Kato, "Development of Quadruped Walking Robot with the Mission of Mine Detection and Removal – Proposal of Shape-feedback Master-Slave Arm," in *Proc. Of the 1998 IEEE Int. Conference on Robotics and Automation*, pp. 1713–1718, 1998.
30 P. Gonzalez, E. Garcia, J. Cobano, and T. Guardabrazo, "Using Walking Robots for Humanitarian De-mining tasks," in *35th International Symposium on Robotics*, pp. 1–6, 2004.
31 National Aeronautics and Space Administration, Jet Propulsion Laboratory, [Available online]: http://www-robotics.jpl.nasa.gov/systems/system.cfm?System=11

3

Locomotion and localisation of humanitarian de-mining robots

P. SANTANA and L. CORREIA, University of Lisbon, Portugal and J. BARATA, New University of Lisbon, Portugal

Abstract: This chapter lays down a set of pragmatic design principles to build sustainable robotic platforms for humanitarian de-mining. The devised set is discussed throughout the chapter, supported by particular aspects of the actual implementation of the sustainable Ares robot. Specific focus is given to the mechanical design, locomotion control and localisation sub-systems, as they are key components of any robot useful for area reduction, survey phases and remotely controlled close-in inspection. Embodiment and exception-driven design are selected as the two macro design principles. Embodiment is taken into account by considering that both software and mechanical design should occur in a highly intricate way, and by taking into consideration the specific environment in which the robot is going to operate. This is essential for the development of smart heuristics leading to simplifications and consequently cheaper, more energetically efficient and robust solutions. Exception-driven design refers to the balanced effort that should be put on both nominal and exceptional behaviour. That is to say that the design focus should be on robustness, rather than on optimality. As is shown, a good methodology to encompass the proposed design principles is the biologically inspired one. Living systems are highly robust, but nevertheless highly adapted to their niche. Field tests with the Ares robot validate the proposed approach.

Key words: humanitarian de-mining, mobile robots, self-localisation, locomotion control, design principles, biologically inspired models.

3.1 Introduction

As technology for landmine detection improves, landmines become correspondingly more advanced. A result of this arms race can be seen for instance in the diminished metallic content of modern anti-personnel landmines, as a response to the newest developments in metal detectors. Hence, the development of more complex landmine detection sensors will most probably induce the development of more complex landmines, which in turn creates a dependence of affected countries (mostly developing ones) on complex and expensive technology. The net result is that those nations become more and more dependent on the will of richer countries, interestingly the ones more involved in the development of both landmines and corresponding counters. The more dependent countries are on others to handle their humanitarian problems, the less ready they are to mitigate their effects. To some extent, it follows that the development of counter-landmine

63

technology is at the same time the result and the cause of much of the suffering inflicted on developing countries.

Following this reasoning, this chapter describes a line of work focused on the development of sustainable robots filling some of the requirements of de-mining robotic portable kits (Santana et al., 2008a), rather than on the development of landmine detection sensors. We believe that, given this arms race scenario, we can at least contribute to delivering something useful to mitigate the number of victims of the de-mining process. This goal is only within reach if robots are: (1) affordable, (2) energetically efficient, (3) modular and made of locally available components, (4) able to traverse highly complex terrains, (5) robust to component field degradation, and (6) user friendly. These highly robust, versatile and sustainable robots can be useful for area reduction, survey phases and remotely controlled close-in inspection, provided that they are equipped with adequate payload.

An instance of a sustainable robot, the Ares robot, has been developed under the aforementioned requirements, and it is partly described here with the purpose of illustrating a set of design principles. With these design principles we aim to provide researchers with pragmatic rules of thumb to build robotic platforms that are useful and deployable in humanitarian de-mining campaigns. Focus is given to some of the basic sub-systems composing any robot useful for humanitarian de-mining, namely: (1) mechanical design, (2) locomotion control, and (3) localisation.

3.2   Mechanical design of humanitarian de-mining robots

The diversity of environments where de-mining is required has motivated fruitful research on robot locomotion. Wheeled, legged, tracked, and hybrid solutions have been developed with different levels of success (refer to Habib, 2007 for a survey). Wheel-based robotic platforms require less maintenance, are typically cheaper, more efficient, and lighter than their tracked and wheeled counterparts. Most solutions are bulky and complex, making their logistics difficult and their cost unbearable. Exceptions are TRIDEM (Mignon and Doroftei, 1999), Shrimp (Estier et al., 2000) and PEMEX (Nicoud and Habib, 1995) robots, whose designs have been driven mostly by simplicity and compliance with the terrain. The two and three wheels of the PEMEX and TRIDEM robots, respectively, limit their performance in demanding environments. The small size of the Shrimp robot greatly constrains its payload.

The wheeled robot described in this chapter, the Ares robot (Santana et al., 2008a), follows the parsimonious design that characterises TRIDEM, PEMEX and Shrimp, but without compromising its ability to handle complex terrains, or carry considerable payloads. To our knowledge, no other middle-sized robot is endowed with the degree of mobility that the Ares robot possesses, especially at such a low cost and low maintenance requirements. As such is well known, these features are essential for the successful deployment of robots in remote developing countries (Habib, 2007).

The robot should be able to dodge obstacles and reposition itself for acquiring a new view of the terrain as fast as possible and with as little ground disturbance as possible. This relieves operators from the burden of planning complex robot trajectories, reduces the mechanical stress and associated power consumption and, most importantly, it reduces the chances of triggering undetected landmines. With its four independently steered wheels, the Ares robot has been developed taking into account all of these issues.

The degrees of freedom Ares possesses allow it to displace in several modes (see Fig. 3.1). In particular, (1) in double Ackerman mode the robot is able to produce circular trajectories, (2) in turning-point mode it rotates around its own geometric centre, (3) in displacement mode it produces linear trajectories along a limited set of directions, and (4) in lateral displacement mode it moves sideways.

The robot is composed of two main blocks (with two wheels each), the front and the rear ones. These blocks can freely and independently rotate around a longitudinal axis. By having this passive joint, the robot is capable of being compliant with respect to uneven terrain (see Fig. 3.3, on p.69). Passively adapting the robot's configuration according to the terrain's unevenness is preferable to an active solution. This passive solution requires less proprioception (i.e. fewer encoders to describe the robot's configuration), is energetically more efficient, and requires no computational power.

The upper bounds of the volume occupied by the robot are $1.5\,\text{m} \times 0.8\,\text{m} \times 1.5\,\text{m}$. By using bicycle wheels, the tires can be easily replaced in order to better comply with the needs of a given terrain. These two characteristics allied to the 0.3 m height to the ground endow the Ares robot with great flexibility. Being able to surmount many natural obstacles, perception requirements for safe navigation are less stringent. Reduced perceptual requirements mean simpler and consequently more robust sensors, less computation and less power consumption. In Santana et al. (2007) we go a step further by showing that the body of the robot can be used as part

3.1 The Ares robot (left) in its four locomotion modes (right): displacement (top-left), double Ackerman (top-right), turning point (bottom-left), and lateral (bottom-right).

of the perceptual algorithms themselves. This view is in line with the concept of morphological computation (Pfeifer and Iida, 2005), which highlights the role the robot's body has in shaping its control system. If the robot is to be affordable, the designer should exploit the characteristics of the environment in which the robot will be deployed, and this can only be fully carried out if mechanical and software design are linked.

For the sake of completeness, the key hardware components are a Videre Design STOC stereo head, a Honeywell HMR 3000 attitude sensor, a Novatel DGPS OEMV-1 system, four RoboteQ AX3500 boards for speed control of the eight Maxon 150 W motors, a Diamond Systems Hercules EBX PC-104 stack as an on-board computer for direct robot control, and a Pentium M 2.0 GHz laptop running stereo vision, mapping, and navigation algorithms. The on-board computer runs a Slackware Linux distribution, whereas the laptop relies on an Ubuntu Linux distribution.

3.2.1 Mechanical design principles

A few design principles can be drawn from our experience designing Ares:

Mechanical compliance: because mechanical loops are more effective and efficient than software ones, delegating as much as possible of the work relative to terrain interaction to mechanical components is advantageous.

Rich manoeuvrability: a rich set of feasible motions reduces the operator's stress, if the robot is being tele-operated, and helps to limit the unavoidable disturbance inflicted on the ground due to robot motion.

Modularity and simplicity: the more modular and simple the robot is, the easier it is to replace damaged parts and to build one operational robot from two damaged ones.

Scaling up: scaling the robot to its natural size (i.e. taking into account the normal size of objects it should be able to surmount) reduces both perceptual and computational requirements

Morphological computation: if software and mechanical design are linked, synergies, simplifications and robustness are more easily achieved.

3.3 Locomotion control of humanitarian de-mining robots

Classical control theory is usually concerned with optimal set points following. When the set points are highly dynamic (i.e. hardly reachable) and the robot's interactions with the environment are highly non-linear and discontinuous, classical solutions collapse drastically. Bearing this in mind, we proposed (Santana et al., 2006) a novel behaviour-based approach focused on robustness instead of optimality for the locomotion control of the Ares robot. In this approach, each wheel has an independent controller composed of a set of behaviours, i.e. units linking perception to action, generating force vectors according to a given criterion, such as respect kinematic constraints, produce a given turning radius or comply with external forces exerted in the wheel. The resulting force is applied to the wheel's steering actuator.

After showing the advantages of using behaviour-based design for locomotion control, we go further (Santana et al., 2008b) by proposing a richer semantic, which in addition to behaviour fusion, also considers behaviour arbitration among other biologically inspired features. In the case of an arbitration node, the action generated by the behaviour with higher priority is passed on. In a fusion node a weighted average of each behaviour output is considered instead, with the weighting being related to each behaviour's priority. Therefore, arbitration nodes are more interesting and allow the changing of the global behaviour in a sudden and qualitative way. On the other hand, fusion nodes are better for enabling cooperation among behaviours and, in particular, to have one behaviour modulating another behaviour's output.

Figure 3.2 depicts the model of each wheel controller. Data flowing from perception to behaviours, from behaviours to coordination nodes, and from behaviours/coordination nodes to actuators are conveyed through information links. In order to allow hierarchical decomposition and run-time adaptation,

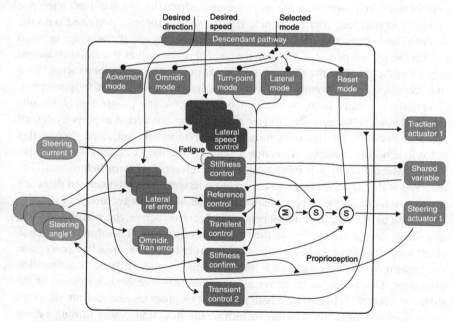

3.2 The behaviour-based wheel controller. Arrows refer to information links. The thicker the arrow, the higher the priority of the behaviour. Lines with dotted and triangular end points refer to excitatory and inhibitory activation links, respectively. Arbitration nodes are denoted by circles with labels S and fusion nodes otherwise.

behaviours can activate (excite) or deactivate (inhibit) other behaviours or information links through activation links. Higher control layers can modulate the locomotion controllers for a given direction of motion, speed, and locomotion mode. Activating a locomotion mode reflects in the activation of the corresponding hierarchical higher behaviour ('Mode' labelled boxes). The desired speed feeds directly the behaviours responsible for controlling the robot's speed ('Speed' labelled boxes), whereas the desired direction feeds into the perceptual entities ('Ref Error' labelled boxes) responsible for computing the wheel's steering error.

As an example of this approach, the double Ackerman mode will be detailed. The reference control behaviour seeks to steer the wheel towards the angle that allows the robot to turn towards the desired motion direction. Hence, each wheel will seek to reach its target regardless of the existence of any other slower or faster wheel. The transient control behaviour adapts the output of the reference control behaviour so as to maintain the Ackerman geometry. If the controlled wheel is drifting from the Ackerman geometry, when compared to the other three wheels, it must be slowed down or speeded up, depending on the situation. As all wheels tend to do the same, eventually they all converge to the Ackerman geometry, provided that the system's parameters have been carefully tuned. Upon reaching an upper threshold of the current in the wheel steering actuator, the stiffness control behaviour suppresses the output resulting from the fusion of both reference control and transient control. Hence, if a too strong force is projected onto the wheel in such a way that contradicts the current steering speed, the wheel is asked to stop for a given period. Each time the current level reaches the aforementioned upper threshold before the stopping period expires, the period is incremented by a time constant and its counting restarted, implementing a kind of fatigue effect. Here fatigue results from an external behaviour, which progressively inhibits other behaviours' outputs for a longer period. When the period expires its default value is restored for the next time the behaviour gets activated. When this behaviour becomes active, it excites a shared variable, i.e. it sets a shared flag to one, which in turn will inhibit all wheels' reference control behaviours. The outcome is that all other three wheels will no longer follow the desired direction in order to focus all the effort in maintaining the Ackerman geometry.

As soon as the stiffness control behaviour becomes active, the wheel stops, and as a result the current applied to the steering motor drops. Therefore, the fatigue effect would never occur in this situation. In order to fix this, an active perception mechanism was developed and implemented in the stiffness confirmation behaviour. This behaviour is active a few seconds before the inactivation of the stiffness control behaviour. When active, it suppresses the output of other behaviours and asks the actuator to turn in the direction it was turning before the activation of the stiffness control behaviour for a while, and then to turn in the opposite direction for the same amount of time. If the obstacle is still present, the current will rise and the stiffness control behaviour will trigger the fatigue effect. Otherwise the wheel will turn in one direction, and then in the opposite one,

3.3 Experimental set-up to test the locomotion control system in which the robot was manually driven along a given path, represented as a black line (left). Snapshots of situations A, B, C, and D (right).

staying roughly in the same place before the stiffness confirmation behaviour was activated. Then, the stiffness control behaviour will timeout and the operations are resumed normally. This swing behaviour of the wheel induces other wheels to follow it in order to maintain the Ackerman geometry. To avoid this, the steering angle percept of the wheel in question is frozen (i.e. gated) during that period, meaning that the other wheels will not be sensitive to the swinging. The transient control behaviour compels the robot to stop when the steering actuator's geometry error reaches a given threshold. This behaviour ensures that the mechanical structure does not collapse in extreme, unexpected, situations.

Fig. 3.3 illustrates the Ares robot moving on highly uneven terrain on which the locomotion controller was tested. The benefits of an exception-driven design become apparent when the forces projected onto the wheels are so high that no other action but a coordinated reflex is a reasonable one. Refer to Santana et al. (2008b) for more details.

Active perception, perceptual gating, fatigue, and behavioural modulation are all ubiquitous mechanisms in nature, and they can be exploited to build up a locomotion control system well adapted to handle exceptional situations. Moreover, these mechanisms reduce the number of sensors required for locomotion control, such as force sensors typically used for force feedback.

As for the mechanical component, let us draw some design principles for locomotion control:

Decoupled design: independent controllers per wheel enable a more intuitive design, which in turn facilitates field system adaptations.

Exception-driven semantics: rather than using techniques for optimal system design, robustness-oriented approaches (e.g. behaviour-based) should be used so as to handle exceptions in a graceful way.

Active perception: acting to influence the sensory flow is a way of reducing the complexity of the sensory apparatus, and consequently the dependency on additional sensors (e.g. force sensors).

3.4   Localisation of humanitarian de-mining robots

Either in tele-operated or autonomous modes, self-localisation is an essential feature for any robot. It is easier to create maps with payload information (e.g. GPR data), autonomous obstacle avoidance can be made more stable, and the operator can be provided with richer situation awareness.

However, in demanding environments, a clear view of the sky is not always available and consequently GPS does not perform with the accuracy required for the aforementioned tasks. Due to this fact, inertial navigation systems (INS) are typically used to complement GPS under the Kalman filtering framework. However, accurate and robust INS are expensive and energy consuming, and so should be avoided, if possible. Thus we are left with odometry, which despite being error prone, has been shown to be able to deliver even in extreme situations (Ojeda et al., 2006). Thus, rather than using expensive inertial systems, odometry has been taken as the primary complement to GPS in the Ares robot.

3.4.1 Wheel odometry

Based on the kinematic model of the locomotion mode in question (e.g. the bicycle model (Wang and Qi, 2001)), the robot's displacement and heading variation, $(\Delta x_i, \Delta y_i, \Delta \psi_i)$, can be estimated by taking into account both steering angle and the travelled linear distance of any wheel $i \in \{1,2,3,4\}$. The redundancy introduced by the possibility of calculating wheel odometry based on a single wheel requires a decision on which wheel to consider. Observations on the dynamical behaviour of the robot suggest that all wheels affect each other strongly, meaning that wheel odometry computation benefits from considering all wheels rather than just a sub-set, as is typically done (Baumgartner et al., 2001; Ojeda and Borenstein, 2004). Bearing this in mind, we proposed (Santana et al., 2008b) a mechanism whereby the estimates of all wheels contribute to the final estimate,

$$(\Delta x, \Delta y, \Delta \psi) = \left(\sum_{i=1}^{4} w_i \cdot \Delta x_i, \sum_{i=1}^{4} w_i \cdot \Delta y_i, \sum_{i=1}^{4} w_i \cdot \Delta \psi_i \right),$$ where the weight

of each contribution, w_i, is given by

$$w_i = \frac{1 - e_i / \sum_{j=1}^{4} e_j}{\sum_{k=1}^{4} \left(1 - e_k / \sum_{k=1}^{4} e_j \right)}$$

being $e_i = |\hat{\delta}_i - \delta_i|$ the difference between the expected angle for wheel i, $\hat{\delta}_i$, and its current steering angle, δ_i. $\delta_i = \frac{1}{3} \cdot \sum_{j \neq i} \Phi(\delta_j)$, where $\Phi(\delta_j)$ returns the steering angle that wheel i should have in order to be coherent (according to the locomotion mode in question) with the steering angle of wheel j.

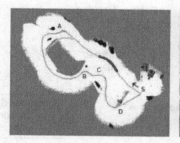

3.4 Resulting 33m × 27m map produced by the robot while following a closed-loop path (left) and snap-shots of situations A, B, C, and D (right).

Intuitively, the odometry computed based on each wheel is weighted for the global estimate according to a function of its estimated error normalised with the estimated errors of the other wheels. Since it is impossible to determine which wheel is failing to meet the kinematic constraints of the locomotion mode, the error of each wheel is computed in relative terms by considering that all other wheels are in the correct position. This simple heuristic helps reduce wheel odometry errors by exploiting, in a simple way, the kinematic and dynamical constraints of the robot.

To test the odometer, the robot was asked to perform autonomously a previously recorded closed-loop path, relying solely on the wheel odometer and a magnetic compass for localisation purposes. In perfect conditions, i.e. with full zero error localisation, the path-following behaviour (Santana et al., 2008b) should take the robot to stop exactly on the place where it started the run (spot 'S' in Fig. 3.4). Actually, after arriving at its final destination, the robot's off-set relative to the position where it departed was measured to be roughly 2 m, which is a remarkably small error, by comparison to the travelled distance. Refer to Santana et al. (2008b) for further details and results.

3.4.2 Visual odometry

In several situations, in particular when wheel slippage is too notorious, wheel odometry is of limited usefulness. Determining whether wheel odometry is still accurate enough (error model) is a challenge in itself. For instance, in a stall situation, wheels may keep rolling despite the fact that the robot is not moving at all. It is thus necessary to have an additional and independent relative localisation mechanism to complement and disambiguate. Since the robot is already equipped with a stereovision system for environment characterisation (Santana et al., 2008c) we can reuse the sensor for localisation purposes as well. For this purpose we have developed a robust visual odometer (Santana and Correia, 2008), which follows the typical three steps decomposition considered since the method's early stages (Matthies, 1989; Agrawal and Konolige, 2006) (see Fig. 3.5):

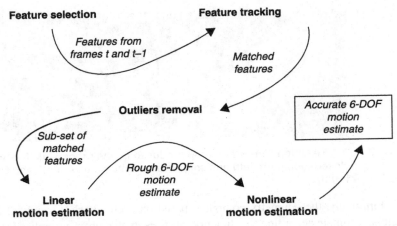

3.5 Visual odometer building blocks.

Feature selection. The first step is to select features, i.e. image regions easily tracked between frames, which could also be called landmarks, in the 2D image, obtained from the left camera of the stereo head. The 3D location of these features, relative to the camera, must be accurately obtainable to be considered. Roughly, they must have a good and unique correspondence to an image region in the right camera.

Feature tracking. In the second step, features from the current frame are matched (i.e. tracked) against the features from the previous frame. The more noisy and homogeneous the environment is, the more matching verification sub-steps are necessary to remove outliers, and consequently the larger the computational load gets.

Motion estimation. Having the 3D positions of the elements of each feature pair, resulting from the stereo computation, the motion estimate can be obtained with a closed-form solution. The numerical instability of such approach requires a preceding robust statistical treatment so as to remove outliers. Then, a fine-grained estimation step is performed in a non-linear minimisation step, taking into account the starting point estimation obtained with the closed-form solution.

The set of matched features generated by the feature-tracking step is certainly populated with a considerable amount of outliers, i.e. mismatched features. The subsequent motion estimation step is robust enough to handle these outliers. However, their anticipatory removal reduces the computational cost of the estimation process, and potentially improves its quality. This section describes our removal algorithm for the outliers, which lies between the feature tracking and motion estimation steps.

The lines connecting each pair of matched features define the scene's optic flow, caused by the robot's motion. From a simple observation on the resulting optic flow (see Fig. 3.6), its homogeneous nature becomes conspicuous. In other

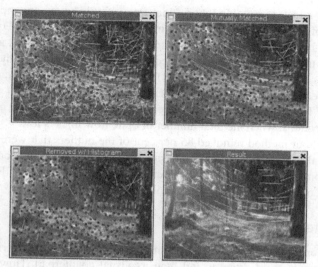

3.6 Feature tracking results. Dots correspond to features in the first and second frames. Basic feature tracking step results (top-left and top-right). Optic-flow lines (outliers) further removed with our method (bottom-left) and final result (bottom-right).

words, optic flow directions change smoothly. This observation can be exploited to distinguish novelties (i.e. outliers) relatively to an optic flow background.

This cue motivated Grinstead et al. (2006) to propose a method to remove optic flow outliers, which follows in the spirit of the approach proposed by Adam et al. (2001). Based on the assumptions that the robot only moves on a surface plane, it was possible to avoid the computational cost of generating rotation hypotheses, as postulated in Adam et al. (2001). The planar motion assumption results in the guarantee that correct optic flow lines share the same direction. Therefore, detecting outliers is done by simply rejecting optic flow lines whose direction differs by a given amount from the average optic flow direction. However, a change in the roll and pitch angles induces a more complex pattern in the optic flow as indicated in Fig. 3.6. That is, there is no longer a common direction. However, since the most dramatic changes in the optic flow are caused by roll, we found out that considering the image's left and right half separately suffices to reduce the problem dramatically (at least in typical all-terrain environments). Although the optic flow lines belonging to each half do not share the same direction yet, their differences are typically much smaller than those of outliers. Consequently, dividing the image (in this case in two equal parts) is a good practice to detect novelties in the optic flow, and consequently outliers.

Our proposal can be summarised as follows. Two histograms capturing the frequency of optic flow directions of each half image, obtained from the left camera, are computed. All paired features whose corresponding optic flow contributed to a histogram's bin whose final frequency is below an empirically defined ratio of the histogram's bin to maximum frequency, are labelled outliers. Intuitively, this process reports paired features as outliers if they exhibit an optic flow direction that is too anomalous when compared to the average. Furthermore, paired features whose corresponding optic flow contributed to a histogram's bin whose final frequency is below a residual empirically defined threshold are also rejected. This latter process removes paired features, which represent residual optic flow directions, when the total number of optic flow lines is low.

On average, our method removed 85 per cent of the outliers that were missed by the original outlier removal mechanism. In addition, the visual odometer managed to provide localisation with a remarkable 5 per cent error. Refer to Santana and Correia (2008) for further details and results.

3.4.3   State estimation

Alone, visual and wheel odometers are of limited utility. In very rough terrain, the wheel odometer is more prone to fail than its visual counterpart. However, in deficient light conditions the visual odometer fails more notably. Moreover, since both approaches are incremental, it is necessary to integrate their output with an absolute localisation system, such as a GPS. Therefore, a fusion mechanism is required and can be described as follows.

The 6-DOF state estimator first predicts the current state based either on visual odometry or on wheel odometry, in case the former fails to produce accurate results. Visual odometer accuracy is estimated by the number of tracked features between previous and current frames. Then, the subsequent correction step uses both GPS and attitude sensor in order to guarantee global convergence of the estimate. The convergence speed towards the global references is a function of the sensor's error estimates. GPS readings are only considered if the robot's speed is above an empirically defined threshold (typically 0.3 m/s) in order to provide added stability to the position estimate.

3.4.4   Localisation design principles

Based on the success of both wheel and visual odometers, a set of design principles can be suggested:

Embodiment: as for the wheel odometer, the exploitation of simple heuristics in the visual odometer (obtained from thorough empirical observation of the effects of action on perception) enabled the reduction of computation and an increase in robustness. Once more we have shown that if the robot's body, sensory apparatus, and algorithms are all seen in an integrated framework, parsimonious design is more easily achieved.

Redundancy: exploit the fact that redundancy exists in physical systems to produce more robust solutions. Histograms of optic flow directions and weighted contribution of local motion estimates obtained from each wheel, are examples of this.

Sensor reuse: reusing sensors, such as stereovision for both environment characterisation and localisation, is a way of reducing both cost and power consumption. In addition, full exploitation of the available resources increases redundancy and consequently robustness.

Intertwined models: wheel odometry, visual odometry, GPS and INS should be mixed in their early stages in a synergetic way (e.g. the heading change obtained from wheel odometry could be used to guide the feature matching process in the visual odometer). This however raises the problem of modularity and synchronisation of modalities with different time scales.

3.5   Conclusions

Developing robots for humanitarian de-mining is a challenging task mainly due to the strict constraints on cost, energy efficiency, robustness and simplicity of the solution. Bearing this in mind, a set of design principles were devised and discussed in this chapter, supported by particular aspects of the actual implementation of a sustainable robot.

Disembodied solutions, in which software and mechanical design are seen as decoupled tasks, are unable to be, at the same time, robust and parsimonious. As we have shown, if body and software are developed in tandem, simplifications through the form of smart heuristics emerge naturally. Simplification is not just about facilitating the design phase; it is also about robustness (fewer components subject to failure) and autonomy (less power consumption).

Robots for harsh environments, such as those where humanitarian de-mining usually takes place, should be designed for robustness rather than for optimality. Therefore, at least as much effort as that invested in the design of the robot's nominal behaviour should be put into making it robust. Exception handling should be embedded in the semantics used to design the system as much as possible. See for instance the behaviour-based locomotion controller, which naturally enables addition of sensorimotor rules to cope with exceptional situations. Fatigue and other biologically inspired mechanisms also introduced some level of creativity into the system, helping it to cope with deadlocks.

In short, embodiment and exception-driven design are two macro design principles that should be kept in mind when building sustainable robots for humanitarian de-mining tasks.

3.6   Acknowledgements

This work was partially supported by FCT/MCTES grant no. SFRH/BD/27305/2006. The developments herein reported were carried out in partnership with the Portuguese company IntRoSys, S.A. under the AMI-02 project for the

Portuguese Ministry of Defence. A special thanks to João Lisboa for his contribution to the design of the Ares mechanical structure and to Carlos Cândido for his support in the development of the work herein presented.

3.7   References

Adam, A.; Rivlin, R. & Shimshoni, I. (2001) ROR: Rejection of outliers by rotations, *IEEE Transactions on Pattern Analysis and Machine Intelligence*, vol. 23, no. 1, pp. 78–84.

Agrawal, M. & Konolige, K. (2006). Real-time Localization in Outdoor Environments using Stereo Vision and Inexpensive GPS, *Proceedings of the 18th International Conference on Pattern Recognition (ICPR '06)*, Vol. 3, pp. 1063–1068.

Baumgartner, E.; Aghazarian, H. & Trebi-Ollennu, A. (2001) Rover Localization Results for the FIDO Rover, *Proceedings of the SPIE Photonics East Conference*, pp. 28–29.

Estier, T.; Piguet, R.; Eichhorn, R. & Siegwart, R. (2000) Shrimp, a Rover Architecture for Long Range Martian Mission, *Proceedings of the Sixth ESA Workshop on Advanced Space Technologies for Robotics and Automation (ASTRA '00)*, December, 5–7, Netherlands.

Grinstead, B.; Koschan, A.; Gribok, A. & Abidi, M. (2006) Improving video-based robot self-localization through outlier removal, *Proceedings of the 1st Joint Emer. Prep. & Response/Robotic & Remote Sys.*, February 11–16, pp. 322–328, Salt Lake City.

Habib, M. (2007). Humanitarian Demining: Reality and the Challenge of Technology – The State of the Arts. *International Journal of Advanced Robotics Systems* (special issue on *Robotics and Sensors for Humanitarian Demining*), Vol. 4, No. 2, pp. 151–172.

Matthies, L. (1989) "Dynamic Stereo Vision," Ph.D. dissertation, School of Computer Science, Carnegie Mellon University.

Mignon, E. & Doroftei, I. (1999) TRIDEM: A Wheeled Robot for Humanitarian Mine Clearance, *Proceedings of the International Symposium on Humanitarian Demining (HUDEM)*, 29–30 April, Brussels, Belgium.

Nicoud, J. & Habib, M. (1995) Pemex-B Autonomous Demining Robots: Perception and Navigation Strategies, *Proceedings of the IEEE/RSJ International Conference on Intelligent Robots and Systems (IROS '95)*, August 1995, pp. 419–424, Pittsburgh.

Ojeda, L. & Borenstein, J. (2004). Methods for the Reduction of Odometry Errors in Over-Constrained Mobile Robots. *Autonomous Robots*, Vol. 16, No. 3, pp. 273–286.

Ojeda, L.; Reina, G.; Cruz, D. & Borenstein, J. (2006). The FLEXnav precision dead-reckoning system. *International Journal of Vehicle Autonomous Systems*, Vol. 4, No. 2, pp. 173–195.

Pfeifer, R. & Iida, F. (2005). Morphological computation: Connecting body, brain and environment. *Japanese Scientific Monthly*, Vol. 58, No. 2, pp. 48–54.

Santana, P.; Cândido, C.; Santos, V. & Barata, J. (2006) A Motion Controller for Compliant Four-Wheel-Steering Robots, *Proceedings of the IEEE International Conference on Robotics and Biomimetics (ROBIO '06)*, 17–2 December, Kunming, China.

Santana, P.; Barata, J. & Correia, L. (2007) Sustainable robots for humanitarian demining. *International Journal of Advanced Robotics Systems* (special issue on *Robotics and Sensors for Humanitarian Demining*), Vol. 4, No. 2, June.

Santana, P.; Correia, L. & Barata, J. (2008a) Developments on an Affordable Robotic System for Humanitarian Demining, in: *Humanitarian Demining: Innovative Solutions and the Challenges of Technology*, M. Habib (Ed.), I-Tech Education and Publishing,

Vienna, Austria.

Santana, P.; Cândido, C.; Santos, P.; Almeida, L.; Correia, L. & Barata, J. (2008b) The Ares Robot: Case Study of an Affordable Service Robot, *Proceedings of the 2nd European Robotics Symposium (EUROS '08)*, 26–28 March, Prague, Czech Republic.

Santana, P. & Correia, L. (2008) Improving Visual Odometry by Removing Outliers in Optic Flow, *Proceedings of the 8th Conference on Autonomous Robot Systems and Competitions*, 2 April, Aveiro, Portugal.

Santana, P.; Santos, P.; Correia, L. & Barata, J. (2008c) Cross-Country Obstacle Detection: Space-Variant Resolution and Outliers Removal, *Proceedings of the IEEE/RSJ 2008 International Conference on Intelligent Robots and Systems (IROS '08)*, 22–26 September, Nice, France.

Wang, D. & Qi, F. (2001) Trajectory planning for a four-wheel-steering vehicle, *Proceedings of the IEEE International Conference on Robotics and Automation (ICRA '01)*, Vol. 4.

4

Sustainable and appropriate technologies for humanitarian de-mining

E. E. CEPOLINA, Snail Aid – Technology for Development,
Italy and M. ZOPPI, University of Genova, Italy

Abstract: The chapter outlines the recent trends in humanitarian de-mining practices, analysing the shift from the concept of absolute need of full clearance to the recently developed idea of land release through technical survey. In this context, the possible use of lower-reliability technologies is discussed and the need to use more sustainable and appropriate technologies raised. The advantages of using locally available machines and agricultural technologies in particular and converting them to assist de-mining operations are outlined after having underlined how much science and technology can contribute to human development. The chapter then looks at the research on sustainable technologies for humanitarian de-mining being undertaken worldwide before presenting the efforts of the PMARlab of the Department of Mechanics and Machine Design of the University of Genova, working in collaboration with the not-for-profit association Snail Aid – Technology for Development. Recent results from PAT and Disarmadillo projects, regarding the development of a machine that costs less than €5000, built around a powertiller, are presented, before introducing a new project called Locostra, co-funded by the Italian Ministry of Economic Development and the Italian Institute for Foreign Trade, that the authors are currently working on. The Locostra machine is built around a small off the shelf 4WD tractor equipped with radio remote control and blast resistant wheels.

Key words: humanitarian de-mining, appropriate technology, agriculture, participatory design, land release.

4.1 Introduction: could 1/100 of Bosnia Herzegovina's tractor population help the country to become landmine impact-free within a year?

In 2007, in Bosnia Herzegovina, 170 km^2 of land were released to public use through area reduction, using 21 accredited de-mining machines (ICBL, 2008).

Area reduction is the process by which the initial area indicated as contaminated is reduced to a smaller area by gathering information (UNMAS, 2008a), either qualitative or by detailed technical interventions, using machines for partially clearing the terrain or verifying the absence of landmines. Therefore, land released after an area reduction process is not fully cleared and contains an element of risk that explosive hazards may have been missed. Although full clearance activities will not guarantee that an area is completely free of mines, land released after area

78

reduction is generally considered to contain a higher residual risk. Nevertheless, this process is increasingly being used to hand over suspected contaminated areas to local populations in a quicker and more efficient manner. Procedures followed by six countries have been analysed by the Geneva International Centre for Humanitarian De-mining (GICHD, 2008a) and standards on how to release land by means other than full clearance are being prepared by United Nations Mine Action Services (UNMAS). Different criteria apply to different scenarios but generally a suspected hazardous area (SHA) that has never been a place of accidents and has already been processed by any ground-engaging machine is released as safe. Therefore, when a physical verification of the presence or absence of mines is needed, it is not necessary to use accredited machines, but virtually every locally available tool can be used to collect data on the real extent of the mined area.

The estimated area that still needs to be cleared in Bosnia Herzegovina is 1738 km^2 (ICBL, 2008). The amount of agricultural tractors in the country in 2007 was approximately 30,000 units (WRI, 2007). If only 300 units, one per cent of the tractors available in Bosnia Herzegovina, could be temporarily equipped with low-cost ground processing tools and light armouring for assessing the presence of landmines, assuming that each one could have the same productivity of one of the 21 machines used for area reduction in 2007, about 8 km^2 per year, the problem of landmines in Bosnia Herzegovina could be potentially solved or drastically reduced to small, confined, highly contaminated areas in less than one year.

Nowadays, when the need for new land is increasing daily due to the world food crisis, we are finally arriving at a very important change: the worldwide acceptance of slightly less accurate humanitarian de-mining operations in order to gain a quicker and more efficient land release to the local population.

This chapter outlines the recent trends in humanitarian de-mining practices, analysing the shift from the concept of absolute need of full clearance to the recently developed idea of land release through general and technical survey. In this context, the possible use of less reliable technologies is discussed and the need for the use of more sustainable and appropriate technologies arose. The advantages of using locally available machines, and particularly agricultural technologies, and converting them to assist de-mining operations is outlined after underlining how much science and technology can contribute to human development. The chapter then looks at the research on sustainable technologies for humanitarian de-mining taking place worldwide before presenting the efforts of the PMARlab of the Department of Mechanics and Machine Design of the University of Genova.

Recent results are presented from tests on the powertiller, a machine costing less than €5000, developed partly at the PMARlab and partly in Jordan with the collaboration of Norwegian People's Aid (NPA), and further work and development are discussed.

The work on participatory agriculture technologies for humanitarian de-mining is now jointly carried out by the University of Genova and the recently formed not-for-profit association, Snail Aid – Technology for Development, using young engineers who have been previously involved into the project as students and led by Emanuela Cepolina.

4.2   2009: deadline in crisis

The year 2009 saw an important appointment for the mine action community: March 2009 was the first deadline for completing clearance in the mine-affected countries that signed the Ottawa Treaty in 1997. Unfortunately, two thirds of them did not make it. Fifteen countries, including Bosnia Herzegovina, asked for a deadline extension of between one and 10 years, leaving a great percentage of their territories unsafe and committing their weak economies to support expensive mine action practices for longer.

The year 2009 also saw many people around the world starving due to the world food crisis, which began in 2007. Different sources (Iacobucci, 2008 and Vadim, 2008) estimated that almost a third of Tajikistan's 6.7 million inhabitants would not have enough to eat during the winter. Many more countries, often those already plagued by landmines left over from recent wars, are facing famine, such as Somalia, Myanmar, Mozambique and Egypt.

In this context, many mine-affected countries around the world are assisting with an informal reappropriation of suspected hazardous areas using the local population before official de-mining activities take place (Fig. 4.1).

4.1 De-miner clearing the field around a house in Cambodia already reoccupied by its owner.

This is often the result of a choice made after balancing different risks: the risk of being injured by a landmine is generally preferred to the risk of dying from starvation. And it is generally an agreeable choice as, on average, less than 3 per cent of cleared land turns out to contain landmines or other explosive hazards after being processed (GICHD, 2008a).

The need of a quick land release to agricultural and grazing use is pressing. If a change toward cheaper and less time-consuming mine action practices has always been desirable, now it finally becomes imperative.

As often happens, it is during crisis (literally meaning judgment in Classical Greek), that solutions come. In fact, we are currently assisting in a dramatic change in mine action practices: the acceptance and standardization of the persistency of a residual risk after clearance (UNMAS, 2008b), in opposition to the traditional requirement of removal and/or destruction of all mine and unexploded ordnance (UXO) hazards from the specified area to the specified depth (UNMAS, 2003).

4.3   Land release by means other than full clearance for humanitarian de-mining

According to the recently developed concept, land release is achieved through one of these, increasingly more costly and more technical, actions: general survey, technical survey or full clearance (Fig. 4.2). While general survey does not involve the use of clearance or verification assets but only collecting and analysing new

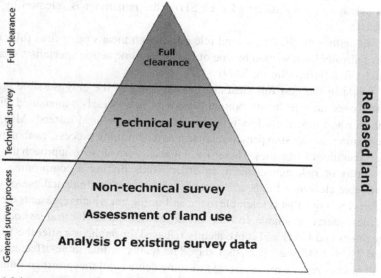

4.2 Land release concept (Source UNMAS, 2008).

and extant information on specific SHA through interviews with local stakeholders and by visual field inspections, technical survey is a detailed technical intervention with clearance or verification assets into a SHA. It aims to confirm the presence or absence of mines and explosive remnants of war (ERW) following the implementation of a general survey. A technical survey may not be required if a general survey suggests that full clearance should be applied and can involve the use of technologies not accredited for proper clearance (UNMAS, 2008b).

Toward the fulfilment of the treaty obligations and the pressing needs of handover of clear land to local population, the wise approach recently adopted is to use current resources more efficiently by managing information better: re-defining the actual size of minefields and spending expensive resources and dedicated equipment only on those areas that have a high probability of actually containing mines. The remaining areas that are released through general survey and technical survey are not physically cleared, or at least not completely, as they are believed not to contain mines. Clearance of a small percentage of land and ground processing of the whole area can occur in a way that gives confidence that there is no reason to believe there are mines in that part of the SHA. If assumptions were wrong and explosive hazards are found, the entire area is fully cleared. Decisions whether to release land by general survey or technical survey are taken in accordance with the rules written in the organizational standard operational procedures (SOPs).

Since 2007, Ethiopia, with technical assistance from the Norwegian People's Aid (NPA), has released several hundred square kilometres through general and technical surveys in more than 1,000 communities in its ongoing land release programme. In 2008, HALO reported that in Angola they only physically clear an average of one-quarter of each SHA (the remainder is released by survey) (ICBL, 2008).

Guidelines on SOPs for land release through means other than full clearance have already been written by one of the major mine action specialists worldwide, Andy Vian Smith (Smith, 2009).

While in the past full clearance was considered the only possible method to hand over land to local communities, it is now widely understood that other quicker and inherently less reliable methods can be used instead. Also, manual de-mining, the most expensive and accurate de-mining process, cannot guarantee the clearance of all mines. Therefore, it makes more sense to approach the problem in terms of risk management, in other words finding a compromise between available clearance funds, technical feasibility and the intended use of the land: achieving a risk that is tolerable to the end-users, but which represents the best use of the resources available. In Western Europe the residual risk in areas contaminated by mines and UXO and subsequently released for public use after the 1914–1918 and 1939–1945 wars persists today. The reality is that mine-affected countries will always remain mine-affected to some degree and they must continue to take measures to deal with that threat (GICHD, 2005).

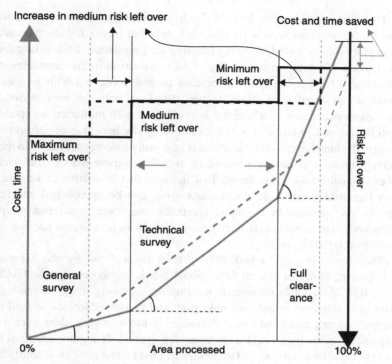

4.3 Land release: analysis of cost, time and risk left over by general survey, technical survey and full clearance.

As risk is defined as a combination of the probability of occurrence of harm and the severity of that harm (ISO Guide 51: 1999 (E)), mine clearance can be seen as a risk reduction process where hazards (mines) in an area are removed to a certain level of reliability (GICHD, 2005). As each one of the actions leading to land release has a certain cost per unit area processed, increasing from general survey, technical survey and full clearance, three straight lines with increasingly higher inclination can be used to approximately represent the cost over area processed (grey lines in Fig. 4.3). As, generally, reductions in risk come at an ever-increasing cost, and the risk left over by general survey will be higher than the one left over by technical survey, that will be higher than the risk left over by full clearance, at the percentage of area released by each action a different level of risk left over (black lines in Fig. 4.3) can be associated. Moreover, considering a behaviour for time per area processed analogous to the behaviour of cost per area processed, attributing an increasingly higher cost to general survey, technical survey and full clearance, the scheme in Fig. 4.3 can be considered to apply, showing benefits induced by using an increased number of machines for technical survey on the total time and cost to release land.

From the scheme, it can be seen that by increasing the number of machines in use in the technical survey process, and therefore increasing the percentage of area processed by technical survey (shifting the boundaries of the technical survey area along the grey arrows in Fig. 4.3), a reduction in cost and time can be achieved. At the same time an increase in area processed with an associated residual risk rated as medium occurs; therefore, both the area released with maximum acceptable risk and the area released with minimum acceptable risk will be reduced. While this shift does not bring any benefit over a short period it does in the long haul. As full clearance is inherently a slower process than technical survey, more land can be released by technical survey than by full clearance over the same time. This means that the area that is waiting to be processed, and therefore contains an unacceptable risk, can be approached and released earlier. An increase in the mine clearance rate, even when that increase is associated with a reduction in reliability, results in accelerated socio-economic benefits (GICHD, 2005).

Therefore, although it had already been pointed out by the Mine Action Equipment: Study of Global Operational Needs carried out by the GICHD in 2002 (GICHD, 2002), the need to develop new and very simple technologies for what was formerly called area reduction, and is now addressed as land release through means other than full clearance, is more evident than ever. Current expensive technologies and work intensive manual de-mining practices can be used to clear land not released otherwise and the largest possible quantity of new, simple and less reliable technologies converted from mature technologies available locally can be used for area reduction.

As long as no machine is expected to conduct clearance without manual or mine detection dogs (MDD) follow-up, a wide range of machines can be used in any way that does not increase risk to staff (Smith, 2009). A quick solution to the landmine problem could already be available in mine-affected countries.

4.4   Local agricultural technologies for land release for humanitarian de-mining

According to the *Mechanical De-mining Equipment Catalogue*, edited by the GICHD in January 2008 (GICHD, 2008b), the total number of de-mining machines working in mine action programmes around the world is less than 650. The market for humanitarian de-mining mechanical technology is small and driven by donors rather than request. Machines are high cost, specialized equipment, mostly heavy as they are designed to destroy mines, and expensively armoured to ensure the safety of the operator on board or equipped with complex control systems when operated remotely. Buyers of these technologies are often donors rather than mine action programme coordinators. Machines are marketed in the same way as military equipment and the price is often part of a package that is negotiated in confidence. Therefore, costs and number of units are not comparable with those of other de-mining technologies, directly bought by programmes, such as sensor technologies (Cepolina et al., 2004).

While these expensive and high resistant machines could be employed where full clearance is needed, other less expensive and more widely available machines need to be conceived for gathering the information required to release land through technical survey. These machines mainly need to verify the absence of mines in the given area. If they encounter an explosion the area needs to be re-categorized and needs to be fully processed by proper clearance. This means that machines need to process the ground and to resist or not be severely damaged by only one explosion at a time, while keeping the operator safe.

As their job is to process the ground, agricultural machines originally conceived to work the soil could be efficiently employed. Agricultural technologies are largely available everywhere and in different sizes. Where they are not already available their presence might be desirable to increase the capability to produce food by farm mechanization. As suggested by the NABARD association (NABARD, 2007), to achieve the desired average farm power availability of 2 kW/ha, necessary to assure timeliness and quality in field operations in India, agro services centres could be established. There, machinery could be provided as and when it is needed on custom hire basis to the small and medium farmers who cannot afford to purchase their own machinery. In the same manner, in parallel to agricultural machines, the agro service centres could also provide machines for de-mining applications, based on agricultural machines. They could develop the modifications required to effectively address the de-mining problem locally, then hire out these machines and provide assistance.

Agricultural technologies are mature and simple, easily reparable in every developing country in local, not specialized workshops. The modularity of agricultural technologies is another advantage; the same tools can be mounted on different tractor units and replaced by dedicated agricultural tools when de-mining operations are over.

While we are assisting with the general re-localization of trades, also occurring in mine action as many operations are being handed over to local commercial companies or NGOs (Filippino and Paterson, 2008), particular importance is placed on not introducing newer technologies dedicated to de-mining, but using local ones. Machines developed or re-adapted locally have a lower initial cost, shorter downtime and lower repairing cost. They would not be under-utilized as much as proper de-mining machines, which are often left next to minefields unused due to the lack of spare parts or the experience needed to fix them, which has to come from abroad. The machines would be much more sustainable than technologies imported from western countries, which are not designed with local conditions in mind.

Moreover, involving local technicians in the re-design of new or improved technology helps to reduce the dependency of local communities on donor help as well as facilitating local human development (Cepolina, 2006).

Human development requires three broad areas of need and capability to be satisfied. First, adequate provisioning for basic human needs – food, shelter, clothing, health and other necessary services – through both public and private effort. Second, development of basic human capabilities: these are, in Amartya Sen's conception, the substantive freedoms that a person needs to lead 'the kind of life he or she enjoys'. They include health, education, knowledge and skills. Third, space for people to apply their innate and acquired assets, individually and communally, to achieve higher welfare outcomes. The defining features of such space include an environment of stability (political, social and economic), of democracy, a human rights culture, and freedom for all to operate as political and economic agents (UNDP, 2005).

Human development is a means towards an even higher ideal, human freedom. An especially important freedom is that of choice. At certain levels of deprivation, people cannot exercise basic choices that are essential for a dignified human existence, choices that every human being should have as a matter of right.

Low incomes constrain access to education and poor people in poor countries lack the basic capabilities that support innovation and the transfer, adaptation and diffusion of technology. And technology is strictly linked to development as it is a tool for, not just a reward of, growth and development. Technology is like education – it enables people to lift themselves out of poverty.

The word technology comes from the fusion of two Classical Greek words: τέχνη (transliterated as techne), art or craft or every kind of knowledge that finds a practical application and λόγος (transliterated as logos), word. In light of etymology, the close relationship between technology and human development becomes clear. Advances in science and technology, in terms of progresses in knowledge on how to do practical things, have been driving the development of human beings from the Stone Age. The desire to innovate and find ways to do labour-intensive activities using less manpower is innate in humankind. The ability to do it, by dealing with problems and finding practical solutions, can be defined as technology, and also the outputs of the process, the practical solutions achieved, can be defined as technologies. When developed by the same people who need it and not driven by a consumerist market but on the basis of real needs, technology is not only sustainable and suitable to the environment where it is designed to work but can also promote the human development of end-users. It does so incrementally, according to a circular path (Fig. 4.4).

When the innovation process starts and a new technology is being produced, people participating in its development acquire knowledge and stimulate their creativity. The technology produced raises the efficiency with which they do things, possibly helping them to achieve a better income and extend their achievement possibilities over time. Time and resources, saved thanks to the increase in efficiency brought by the new technology, can be invested to meet higher needs such as leisure or the desire to do research, experiment and discover more knowledge. The ability of people to participate actively in social and political life increases as well.

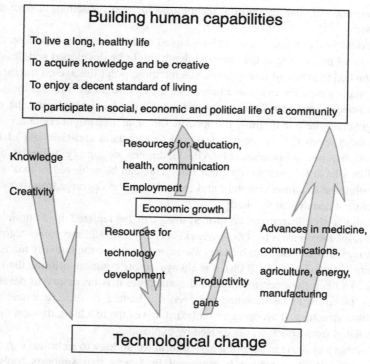

4.4 Links between technology and development (Source: Human Development Report 2001, UNDP).

Unfortunately, the innovation process leading to the development of new technologies and the enhancement of human development, described by the box in Fig. 4.4, cannot start in many developing countries. Poor people, living in insecure, conflict environments, are seldom innovators. They lack the basic resources needed to develop new technologies that could help them solve their own problems and start the innovation incremental process.

An external, input is needed (possibly once only) to start the process. This could be in the form of a participatory design process, bringing together researchers of western countries and local people, through all stages of the technical design.

Empowerment is an integral part of many poverty reduction programmes. It is seen as essential to promote human development and human freedom to help individuals and communities to function as agents for the improvement of their own wellbeing. Empowerment is not only about the state providing resources and opportunities, it is about citizens taking responsibility for self-improvement (UNDP, 2005).

The handover of all mine action activities to local entities, who can perform the majority of the work and can gain skills while participating in the creation and maintenance of new agricultural technology for area reduction, is desirable and necessary.

Agricultural machines need to be adapted to the de-mining task. Special tools for ground processing at the required depth might be developed and designed to be attached to standard linkages such as the three-point linkage on tractor units.

In many cases, the explosive threat that a SHA might be affected by is known before operations start. Information collected from local people and the military can generally help in defining the specific threat an area might contain.

Besides, even if they are not designed to withstand anti-tank (AT) landmine explosions, just anti-personnel (AP) landmines or simply to resist damage, machines have to keep the operator safe. This can be achieved in two ways: by operating the machine remotely and by isolating the operator from the machine structure when driven manually.

While a simple remote control system can be realized in a modular way, relatively inexpensively (Kostrzewski et al., 2007), and semi-autonomous machines are considered to be a key element to improve total quality management in mine action (Eriksson, 2008), it is always a more complex solution than manual drive. To keep the operator near to the machine, it is necessary to devise some shock isolators to be mounted between the handle or driving wheel and the machine structure. If an operator on board drives the machine, the seat must also be isolated from shock waves caused by explosions.

Another key issue in adapting agricultural technology to technical survey is the armouring. If the machine is equipped in a way that supports tools at the front, there is no need to spend too many resources in armouring and only a light shield is needed to protect the delicate parts. Otherwise, if the machine was originally conceived to support tools at the back, as is the case the majority of times, then a system to protect the structure from damage caused by the possible explosion of mines underneath has to be developed. A good approach in this case is to design special blast resistant wheels that do not transmit the shock associated with an explosion to the chassis, either by deforming flexibly or by releasing energy through frictional pins. Research on blast resistant wheels, shock isolators and modular remote control systems, if flexible enough to be adapted to different agricultural machines, would enormously benefit technical survey processes.

Where animals are already employed in agriculture and no machines are available, agricultural implements generally drawn by animals could be adapted to process the soil as required by technical survey. In this case, to protect the animal it is necessary to place the tool at the front; implements have to be re-designed to embed wheels to support the tool frontally.

Examples of agricultural tools that could be used in technical survey, with little adjustment, are shown in Fig. 4.5(a)–(d).

We argue that technology developed on the basis of real needs, in a participatory way together with people who expressed these needs, contributes significantly to

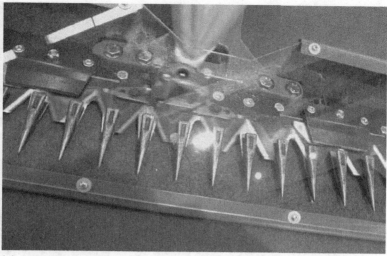

(a)

(b)

4.5 **Agricultural implements suitable for technical survey use: (a) vegetation cutter; (b) small machine with hydraulic driven front loader; (c) potato digger; (d) potato digger with vibrating system.**

(c)

(d)

4.5 Continued

their human development, by enhancing their knowledge and creativity and hopefully by solving a real problem and improving their lives. And such technology is appropriate and sustainable because it is developed with end-users who live in and know the environment in which it will operate, using local available materials and resources. Not only that, but it can be upgraded further, when and if it is needed, without any more help from outsiders. This technology and the knowledge behind it are actually owned by the users.

4.5 Participatory design and appropriate technology for humanitarian de-mining

The concept of participatory design was born in the context of research and development in the field of agriculture. It arose as an answer to the Green Revolution, which in the 1960s and 1970s brought a dramatic increase in world food production, despite a massive industrialized, monoculture agriculture implementation, made possible by the transfer of new technologies, such as fertilizers and new types of more productive breeds, from western countries to developing countries. This industrialized type of agriculture led to the loss of local traditional practices and biodiversities as well as to pollution and erosion of the soil, due to the introduction of chemical products, in a way that by many was seen as unsustainable (Chambers, 1997).

Participatory research and development is an approach to learning and innovation, in which a wide range of actors, users and stakeholders are required to participate. It redefines the role of local people from being merely recipients and beneficiaries to actors who influence and provide key inputs to the process and at the same time enhance their knowledge (IDRC, 2005). In the agricultural field, the approach involves collaboration between researchers and farmers in the analysis of agricultural problems and testing of alternative farming practices. It makes use of participatory tools, which are very simple and creative communication tools, designed to help two-way communication by largely exploiting visual techniques, such as ranking and rating tools, where end-users are asked to rank and rate images representing possible choices. The final aim is to improve farmers' livelihoods by proposing sustainable solutions suitable to the local environment and exploiting local resources. At the end, solutions achieved by participatory research and development are intellectually owned by all participants who become less dependent on outsider help and are later encouraged to start their own innovation process.

Following the move towards the idea of human centred development, formalized by the United Nations Development Programme (UNDP) in 1990, the concept of appropriate technology started forming. It was introduced by the economist E. F. Schumacher in the book *Small is Beautiful* (1973).

From the acknowledgment that our economy is not sustainable, Schumacher proposes a shift from technology transfer to the design of appropriate technology, which is defined as technology that is designed with special consideration to the environmental, ethical, cultural, social and economic aspects of the community it is intended for (Darrow and Saxenian, 2008). With these goals in mind, appropriate technology typically requires fewer resources, is easier to maintain, and has a lower overall cost and less of an impact on the environment. A related term is also intermediate technology, used to define technology that costs more or is more sophisticated or complex than those currently in use in developing nations but still much less costly, or more accessible, than those tools that would be used in developed nations. According to a definition elaborated by the British architect J. Turner a truly appropriate technology is a technology that does not make users dependent on the system over which they have no control (Darrow and Saxenian, 2008). A very good example of what can be defined as appropriate technology is the 100 dollar laptop, developed within the One Laptop per Child project, carried out at the Media Lab of the Massachusetts Institute of Technology (MIT). The outcome is a very cheap laptop specifically designed for children in developing countries, robust, consuming low energy and rechargeable by human power.

Although addressing problems of sustainability and dependency from external donors, appropriate or intermediate technologies do not specifically involve a full participatory design process. End-users are involved at the initial stage in the definition of requirements but are not generally asked to give their contribution in following design choices. Instead, an increased participation in technical design can be found in the development of commercial products. Some companies use techniques known as quality functional deployment (QFD) or user centered design (U-CD) to improve their products by better identifying and meeting customers' needs. These processes are aimed at producing more usable and easier to sell products but are not aimed at enhancing end-users' knowledge of the technology or the concepts behind it.

Another field, in which participation finds space is the information and communication technologies (ICT) field, where collaborative software is being implemented. It consists of software providing different users with the ability to create and manage information through synchronous or asynchronous communication using different channels, mainly supported by the Internet. It has been suggested that Metcalfe's law — the more people who use something, the more valuable it becomes — applies to such software. Collaborative software represents a good tool to favour participatory design, by providing a means through which knowledge of different users, not necessarily located in the same place, is shared and improved.

An example for all, of successful use of collaborative software is Wikipedia, the encyclopedia where definitions are inserted by the same users, which has more than two million articles and currently ranks among the top 10 most-visited websites worldwide. Another interesting collaborative project carried out at the MIT Media Laboratory is ThinkCycle, a digital platform where end-users in developing countries meet with university students in Western countries and provide requirements to them for designing some useful technologies they need. The project was born from a student idea of creating an online database of well-posed problems and evolving design solutions (Sawhney et al., 2002). In practice, the platform supports problem formulation and design exploration through the extensive use of forums, allowing ongoing dialogue among many dispersed participants, plus file sharing. At the moment, there is no suggested structured design cycle to follow and little use of visual communication, and less educated end-users are not able to take part in the design.

Although differing in audience, all participatory design methodologies analysed are iterative processes. Sometimes, as in the case of user-centred design, low fidelity prototypes are used to give end-users a physical representation of that particular aspect of the final product that is under investigation, and allow them to make relative design choices.

Like every process involving discussion and consensus building, participatory design activities take longer than traditional ones; instead of the time to market, functionality, intellectual ownership and sustainability are key aspects of the final product. Building trust and developing community capacity takes time. Moreover, time is needed both at the planning stage and at the implementation stage of participatory sessions: use of imagination is essential both to maintain a high level of attention and also to enhance creativity by suggesting new ways of thinking.

Participatory tools are already used in mine action activities. Mostly, they are employed in mine risk education (MRE), to enhance a two-way communication between practitioners and listeners: active, dynamic and creative communication is considered very important to convey a message more efficiently and promote behavioural change (GICHD, 2004). While the contribution of participatory techniques to MRE is more obvious, it is not the only mine action activity in which end-users are involved. The contribution of local communities to clearance is as old as de-mining itself. In Afghanistan, where the first humanitarian de-mining operations began in the 1980s, local people were trained as de-miners, given basic tools and sent back to their villages to start clearing. The programme was considered unsuccessful and the experience was not repeated in future projects. Nevertheless, in some countries such as Cambodia, informal de-mining by villagers with no proper tools or training continued on an as-needed basis, covering the inability of the formal mine action sector to meet the needs and priorities of communities living in mine-affected areas. Only recently have attempts started to bring informal de-miners up to the standards of professional teams through training. In Cambodia, two similar projects started in 2004, under the Mine Advisory Group (MAG) and the Cambodian Mine Action Centre (CMAC). They created de-mining teams by recruiting the poorest members of mine-affected communities to work as de-miners in their own communities under the formal supervision of the organization. De-miners are trained for several weeks by experienced staff of the organization and work following the organizational standard operational procedures (SOPs). Usually they are hired for a short term and paid less than a traditional de-miner, as they live at home, not in campsites like traditional de-miners.

Incident records are promising, being zero for MAG and a minor one for CMAC (Bottomley, 2005); they suggest that the approach leading to an increase in involvement of local communities and a handover of responsibilities to them is a possible way forward.

In the case of the rake system in Sri Lanka, local communities also participated in the development of a new de-mining technology. Simple rakes, locally sourced, were used by both government and Tamil forces in de-mining operations, before Norwegian People's Aid (NPA) was brought into the country to advise the local de-mining agency. Rather than changing the approach, NPA refined the tool and introduced it into their SOPs. The rake system is one of the simplest and most efficient de-mining methods implemented worldwide. Like all techniques born in the field at community level, it is well adapted only to the environment where it was designed or to similar ones; it is not a universal clearance method.

4.6   Agricultural technologies on the humanitarian de-mining technology market

Between technologies available on the market and reported in the *Mechanical De-mining Equipment Catalogue* (GICHD, 2008b), we have analysed with more attention those derived from the agricultural sector. Even if their price is relatively high, as they are still sold on the small humanitarian de-mining market, it is very useful to see in which application different agricultural tools are used to address mine clearance. Except for the Pearson minefield tractor, all other prime movers do not come from the agricultural sector, being either specially-designed remotely controlled units or commercial excavators and front-loaders.

The Pearson tractor represents an exception because of the idea that drove its conception: minefields often occupy productive land, which can be exploited even if contaminated by landmines if a sufficiently armoured tractor is available. Therefore, the first Pearson tractor, called the Pearson survivable demining tractor (Fig. 4.6), was a ten tonne 4×4 tractor, armoured and equipped with open cage wheels, designed to resist detonation of anti-personnel landmines. Similar to the previous version, which was based on a John Deere medium-sized 6920 tractor with a 110 kW powered engine, the Pearson survivable demining tractor (Fig. 4.7)

4.6 Pearson survivable demining tractor (GICHD catalogue, 2003).

4.7 Pearson survivable demining tractor (GICHD catalogue, 2006).

4.8 Pearson sifter (GICHD catalogue, 2006).

can support any kind of agricultural implement, attachable at the front or at the rear to the three-point linkage attachments which are standard for every type of tractor. At the same time, it also supports specialized tools, designed by the Pearson Engineering company appositely for humanitarian de-mining, some more traditional such as the mine roller or the magnet and others more similar to agricultural tools commonly used for different applications, such as the sifter, the comb, the heavy soil loosener and the lighter spring tine cultivator.

The Pearson sifter (Fig. 4.8) is designed to be pulled by the tractor or other suitable prime mover of minimum 50 kW power with mechanical and hydraulic power take off. It is based on a commercial agricultural de-stoner. It cuts the soil at the selected depth with a set of horizontal shares; these lift the soil on to a slatted vibrating conveyor. The soil falls through the slats leaving mines, stones and large clods deposited in a windrow behind or to one side of the sifter for manual removal. Before using it, the ground must be prepared by removing vegetation and rolling to break up clods.

Similar kinds of sifters are also produced by other manufacturers of humanitarian de-mining equipment such as Armtrac Sifter and the KZ Sifter produced by the Iraqi company Khabat Zangana Company (KZC) (Fig. 4.9 and 4.10). Hendrik Ehlers Consult (HEC) in Namibia produces a slightly different sifter called the HEC Rotar Mk-I sifter system (Fig. 4.11), which is a front-mounted sifter drum used to scoop a bucket load of soil and process it. Processing happens when the

4.9 Armtrac sifter (GICHD catalogue, 2006).

4.10 KZ sifter (GICHD catalogue, 2006).

4.11 HEC Rotar (GICHD catalogue, 2006).

sifter drum is rotated: particles smaller than 40 mm × 40 mm fall from the drum through a round steel bar mesh. Material that remains in the drum is visually checked for the presence of anti-personnel landmines by de-miners.

All sifters are designed for anti-personnel landmines. The Pearson mine comb (Fig. 4.12), instead, is an anti-tank mine clearing tool. It is mounted in front of the prime mover and operates by combing large objects including mines gently to the ground surface from which they can be disposed of in an appropriate manner. The mine comb is very similar to the mine plough (Fig. 4.13 and 4.14), another tool designed by the same company, Pearson Engineering. The mine plough undertakes military de-mining and it is pushed by a tank. A raking action brings mines to the surface and moves them to each side of the vehicle. Another tool called the mine clearing cultivator, which uses the same principle of raking the ground and lifting up landmines in front of a prime mover, has been developed by the US Army CECOM Night Vision. Also, different models of fork shaped tools exist, both for anti-personnel and for anti-tank landmines. Generally they are attached to excavators.

Details of the research and development of a machine for humanitarian de-mining employing a sifter system similar to the ones presented above are found in the paper 'The MiSa 1, an agricultural machine with demining capabilities', written by Detlef Schulz and published in the *Journal of Mine Action* (Shultz, 1999). Unfortunately no other reference or picture is available. The core of the machine is the vibrating sifter, which performs an almost automatic mechanical separation of soil and metal parts, through the use of a metal detector and a few other sensors. The sifter is mounted on a tracked chassis, which is

4.12 Pearson mine comb (GICHD catalogue, 2006).

4.13 Pearson mine plough (GICHD catalogue, 2006).

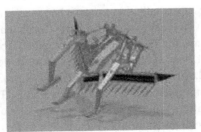

4.14 Pearson mine plough mock-up (www.pearson-eng.com).

operated by radio remote control. It was specifically designed for clearing military areas in Germany contaminated with ammunition, but also takes into consideration requirements typical of Bosnia, Cambodia and Africa.

4.7   Development Technology Workshop and other sustainable ideas for humanitarian de-mining

The Development Technology Workshop (DTW) produces the cheapest machine on the humanitarian de-mining technology market. It is a small non-profit company undertaking product design and technology for developing countries, operating in Cambodia. It is affiliated to the Development Technology Unit of the University of Warwick in the United Kingdom (UK), which is a research centre specialized in technologies for the rural development of developing countries. The philosophy adopted by DTW is different from those driving other companies who are producing de-mining equipment at a profit (Gasser and Terry, 2000), whereas sustainability and suitability to local production are fundamental characteristics of the technology DTW produce. They attempt to develop technology that can promote development and is suitable for developing countries. Recently they started working on new projects regarding the production of low-cost technologies for blind people, such as mobility canes and Braille writing machines, and work on irrigation and sanitation as well as energy production.

The DTW's stated mission is to improve the livelihoods of poorer people through job creation and by introducing locally manufactured equipment into the small industries sector. DTW products go through the six stages represented in Fig. 4.15.

The major advantages of this approach are the production of cheaper equipment and the creation of a long-term, sustainable indigenous capacity specialized in the production of de-mining equipment.

At the moment there is one Development Technology Workshop in Cambodia. There, 33 Cambodians work, of whom half have disabilities, together with three expatriate engineer/technicians. Local (Cambodian) end-users are involved in the development of new technologies in two ways: at the first important stage of the design process, the needs assessment, and during production, as they are hired for the actual manufacturing of the technology. This process was not developed when work on Tempest began, with its first design being conducted in the UK. Although

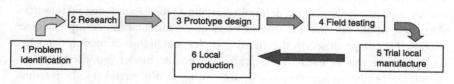

4.15 DTW product stages.

4.16 TEMPEST (Source: GICHD mechanical equipment catalogue 2006).

designed for the Cambodian environment, the Tempest machine (Fig. 4.16) has been donated to Bosnia and Africa. This proves two important facts: the common assumption that locally made low-cost technology produced in developing countries is of low quality is wrong and the idea of involving end-users in the design process works.

Another project that is worth investigating because it shows the relationship between landmine clearance and other development initiatives is the Modular De-Mining project undertaken by De-mining Systems UK Ltd, represented by Roy Dixon. It is at a less advanced stage than DTW, being mainly a proposal waiting to be funded, but it has some interesting ideas behind it. The project aims at developing self-sustainable technology (Dixon, 2004) for humanitarian de-mining. The assumption behind it is that with the current method of foreign government donation the de-mining industry can only expand relative to the amount of funding that is available to it. Considering the problem relative to other development problems, such as HIV or other pressing needs, funding for the de-mining industry seems to be peaking and starting to decrease in size.

Therefore, De-mining Systems suggests applying engineering business principles to humanitarian de-mining by proposing a new technology that generates its own funds. The modular de-mining machines (Fig. 4.17) would simultaneously clear vegetation and landmines, cultivate the ground and plant crops. After harvesting, these crops could be sold on the open market and the money used to pay the costs of de-mining, giving a return to the original investors. Although, in our opinion, the sustainability of the modular de-mining machines proposed is compromised by the highly complex and high-tech modules attached to the tractor that is used as the prime mover, the model proposed for self-sustaining de-mining activities is very interesting as it is intended to help reduce the dependency of local communities on outside donors.

4.17 Modular de-mining (Source Roy Dixon).

4.18 ARJUN machine (Source Andy Smith).

Finally, another project we analysed was the infield development of the Arjun machine (Fig. 4.18) by the Indian NGO Sarvatra in collaboration with NPA Sri Lanka and Andy Vian Smith, in Sri Lanka. This is interesting because the overall cost of each machine is approximately US$30,000, less than a good four-wheel drive vehicle.

For more than a year, Sarvatra has been using adapted construction-site machines to provide the platform for vegetation cutters. The platforms are low-cost earth-moving machines with hydraulic arms designed to carry excavation buckets for use on building sites. The hydraulic arm reaches out into the minefield while the machine stays on safe ground.

With NPA assistance, the platforms have been armoured and new ground preparation tools have been made. The platforms are being used in advance of the manual de-miners. The tools remove dense undergrowth and scarify the ground, raking it to depths beyond that needed for confident mine clearance. The mechanized vegetation cutters and rakes are not designed to expose or detonate mines, but merely to break up the ground so that manual de-mining can be rapidly conducted behind the machine.

The use of widely used plant machinery means that spare parts are readily available, servicing is simple, operation is straightforward and the machine can be converted back to conventional uses by retrofitting its original tools in a matter of minutes. This versatility is unique and guarantees that the machines will not have to be scrapped when their de-mining roles are over.

Arjun is another good example of simple technology, built around commercially available mature technology, designed by end-users.

4.8   Participatory agricultural technology for humanitarian de-mining: PAT and DISARMADILLO machines

The work being undertaken at the PMARlab of the University of Genova, in collaboration with Snail Aid, is aimed at expanding the approach used by DTW by involving local end-users in the whole design process of new technologies. We ensure the integration of end-users' contribution to the first needs assessment stage and try to involve them in proper design choices, after having explained the main underlying concepts. This is easier when dealing with mature technology that most of the people already know and possibly use, such as agricultural machines.

Moreover, we include other stakeholder contributions in the design, such as technicians working in workshops, blacksmiths, researchers studying agricultural subjects, people experienced in earth moving machines and everybody else who believes that he or she could contribute. The final result is a technology that end-users own both physically and intellectually.

To do this, we developed a new design methodology representing a further step from the concept of appropriate technology, promoting participatory technology. It consist of a structured iterative design cycle, called the Snail system, in which end-users are involved at all choice-making stages; the final product of such a process is sustainable, appropriate, reproducible and upgradeable with no need of more external help.

As face-to-face communication is not always possible we set up different collaborative tools to use through the project website (www.dimec.unige.it/PMAR/demining). By using these tools it is possible to store documents such as presentations, Excel sheets or Word documents online and to view them either on the website where they are stored, or on any other website where they have been embedded. Different users can access the documents and modify them; all changes appear in real time on the website and can be seen by all visitors. We use PowerPoint-type presentations, provided by Zoho (http://www.zoho.com/) service's free collaborative online tools, capable of including text and simple drawings, to explain as easily as possible all design steps, and we have linked them to the main pages describing the work. Comments, errors or sketches can be added by any visitor, accessing the presentations by using the account created for contributors on Zoho. A good portion of inputs to the design come from members of the Italian Machine for Soil Movement (MMT) forum. Through the use of the online forum, people with first-hand experience in agricultural equipment get involved with the project, which seems attractive probably because of its humanitarian application and diversity from other problems traditionally posted. Not only do they give us precious suggestions but also sketches of their own ideas.

Between more traditional participatory tools, which can be used only in face-to-face sessions, we carry out interviews with groups of de-miners, using ranking tools and simple presentation slides, clearly introducing the problem and trying to generate ideas.

By using this participatory approach, we developed PAT, the first prototype of a small machine built around a commercial small agricultural machine widely available in south east Asia, also second hand: the power tiller (Fig. 4.19).

Also called two-wheel tractors, walking tractors or iron buffaloes, power tillers have great importance in their nations' agricultural production and rural economies. They are very versatile machines, having many attachments for performing different ground processing work, such as rotovators, moldboards,

4.19 PAT machine: original powertiller.

disc-plows, seeders, planters, and harvesters. Also very important is their ability to pull trailers with two ton plus cargoes.

PAT is specifically tailored to the environment of the Vanni region of Sri Lanka, the country for which it was developed, and therefore it is designed to process soft soil contaminated by small plastic anti-personnel landmines containing not more than 50g of TNT. The contextualization of the project allowed the design to be really participatory and helped to develop a simpler and more effective solution, even if it is not universal. It is commonly recognized that a solution pretending to address the general de-mining problem, referred to as a 'silver bullet', is not feasible due to the extreme variety of scenarios, differing either in environmental or landmine contamination characteristics. Nevertheless, a solution developed for a specific area of a specific country can be employed in other regions presenting similar characteristics. This is the case of the southern minefields between Jordan and Israel, where sand and small plastic landmines are prevalent as they are in the Vanni. It was there that we tested the ground-processing tool of the machine, as it was nearly impossible at that time to bring equipment to the Vanni due to the closure of the border. The *Study of Global Operational Needs*, carried out by the GICHD in 2002, classifies humanitarian de-mining scenarios in 12 types. Soft ground, the type present in the Vanni region of Sri Lanka, is generally found in three de-mining scenarios: woodland, desert and paddy fields. From annex H of the same document reporting the 'spread' of de-mining scenarios found in each region it can be seen that woodland, desert and paddy fields scenarios are the dominant scenarios in many regions throughout the world.

The PAT machine (Fig. 4.20) uses a modular structure consisting of three different modules:

- A tractor unit (Fig. 4.21) based on a 7.5kW powertiller opportunely modified to incorporate rubber tracks and to support different front end-effectors (Cepolina, 2008)
- The ground processing tool (Fig. 4.22), a rake-like system designed to sift the soft soil in front of the machine, lift mines up on soil surface and leave them beside it for later manual removal (Cepolina & Snobar, 2008)
- The remote control (Fig. 4.23 and 4.24), allowing differential skid steering of the vehicle, operated pneumatically through an air compressor coupled with the power tiller engine shaft (Kostrzewski et al., 2007).

The task of the machine is to smooth the soil up to the required depth of 150mm and expose landmines by lifting them to the soil's surface, possibly without actuating them. The requirements that the ground-processing tool has to satisfy therefore are to process the soil at a constant depth and, as it is placed at the front of the machine, to remove landmines before the tractor unit passes over them. The machine is designed to support manual de-miners in area reduction operations. In fact, where landmines have been laid in well-defined patterns along 'mine-belts', usually for defending trenches during conflicts, such as in Sri Lanka, the most

4.20 PAT: digital mock-up of the assembled machine.

4.21 PAT: tractor unit prototype.

efficient way to employ the machine would be simply to locate the beginning of the mine-belt. When the first landmines are exposed, and the belt therefore located, de-miners can proceed to manually clear the belt without wasting energy and time working where there are no mines, while the machine can be employed to locate the belt in the next mine field.

A slightly different ground-processing tool could be designed to collect landmines at the same time as raking them out of the ground, to lesser de-miners' work. This solution is more suitable to environments where landmines are found at random locations. In fact, where a high concentration of mines is located, it is

4.22 PAT: ground processing tool prototype.

4.23 Remote control interface.

faster and cheaper to proceed with manual de-mining than with a machine that can be damaged by too many explosions.

Even if other bigger and more dangerous landmines can also be found, including fragmentation types and anti-tank mines, the machine targets only small plastic

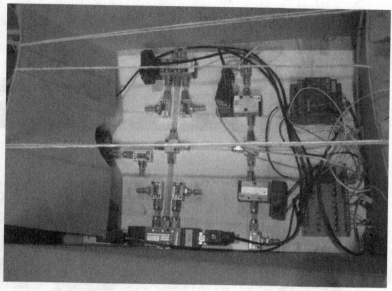

4.24 PAT: remote valve bay.

landmines, which are also the most common types. A major requirement of the machine is for it to be low cost and therefore with limited power. Before starting clearance it is generally possible to say in which minefields bigger and more dangerous mines can be found: the machine will not be used in those areas.

The PAT machine is much simpler and lower cost (the prototype cost is less than 5000€) than other machines available now, even Tempest, and can be built in any unspecialized workshop that can be found (or easily built) in developing countries. The participatory approach we used led to the development of a technology whose technical drawings are freely available on the Internet and can be built everywhere. It is not intended to be commercialized.

While the tractor unit and the remote control have been developed and tested in Italy at the PMARlab of the University of Genova and the ground processing tool has been developed and tested in Jordan with the collaboration of the University of Jordan and Norwegian People's Aid (NPA), a complete test of the whole assembled machine is still to be carried out. In order to assess machine capabilities and propose it on the humanitarian de-mining technology market as a realistic local and participatory developed alternative to current expensive technologies, we have decided to build a second prototype called DISARMADILLO (Fig. 4.25). It will be based on a new, more powerful powertiller, the G131 model with 10 kW power, kindly supplied by the Italian producer Grillo SpA (Fig. 4.26) and it will embed the structure, wheels, tracks and the remote control system already developed for PAT.

The aim of the second prototype is also to demonstrate the modularity of the design with little adjustments that can be mounted on different models. The new

4.25 Disarmadillo logo.

4.26 Disarmadillo new core prototype.

prototype will be funded by the Rotary Club in Genova and will be made with the collaboration of the students of a technical secondary school where Emanuela Cepolina is working part time as a teacher. A new website is under construction and will embed collaborative tools for the participatory re-design.

4.9   Low cost tractor for humanitarian de-mining: Locostra

Most of the partners of the PAT project have agreed to come together again for the development of the Locostra (Low-cost Tractors for Humanitarian Demining) project. Encouraged by that successful experience, and motivated by the same principles of producing a sustainable, modular and agricultural derived technology, we aim to exploit the skills acquired and develop a new machine, slightly more costly than the powertiller, but also more powerful and more versatile. One that, we believe, can greatly improve humanitarian de-mining activities worldwide by speeding up technical survey processes.

Using a commercially available tractor as a drive unit combined with converted off-the-shelf agricultural tools, the machine will be able to cut and clear undergrowth before processing the ground with specialized rakes or harrows. It could be used in technical survey or in full clearance for preparing the ground in advance of de-miners entering the minefield. The only entirely innovative parts are blast-resistant wheels designed to withstand the forces associated with an anti-personnel mine detonation, so allowing the machine to continue to work if it inadvertently detonates a mine.

The remotely controlled machine is based on a small, lightweight, four wheeled, agricultural mini-tractor designed to be equipped with one or more of a range of proven agricultural tools. Power to the attachments is drawn from a power-take-off (PTO), available both at the rear and front of the vehicle.

The design allows the use of relatively heavy attachments, such as brush-cutting implements. A standard, category I, three-point linkage attachment at the rear allows hydraulic lifting and positioning of most of the shelf agricultural tools. The machine is designed to be easily transported over unimproved terrain without the need for a dedicated transporter. The overall vehicle dimensions allow it to be loaded directly into the bed of a Toyota 'Hi-Lux' (with no attachments) or larger pick-up truck load-bay. The overall weight is designed to be less than 1500 kg, allowing the transport of all parts by pick-up and, if necessary, a trailer can be attached. In hazardous areas, the machine can traverse uneven and steep ground with ease. Its low centre of gravity and relatively high ground clearance make it suitable for use in rough terrain.

The vehicle is four-wheel drive with articulated skid-steering. A hydrostatic differential transmission provides power to the drive. Intended for use in areas where there is a risk from explosive devices, its wheels are designed to withstand the detonation of 200 g of TNT without damage that would halt operations. The same blast resistant wheels are mounted at the front and rear. These large wheels are heavily braced dishes of 6.35 mm plate with a tread attached and a large solid rubber wheel in the hub. Designed to allow blast-ventilation and prevent shock transfer to the bearings and chassis, these wheels are original and highly robust. Integral shear-points minimize the potential for shock-transfer to the hull in a major blast. On soft ground, it is anticipated that the tread on these wheels may need to be replaced after five blasts. On hard ground, the loss of the tread should not have any great effect on mobility.

The design of the wheels may change after blast testing. If needed, the use of specially designed magnetic joints, which withstand the normal working load and detach in case of explosion and are placed between the wheels and axles, will be considered.

A robust radio control system controls the petrol engine, the wheels and the electrical switching for the implements attached. To maintain a low weight, the minimal armouring required would be a composite of ballistic aramids and polycarbonate with critical areas further protected by steel plate. The armouring will be easily removed for servicing and when working in extreme temperatures (over 40°C). Deflection angles will be maximized to allow blast ventilation and to protect critical components from high-speed fragmentation (over 550 m/s). The machine is too small and light to allow any significant protection against large explosive threats (such as AT mines) and no effort will be made to provide this. The machine is designed for use in areas with an AP mine threat.

Throughout the programme it will be borne in mind that the system must be prepared complete with operating instructions that include draft SOPs for safe use in a hazardous environment. A training programme will also be structured and the range of spare parts and the equipment necessary for field repairs evaluated and maintained. It is intended that the final purchase price will be kept below €50,000 per unit, with operating costs comparable to that of a road vehicle.

Innovative design solutions, blast-resistant wheels and radio control from the Locostra project will also benefit other areas. Blast-resistant wheels can be mounted on the powertiller developed during the PAT project, to make that platform more versatile and capable of driving safely over landmines while carrying tools at the rear, in areas where genuinely sustainable technologies are needed and programmes cannot afford the slightly more expensive Locostra. The radio control, modular enough to be mounted on every small agricultural machine, could be used to make heavy agricultural works suitable also for weaker subjects such as disabled people and women.

The Locostra project is at the moment only at the proposal stage, being under evaluation for funding by the Italian Ministry of Economic Development. The collaborators include NPA, the University of Jordan, Pierre Trattori (a small Italian producer of mini tractors), the University of Melbourne and Rama agricultural equipment. The proposal has been written with the help of Andy Vian Smith.

4.10   References

Bottomley, R. (2005). Community Participation in Mine Action. Norwegian People's Aid December 2005. Available from: http://www.npaid.org/?module=Articles;action=Article.publicShow;ID=3103, Acceded 3/2/2009.

Cepolina, E.E. (2008). Humanitarian Demining, Ch. 16. *Power Tillers for Demining in Sri Lanka: Participatory Design of Low-cost Technology*, I-Tech Education and Publishing, ISBN: 978-3-902613-16-5, Vienna, Austria.

Cepolina, E.E. (2006). Power Tillers and Snails for Demining in Sri Lanka. *Journal of Mine Action*, Issue 10.1, Summer 2006, pp. 76–79, ISSN: 1533–9940.

Cepolina, E.E.; Bruschini, C. & De Bruyn K. (2004). Field Survey Results. EUDEM2 Internal publication. Available from: http://www.eudem.vub.ac.be/files/FieldSurvey_Results_V1.0.0.pdf. Accessed: 01/02/2009.

Cepolina, E.E. & Snobar, B. (2008). Agricultural derived tools for ground processing in humanitarian demining operations: set up of testing facility in Jordan. *Proceedings of the VII International Workshop HUDEM08*, Cairo, Egypt, 28–30 March 2008.

Chambers, R. (1997). *Whose Reality Counts? – Putting the First Last*, Intermediate Technology Publications, ISBN: 1-85339-386-X, London, UK.

Darrow, K. & Saxenian, M. (2008). *Appropriate Technology Sourcebook*. Village Earth, 2008. Available from: http://www.villageearth.org/pages/Appropriate_Technology/ATSourcebook/index.php. Accessed: 03/02/2009.

Dixon, R. (2004), A concept of implementing technology to encourage economic growth in de-mining, *Proceeding of the International Workshop on Robotics and Mechanical Assistance in Humanitarian Demining and Similarly Risky Interventions*, Brussels, 16–18 June 2004.

Eriksson, D. (2008). Total Quality Management in Mine Action. *Journal of Mine Action*, Issue 12.1, Summer 2008, pp. 80–83 ISSN: 1533–9940.

Filippino, E. M. & Paterson T. (2008). Local NGOs and Firms in Mine Action. *Journal of Mine Action*, Issue 12.1, Summer 2008, pp. 50–52 ISSN: 1533–9940.

Gasser, R & Terry, T. (2000) Developing New Technology for Humanitarian Demining, *Journal of Mine Action*, Vol. 4. No. 1, 2000. ISSN 1533–9440.

GICHD (2008a). *A Guide to Land Release: non-technical methods*, Geneva International Centre for Humanitarian Demining (GICHD), ISBN: 2-940369-20-8, Geneva, Switzerland.

GICHD (2008b). *Mechanical Demining Equipment Catalogue 2008*, Geneva International Centre for Humanitarian Demining (GICHD), ISBN: 2-940369-11-9, Geneva, Switzerland.

GICHD (2005). *A Study of Manual Mine Clearance – 4*. Risk Assessment and Risk Management of Mined Areas, Geneva International Centre for Humanitarian Demining (GICHD), ISBN: 2-88487-032-6, Geneva, Switzerland.

GICHD (2004). *A Guide to Improving Communication in Mine Risk Education Programmes*, Geneva International Centre for Humanitarian Demining (GICHD), ISBN: 2-88487-022-9, Geneva, Switzerland.

GICHD (2002). *Mine Action Equipment: Study of Global Operational Needs*, Geneva International Centre for Humanitarian Demining (GICHD), ISBN: 2-88487-004-0, Geneva, Switzerland.

Iacobucci, A. (2008). Tajikistan: Almost one-third of the population is in danger of going hungry this winter. *Eurasia Insight*, October 2008. Available from: http://www.eurasianet.org/departments/insight/articles/eav100708.shtml. Accessed: 03/02/2009.

ICBL (2008). *Landmine Monitor Report*, International Campaign to Ban Landmines (ICBL), Available from: http://www.icbl.org/lm/. Accessed: 24/01/2009.

IDRC (2005). *Participatory Research and Development for Sustainable Agriculture and Natural Resource management – A Sourcebook*. International Development Research Centre, 2005. ISBN: 1-55250-181-7, Laguna, Philippines.

Kostrzewski, S., Apputhanthri, R. D., Masood, J. & Cepolina, E. E. (2007). Portable mechatronic system for demining applications: control unit design and development. *Proceedings of the International Symposium on Humanitarian Demining 2007 – Mechanical Demining*, 24–27 April 2007, Sibenik, Croatia.

National Bank for Agriculture and Rural Development (NABARD) (2007). *Indian National Strategy, Farm Machinery*. Available from: http://www.nabard.org/modelbankprojects/farmmachinery.asp, Accessed: 03/02/2009.

Sawhney, N., Griffith, S., Maguire, Y. & Prestero, T. (2002). *ThinkCycle at M.I.T*, TechKowLogia, January–March 2002.

Schulz, D. (1999). The Misa 1, an Agricultural Machine with Demining Capabilities, *Journal of Mine Action*, Vol. 3, No. 2, Summer 1999. ISSN 1533–9440.

Smith, A.V. (2009). *Generic SOPs*. Available from: http://www.nolandmines.com/Generic%20SOPs.htm. Accessed: 28/01/2009.

UNDP (2005). *Harnessing Science and Technology (S&T) for Human Development*, United Nations Development Programme (UNDP), Botswana.

UNMAS (2008a). IMAS 04.10 – *Glossary of mine action terms, definitions and abbreviations*, United Nations Mine Action Services (UNMAS), New York, USA.

UNMAS (2008b). IMAS 08.20 – *Land Release – Draft edition*, United Nations Mine Action Services (UNMAS), New York, USA.

UNMAS (2003). IMAS 09.10 – *Clearance Requirements*, United Nations Mine Action Services (UNMAS), New York, USA.

Vadim. (2008). Hunger to replace energy crisis. *Neweurasia*, March 2008. Available from: http://tajikistan.neweurasia.net/2008/03/04/hunger-to-replace-energy-crisis/. Accessed: 03/02/2009.

Wikipedia (2009). *2007–2008 World Food Price Crisis*, Wikipedia. Available from: http://en.wikipedia.org/wiki/Food_crisis#cite_note-129. Accessed: 26/01/2009.

WRI (2007). *Earth Trends: Environmental Information*, World Resource Institute (WRI), Washington DC, USA. Available from: http://earthtrends.wri.org. Accessed: 26/01/2009.

5
Some problems of robotic humanitarian de-mining evolution

V. G. GRADETSKY, Institute for Problems in
Mechanics, Russia

Abstract: The chapter discusses some problems of the effectiveness of
sensor-based technology and mobile robotic systems related to estimation and
evolution of robotic humanitarian de-mining. The quality data of different
robotic automation technologies and relevant parameters are compared.
Discussing the quality characteristics of de-mining technologies and comparing
them shows the evolution of humanitarian de-mining by means of mobile robots.

Key words: humanitarian de-mining, mobile robot, sensors technology
evolution.

5.1   Introduction

The evolution of robotic humanitarian de-mining depends on the effectiveness
of sensor-based technology and mobile robotic systems, including transport
mechanics and control.

Various solutions have been suggested, and some of them include the analysis
of mechanical, sensor and control systems of different countries (Rachkov et al.,
2002; Hofmann et al., 1998; Noro et al., 1999; Gradetsky et al., 2001a; Verlinde
et al., 2001; Baudoin and Acheroy, 2002; Almeida et al., 2002; Grand et al., 2002;
Gradetsky et al., 2001b).

In this paper several problems related to the estimation and evolution of robotic
humanitarian de-mining of fields have been considered. Some ideas may be used
only as references, other as suggestions.

The characteristics of different robotic automation technologies and relevant
characteristics are compared. The quality data are summarized and presented in
tables.

5.2   Evolution of sensor systems for de-mining

The main sensor-based technologies include the following methods:

Metal detection. The metal detector is the most popular sensor for detecting
mines. It consists of a coil generating a magnetic field that may be disturbed
by a metal object. The consequence is higher power consumption or a change
in the magnetic field induced into another coil. Unfortunately this method is
often unreliable or time consuming, as some mines have minimum metal to

113

detect and they are unrecognizable amongst all the various kinds of metallic objects on a former battlefield.

Thermal imaging. Advanced aerial photographic techniques can be used to identify temperature anomalies in the ground.

Ground penetrating radars (GPR). The measure of the resonance frequencies of the ground to a radar impulse depends on buried objects.

At frequencies where the wavelengths are comparable to the overall size of the object, the smaller details of the structure are irrelevant, making it possible to describe the shape and the material properties of a body with only a limited number of parameters.

Odor sensing. Artificial odor sensors can be used to detect the evaporation of explosives. However, in these cases concentrations must be very high for a few trinitrotoluene (TNT) molecules to be detected.

Antibodies. Antibodies are very selective to other molecules and are therefore suitable for building sensors that react to specified molecules.

Biosensing. A particular molecule belonging to the trinitrotoluene (TNT) family relaxes a tense muscle. However, the molecule's lifetime is limited and therefore difficult to detect.

Laser sensing. Another possibility for TNT sensors is to use light absorption. Light energy triggers the explosive reaction of the molecule. The heat increase can then be detected.

Ion mobility spectrometry (IMS). IMS separates ionized molecular compounds on the basis of their transition times when subjected to an electric field in a tube. This time is then compared to stored transition times of known compounds. Therefore it is possible to distinguish TNT from other molecules. This technique is fast and makes a compact device possible. Unfortunately the sensitivity is not very high, in particular for compact designs.

X-ray tomography. In X-ray tomography, the emitted radiation, either in the form of neutrons or gamma rays, is designed to react with different elemental components of the object of interest to produce a reaction particular to the specific detector application. To detect plastic explosives, it is necessary to produce the particular energy that reacts with the subject chemical element in the explosive. In general, plastic explosives contain several elements, such as nitrogen, which has unique characteristics that lend itself to a host of nuclear detection methods, such as thermal neutron analysis and others. A radiation of neutrons reacts with nitrogen nuclei to produce specific detectable gamma rays.

Plant indicators. An as yet unexplored idea to find buried mines is to imagine that plants are sensitive to the presence of TNT in the soil. Hence it could be possible to genetically manipulate plants to have them change their behavior in presence of TNT, for example, changing color, growing very high or any other detectable sign. Other signs such as changes in UV reflection are also usable and measurable by simple tools.

The humanitarian de-mining mission consists of the detection of mines and their removal. To fulfill these tasks automatically it is necessary to have a mobile platform, which can move across the rough terrain. The platform should be tele-operated and have on-board sensors to detect mines and a manipulator to remove them.

The de-mining robots are responsible for clearing a direct vertical path between the top and bottom edges of the minefield. At each time step, robots must select one of three possible actions:

- Move: the robot moves towards a target cell, aiming to perform either a scan or a defuse action on arrival. A behavior-based robot controller plans the path, and obstacle avoidance is achieved through a swirling behavior.
- Scan: the robot locates mines concealed in the current cell. Scanning is not instantaneous, and the difficulty is abstractly modeled as a time cost. The robot may elect to abort its scan based upon communication with its peers, in which case no information about the current cell is obtained.
- Defuse: the robot removes the mines in the current cell. Defusing is typically more time-consuming than scanning. As for scan, an aborted defuse action is assigned no partial credit. Thus, robots must weigh their desire to complete the current task with the goal of helping their peers.

By varying the costs associated with each action, it is possible to model the characteristics of different robotic de-miners. For instance, when the cost of scanning is high, strategies that involve a thorough exploration of the minefield become less feasible. Conversely, when defusing is expensive, robots are encouraged to explore their environment before focusing on a common path containing few mines. The risk of accidental mine detonations, leading to robot loss, can also be represented in the simulation.

The robotic de-mining problem shares characteristics with consume and graze tasks. Robots individually explore unknown areas to identify low-cost paths (paths with fewer cells to be defused), before focusing on the most promising candidates and "consuming" the mines.

The motion simulator is an interactive, high resolution, entity level simulation that represents combined arms tactical operations up to the battalion level. One can define simulated sensors and actuators to suit the real robot system. It should be a cinematically realistic simulator, capable of modeling sensor noise and positional uncertainty if the simulated sensors are defined appropriately.

De-mining robots are modeled as a subclass of the foraging robot in the Java based minefield environment. All robots are assumed to have the following capabilities:

- Positional encoders: gives robot position in absolute world coordinates.
- Kin sensor: detects other robots within a certain range.
- Obstacle sensor: detects obstacles within a certain range.
- Mine sensor: detects mines within a specified local area.

Each robot translates its sensor readings into a grid-based coordinate system for mapping the minefield. Cells are marked as being unexplored, safe, mined, or defused, and each robot maintains its own version of this map.

Inter-robot communication is handled through a socket-based communication server external to the simulation environment. Through the server, robots can broadcast or unicast serialized Java objects to each other. The communication medium is identical to that used by real robots, and failures occasionally occur if the server is accidentally shut down or buffers overflow. At periodic intervals, robots broadcast their internal maps to the other robots and to the robot manager. Communicated maps are merged with information derived from personal exploration (internal maps) to form more comprehensive world maps.

At every time step, robots identify target positions based on their current knowledge about the minefield. Once the robot arrives at its target location, it proceeds to either scan the cells for mines, or to defuse previously marked mines in its cell. Obstacle avoidance during path execution is implemented as a swirling behavior.

The multi-agent system is composed of communicating software entities, designed to assist the human commanders in playing game scenarios in the robot motion simulation environment. The software modules can be grouped into four basic categories:

- middle agents
- human interface agents
- robot motion related agents
- specialized task agents.

The middle agent's architecture includes a group of services useful for developers of distributed systems. These resources include: Agent Name Server, Matchmaker, DemoDisplay, Logger, and Launcher. Agent lookups can be performed using the agent name server, which maintains a simple address listing for all agents, and the more sophisticated Matchmaker, which allows agents to perform lookups based on services, rather than name, using the LARKS advertisement language.

5.3   Comparison of different de-mining automation technologies

As an example, it is possible to list systems such as an off-road mobile robot, radar mobile system, wheeled robot with chains detonation, mine detonation by a wheel, lightweight de-mining system, and backhoe excavator attachment system with vegetation cutting attachment, marshland robot, caterpillar system with weights detonation, tele-operated de-mining robot with TV camera, screws, leg locomotion and hybrid system.

Tables 5.1, 5.2 and 5.3 illustrate the quality characteristics of Russian mines, show a quality comparison of automatic technologies, and give a comparison of relevant characteristics of de-mining automation technologies.

Table 5.1 The quality characteristics of Russian mines

Name	Case material	Shape	Length (mm)	Height (mm)	0 (mm)	Effect
KHF-1 Chemical Antipersonnel Mine	Sheet metal	Cylinder		245	150	Chemical
MON-100 Antipersonnel Mine	Sheet steel	Circular		80	220	Directed fragmentation
MON-50 Antipersonnel Mine	Plastic with embedded cylindrical fragments	Rectangular	220	105		Directed fragmentation
MZ Antipersonnel Mine	Steel	Cylinder		254	127	Fragmentation
OZM-160 Antipersonnel Mine	Metal	Cylinder		930	170	Bounding fragmentation
PFM-1 Antipersonnel Mine	Plastic (low density polyethylene)	Irregular	120	61		Blast
PMD-6 Antipersonnel Mine	Wood	Rectangular	191	64		Blast
PMK-40 Antipersonnel Mine	Waxed cardboard	Cylinder		38	70	Blast
PMM-3 Antipersonnel Mine	Sheet metal	Cylinder		37	100	Blast
PMM-5 Antipersonnel Mine	Cast iron, sheet metal	Rectangular	150	50		Blast/ fragmentation
PMN Antipersonnel Mine	Plastic, rubber, metal	Cylindrical		56	112	Blast
PMP Antipersonnel Mine	Metal	Cylindrical		120	36	Cartridge projectile
POM-IS Antipersonnel Mine	Steel	Spherical			60	Fragmentation

Table 5.2 Quality comparison of de-mining automatic technology classes

Class	Subclass	Advantages	Disadvantages
Brute force	Mechanical	Simple and effective	Requires heavy and large vehicles, only suitable for some terrains
	Explosive	Simple and economical	Limited applications, displaces some kinds of mines rather than exploding them
Prodding		Accurate	Slow, lots of false alarms
Induction coil	Metal detector	Fast, reliable metal detection	Lots of false alarms, useless for plastic mines
	Imaging	Reliable metal detection	Lots of false alarms, heavy, useless for plastic mines
Ground penetration radar		Possible compromise solution between depth and resolution, suitable for all kind of mines, scanning in real time	Hard to interpret data, lots of false alarms, expensive
Infrared imaging	Passive	Scanning in real time	Hard to interpret data (pattern resolution), works well only in special moments of the day, limited depth of penetration, lots of false alarm
	Active	Scanning in real time	Hard to interpret data (pattern resolution), doesn't works in special moments of the day, limited depth of penetration, lots of false alarm, more expensive
Electro-chemical	Antibodies	High sensitivity	Sensing devices must be replaced periodically
	Artificial noses	Detect non-metal mines	Low sensitivity, slow, large
Nuclear quadrupole resonance		Few false alarms, high detection rate, very reliable, good for small and non metallic mines	Affected by electro-magnetic pollution
Ultrasonic methods		Inexpensive	Works well only in some terrains and in water
Optical methods (multispectral)		Distant	Works well only in some terrains
X-ray backscatter		Provides accurate images of the mines, simple to interpret data	System complexity, limited depth of penetration, sensitivity to environmental condition
Conductivity meters	Frequency domain	Suitable for all kinds of mines, the best for small and shallow buried mines	Seriously degraded in some terrains, limited depth of penetration

Table 5.2 Continued

Class	Subclass	Advantages	Disadvantages
	Time domain	More exact	Seriously degraded in some terrains
Ion mobility spectrometers		Fast, compact	Low sensitivity
Magnetometers	Fluxgate	Inexpensive, reliable, low energy consumption	Lots of false alarm
	Proton precession	More sensitivity	Lots of false alarm, slower
	Optically pumped	More sensitivity	Lots of false alarm, more expensive
Photoacoustic spectroscopy		Exact	Works well only in some terrains, slow, limited depth of penetration
Thermal Neutron Analysis		Very reliable	Not for all kind of mines, system complexity, limited depth of penetration
Passive millimeter wave detection		Possible compromise solution between depth and resolution, simpler than GPR	Works well only in dry soil and for metallic mines

Table 5.3 Comparison of relevant characteristics of de-mining automatic technologies

Class	Subclass	Terrain	Speed	False alarms rate	Cost and complexity	Depth	Kind of mines
Brute force	Mechanical	Roads, lands	High	No alarms	Low	Changeable	All
	Explosive	All	High	No alarms	Low	Changeable	All
Prodding		All	Low	High	Low	Deep	All
Induction coil	Metal detector	All	High	High	Low	Deep	Metallic
	Imaging	All	High	High	Medium	Deep	Metallic
Ground penetrating radar		All	High	High	Medium	Changeable	All
Infrared imaging	Passive	All	High	High	Medium	Shallow	All
	Active	All	High	High	Medium	Shallow	All
Electro-chemical	Antibodies	All	Low	Low	High	Deep	All
	Artificial noses	All	Low	Low	High	Changeable	All

Table 5.3 Continued

Class	Subclass	Terrain	Speed	False alarms rate	Cost and complexity	Depth	Kind of mines
Nuclear quadrupole resonance		All	Low	Low	High	Changeable	All
Ultrasonic methods		Water, wet ground	High	High	Medium	Shallow	All
Optical methods (multispectral)		Some terrains	High	High	High	Changeable	All
X-ray backscatter		Dry terrains	High	Low	High	Shallow	All
Conductivity meters	Frequency domain	Dry terrains	High	High	Medium	Shallow	AH
	Time domain	Dry terrains	High	High	Medium	Shallow	All

5.4   Conclusion

Describing the quality characteristics of de-mining technologies and comparing them enables us to analyze the evolution of humanitarian de-mining tools, compare the possibilities of the applied robots and indicate the advantages and disadvantages for future research and development in this field.

5.5   References

Almeida A, Marques L, Rachkov M and Gradetsky V, 'On-Board Demining Manipulator', *Proceedings of IARP Workshop on Robots for Humanitarian Demining*, Vienna, Austria, 2002.

Baudoin Y and Acheroy M, 'Robotics Systems for Humanitarian Demining: Modular and Generic Approach and Cooperation Under IARP/ITER/ERA Networks', *IARP Workshop on Robotics for Humanitarian Demining*, Vienna, Austria, HUDEM'02, 2002.

Gradetsky V, Veshnikov V, Kalinichenko S and Kravchuk L, Mobile Robot's Control Motion Over Arbitrarily Oriented Surfaces in the Space, Moscow, Nauka publisher, 2001a.

Gradetsky V, Veshnikov V, Kalinichenko S, Rizzotto G and Italia F, 'Fuzzy Logic Control for the Robot Motion in Dynamically Changing Environments', *Proceedings of 4th CLAWAR 2001 Conference*, Germany, FZI, Professional Engineering Publishing, 2001b.

Grand C, Ben Amar F, Plumet F and Bidaud Ph, 'Simulation and Control of High Mobility Rovers for Rough Terrains Exploration', *Proceedings of IARP Workshop on Robots for Humanitarian Demining*, Vienna, Austria, 2002.

Hofmann B, Rockstroh M, Gradetsky V and Rachkov M, 'Acoustic navigation and inspection methods for underwater mobile robots', In *Proceedings of 1st International Workshop on Autonomous Underwater Vehicles for Shallow Waters and Coastal Environments*, Lafayette, USA, 1998.

Noro D, Sousa N, Marques L and Almeida A, 'Active Detection of Antipersonnel Landmines by Infrared', *Annals of Electrotechnical Engineering Technology*, Portuguese Engineering Society, 1999.

Rachkov M, Marques L and Almeida A, *Automation of Demining*, Portugal, University of Coimbra, 2002.

Verlinde P, Acheroy M and Baudoin Y, 'The Belgian Humanitarian Demining Project and the European Research Context', *Thematic Network on Climbing Walking Robots*, Clawar Royal Military Academy VR-Mech'01, Brussels, Belgium, BSMEE, 2001.

Part II
Sensors for mine detection and robotics

6

Sensing capabilities for mobile robotics

L. NOMDEDEU, J. SALES, R. MARÍN, and
E. CERVERA, Jaume I University, Spain and
J. SAEZ, Sheffield Hallam University, UK

Abstract: This chapter provides an overview of sensing capabilities for the estimation of distances and localization in a team of mobile robot platforms. This work, similar to others in this book, has been carried out in the context of the EU GUARDIANS project (Group of Unmanned Assistant Robots Deployed In Aggregative Navigation supported by Scent detection). There is a need for a localization method not only for robot platforms but also for fire fighters. This chapter will present an overview of the current sensing capabilities available, both commercial, and those not yet commercialized.

Key words: GUARDIANS project, sensing capabilities, localization, mobile robotics, ego-motion, environment perception.

6.1   Introduction

Nowadays mobile robotics and more specifically autonomous mobile robotics is a very active topic all over the world. It is difficult to find any medium-to-large research institution that does not have a special division devoted to this topic. Algorithms for localization, autonomous navigation, obstacle avoidance, path planning, map building, and other important issues are available. Also new techniques are being developed while classical approaches are being refined every day. But to make them work, autonomous mobile robots need to collect environmental and self-state data.

The availability and accuracy of sensor data is crucial and fully constrains the algorithms and techniques, according to the platform and the environment. We often find a close relationship between the algorithms developed and the robotic platforms used, and specifically the types of sensors employed, up to the point that often the algorithms only work with a particular model of sensors and specifically with a specific configuration. This gives us a glimpse of the crucial importance that sensors have.

In this chapter we will try to provide an insight to the sensor capabilities commonly found or at least available for mobile platforms. We draw special attention to the limitations of each type of sensors and techniques.

Since this study lies within the context of the GUARDIANS European Project, we have several constraints due to the dynamic and hazardous unstructured and unknown environment.

In the GUARDIANS project, multiple robot platforms and sensors are used for navigating and surveying in fire-fighting scenarios indoors. The problem faced in

125

these environments is that smoke blinds typical laser range finder technologies (Pascoal, 2008). As a result of this, we performed research in technologies available for the purposes of our project, keeping in mind the characteristics of our indoor environments.

We present a survey of sensing technologies available in commercial or development kits nowadays, and their use in these kinds of environments. We also provide the results of our tests of some of the technologies and our conclusions.

From simple infrared (IR) sensors to heavy hi-tech radars, we looked at the physical principles, software techniques, and enhancing practises for each type of sensor.

In the following sections, we show various technologies and the different uses of sensors and techniques for upper layers of the global system.

In section 6.2, an overview of the layers and elements is presented. Then, in sections 6.3 to 6.5 each layer is detailed as well as its components. The principles of sensing are also covered and we also have a look at new approaches like sonar wide beam based triangulation, etc.

Finally, section 6.6 presents some final remarks and lines of ongoing research.

6.2   Sensing technologies taxonomy

While navigating through unknown environments, mobile platforms need to collect information about the environment to be able to carry out the task we build them to perform. For this purpose these are some of the capabilities that we would like to have in a mobile platform:

- some information about 'ego-motion' of the platform is needed
- 'environment-perception' data is required
- 'localization' information that is as accurate as possible, to be able to navigate to a target position.

So, basically these are also the groups into which we have split our summary of sensing technologies. In Fig. 6.1 we present a possible taxonomy according to the abovementioned characterization.

In the first group, named 'ego-motion', we have put sensors that will give us information about the platform displacement. Some of them calculate this movement without measuring external environmental features, while others work out this information by observing the environment. GPS and other outdoor technologies have not been included due to our indoor approach in the GUARDIANS project.

The second group, named 'environment-perception', includes sensors that typically give us information about obstacles, walls, etc. This information on items detected in the environment is used, for example, to navigate avoiding obstacles. Vision based sensors (cameras) have deliberately been excluded

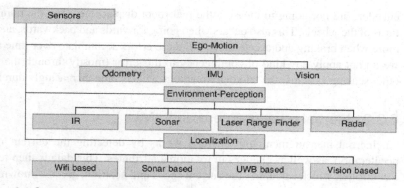

6.1 Sensing technologies schema.

from this group for two main reasons: on one hand the complexity of these devices makes them deserving of a whole specific paper, but, on the other hand, in the context of the GUARDIANS project, they are useless for distances above 10 cm.

The third group, named 'localization', incorporates techniques and technologies used for localizing robots or tagged items in the environment, so we can, at any time, go from our position to a new desired position where we know there is a power supply, or we can take actions knowing where other team-mate robots are.

6.3   Ego-motion capabilities for mobile robots

6.3.1 Odometry

The word odometry is composed from the Greek words *hodos* (meaning 'travel', 'journey') and *metron* (meaning 'measure').

Odometry[1] (Siciliano, 2008) is the study of position estimation during vehicle navigation. The term is also sometimes used to describe the distance travelled by a vehicle. Odometry is used by robots to estimate (not determine) their position relative to a starting location. Basically odometry is the use of data from the movement of actuators to estimate change in position over time. This method is very sensitive to error. Rapid and accurate data collection, equipment calibration, and processing are required for odometry to be used effectively.

Suppose a robot has rotary encoders on its wheels. It drives forward for some time and then would like to know how far it has travelled. It can measure how far the wheels have rotated and, if it knows the circumference of its wheels, compute the distance.

But there is a big issue here: slippage. When moving, and especially when turning, the robot wheels tend to slip slightly and slide on the floor, and so the

[1] http://en.wikipedia.org/wiki/Odometry

encoders are not going to measure the real robot displacement but the number of turns of the wheels. This also occurs when going forwards and backwards, and even more when braking and accelerating. All these errors accumulate over time and, if we do not apply any kind of contingency to this error (mostly through the use of other sensing systems) the odometry readings will become increasingly unreliable.

6.3.2 Inertial measurement unit (IMU)

An inertial measurement unit (IMU[2]) works by detecting the current rate of acceleration, as well as changes in rotational attributes. This data is then fed into a controller, which calculates the current speed and position, given a known initial speed and position, by integrating the accelerations.

A major disadvantage of IMUs is that they typically suffer from accumulated error. Because the guidance system is continually adding detected changes to its previously calculated positions (see dead reckoning), any errors in measurement, however small, are accumulated from point to point. This leads to 'drift', or an ever-increasing difference between where the system thinks it is located, and the actual location.

Some IMU units also provide compass information, usually with a magnetic device. This compass information can be useful in some cases but we have to take care, as it is very sensitive to magnetic fields.

In Fig. 6.2 we can see the variation in measures of the IMU device mounted on the back of a pedestrian.

6.3.3 Vision

Although vision is usually an 'environment-perception' mechanism, here we describe some techniques to estimate 'ego-motion' information from vision data.

Visual odometry

When traditional odometry techniques cannot be applied to robots due to non-standard locomotion methods, or simply because of the universal precision problems of odometry, we have to find other ways of getting this data.

Visual odometry[3] is the process of determining equivalent odometry information using only camera images. Compared to traditional odometry techniques, visual odometry is not restricted to a particular locomotion method, and can be utilized on any robot with a sufficiently high quality camera. Most existing approaches to visual odometry are based on the following stages:

- Acquire input images: using single cameras, stereo cameras, or omni-directional cameras (Scaramuzza and Siegwart, 2008; Corke et al., 2004).

[2] http://en.wikipedia.org/wiki/Inertial measurement unit
[3] http://en.wikipedia.org/wiki/Visual odometry

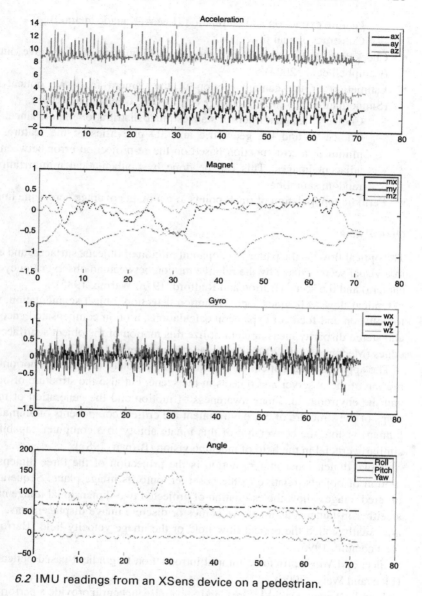

6.2 IMU readings from an XSens device on a pedestrian.

- Image correction: apply image processing techniques for lens distortion removal, etc.
- Feature detection: define interest operators, and match features across frames and construct optical flow field.
    - Use correlation to establish correspondence of two images, and no long-term feature tracking.

  - Feature extraction and correlation (Lucas–Kanade method).
  - Construct optical flow field.
- Check flow field vectors for potential tracking errors and remove outliers (Campbell et al., 2004).
- Estimation of the camera motion of the camera from the optical flow (Sunderhauf et al., 2005; Konolige et al., 2006; Cheng et al., 2006).
  - Choice 1: Kalman filter for state estimate distribution maintenance.
  - Choice 2: find the geometric and 3D properties of the features that minimize a cost function based on the re-projection error between two adjacent images. This can be done by mathematical minimization or random sampling.
- Periodic repopulation of track-points to maintain coverage across the image.

Optical flow

The optical flow[4] is the pattern of apparent motion of objects, surfaces, and edges in a visual scene caused by the relative motion between an observer (an eye or a camera) and the scene (Burton and Radford, 1978; Warren, 1985).

Optical flow techniques such as motion detection, object segmentation, time-to-collision and focus of expansion calculations, motion compensated encoding, and stereo disparity measurement utilize this motion of the object's surfaces and edges (Medeiros et al., 2008; Barron and Beauchemin, 1995).

The application of optical flow includes the problem of inferring not only the motion of the observer and objects in the scene, but also the structure of objects and the environment. Since awareness of motion and the generation of mental maps of the structure of our environment are critical components of animal (and human) vision, the conversion of this innate ability to a computer capability is similarly crucial in the field of machine vision (Brown, 1987).

A two-dimensional image motion is the projection of the three-dimensional motion of objects, relative to the observer, onto its image plane. Sequences of ordered images allow the estimation of projected two-dimensional image motion as either instantaneous image velocities or discrete image displacements. These are usually called the optical flow field or the image velocity field (Barron and Beauchemin, 1995).

Fleet and Weiss provide a tutorial introduction to gradient-based optical flow (Fleet and Weiss, 2006).

John L. Barron, David J. Fleet, and Steven Beauchemin provide a performance analysis of a number of optical flow techniques. It emphasizes the accuracy and density of measurements (Barron et al., 1994).

Optical flow was used by robotics researchers in many areas such as: object detection and tracking, image dominant plane extraction, movement detection, robot navigation and visual odometry (Medeiros et al., 2008).

[4] http://en.wikipedia.org/wiki/Optical flow

6.4   Environment-perception capabilities for mobile robots

6.4.1   Infrared sensors technology

Infrared light has been used for years in a wide range of topics[5] from thermo-graphical cameras and night vision cameras to communications devices, meteorology, health care, etc.

Basic principles

The basic idea is this: a pulse of infrared (IR) light is emitted by the emitter. This light travels out into the field of view and either hits an object or just keeps on going. If there is no object, the light is never reflected and the reading shows no object. If the light reflects off an object, it returns to the detector and creates a triangle between the point of reflection, the emitter, and the detector, as shown in Fig. 6.3.

The angles in this triangle vary based on the distance to the object. The receiver portion of the detectors[6] is actually a precision lens that transmits the reflected light onto various portions of the enclosed linear charge-coupled device (CCD) array based on the angle of the triangle described above. The CCD array can then determine what angle the reflected light came back at and, therefore, it can calculate the distance to the object.

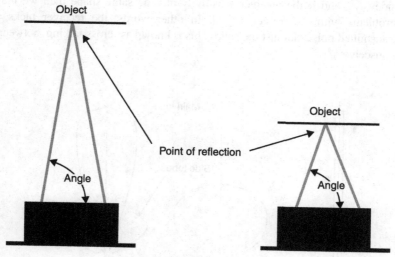

6.3 Different angles with different distances.

[5] http://en.wikipedia.org/wiki/Infrared
[6] http://www.acroname.com/robotics/info/articles/sharp/sharp.html

6.4.2 Sonar technology

What is sonar?

Sonar (Marshall, 2005; Fraden, 1996) is a method of finding the distance to an object by measuring the time it takes for a pulse of sound (usually ultrasound) to make the round trip back to the transmitter after bouncing off the object. At sea level, in air, sound travels at about 344 metres per second (1130 feet per second). In practical terms this means 2.5 cm are covered in about 74 microseconds. These sorts of numbers are easily managed by a simple microcontroller system.

In principle, all you do is send a burst of ultrasound from a suitable transmitter, setting a clock or timer running at the same time. When the receiver picks up the reflected signal or 'echo', the clock is stopped and the elapsed time is proportional to the distance. Although it seems to be simple, there are some practical problems that need to be addressed.

Transducer characteristics

An ultrasonic transducer is basically a miniature loudspeaker tuned to emit or receive a single frequency, usually 40 kHz. The sound does not emerge from the front of the device in a nice pencil-shaped beam, but has a shape like that in Fig. 6.4.

We should have in mind that these shapes are three-dimensional, so Fig. 6.4(a) would be a cylinder shape. The wide beam transducer puts out a lot of energy sideways, and if the receiver sensitivity has the same shape then we may have problems with the 'direct' signal. In other words, the receiver picks up the transmitted pulse, not just the echo. This is known as 'cross-talking' between sonar transceivers.

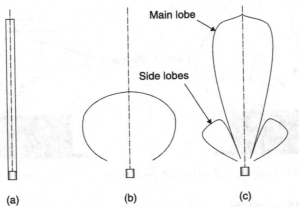

6.4 Transducer beam/sensitivity patterns (a) ideal; (b) wide beamwidth; (c) narrow beamwidth.

Reflection problems

When the waves hit an object, part of their energy is reflected. In many practical cases, the ultrasonic energy is reflected in a diffuse manner. That is, regardless of the direction where the energy comes from, it is reflected almost uniformly within a wide solid angle, which may approach 180 degrees. If targets were always at right angles to the axis of the beam, distance measurements would be reliable and accurate. Unfortunately they seldom are, and to make it worse the target material may be such that no signal is bounced back at all. Figure 6.5 illustrates some problematic situations. For simplicity, the signal path is shown as a single line. If an object moves, the frequency of the reflected waves will differ from the transmitted waves. This is called the Doppler effect.

If the robot is heading towards the wall at an angle, then you can see from Fig. 6.5(a) that the measured distance will be too long. A strange effect occurs when the angle with the wall becomes shallower as in Fig. 6.5(b). At a certain critical angle, specular reflection takes place and all the signal is reflected away from the robot. In other words, the wall disappears because there is no echo. This critical angle depends on the wall material and its surface: a very smooth gloss finish on a skirting board can lead to specular reflection. Slightly rough surfaced materials, such as cardboard, cause the signal to be scattered in all directions, so at least some makes it back to the receiver. Figure 6.5(c) shows how specular reflection in a corner can fool the sonar into making a very large error indeed.

The hardness of the target material is also an important factor. A skirting board will reflect most of the energy, absorbing very little, but soft furnishings will do the reverse. This means that the maximum detection range is large for walls, but soft objects may not be 'seen' until the robot is nearly touching them. Note that this does not affect the accuracy of distance measurements, only the maximum range at which a particular object can be detected.

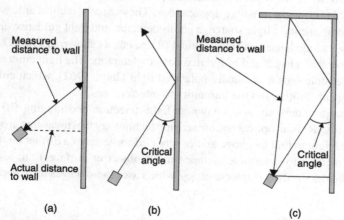

6.5 Sources of error (a) triangle error; (b) specular reflection; (c) corner error.

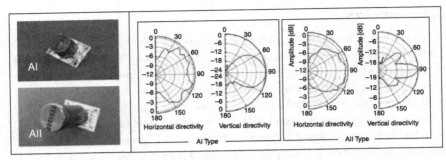

6.6 AniBat wide-beam sonar sensors.

New wide-beam approach

Recently, a new wide-beam[7] based approach has been introduced in contra position to the narrow-beam classical approach. Using several wide-beam sensors and ringing them in the correct order we are able to detect small objects, and work out their correct relative positions, as shown in Fig. 6.6. This is accomplished through the use of techniques like 'different time of arrival'. The principle is that knowing the geometrical distribution of sensors and the order of firing, the different readings in each sensor will give us an approximation of the position of the obstacle by triangulation.

6.4.3 Laser range finder

Optical sensors

Nowadays, optical sensors are probably the most popular for measuring distances, and also for position and displacement measurements. Their main advantages include, among others, simplicity, relatively long operating distances, accuracy, and insensitivity to magnetic fields and electrostatic interferences, which makes them suitable for many sensitive applications. They usually require at least three essential components: a light source, a photo-detector, and light guidance devices, which may include lenses, mirrors, optical fibres, etc, as shown in Fig. 6.7. Light is guided towards a target and is diverted back to detectors. The light emitted from an optoelectronic sensor is usually polarized light (Juds, 1962), which enhances the sensing capabilities and the immunity to interferences.

Laser range finders are also known as light detection and ranging (LIDAR) devices. LIDAR[8] is an optical remote sensing technology that measures properties of scattered light to find the range and/or other information of a distant target. The prevalent method to determine distance to an object or surface is to use laser pulses. Like the similar radar technology, which uses radio waves instead of light,

[7] http://www.hagisonic.com/
[8] http://en.wikipedia.org/wiki/LIDAR

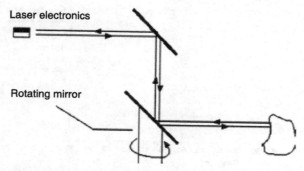

6.7 Laser range finder schematics.

the range to an object is determined by measuring the time delay between transmission of a pulse and detection of the reflected signal. LIDAR technology has applications in archaeology, geography, geology, geomorphology, seismology, remote sensing and atmospheric physics (Cracknell and Hayes, 2007).

A laser typically has a very narrow beam, which allows the mapping of physical features with very high resolution compared to radar. In addition, many chemical compounds interact more strongly at visible wavelengths than at microwaves, resulting in a stronger image of these materials. Suitable combinations of lasers can allow for remote mapping of atmospheric contents by looking for wavelength-dependent changes in the intensity of the returned signal. The beam densities and coherency are excellent. Moreover the wavelengths are much smaller than can be achieved with radio systems, and range from about 10 micrometers to the UV (ca. 250 nm). At such wavelengths, the waves are 'reflected' very well from small objects (see Fig. 6.8). This type of reflection is called backscattering. Different

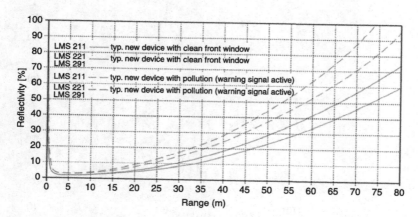

6.8 SICK LMS 211, LMS 221, LMS 291, relationship between reflectivity and range with good visibility (SICK Manual).

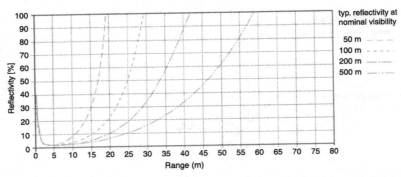

6.9 SICK LMS 211 – relationship between reflectivity and range in fog (SICK Manual).

types of scattering are used for different LIDAR applications, most common are Rayleigh scattering, Mie scattering and Raman scattering as well as fluorescence. The wavelengths are ideal for taking measurements of smoke and other airborne particles (aerosols), clouds and air molecules (see Fig. 6.9) (Cracknell and Hayes, 2007).

Although LIDAR is the most accurate system we can use now, in the field of the GUARDIANS project it is almost useless, as we face environments full of smoke (Pascoal, 2008), where it is proven that the typical laser range finder devices commonly available are useless in the density of smoke and over small distances.

In the sequence of images in Fig. 6.10 we can briefly see how the laser range finder is blinded when a high density of smoke is reached.

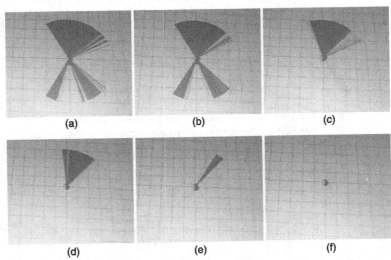

6.10 (a)–(f) Sequence of laser range finder readings while inserting smoke.

6.4.4   Micro and millimetre wave radar

Basic principles

The basic principle[9,10] of the operation of primary radar is simple to understand. However, the theory can be quite complex. An understanding of the theory is essential in order to be able to specify and operate primary radar systems correctly. The implementation and operation of primary radars systems involve a wide range of disciplines such as building works, heavy mechanical and electrical engineering, high power microwave engineering, and advanced high-speed signal and data processing techniques. Some laws of nature have a greater importance here.

Radar measurement of range, or distance, is made possible because of the properties of radiated electromagnetic energy. The electromagnetic waves are reflected if they meet an electrically leading surface. If these reflected waves are received again at the place of their origin, then that means an obstacle is in the propagation direction.

The primary difference between LIDAR and radar is that with LIDAR, much shorter wavelengths of the electromagnetic spectrum are used, typically in the ultraviolet, visible, or near infrared. In general it is possible to image a feature or object only about the same size as the wavelength, or larger. Thus LIDAR is highly sensitive to aerosols and cloud particles as well as smoke (Cracknell and Hayes, 2007). On the other hand, radar has less accuracy.

An object needs to produce a dielectric discontinuity in order to reflect the transmitted wave. At radar (microwave or radio) frequencies, a metallic object produces a significant reflection. However, non-metallic objects, such as rain and rocks, produce weaker reflections and some materials may produce no detectable reflection at all, meaning some objects or features are effectively invisible at radar frequencies. This is especially true for very small objects (such as single molecules and aerosols) (Cracknell and Hayes, 2007).

See Fig. 6.11, which shows the measuring of a round trip time of a microwave. Electromagnetic energy travels through the air at a constant speed, at approximately the speed of light, 300,000 kilometres per second. This constant speed allows the determination of the distance between reflecting objects (airplanes, ships or cars) and the radar site by measuring the running time of the transmitted pulses.

This energy normally travels through space in a straight line, and will vary only slightly due to atmospheric and weather conditions. By use of special radar antennas this energy can be focused in a desired direction. Thus the direction and elevation of the reflecting objects can be measured. These principles can basically be implemented in a radar system, and allow the determination of the distance, direction and height of the reflecting object.

---

[9] http://www.radartutorial.eu/index.en.html
[10] http://en.wikipedia.org/wiki/Radar

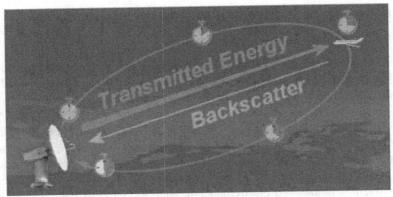

6.11 Radar principle: the measuring of a round trip of a microwave.

Millimetre wave technologies have proved to work under the worst weather circumstances (Vane, 1986) and even with heavy dust. Therefore, it must be taken seriously as an option for ensuring ranging capabilities in the field of the GUARDIANS project or similar environments.

A review of the use of millimetre wave radar for robotics is presented in Brooker et al. (2001). The paper also includes a brief overview of the different radar and scanner technologies.

Figure. 6.12 shows the results of a prototype. The system consists of a radar and a sectorial antenna, mounted on a rotary joint.

Radar technologies have caught the eye of politicians and car manufacturers as they enhance the security of drivers throughout Europe, especially short-range radar systems (SRR).[11] The long-term goal is to have short-range radar systems available in a harmonized way on 1 July 2013. This will provide us, the research community, with a new source of low cost market-ready reliable radar devices to work with.

6.5   Localization capabilities for mobile robots

Localization involves one question: Where is the robot now? Although a simple question, answering it is not as easy, as the answer is different depending on the characteristics of your robot. Localization techniques that work fine on an outdoor robot would not work very well or even at all for an indoor robot.

All localization techniques have to provide two pieces of information:

- What is the current position of the robot?
- Where is it heading to?

The first could be in the form of Cartesian or polar coordinates and the second as compass headings.

[11] http://ec.europa.eu/information_society/newsroom/cf/itemdetail.cfm?item%20id=4629

6.12 Prototype radar test.

The current position of a robot can be determined in several very different ways. Those that will work well outdoors (like GPS) are useless indoors.

In the GUARDIANS project we point to an indoor environment, so we will focus on indoor technologies rather than outdoor systems. Most of the indoor technologies involve using some kind of signal spreading through the air.

Measuring the 'time of flight' or the 'time of arrival'[12], the 'time difference of arrival'[13] and the 'angle of arrival'[14], and knowing the spreading speed of the specific signal as well as the geometrical position of the receivers, the position of the emitter can be determined with more or less accuracy depending on the technology, the number of beacons, the conditions of the environment, etc.

6.5.1 Wifi

In this case, common Wifi ethernet devices are used. While packets are sent and received through the wireless network, we can also measure the 'strength' of the signal for each packet. Doing some tests for each type of ethernet device, we can build a regression table that will allow us to estimate the distances the emitter is from us in further experiments.

Using a trigonometric approach, with three or more network partners we can estimate a single planar position, where we are. See Fig. 6.13.

Multilateration, also known as hyperbolic positioning, is the process of locating an object by accurately computing the time difference of arrival (TDOA) of a

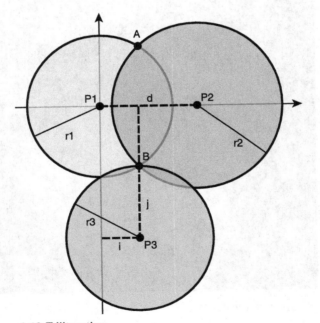

6.13 Triliteration.

[12] http://en.wikipedia.org/wiki/Time of arrival

[13] http://en.wikipedia.org/wiki/Time difference of arrival

[14] http://en.wikipedia.org/wiki/Angle of arrival

signal emitted from the object to three or more receivers. It also refers to the case of locating a receiver by measuring the TDOA of a signal transmitted from three or more synchronized transmitters.

Multilateration should not be confused with trilateration, which uses distances or absolute measurements of time-of-flight from three or more sites, or with triangulation, which uses a baseline and at least two angles measured e.g. with receiver antenna diversity and phase comparison.

These approaches can also be considered while using other communications technologies, like Bluetooth or ZigBee. More in-depth results can be found in our work (Sales et al., 2009).

6.5.2 Sonar

Ultrasounds do not suffer from electromagnetic interference with or from other equipment. On the other hand, ultrasounds can be 'blinded' if they reach some specific materials, as reflection varies depending on the characteristics of the materials.

The same techniques described in Section 6.5.1 can also be applied while using sonar (MacMullen et al., 1996), as it also travels through the air with a known speed and attenuation. In addition, some of the techniques described in Section 6.5.4 will also allow the use of sonar range data, bearing in mind the different accuracy of this ranging data and laser-based ranging data.

There are several companies that provide sonar-based location systems, as for example, Sonitor.[15] They provide ultrasound real time location systems (RTLS) that automatically track the real-time location of movable equipment or people in complex indoor environments with high room-level, or zone-level accuracy (such as bed-level) within a room.

For this technique to work, it usually needs wireless detectors installed throughout the environment, and tags attached to every item that we want to track. The motion-activated tag transmits its unique identification signal using ultrasound waves as the item moves through the environment area. Detectors transmit signals in digital format via the existing wireless LAN to a central computer unit that stores the information about the tag's location and the time of the receipt of the tag signal. This information enables the retrieval of a tag's position and/or movement and allows us to determine precisely by room where the item with that specific tag is located.

The main problem with this approach is that a pre-setup of the environment is needed to be able to locate the robots, and so, robots will not be able to be located except in very specific buildings, and only if the electric power is working properly. It also needs an infrastructure network, which is an extra constraint.

[15] http://www.sonitor.com

6.5.3 UWB

What is UWB? Ultra wideband systems transmit signals across a much wider frequency than conventional systems and are usually very difficult to detect. The amount of spectrum occupied by a UWB signal, i.e. the bandwidth of the UWB signal, is at least 25 per cent of the centre frequency. Thus, a UWB signal centred at 2 GHz would have a minimum bandwidth of 500 MHz and the minimum bandwidth of a UWB signal centred at 4 GHz would be 1 GHz. The most common technique for generating a UWB signal is to transmit pulses with durations less than 1 nanosecond.[16]

Ultra wideband technology[17] (Siwiak and McKeown, 2004; Di Benedetto et al., 2004) is quite an old technology, but nowadays is a new trend in radio signal based technologies for mass data transfer mainly, but also for localization and tracking of goods, materials, and any kind of item that can carry a tag. In fact, a European industrial driven project[18] has been created to carry on ultra-wideband technology research.

UWB is used as a part of location systems (Sahinoglu et al., 2008; Falsi et al., 2006) and real time location systems. The precision capabilities combined with the very low power make it ideal for certain radio frequency sensitive environments such as hospitals and healthcare. Another benefit of UWB is the short broadcast time, which enables implementers of the technology to install a vastly greater number of transmitter tags in any given environment relative to competitive technologies.

UWB[19] is also used in 'see-through-the-wall' precision radar imaging technology, precision locating and tracking (using distance measurements between radios), and precision time-of-arrival-based localization approaches (Saeed et al., 2006).

There are now several projects, companies, and research labs working in this field to get a new high accuracy, high reliability system. For instance, the EUROPCOM concept (Harmer, 2008) and project, which aims to investigate the UWB technology and the positioning system fields (see Fig. 6.14).

The Ubisense[20] precise real-time location system based on UWB technology is able to locate, track, record and analyse the movements of goods and people to an accuracy of up to 15 cm in 3D.

6.5.4 Visibility-based

Vision-based localization (Thompson et al., 1993) has been an important topic for a very long time. Localization based on visual landmarks (Yuen and MacDonald,

---

[16] http://www.palowireless.com/uwb/tutorials.asp

[17] http://en.wikipedia.org/wiki/Ultra wideband)

[18] EUWB (FP7-ICT-215669): http://www.euwb.eu/

[19] http://www.uwbforum.org/

[20] http://www.ubisense.net/

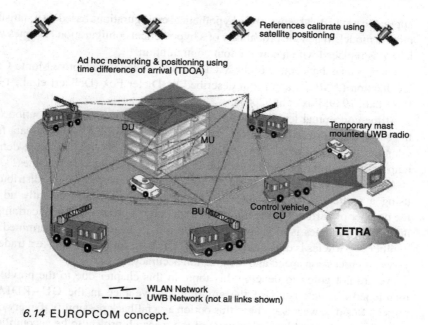

References calibrate using
satellite positioning

Ad hoc networking & positioning using
time difference of arrival (TDOA)

DU

MU

Temporary mast
mounted UWB radio

BU

Control vehicle
CU

TETRA

⌐⌐⌐   WLAN Network
—·—·— UWB Network (not all links shown)

6.14 EUROPCOM concept.

2005; Wolf and Burgard, 2002) requires feature extraction from views and a map, and the matching of features between views and a map. There are several hard to solve issues, starting at the very beginning with procedures for extracting features, and ranging to higher-level problems like solving methods for establishing feature correspondences.

Other visibility-based localization techniques involve the use of sensors that need visibility to work, like the laser range finder sensors. These sensors on their own do not provide localization, but there are well-known techniques to use their ranging data to provide an estimation of the position and orientation of the robot that holds the sensor, given a good enough knowledge of the environment.

The Monte Carlo localization algorithm

Monte Carlo localization, or MCL, is a Monte Carlo method to determine the position of a robot given a map of its environment based on Markov localization.

It is basically an implementation of the particle filter applied to robot localization, and has become very popular in robotics literature. In this method a large number of hypothetical current configurations are initially randomly scattered in configuration space. With each sensor update, the probability that each hypothetical configuration is correct is updated based on a statistical model of the sensors and Bayes' theorem. Similarly, every motion the robot undergoes is applied in a statistical sense to the hypothetical configurations based on a statistical motion model. When the probability of a hypothetical configuration becomes very low, it is replaced with a new random configuration.

There is an enhancement of the MCL method, called the adaptive Monte-Carlo localization (AMCL) algorithm described by Dieter Fox (Dellaert et al., 1999; Fox et al., 1999; Fox et al., 2001).

At the conceptual level, the AMCL maintains a probability distribution over the set of all possible robot poses, and updates this distribution using data from odometry, sonar and/or laser range-finders. The AMCL also requires a pre-defined map of the environment against which to compare observed sensor values.

At the implementation level, the AMCL represents the probability distribution using a particle filter. The filter is 'adaptive' because it dynamically adjusts the number of particles in the filter: when the robot's pose is highly uncertain, the number of particles is increased; when the robot's pose is well determined, the number of particles is decreased. The AMCL is therefore able to make a trade-off between processing speed and localization accuracy.

We are not going to develop this topic in this chapter due to the previously mentioned characteristics of the environment we face in the GUARDIANS project. Besides, we always have this option as a fallback possibility if every other technology fails, enabling other parts of the research project to be accomplished while waiting for a better localization technique in the near future.

6.6   Future trends

In this chapter we have presented a survey of sensing technologies for mobile platforms, their strengths, and their weaknesses from the point of view of an environment filled with smoke, as in the GUARDIANS project.

In order to develop a fully functional system it is clear that a very well designed integration is needed. No single sensing technology is able to solve every problem. Thus finding and using the strengths of each technology will complement the weaknesses of the others, giving us much more accurate and useful information in the environment, and so providing us with a base and well-informed layer to build new algorithms upon.

6.7   References

Barron, J. L. and S. S. Beauchemin (1995) *The Computation of Optical Flow*. ACM New York, NY, USA.

Brooker, Graham, Mark Bishop and Steve Scheding (2001) "Millimetre waves for robotics." In Australian Conference in Robotics and Automation.

Brown, Christopher M. (1987) *Advances in Computer Vision*. Hillsdale, NJ: Lawrence Erlbaum Associates.

Burkhardt, Hans, J. Wolf and Wolfram Burgard (2002) *Robust Vision-based Localization for Mobile Robots*. Unknown.

Burton, Andrew and John Radford (1978) *Thinking in Perspective: Critical Essays in the Study of Thought Processes*. London: Routledge.

Campbell, J., R. Sukthankar, I. Nourbakhsh and I.R. Pittsburgh (2004) "Techniques for evaluating optical flow for visual odometry in extreme terrain." In *Proceedings of Intelligent Robots and Systems (IROS)*.

Cheng, Y., M.W. Maimone and L. Matthies (2006) Visual odometry on the Mars exploration rovers. *IEEE Robotics and Automation Magazine*.

Corke, P., D. Strelow and S. Singh (2004) "Omnidirectional visual odometry for a planetary rover." In *Proceedings of Intelligent Robots and Systems (IROS)*.

Cracknell, Arthur P. and Ladson Hayes (2007) *Introduction to Remote Sensing*. London: Taylor and Francis.

Dellaert, F., W. Burgard, D. Fox and S. Thrun. (1999) "Monte Carlo localization for mobile robots." In IEEE International Conference on Robotics and Automation (ICRA 1999).

Di Benedetto, Maria-Gabriella and Guerino Giancola. (2004) *Understanding Ultra Wide Band Radio Fundamentals*. Prentice Hall PTR.

Falsi, Chiara, Davide Dardari, Lorenzo Mucchi, and Moe Z. Win. (2006) Time of arrival estimation for uwb localizers in realistic environments. EURASIP Journal of Applied Signal Processing.

Fleet, David J., John L. Barron and Steven Beauchimen (1994) Performance of optical flow techniques. *International Journal of Computer Vision*.

Fleet, David J. and Yair Weiss. (2006) Optical Flow Estimation. *Handbook of Mathematical Models in Computer Vision*. Springer.

Fox, D., W. Burgard, F. Dellaert and S. Thrun (1999) "Monte Carlo localization: Efficient position estimation for mobile robots." In Sixteenth National Conference on Artificial Intelligence (AAAI'99).

Fox, D., W. Burgard, S. Thrun and F. Dellaert (2001) *Robust Monte Carlo localization for mobile robots*. Amsterdam: Elsevier.

Fraden, Jacob (1996) *Handbook of modern sensors*. Thermoscan.

Harmer, David (2008) "Europcom, ultra wideband (uwb) radio for rescue services." In *Proceedings of the EURON/IARP International Workshop on Robotics for Risky Interventions and Surveillance of the Environment*. Benicassim, Spain.

Juds, S. (1962) *Thermoelectricity: an introduction to the principles*. John Wiley & Sons.

Konolige, K., M. Agrawal, R.C. Bolles, C. Cowan, M. Fischler and B.P. Gerkey (2006) "Outdoor mapping and navigation using stereo vision." In *Proceedings of the International Symposium on Experimental Robotics* (ISER).

MacMullen, W. G., B. A. Delaughe and J. S. Bird (1996) A simple rising-edge detector for time-of-arrival estimation. Instrumentation and Measurement. *IEEE Transactions on instrumentation and measurement*.

Marshall, Bill (2005) *An Introduction to Robot Sonar*. Robotbuilder.

Medeiros, Adelardo A. D., Kelson R. T. Aires and Andre M. Santana. (2008) *Optical Flow Using Color Information*. ACM New York, NY, USA.

Pascoal, José, Lino Marques and Aníbal T. de Almeida (2008) "Assessment of laser range finders in risky environments." In *Proceedings of the EURON/IARP International Workshop on Robotics for Risky Interventions and Surveillance of the Environment*, Benicassim, Spain.

Saeed, Rashid A., Sabira Khatun, Borhanuddin Mohd. Ali, and Mohd. A. Khazani. (2006) Performance of ultrawideband time-of-arrival estimation enhanced with synchronization scheme. *Transactions on Electrical Eng., Electronics, and Communications*.

Sahinoglu, Z., S. Gezici, and I. Guvenc. (2008) *Ultra-wideband Positioning Systems: Theoretical Limits, Ranging Algorithms, and Protocols*. New York: Cambridge University Press.

Sales, Jorge, Raul Marín, Leo Nomdedeu, Enric Cervera, and J.V.Martí (2009) "Estimation of the distance by using the signal strength for localization of networked mobile sensors and actuators." In RISE 09.

Scaramuzza, R. and D. Siegwart (2008) Appearance-guided monocular omnidirectional visual odometry for outdoor ground vehicles. *IEEE Transactions on Robotics*.

Siciliano, Bruno and Oussama Khatib (2008) *Handbook of Robotics*. Springer.

Siwiak, Kazimierz and Debra McKeown (2004) *Ultra-wideband Radio Technology*. Wiley: UK.

Sunderhauf, N., K. Konolige, S. Lacroix and P. Protzel (2005) Visual odometry using sparse bundle adjustment on an autonomous outdoor vehicle. Tagungsband Autonome Mobile Systeme.

Thompson, William B., Thomas C. Henderson, Thomas L. Colvin, Lisa B. Dick and Carolyn M. Valiquette (1993) "Vision-based localization." In *Proceedings of the 1993 Image Understanding Workshop*.

Vane, M. R. and J. K. E. (1986) *Multifunction millimetre-wave radar for all-weather ground attack aircraft*. Tunsley.

Warren, David H. and Edward R. Strelow. (1975) *Electronic Spatial Sensing for the Blind: Contributions from Perception*. Springer.

Yuen, David C. K. and Bruce A. MacDonald. (2005) Vision-based localization algorithm based on landmark matching, triangulation, reconstruction and comparison. *IEEE Transactions on Robotics*.

7

Sensor fusion for automated landmine detection on a mobile robot

S. LARIONOVA, PromAutomation Ltd., Russia and
A. T. DE ALMEIDA and L. MARQUES,
University of Coimbra, Portugal

Abstract: This chapter describes an approach of feature-based sensor fusion for landmine detection on a mobile robot using several non-selective sensors. A strategy of step-by-step reduction of the false alarm rate is used with newly developed techniques of *selective training* and *dominant class*. The proposed ideas are tested on experimental data from testing minefields obtained by a scanning device and a mobile robot.

Key words: feature-based sensor fusion, landmine detection, de-mining robot, landmine classification, de-mining experiments.

7.1   Introduction

Among all types of landmines, antipersonnel landmines (APLs) pose the biggest problem because they are specially designed to kill or maim people. An APL can be a very simple device which costs on average only US$3. Normally, its size is not larger than 15 cm and it triggers from a pressure of several kilograms. To achieve their purpose during wartime, antipersonnel landmines are hidden in the ground and the minefields are not marked (especially minefields created by terrorists).

In contrast to military de-mining, humanitarian de-mining should lead to a complete clearance of the specified area, so it can be safely used for civilian purposes. The current standard of the clearance maintained by Geneva International Centre for Humanitarian Demining (GICHD) recommends considering the area to be clean if all mines and unexploded ordnance (UXO) hazards are removed and destroyed (International Mine Action Standard (IMAS) 09.10. Clearance requirements). In other words, a clearance rate of 100 percent is required. This situation is further complicated by the following:

- Minefields are located in very different environmental conditions, often in difficult to access areas.
- There are a lot of types of landmines worldwide. They differ in size, shape, materials used (see, for example, Fig. 7.1) which complicates the development of standard detection methods.
- The landmine problem greatly affects countries where the cost of removing them is often the main concern. This limits the use of new and developing technologies.

147

7.1 Examples of AP landmines.

Human inventiveness has produced hundreds of different antipersonnel landmines (see, for example, the ORDATA database). However, it seems that currently the only effective way to remove and destroy a landmine is by using manual techniques. Landmines usually contain an amount of metal, thus a widely used tool for manual de-mining is a metal detector. However, only a few types of landmines have metallic cases, and others may include only a few grams of metal (a fuse), which makes them undistinguishable from metal debris left in the ground after the conflict. In this situation the data from a metal detector is ambiguous, providing a lot of false alarms and resulting in the need for careful probing to discover landmines.

The manual procedure is the basic tool of humanitarian de-mining nowadays. Even so, other more advanced techniques exist but they are seldom used because their cost is not affordable in most situations. The most common helpers of the de-miners are dogs, which have a highly developed olfactory sense, allowing them to detect the explosives leaking from the landmines. However, a specially trained de-mining dog is still too expensive. From the above information it can be seen that the technology of humanitarian de-mining needs to be advanced in terms of speed and safety to make the complete clearance of the existing minefields an affordable task. It was estimated that one de-miner is killed for every 5000 landmines removed (Landmines must be stopped, 1997).

7.1.1 Automation of humanitarian de-mining

The problems related to manual humanitarian de-mining have provoked many attempts to automate this process. There are two main objectives for this effort: decreasing the time of mine clearance and eliminating human participation on the minefield.

Following clear goals, an automated system must still provide the same quality of mine clearance achieved by the manual technique. Automation of humanitarian de-mining should be understood here in a broad sense meaning any tool that enables the diminishing or complete elimination of the participation of humans on the minefield. It is important to mention that there is no automated de-mining system that can provide the required clearance rate yet. The research in this area is ongoing in several directions, which are presented below.

A naive technique for mine clearance consists of the safe activating of landmines situated on a minefield. This is particularly useful in the case of antipersonnel landmines because their destroying force is not sufficient to damage a relatively simple vehicle. The field of mechanical mine clearance is the most developed one and there are commercially available vehicles. However, these machines cannot guarantee a complete clearance, thus they have to be used in combination with manual techniques. Moreover, the cost is still a limitation for their wide use.

One step toward an automatic system is the teleoperated method, which allows humans to be substituted by a robot in dangerous conditions. A teleoperated vehicle is controlled by the operator from a safe distance using video cameras. Changeable manipulators are used to perform the necessary operations, which include, for example, target deactivation by high-pressure air. Such machines are available in the market and are widely used by bomb disposal units to assist in their missions. They are mostly suitable for short operations when the location of the dangerous object is more or less known. However, teleoperated robots for humanitarian de-mining on a large scale are still being developed. Several new designs of semi-autonomous robots are proposed in Ishikawa et al. (2005). These vehicles can move to a specified location and then scan a small area using an advanced manipulator. The advantage of the teleoperated concept is the ability to develop a relatively simple vehicle and give the human operator the complex task of decision making. Then, the robot itself can be made very reliable, which is important because reliability is one of the main concerns on a minefield.

The largest degree of automation for humanitarian de-mining can be implemented in a fully automated system, which should be able to perform all stages. However, this is a very complex and probably unaffordable task. So, currently research is mostly concentrated in the direction of automated mine clearance because it is the most dangerous stage of humanitarian de-mining. An automated vehicle has to perform the following main tasks in order to assist the mine clearance task:

- landmine detection
- removal or marking (for further removal by humans) of the detected landmines
- planning a safe path through the minefield to ensure its full exploration
- avoiding obstacles in the way.

In the area of automated mine clearance there are several platforms being developed. Most of the research projects are dedicated to the mechanical design of the platform to make it reliable in the harsh outside conditions. A few projects also incorporate landmine detection and path planning tasks. There are also a few developments in the area of automated landmine removal.

A quadruped walking mechanical structure for the de-mining robot is proposed in Kato et al. (2000), Hirose et al. (2005) and Hirose et al. (1997). The implemented TITAN-IX robot shown in Fig. 7.2(a) is able to change working tools in the end of one leg allowing it to operate as a landmine detector (specifically, a metal detector). The working principle of the robot consists of scanning a small area while staying stable and then moving to the next location.

Another spider-like robot SILO4 proposed for use in humanitarian de-mining is described in de Santos et al. (2003). The robot has a good adaptability for outdoor terrain (see Fig. 7.2(d)). The developments also include an adaptable manipulator for a landmine detector.

7.2 Examples of detection robots. (a) Titan-IX; (b) Finder; (c) Ladero; (d) Silo4.

Figure 7.2(b) shows a simple wheeled platform for the testing of de-mining path planning algorithms (Acar et al., 2001). This project is concentrated mostly on the path planning and navigation algorithms. The proposed unknown coverage algorithm is based on the Morse decomposition of the area (Acar, Choset, Rizzi, Atkar & Hull, 2002; Acar, Choset, Zhang & Schervish, 2002). The main disadvantage of this algorithm is its high dependence on the precise localization of the robot, which requires improvements in the mobile platform localization sensors. There are also developments of probabilistic path planning algorithms specially designed for fast landmine search (Acar et al., 2003; Zhang et al., 2001), which are appropriate for the military but not for humanitarian de-mining.

A pneumatic cartesian platform is proposed for landmine detection in Rachkov et al. (2005) and Neves et al. (2003) (see Fig. 7.2(c)). The platform has a simple structure and scans the area during movement. Several sensors can be installed on the front of the robot to perform the landmine detection.

Summarizing the above overview it can be concluded that currently the most widely used technologies are mechanical mine clearance and teleoperated vehicles. There is also some work being done in the area of airborne assistance of landmine detection and minefield reduction (Clark et al., 2000; Kempen et al., 1999). However, the domain of automated mobile platforms for landmine detection (and, eventually, removal) is developing and can be considered the most promising for future applications.

Present work is focused on the development of a sensor fusion approach for landmine detection assisted by an autonomous mobile de-mining robot. The main goal is the creation of new sensor data processing algorithms and their implementation for a prototype scanning platform. It is assumed that the de-mining robot can be equipped with available sensor technologies, such as a metal detector, infrared (IR) sensors, and ground penetrating radar (GPR). The algorithms are implemented on a prototype de-mining robot LADERO, which has a cartesian mechanical structure and is controlled by pneumatic actuators (see Fig. 7.2c). This simple scanning platform is in accordance with the idea of simplicity to be used throughout the developed algorithms, which should not be computationally expensive, that are to be implemented on ordinary hardware.

7.2   Landmine recognition on a mobile robot

An antipersonnel landmine is a complex object composed from several heterogeneous parts: explosive charge, fuse system and case (see Fig. 7.3). These parts have physically different properties, thus there is no general "landmine sensor" which would allow the sensing of a landmine as a whole object. However, the parts of the landmine can be sensed separately, for example, as shown in Fig. 7.3.

Each of the characteristics provides a disturbance of some physical property in relation to the surrounding soil. Any of these characteristics can be used for the detection of the whole landmine, if it is unique for the landmines in comparison to other objects located on the minefield. However, in general, all the characteristics shown in Fig. 7.3, except the explosive material, are not selective. In comparison, sensors that are suitable for sensing of the explosive materials do not have high selectivity themselves. In this situation, it is usually not possible to provide the required quality of landmine detection by sensing only one of the landmine characteristics. Thus, there have been developments of sensing devices using different physical principles (appropriate to the landmine characteristics) (Dawson-Howe and Williams, 1998), in order to provide as much information as possible. A combination of the characteristic and the sensing method defines a type of signature that a specific detector acquires from the landmine. Landmine signature is a general term which can be used both in the case of manual and automated landmine detection: it is a set of signals provided by the detector together with their relation to some spatial information. It should be noted that the degree of non-selectivity of a sensing method also depends on the approach used for the discrimination of landmines and other objects. For example, the amplitude of the signal may not be a selective property, but the shape of its spatial signature may provide enough information for the discrimination.

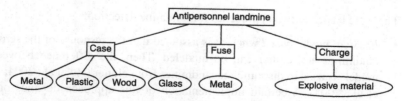

7.3 Characteristics of an antipersonnel landmine which can be sensed.

The quality of the sensing method is affected both by the selectivity of the sensor and by the selectivity of the sensed property. It must also be mentioned that the final result of landmine detection is a binary decision taken by some algorithm using the sensor data. Thus, the parameters of the data processing algorithms also affect the detection quality. The whole process of landmine detection is usually characterized statistically using the notations of the signal detection theory. As applied to the case of landmine detection the following terms can be defined:

- N_M: number of landmines (or, more generally, dangerous objects) to be examined, a number of landmines in a specified area S.
- $N_{\overline{M}}$: number of non-hazardous objects in S.
- *False alarm*: a decision that a landmine is present in its absence.
- *Probability of detection* or *detection rate* is the ratio of the number of correctly identified landmines to the total number of existing landmines

$$P_d = \frac{N_{M+}}{N_M},$$ [7.1]

where N_{M+} is the number of correctly identified landmines.
- *Probability of false alarm* or *false alarm rate* is the ratio of the number of false alarms to the number of non-hazardous objects

$$P_{fa} = \frac{N_{\overline{M}+}}{N_{\overline{M}}},$$ [7.2]

where $N_{\overline{M}+}$ is the number of false alarms. This measure is convenient for estimating the quality of the detection algorithms. However, it is not very meaningful for representing the quality of the landmine detection overall because every false alarm causes a costly operation of attempted landmine removal. Thus, not only the number but also the area of false alarms is important. A widely used measure in this case is the number of false alarms per square meter

$$P_{fa/m^2} = \frac{N_{\overline{M}+}}{S},$$ [7.3]

where every false alarm is allowed to cover an area not larger than a specified size (otherwise it is considered as multiple false alarms).

7.2.1   State of the art

There are basically two approaches for landmine detection:

- *Physical model based* where it is assumed that the response of the sensor for landmines and clutter can be modeled. Then, the difference between the models is used to discriminate landmines from clutter (Ho et al., 2004; Zhang et al., 2003). A certain implementation of this approach depends on each sensor technology.

- *Learning based* approach does not assume any specific model for the sensor response (the model is hidden). Instead, a set of classification features that characterize the response are used. The relations between the features and the detected object are determined by learning the hidden pattern from training data. Such approaches are based on the techniques of pattern recognition (Duda et al., 2000). Then, the landmine detection is usually considered as a classification with two classes: landmines and background. Different classifiers and feature extraction methods were used in previous works: Bayesian classification is used in Cremer, Jong, Schutte, Yarovoy and Kovalenko (2003), Zhang et al. (2003), Collins et al. (2002), Collins et al. (1999), Gao et al. (1999); support vector machines in Torrione and Collins (2004), Zhang et al. (2003); Dempster-Shafer classification in Perrin et al. (2004), Milisavljevi et al. (2003), Milisavljevi et al. (2000); neural networks in Stanley et al. (2002), Filippidis et al. (1999) and Sheedvash and Azimi-Sadjadi (1997), etc. In some works the different methods are compared, however, a definite conclusion about the most suitable approach is not made. This can be explained by the fact that the performance of landmine detection is affected by many parameters including the techniques for estimation of classification features and preprocessing of the sensor data.

There are also approaches that combine both the mentioned strategies. For example, the parameters of the physical model can be estimated using one of the learning techniques (Ho et al., 2004). Analyzing the previous work that uses both approaches it can be concluded that the model based techniques are less appropriate for the detection of landmines in real conditions and at least some amount of learning should be incorporated in the algorithm. Moreover, the physical modeling is well applicable only for some of the sensor technologies (mostly, metal detection) and is usually not appropriate for the use of multiple sensors together.

The basic two-class learning-based strategy is usually not effective enough due to the great complexity of the landmine detection task: the division of all the objects into categories of landmines and background is not logically sound because of the presence of other objects which significantly differ from the background. There are few studies that pay attention to this issue. A two-step strategy for landmine detection using handheld detector is used in Ho et al. (2004). Several works consider a preprocessing step of region-of-interest extraction (Roughan et al., 1997; Cremer, Schavemaker, Jong and Schutte, 2003; Clark et al., 2000; Frigui et al., 1998).

The currently existing algorithms for landmine detection do not achieve the required high quality yet; research in this area is ongoing in several directions. It is widely accepted that the integration of data from several sensors, called sensor fusion, should improve landmine detection in comparison to the performance of each sensor being used alone. This fact was also confirmed experimentally in several previous works (Filippidis et al., 2000; Milisavljevi et al., 2003; Breejen et al., 1999). In this situation the development of learning-based algorithms is an important step forward to develop a reliable landmine detection system.

In general, the process of multisensor integration can be implemented in different ways (Luo and Kay, 1995):

- separate operation for each sensor
- controlling one sensor by the information from another one
- fusion of data from several sensors.

The sensor fusion approach is widely accepted as the most promising approach to improve the landmine detection process. The term sensor fusion has several definitions, which can be found for instance in Waltz and Llinas (1990) and Luo and Kay (1995), and basically means any technique which assists the integrated usage of several sensor sources to achieve a common goal or decision. Four main types of sensor fusion can be distinguished: signal-level, pixel-level, feature-level and decision-level fusion (Luo and Kay, 1995). The signal and pixel-level approaches provide the highest degree of integration. However, they can only be used for sources with equal working principles, while the feature-level fusion can be used for any combination of sensors: the integration is performed using features which can be extracted from heterogeneous sources but have the same structure themselves. The decision-level fusion represents the highest level of abstraction from the sources used. The integration is performed after the decisions have been obtained for each source separately. The structure of the most appropriate method is determined by the specifics of the used sensors and the final goal of sensor fusion. In the case of landmine detection the following properties should be noticed:

- sensors have different physical principles
- sensors are not selective for the landmines, leading to a high false alarm rate.

The final stage of sensor fusion should always be a decision or feature-level algorithm because the goal of the landmine detection is a binary decision about the presence of a landmine. The decision-level sensor fusion combines the decisions made separately by each sensor, while in the feature-level fusion the raw data (classification features) from all sensors are processed together. This processing can be performed on the raw data or on the result provided by a lower level fusion algorithm (signal- and pixel-level fusion).

Examples of using the feature-level sensor fusion can be found in previous works on different combinations of sensor sources: metal detector and GPR are used in Perrin and Du (2004) and Stanley et al. (2002), IR camera and GPR are used in Cremer, Jong, Schutte and Bibaut (2003) and Cremer, Jong, Schutte, Yarovoy and Kovalenko (2003), and a combination of three sensors (metal detector, IR camera and GPR) are used in Gunatilaka et al. (2001), Gunatilaka et al. (1999) and Baertlein et al. (1998).

An example of pixel-level sensor fusion used for IR cameras with different bands can be found in Clark et al. (1993) and Clark et al. (1992). The decision-level fusion is the simplest approach in terms of implementation. It is explored for a combination of metal detector and GPR in Ho et al. (2004), Stanley et al. (2002), Bruschini et al.

(1996); for IR camera and GPR in Cremer et al. (2000); for a combination of IR cameras in Filippidis et al. (2000); and for the three sensors in Milisavljevi et al. (2003), Cremer et al. (2001), Schavemaker et al. (2001), Gunatilaka et al. (2001), Milisavljevi et al. (2000) and Breejen et al. (1999). There is also some research in the domain of sensor fusion for landmine detection without considering any specific sensors (Kacelenga et al., 2003; Kacelenga et al., 2002). In some works these two concepts are compared (Stanley et al., 2002; Cremer et al., 2001; Gunatilaka et al., 2001; Milisavljevi et al., 2000), but there is no experimentally proven conclusion as to which strategy is better. However, it is considered that the feature-level fusion is more preferable because it provides a deeper integration of data.

Having features obtained from landmine detection sensors, the task of the consequent steps is to assign a class label to it (classification task). Using different classifiers in previous reports did not lead to a clear conclusion on a preferable one. This is probably due to the fact that the performance of the classification depends on many other factors, including classification features, training strategy, and the availability of data for training.

It must be noted that landmine recognition is a very challenging classification task due to the high ambiguity of the sensor data. As the classifier represents the top structure in this process, the best strategy seems to be the implementation of a simple classifier and using it for analyzing other factors. Particularly, the analysis of classification features is one of the most important tasks to be done before a conclusion about the best configuration can be made. The problem of selection and estimation of classification features has not received the required attention, as can be concluded from the analysis of previous works. The most frequently used features are statistical measures (Roughan et al., 1997; Cremer, Schavemaker, Jong and Schutte, 2003; Clark et al., 2000), such as mean value inside the region-of-interest (ROI), and measures related to simple shapes (Roughan et al., 1997; Milisavljevi et al., 2003; Frigui et al., 1998; Perrin and Du, 2004). More complex features like entropy, contrast and correlation are considered in Clark et al. (2000) for the processing of IR data, but the results are difficult to analyze due to the small number of samples in the available data sets. In Stiles et al. (2002) the possibility of using shape symmetry features for landmine detection is analyzed. However, the presented results were obtained only for well-controlled conditions. Texture features are used in Filippidis et al. (2000) for surface landmine detection.

In summary, there is definitely a need for detailed research that clarifies the choice of classifiers and classification features. Moreover, the existing solutions do not provide the performance required by humanitarian de-mining standards yet, so new ideas are needed.

7.2.2 Approach for landmine detection

The most developed sensing technologies for landmine detection currently include metal detectors, infrared sensors and ground penetrating radar. They are considered here as typical detectors whose data should be combined by means of sensor fusion techniques. The goal of this process is to provide a single decision about presence or absence of a landmine. The approach developed in this work is based on feature-level sensor fusion and has several stages. The multi-stage nature of the approach is necessary to address the specific challenges associated with the landmine detection task.

First of all, it is important to mention that the sensors used do not detect the landmines directly. Instead, they can only distinguish heterogeneity of some physical parameter against the background. This physical property found in a landmine can also be present in other objects, called clutter. The presence of heterogeneity signifies only that there may be an object suspected of being a landmine. The difference between the background and the objects (heterogeneities) gives the largest variation in the sensor signal, while the difference between landmines and clutter is much lower. An attempt to separate the three categories (background, landmines, and clutter) in one classification process is a complicated task because the signatures of landmines and clutter are nearly the same in relation to the background. Thus, the most effective first stage of the landmine detection process seems to be the classification with two classes: *background* and *suspicious objects*.

Here suspicious objects include the actual landmines, other man-made objects, natural debris, and can also represent just heterogeneities of the soil properties.

The algorithm for detection of a suspicious object incorporated in this work uses the difference of the sensor value and the spatial information to detect a region, called region-of-interest (ROI), where the signature of the object is located. For this purpose a novel online algorithm was developed in Larionova (2007). It is able to detect the object right after being fully scanned by the scanning device, thus enabling online processing of the data while the mobile scanning platform is exploring the terrain. The algorithm is hardly sensitive to the quality of the sensor data. A sample result of a ROI detected by this algorithm can be seen in Plate I (in colour section between pp 166–7). In the next step, the ROIs which represent the signatures of the same object for different sensors are combined together forming a complete signature of the suspicious object.

The next stage of landmine detection should finally distinguish landmines from the previously detected suspicious objects. An important issue at this point is the classification features that can be used for this purpose.

While the difference between the background and the objects is clear on the spatially mapped sensor data, the signatures of the landmines contain almost no visual tokens that would help to identify them from other objects. As there is no single "selective" classification feature, a large number of unselective features can be used to perform the classification. This situation is similar to the usage of sensor fusion for several non-selective sensors each of which cannot be used alone.

The classification features may reflect the difference between signatures based on their shape, statistical and information measures, etc, as described later in this section. The common idea behind the classification features is an attempt to recognize some regularity in the object signature because a landmine has a regular structure with parts in certain locations and some symmetry. This property is probably the only possibility that can be used to separate the landmines from clutter. However, regularity is also present in many other man-made objects. If the landmines were laid on the ground surface their structural signatures could be easily recognized from other artificial objects by a vision system, but the subsurface detectors cannot usually provide such rich information.

The recognition stage is further divided into sub-stages which can be performed or not, depending on the data quality: the suspicious objects are classified in classes: *man-made objects* and *natural clutter*, and then the landmines are identified among the man-made objects. The new algorithms for these two stages are developed in the present work. The quality of the sensor data obviously limits the possibilities of the recognition method. By having better sensor data, this process could even be extended to the identification of certain types of landmines. However, such data is not currently available.

Plate II (in colour section) shows the process of landmine detection described above. One more stage is added to it for research purposes: the signatures of detected suspicious objects are collected in a database. Each object in the database is associated with the experiment where the data were obtained, the coordinate data, the ground truth, etc., providing a unified approach for representation of landmine signatures. The stored signatures are later used offline for the testing of classification algorithms.

One of the advantages of the proposed strategy is its ability to account for the quality of the sensor data as shown in Fig. 7.4. The landmine detection process can be terminated at any stage if the data quality is not high enough to perform a more precise classification. Then, all objects of the appropriate class (shown with gray background in Fig. 7.4) are considered to be landmines, allowing detection to be

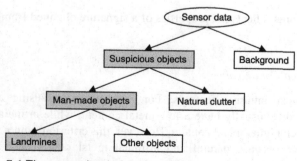

7.4 Three-step landmine detection.

performed but with a higher false alarm rate. Thus, the developed landmine detection strategy naturally appears as a step-by-step reduction of the false alarm rate.

The last two tasks on Fig. 7.4 are naturally the same in terms of classification strategy but should be accomplished separately in order to improve the separability of the classes and to account for the quality of sensor data.

The basic classifier considered in this work is a Bayesian classifier which models the joint probability as a multi-variate Gaussian distribution. Based on this model, effective classification features should be developed. The classifier performs as a fusing algorithm, which accepts the feature vector created from the features calculated for each signature of the object. Thus, the fusion is performed on the feature-level by using this combined feature vector. Taking into account that landmine recognition is characterized by high overlapping of the classes, suitable solutions based on the simple classifier are developed.

7.2.3 Classification features

The first stage of any classification process is the estimation of the classification features. In general, the raw sensor data can play the role of the features themselves. However, such strategy is prone to over training. It is considered here that a classification feature is a single number calculated for a ROI (its data map, segmented map or object area, see Plate I) or a collection of ROIs.

Such features are suitable for use in the framework of feature-level sensor fusion forming the combined feature vector. Some of the features described below were adapted from different fields of pattern recognition mostly related to computer vision, and others (object skewness, fractal dimension and golden ratio measure) were specially developed here. The main property that a feature should reflect is the regularity of the signature, which may signify the regularity of the object itself and provide useful information for landmine recognition. As long as there is not one perfect feature that provides this information, several less specific features can be used instead.

Sensor-based features

This group of features reflects the properties of a signature obtained from data of a single sensor.

Features based on absolute value

These are the most intuitive features. For example, when using a metal detector the landmines usually have a low metal content, while artificial metal objects have a much higher metal content. However, this criterion cannot provide good results alone because natural and artificial metal debris may also be low-metal objects.

Statistical measures: mean, standard deviation, skewness, kurtosis.

Contrast

$$C = \max(C_{i,j}),$$

where $C_{i,j}$ is the local contrast

$$C_{i,j} = |x_{i,j} - (x_{i-1,j} + x_{i+1,j} + x_{i,j-1} + x_{i,j+1})/4|$$

Features related to object shape

Sensors like IR cameras and GPR may provide a good estimation of the object shape.

Size, aspect ratio

Vertical skewness of object

$$VS = \frac{|\max(D_i) - \min(D_i)|}{width},$$

where D_i is the distance between the center of the object along X axis and the local maximum along X axis.

Horizontal skewness of object

$$HS = \frac{|\max(D_i) - \min(D_i)|}{height}$$

where D_i is the distance between the center of the object along Y axis and the local maximum along Y axis.

Occupied part

$$OP = \frac{Area\_of\_object}{Size\_of\_ROI}$$

Compactness

$$M_{cmp} = \frac{\mu_{00}}{\mu_{20} + \mu_{02}},$$

where $\mu_{00}, \mu_{20}, \mu_{02}$ are central shape moments.

Eccentricity

$$M_{ect} = \frac{\sqrt{(\mu_{20} + \mu_{02})^2 + 4\mu_{11}^2}}{\mu_{20} + \mu_{02}},$$

where $\mu_{11}, \mu_{20}, \mu_{02}$ are central shape moments.

Circularity

$$F = \frac{4\pi S}{P^2},$$

where S is area and P is perimeter of the object.

Features related to object nature

These features are the most challenging ones because they attempt to analyze if the object is natural or artificial. One of such features is the fractal dimension: in most cases its value for natural objects must be higher than for artificial ones. A novel feature introduced here is based on the idea of the golden ratio.

Fractal dimension (measure of self-similarity)

$$FD = \frac{\log(N_r)}{\log(1/r)},$$

where N_r is a number of copies of the object scaled down by ratio r. FD is estimated using a differential box-counting approach (Sarkar and Chaudhuri, 1994).

Entropy (measure of disorder)

$$H(X) = -\sum_x P(x)\log P(x),$$

where $P(x)$ is the probability that X is in the state x

GR measure (measure of golden ratio)

The segmented map is analyzed as follows in order to determine how different the averaged ratio between two neighboring segments is from the golden ratio:

$$GRM = \sum_{i=1}^{N-1} grm_i / N,$$

$$grm_i = \left| \frac{\min(S_i, S_{i-1})}{\max(S_i, S_{i-1})} - \frac{1}{\phi} \right|,$$

where S_i is the size of segment i, N is a number of segments and $\phi = \frac{1}{2}(1+\sqrt{5})$ is the golden ratio.

Multi-sensor features

Besides using the sensor-based features it is possible to analyze the relationship between the signatures obtained from different sensors. This feature can be calculated directly using several ROIs, as correlation features, or a combination of several sensor-based feature and combined features.

Correlation features

Distance between ROIs

Aspect ratio correlation

$$HWC = \frac{HW1}{HW2}$$

if

$$HW1 < HW2$$

and

$$HWC = \frac{HW2}{HW1}$$

otherwise. Where *HW* is aspect ratio.

Correlation (image correlation)

$$Corr = \max_{x1, x2} |Corr(s1, s2)|,$$

$$Corr(s1,s2) = \frac{\sum\limits_{p1=x1, p2=x2}^{N1, N2} Vn1_{p1} \cdot Vn1_{p1}^{2}}{\sum\limits_{p1=x1}^{N1} Vn1_{p1}^{2} \cdot \sum\limits_{p2=x2}^{N2} Vn2_{p2}^{2}},$$

where *Vn*1 and *Vn*2 are normalized values, and *s*1 and *s*2 are the coordinates of the starting points for the maps. Changing *s*1 and *s*2 enables the shift of one map relatively to the other allowing us to search for the best correlation.

Shape correlation

$$SC = \max_{x1, x2} |SC(s1, s2)|,$$

$$SC(s1,s2) = \sum\limits_{p1=x_1, p2=x_2}^{N1, N2} OA1_{p1} \cdot OA2_{p2} / N(s1,s2),$$

where *s*1 and *s*2 – see Correlation, and N(*s*1, *s*2) is the number of common points for both maps.

Contour correlation

The object area is processed to obtain its contour chain code (represented by numbers from 0 to 7). Then the correlation between the chains is calculated as follows:

$$ContourCorr = \sum\limits_{i1=0, i2=0}^{N1, N2} \frac{7 - |C1_{i1} - C2_{i2}|}{7},$$

where *C*1 and *C*2 are values of the chain codes, and $|C1_{i1} - C2_{i2}|$ are adjusted to be always ≤ 4.

Combined features

The combination of several sensor-based features can improve the feature performance if they represent a general property of the object reflected in signatures of different sensors in a similar way. In other words, the feature should not highly depend on the sensor from which the signature was obtained. The process of feature combination is identical to the signal-level sensor fusion when the signals are fused, for example, by averaging. In this work the combined features are calculated by averaging the values of the same feature calculated for signatures of the object obtained from different sensors. In this case the confidence in the feature value should increase with the number of sensors detecting the object. Moreover, even if there is only one sensor detecting the object, the feature is still present. Of course, this operation cannot be considered for very simple features like mean or size, but it can show a good performance for the more complex ones, like entropy.

Feature selection

The large number of features analyzed in this work cannot be directly used by a classifier because the amount of training data is usually not enough to estimate the joint distribution with so many parameters. Therefore, using too many features decreases the performance of a classifier. Moreover, some features can even confuse the classifier because they do not represent useful information for the separation of classes. Each feature should be evaluated in terms of relevance, and only the most relevant ones should be considered for the classification. This process is usually called feature selection and is commonly used in pattern recognition tasks to reduce the dimension of the feature vector in order to reduce the processing time.

Classification features are evaluated in this work using two evaluation criteria: mutual information and Hausdorff distance. Mutual information gives a measure of how much information a random variable contains about another (Kwak and Choi, 2002). Two random variables are considered: continuous value of the feature f and discrete class c. The Hausdorff distance characterizes the position of each point of one set, relatively to the points of another set (Piramuthu, 1999). In order to benefit from both evaluation measures, their product is used as an evaluation criterion for the selection of the most relevant features. The feature selection was performed manually by choosing the features with the largest value of the evaluation criteria.

7.2.4   Multi-stage classification

As it is defined by the strategy of landmine detection the recognition of landmines should be performed in several stages according to the quality of sensor data. After each stage, only the objects of a landmine-suspicious class are taken for the further stages. In this work the process is limited to a two-stage classification, as it does not seem reasonable to extend it further with the available sensor data. Each stage is implemented as a standard Bayesian classifier (Duda et al., 2000). The Bayesian classifier is based on the optimal decision rule, which is in the case of two classes:

$$decide \; class_1 \; \text{if} \; P(class_1 \,|\, x) > P(class_2 \,|\, x),$$

otherwise decide ^{class_2},

where x is a feature vector representing the state of the object to be classified, and $P(class_i \mid x)$ is the probability of the object represented by x belonging to the class $class_i$, also called posterior probability. The posterior probability can be obtained by the Bayes' formula:

$$P(class_i \mid x) = \frac{P(x \mid class_i)P(class_i)}{P(x)},$$

where $P(x \mid class_i)$ is a probability of observing feature vector x from an object of class $class_i$. This probability can be learned from the training data set using one of the techniques for parameters estimation (Duda et al., 2000).

Then, the decision rule can be modified as follows:

$$decide \ ^{class_1} \ if \ \frac{P(x \mid class_1)}{P(x \mid class_1)} > \lambda,$$

otherwise decide ^{class_2},

where λ is a constant which includes the unknown $P(class_i)$ and risk factors. Varying λ the decision can be shifted in favor of one class, which can be used in the case of landmine detection in order to achieve a high detection rate. To obtain an optimal value of λ the training data set can be used.

The analysis made in this work showed that the probability distribution of most classification features can be modeled by a Gaussian function. Thus, $P(x \mid class_i)$ is modeled as multivariate Gaussian:

$$p(x) = \frac{1}{(2\pi)^{n/2} \sqrt{\hat{\Sigma}}} \exp \left[-\frac{1}{2}(x - \hat{\mu})^t \hat{\Sigma}^{-1}(x - \hat{\mu}) \right], \quad\quad [7.4]$$

where n is the dimension of the feature vector, and $\hat{\Sigma}$ is an n × n covariance matrix. A maximum likelihood estimation is used to compute the parameters of Equation [7.4] from the training data set (Duda et al., 2000).

$$\hat{\mu} = \frac{1}{N} \sum_{k=1}^{N} x_k \quad\quad [7.5]$$

$$\hat{\Sigma} = \frac{1}{N-1} \sum_{k=1}^{N} (x_k - \hat{\mu})(x_k - \hat{\mu})^t, \quad\quad [7.6]$$

where N is the size of the training set.

7.2.5 Classifier training

One important challenge in landmine detection is the absence of the ability to train the classifier on the data obtained directly from the minefield. In most cases it is possible to acquire some data from the background and non-hazardous objects, which can be useful for adjusting parameters for the detection of suspicious objects. However, the availability of the data from real buried landmines should not be considered. In this situation the algorithm has to be pre-trained beforehand with data obtained from other experiments made on test fields. To address this issue, different training scenarios are considered in this work:

- Randomly chosen training set. The whole data set is randomly divided into two equal sets, one of which is used for training, and the other for evaluation.
- The training set only contains data from one field with a certain soil type. Each soil type can be considered as a separate training experiment, and then other fields are used for evaluation. This scenario models the case when the algorithm is trained in one type of condition and then is used in others.
- The training set is chosen manually to contain only "good" signatures. Such a scenario models a case when only well-controlled experiments are available for training. If it were possible to train the algorithms in such a way, then it would simplify the experiments, and thus increase the amount of training data.

Table 7.1 shows the number of samples of the considered training and evaluation sets. It is logical to assume that among the possibilities mentioned the use of randomly chosen training sets may give better classification results. However, in practice it is not possible to implement this scenario because the full data set (from which the random samples would be chosen) is usually not available.

7.2.6 Missed features

A practical problem of any object recognition system is the fact that in real conditions some features can be missed. A Bayesian classifier can solve this problem by integrating the posterior probabilities over the missed features (Duda et al., 2000). However, in the case of landmine detection, the features are missed only when a sensor does not detect the object at all (for example, a metal detector does not sense a plastic object). In this situation, all features related to that sensor

Table 7.1 Number of samples of training and evaluation sets used in experiments (LM – landmines, Others – not landmine objects)

Experiment description		Training set			Evaluation set		
		All	LM	Others	All	LM	Others
3 sensors	1. Random	1287	183	1104	1287	183	1104
	2. Soil 2	666	82	584	1908	284	1624
	3. Manual	404	168	236	2170	197	1972
4 sensors	4. Random	965	129	836	965	129	836
	5. Soil 2	472	52	420	1458	205	1253
	6. Manual	351	98	252	1579	159	1420

are missed. Thus, there are a limited number of feature combinations determined by the number of sensors (for example, there are 15 combinations for four sensors: 1, 2, 3, 4, 1–2, 1–3, 1–4, 2–3, 2–4, 3–4, 1–2–3, 1–2–4, 1–3–4, 2–3–4, 1–2–3–4). It is practically possible to consider all the combinations and train several classifiers. Then, during the process of object classification, a classifier with the largest possible number of features is chosen (Fig. 7.5(a)). Another solution comes automatically if only combined features are used as shown in Fig. 7.5(b) because all features are always present if at least one sensor detects the object. However, the last approach is probably limited because there are not enough features that show good performance in the combined version. Thus, the strategy implemented in this work is a combination of both techniques.

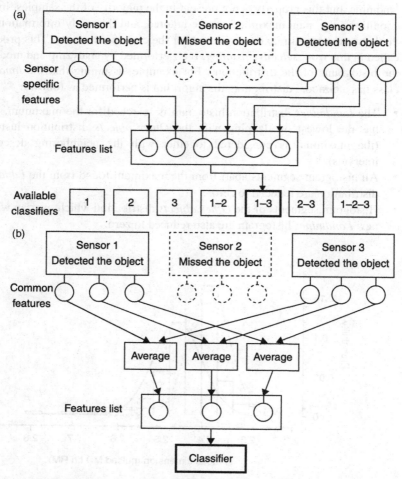

7.5 (a) Possible solution for missing features problem. (b) Possible solution for possible features problems.

7.2.7 Concept of selective training

The classification features implemented in this work basically show high overlapping of the classes, which is represented by two types of behavior:

- Distributions for different classes have shapes close to the Gaussian distribution and maxima are poorly separated (see, for example, Fig. 7.6).
- One of the distributions is bimodal: the main maxima are overlapped, and another weaker maximum is separated from the main one (see Fig. 7.7(a)).

The usual strategy followed in the second case is to consider only the main maximum. However, if the feature has a weak (or equally sized) but better separated maximum, it may signify that it reflects a specific property of the landmine, but this property is not present in the majority of the samples. In these conditions, the main maximum can be ignored, and then only information from the least overlapped maximum is used for the classifier training. This process is called in this work *selective training*. It is performed by analyzing and modifying the histograms of the distributions. For example, assuming that the *landmines* class has a bimodal distribution, the algorithm is performed as follows:

- The *Landmines* distribution histogram is searched for the maximum, which has the lowest overlapping with the *Other objects* distribution histogram (the maximum is considered together with the neighboring descending intervals).
- All histogram segments apart from the maximum found from the *Landmines* distribution are reduced to zero.
- Histogram segments of the *Other objects* distribution which overlap with the new *Landmines* histogram are also reduced to zero.

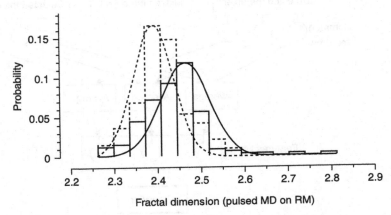

7.6 Example of feature whose distributions for different classes are poorly separated.

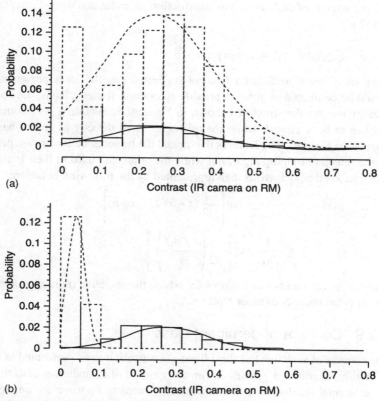

7.7 (a) Example of feature which has bimodal distribution for other objects class. (b) Example of feature which has bimodal distribution for other objects class.

The distributions obtained (see, for example, Figure 7.7(b)) are then used for training. If there are enough features with such property able to separate different samples, this process may improve the separability between the classes.

Assuming that only a well-separated part of the distribution of each feature is used, the relationships between the features change. It becomes inappropriate to represent $P(x \mid class_i)$ as a multivariate Gaussian because it reflects an AND relation between the features. However, the features processed by the selective training algorithm have an OR relation to each other: the effect of one feature should be added to the others to assure that the object is detected if values of some of its properties are in the selected ranges (but not values of all properties, as would be the case for AND relation). In this case a more suitable model for $P(x \mid class_i)$ is an additive Gaussian model which can be represented as follows:

$$P(x) = \sum \frac{1}{\sqrt{\pi}\hat{\sigma}} \exp\left[-\frac{1}{2}\left(\frac{(x-\hat{\mu})^t}{\hat{\sigma}}\right)^2\right], \tag{7.7}$$

The parameters of each univariate distribution are estimated using Equations [7.5] and [7.6].

7.2.8 Combined strategy

Analysis of the classification features suggested that the ideas presented above should be combined in order to provide an optimal strategy. Thus, the combined strategy merges the structures shown in Figure 7.5, assuming only some of the features to be averaged, while the others are used directly as sensor-based and correlation features. The features that reveal the bimodal behaviors are processed by the selective training algorithm while the others are used in their initial form. The combined p(xjclassi) is then represented by the following equation:

$$p(x) = \frac{1}{(2\pi)^{n_1/2}\sqrt{\hat{\Sigma}}} \exp\left[-\frac{1}{2}(x-\hat{\mu})^t \hat{\Sigma}^{-1}(x-\hat{\mu})\right]$$
$$+ \sum_{j=1}^{n_2} \frac{1}{\sqrt{\pi}\hat{\sigma}_j} \exp\left[-\frac{1}{2}\left(\frac{(x-\hat{\mu}_j)^t}{\hat{\sigma}_j}\right)^2\right]$$

[7.8]

where n_2 is the number of features for which the selective training is applicable, and n_1 is the number of other features.

7.2.9 Concept of dominant class

The concept of selective training allows classification to be performed in the case of highly overlapped classes as in the case of the landmine detection task. Experimental results show that this strategy helps to improve the quality of the recognition (see Section 7.2.10). However, there is one specific feature of the landmine detection task that is not fully accounted for in this concept: the recognition process should provide the highest possible detection rate (DR) (which is specified to be 100 percent by the landmine detection standards). The usual solution for achieving a high detection rate is to choose the classification constant λ appropriately. This allows shifting of the working point on the receiver operating characteristic (ROC) curve to the area of higher DR (and consequently higher false alarm rate (FAR)). However, the problem still remains because for high detection rates a small increase of DR causes a much higher increase of FAR and in most cases the DR of 100 percent cannot be reached at all. This follows consequently from the fact that the classification process intends to obtain the best solution in respect of the classification error, which includes the quality of detection for both the classes considered. On the other hand, in the case of landmine detection one class (*Landmines*) should dominate allowing the 100 percent DR to be achieved (which cannot be achieved by simply shifting the λ). It is proposed in this section to develop the concept of selective training further to allow the domination of the *Landmines* class. Following the same principle the feature distributions are modified before they are used for the classifier training.

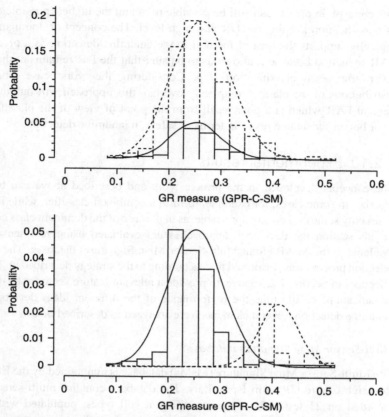

7.8 Illustration of the training process to account for a dominant class.

The goal of the modification is to consider only the dominant class where the distributions overlap and let the distribution of the other class be present only outside of the main class. Here, all types of distributions are considered, both unimodal and bimodal. The preprocessing of the distributions is performed in the following steps (see Fig. 7.8):

- The *Landmines* distribution histogram is searched for the main maximum (the maximum is considered together with the neighboring descending intervals).
- Histogram segments of the *Other objects* distribution which overlap with the found maximum are reduced to zero.

The obtained distributions are then used for the training of the classifier which later processes the data using the model in Equation [7.4]. This process obviously leads to loss of information about the *Other objects* class. However, it might not be possible in general to perform a classification that achieves good quality in relation to both classes in the case of highly overlapped distributions. By using this concept, in turn, it can still be possible to obtain the highest possible quality of classification keeping the DR at a high level. The concept of dominant class logically supports the idea of the multistage landmine detection as a process of FAR reduction because it allows us to be sure that the DR remains at the same lever after each classification step. Considering the worst case when the distributions of the classes completely overlap, this approach will output a 100 percent FAR which is a poor result from the point of view of the classification error but an acceptable result in terms of safety in landmine detection.

7.2.10 Experimental results

The concepts developed in the present work and described above can be used together in some cases, for example, to form a combined classifier, while in other cases only some of them are applicable, as in the case of the dominant class concept. In this section the developed approaches are evaluated using real sensor data available in the MsMs (Joint Multi-Sensor Mine Signature) database. The feature selection process was performed first, according to the strategy described in 'Feature selection' in Section 7.2.3 above, to provide a relevant feature vector to be used for classification. To illustrate the performance of the different ideas developed for landmine detection, several elements were analyzed as described below.

Multi-Sensor Mine Signature database

The Multi-Sensor Mine Signature (MsMs) database is maintained by the EU Joint Research Centre (JRC) in Ispra, Italy. The database contains multi-sensor data recorded on 21 test fields of seven different soil types, populated with mine surrogates and other objects. In this work the data from four sensors were used: Vallon ML 1620C pulsed metal detector, Foerster Minex 2FD 4.500 continuous metal detector, AGEMA 570 infrared camera and experimental ground penetrating radar C scans. Data from only nine fields are available for all four sensors: A1, A2, A3, C1, C2, C3, C4, C5, C7 (MsMs database). Thus, a combination of three sensors (without the GPR) was also used for experiments in order to maximally use all data sets obtained from the 21 fields. Each field measures about 6 m × 2 m.

One-stage vs multi-stage classification

It is not obvious that for landmine detection a multi-stage classification is required. To analyze this issue, the feature space was processed in Larionova (2007) by principal component analysis (PCA) (Duda et al., 2000). The variance in the data revealed by PCA does not reflect the difference between the *Landmines* and *Other objects* classes, confirming the concept of highly overlapped classes in the case of landmine detection. Moreover, during such classification most false detections consist of other man-made objects: the classifier is not able to distinguish landmines from other man-made objects. The classes *Man-made objects* and *Natural clutter*

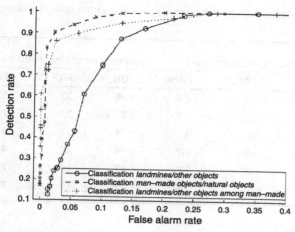

7.9 Comparison of different classifications.

are better separated than the previous pair. Another possibility to support the two-stage approach is to compare the classifications involved in the one-stage and two-stage classifications in terms of detection and false alarm rates. A comparison of ROCs for different classifications shown in Fig. 7.9 also confirms this idea. It can be seen that, as expected, the classification with *Man-made objects/Natural clutter* classes has better performance than the classification with *Landmines/Other objects* classes. Moreover, the classification *Landmines/Other objects* performs better when it is used to distinguish between the *Man-made objects*.

Influence of the training set

The problem of classifier training is important due to the difficulties in obtaining the sensor data for landmine detection. To analyze this issue a classifier which contains only sensor-based features was constructed eliminating the influence of the combined features and selective training on the results. Table 7.2 shows the results of classification using different combinations of training/evaluation data sets and the mentioned classifier with fixed λ. Three possibilities for the forming of the training set were considered: random, manually chosen 'good' signatures and data from a selected type of soil (Table 7.1). As expected, using the randomly created training set leads to the best classification results (Table 7.2 with N=1 and N=4). From the two other possibilities, the training based on the selected experimental field is the most practical because it shows better results (Table 7.2, N=2, 5) than in the case of manual selection (Table 7.2, N=3, 6). This signifies that the manually chosen training set does not contain enough information due to, for example, the absence of clutter. The ability to train the classifier on one type of soil is important for the practical implementation of the automated landmine detection system in which neither random nor leave-one-out training is possible.

Table 7.2 Sensor fusion results for different training/evaluation sets, N = number of sets from Table 7.1

N	Performance on training set			Performance on evaluation set		
	DR, %	FAR, %	FA/m^2	DR, %	FAR, %	FA/m^2
1	98	36	3.3	96	37	3.3
2	90	13	2.7	70	23	2
3	90	35	3	74	37	4.9
4	96	35	4.7	95	35	4.9
5	100	13	2.8	77	20	2.8
6	93	23	2.6	74	19	3

Table 7.3 Comparison of sensor fusion results for different feature sets and different types of training, N = number of sets from Table 7.1

N	Training	Performance on evaluation set		
		DR, %	FAR, %	FA/m^2
1	Normal	90	30	2.7
	Selective	95	32	2.8
4	Normal	90	32	4.1
	Selective	95	32	3.9

Selective training

In order to show the benefits of the concept of selective training, the performance of the classification with and without it was compared. A combined classifier containing sensor-based and combined classification features was created for this purpose. In one case the classification features with bimodal distributions were trained normally (without ignoring the parts of the distribution), in the second, according to the selective training concept. Table 7.3 shows the results of applying this classifier with fixed λ for both cases. It can be seen that the introduction of selective training improves the results.

Dominant class

The dominant class concept is tested here using the experimental data from the MsMs database. These data are extensive allowing the selective training to be performed as it was confirmed in the previous section. While the dominant class concept is expected to be especially effective in the case of poor sensor data, the results of applying both concepts are very similar in the case of rich sensor data.

To perform a clear comparison of the training concepts, the performance of the classifier was intentionally degraded by using only a few combined features, simulating a situation when the sensor data are not enough to perform the training. Training and evaluation of the classifier were carried out with the sets under number 5 in Table 7.1. The results of the comparison of three training techniques in this case are shown in Figure 7.10. The dominant class training outperforms the selective training and normal training (where the distributions of the classification features are not changed). The selective training, in turn, outperforms the normal training, again confirming its usefulness.

A more appropriate case where the dominant class training is vital is considered in Section 7.3 in which landmine recognition is performed on the sensor data obtained by the mobile platform LADERO.

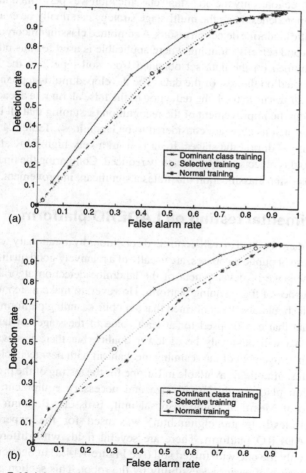

7.10 (a) and (b) A comparison of ROC for different training strategies.

Table 7.4 Changing of false alarms per m² for detection rate 95% over the steps of
landmine detection

	ROIs extraction	Two-step classification	
		Man-made objects/ natural clutter	Landmines/other objects
FA/m²	12.1	5.4	4.5
		One-step classification	
		7	

Step by step reduction of false alarm rate

In the previous subsections only the one-stage classification was performed in order
to evaluate other approaches. Here the multi-stage strategy can finally be evaluated
by providing specific landmine detection results. A combined classifier incorporating
combined features and selective training where applicable is used for this purpose.
The classifier is trained on the data set obtained from soil type 2 in the MsMs
database and evaluated on the rest of the data. The developed multi-step landmine
detection strategy allows us to track the reduction of the false alarm rate at each step
therefore monitoring the improvement of the recognition (assuming that all objects
of the landmine-suspicious class are considered to be landmines). Table 7.4 shows
how the FAR changes during the stages. It starts from a very high FAR after the
detection of suspicious objects, and is constantly reduced. Comparing the final FAR
with the one for one-step classification illustrates a significant improvement.

7.3   Experimental testing on LADERO platform

The experimental data obtained in well-controlled conditions by a stationary scanning
device, as in the case of public databases, are usually of a relatively good quality. This
is important in the stage of development of the landmine detection algorithms to
eliminate the influence of the scanning platform. However, to make a step towards
the real situation it should also be confirmed that a mobile scanning platform is able
to provide the data that can be used for at least some of the stages of landmine
detection. Such data will obviously be of lower quality, but there is always the
possibility of the improvement of the scanning mechanism with new technologies.

There are few test minefields available in Europe for the testing of the prototype
landmine detection platform due to its size and necessary equipment. A test
minefield located at Meerdaal bomb disposal unit, Brussels, Belgium (http://
www.itep.ws/facilities/Belgium/belgium.htm) was used for the experimental
testing of the LADERO platform. There are several fields with different soil
types. However, only the one with mixed soil (see Figure 7.14(a) in Section 7.3.2)
was used here because of easier accessibility by the robot. This section presents
the results of the experimental testing of the mobile scanning platform LADERO

performed on a test minefield in real conditions. The testing included the gathering of sensor data from a test minefield and landmine detection data processing.

7.3.1 De-mining robot prototype

The previously developed prototype de-mining platform LADERO (Rachkov et al., 2005) has a Cartesian mechanical structure with four axes supported by eight legs (see Fig. 7.11). The movements of the robot are performed using discrete pneumatic actuators powered by an external air compressor.

The robot can perform two types of Cartesian motions in order to scan a specified area: a scanning step performed along the X-axis in order to acquire data from the current scanning line, and an advancing step along the Y-axis in order to move to the next scanning line and start the motion in the opposite direction. The rotations are theoretically possible (Rachkov et al., 2005), however, experiments show that the performed motion is unreliable due to the slippage and deformation of the robot body, so it cannot be used in practice to achieve a predictable rotation.

The basic scanning step is performed in several stages shown, for example, in Fig. 7.12 for a left-to-right step:

1. The feet of the supporting axes are landed, and the moving axes are shifted to the right.
2. The feet of the moving axes are landed and the feet of the supporting axes are lifted.
3. The moving axes are moved left, making the robot body drive to the right.

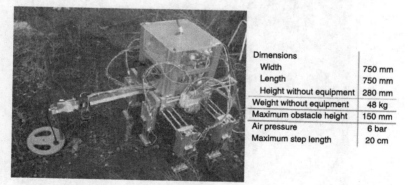

Dimensions	
Width	750 mm
Length	750 mm
Height without equipment	280 mm
Weight without equipment	48 kg
Maximum obstacle height	150 mm
Air pressure	6 bar
Maximum step length	20 cm

7.11 LADERO Prototype and its characteristics.

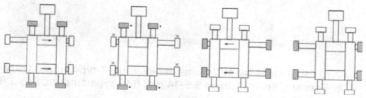

7.12 Basic scanning step of the robot.

Depending on the step direction the supporting and the moving axes change. The advancing step is performed in a similar manner, but the moving axes are shifted by a small distance providing a small step. Landmine detection sensors are installed in front of the robot providing the possibility of detecting landmines in advance in order to avoid them with the robot. The sensors currently installed are one metal detector and two infrared sensors. The Schiebel ATMID metal detector installed on the robot is a professional metal detector certified for manual de-mining (Fig. 7.13(a)). Its control electronics are able to control one of the two available search heads: continuous (ATMID) and pulsed (AN 19/2). The

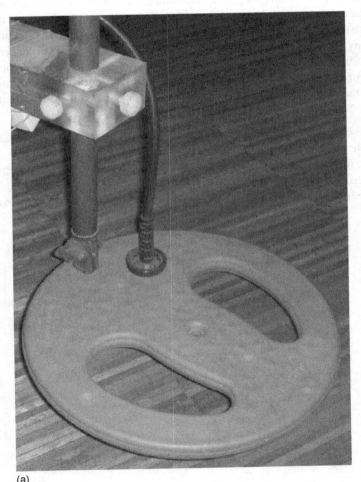

(a)

7.13 (a) Schlebel ATMID metal detector; (b) K-type output OS36-10-K-80F sensor. Spectral band 6.5–14 μm; (c) J-type output OS65-J-R2-4 sensor. Spectral band 8–14 μm.

(b)

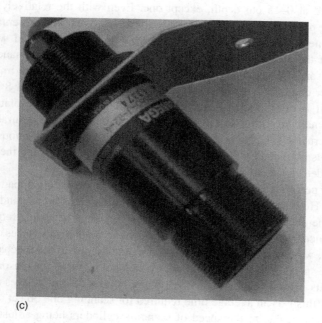

(c)

7.13 Continued.

continuous mode is known to be more sensitive. Thus, the ATMID search head is used in order to make the detection of small metal objects possible.

Two infrared sensors shown in Fig. 7.13(b) and 7.13(c) are based on the thermocouples with different characteristics: K and J. The sensors provide signals proportional to the temperature of the soil at the current location. To obtain an IR image of a larger area the sensors should be moved over the area by a scanning device. It should be noted that these signals are very noisy due to the natural fluctuations of the thermal radiation from the soil.

7.3.2 Data acquisition

Figure 7.14(a) demonstrates the robot on the test field during the scanning. The data were acquired from 27 objects using a metal detector and two IR sensors installed on the platform. Fig. 7.14(b) shows the layout of the test minefield. Each object on the field is located in a separate rectangle of size $50 \times 100cm$ and surrounded with an empty area to allow easy access. Not all of the available objects could be scanned by the robot because of its relatively large size. Each object was scanned in a separate experiment thus reducing the positioning errors that would be accumulated on a larger area. Examples of the data can be seen in Plates III, IV and V (in colour section).

The metal detector used on the robot showed a good performance with the landmine targets located on the test field. It detected all low-metal plastic landmines buried at 0–15 cm depth, except one. Even with the relatively low quality scanning provided by the platform, this allows us to obtain representative data from the metal detector due to the large size of the sensor head which compensates for the positional inaccuracy of the platform. The performance of the infrared sensors is acceptable in only a few cases depending on the time of day and weather conditions (the tests were performed from 8 am until 8 pm). Unfortunately, it is questionable if such sensors can be used on the platform because of the small area of the soil that they cover, meaning that the accuracy of the platform is crucial. It has to be concluded that the IR data are ambiguous due to the small spots provided by the sensors on the ground and, thus, only the data from the metal detector were used for the classification.

The scanning performed by the robot is very unreliable in terms of slippage and small rotations. However, the possibility of using a similar platform for landmine detection in general can be tested using the current version. The testing requires constant assistance by a researcher in order to correct the small disturbances of the robot trajectory, which cannot be corrected automatically due to the mechanical limitations. The speed of the robot is limited by the time required for sensor data acquisition. In its current version an average scan of 100×50 cm requires about 40–50 minutes (this was an average time required for scanning one object on the Meerdaal test fields), due to the speed of scanning cylinders being adjusted to approximately 40 mm/s in order to provide reliable motion.

(a)

D	C	B	A
		B1 APL 0mm	A1 APL 0mm
	C2 APL 0mm	B2 APL 0mm	A2 APL 0mm
	C3 APL 50mm	B3 APL 50mm	A3 APL 50mm
	C4 APL 100mm	B4 APL 100mm	A4 APL 100mm
	C5 APL 100mm	B5 APL 100mm	A5 APL 100mm
D6 Nails 60mm		B6 APL 150mm	A6 APL 150mm
	C7 ATM 125mm	B7 Brick 75mm	A7 Bottle 75mm
D8 APL 30mm	C8 Grenade 30mm		

(b)

7.14 (a) Robot on the test field. (b) Layout of the test field.

The data obtained during the tests are unique due to the fact that they are acquired by a mobile platform from two sensors at the same time in real conditions. The condition of the field is close to natural because it was not used for a long time. The usage of these data represents a real skill in employing a mobile robot for landmine detection in contrast to using data obtained with high precision scanning devices.

7.3.3 Landmine detection

The data obtained by the mobile platform during scanning were used for landmine detection. Taking into account the low spatial resolution of the data it is reasonable to consider only a two-stage classification as shown in Fig. 7.15. The first stage of the detection, namely the detection of suspicious objects, showed good results providing the detection of all objects. If the detection process is terminated after this stage (all the detected objects are considered to be landmines), the value of the false alarm rate would be 5.4 FA/m^2.

The possibility of performing further classification is tested using the concept of dominant class. This concept assures that the failure of the classification step does not lead to a significant change of the detection rate leaving the false alarm rate also unchanged. After the ROI detection stage for these experimental data the obtained data set has 70 objects, including 27 real objects, several double detections and clutter. This data set is obviously not large enough for performing training and evaluation of a classifier. Therefore, it was decided to investigate the possibility of training the classifier using experimental data from the MsMs database. This task is especially important because the situation of not having enough training data can easily happen in reality. The main obstacle to this approach is the fact that the sensors installed on LADERO are not considered in the MsMs database. This means that the classifier has to be trained on the data obtained from different sensors.

The only possibility of performing such training is to reveal classification features that have weak dependency on the nature of the sensor. Such features could be fused for signatures of the same object obtained from different sensors using the idea of combined features. To choose the classification features that are

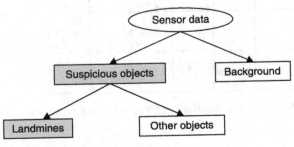

7.15 Two-step landmine detection.

general enough to be used for the created classifier, a measure of feature generality in respect of the dominant class is introduced. This measure will estimate the difference between the distributions of the feature for the dominant class obtained from different sensors. The closer the distributions of the same feature for different sensors are to each other, the more general the feature is. Only the distributions for the dominant class are considered here, allowing the feature not to be general in respect to the other class. The measure is calculated as follows:

$$FG = \frac{1}{N} \sum_{k=1}^{N} \frac{\Delta \mu_{ref} - \left| \mu_{comb(class1)} - \mu_{k(class1)} \right|}{\Delta \mu_{ref}}$$

$$\Delta \mu_{ref} = \left| \mu_{comb(class1)} - \mu_{comb(class2)} \right|,$$

where N is the number of sensors, $\mu_{comb(class1)}$ is the mean of the combined feature distribution for class1, $\mu_{k(class1)}$ is the mean of the feature distribution for class1 obtained by sensor k, and class1 is the dominant class. The measure shows how far the individual features calculated for each sensor are from the averaged feature normalized by the distance between the means of the classes (after processing the distributions to account for the dominant class). It should change from 0 to 1, but can also be negative showing very high non-generality.

Only the features which have the value of FG close to 1 are considered in the construction of the classifier. Three features with highest values of FG are chosen:

- combined Occupied part (OA)
- combined Entropy (SM)
- combined GR measure (SM).

The created classifier is then trained on the database experimental data from the MsMs database using the largest available data set, which includes data from four sensors and counts 2574 samples. The evaluation of the classifier is performed on the data obtained by LADERO. After the classification one of the previously detected landmines is missed and the FA/m2 is reduced from 5.4 to 2.9. The results of landmine detection through the classification stages are shown in Table 7.5.

Table 7.5 Landmine detection results over the stages of two-step detection process (the evaluation set contains 19 landmines)

	Classification step	
	ROIs extraction	Classification landmines/ other objects
Landmines detected	19	18
DR	100%	95%
FA/m2	5.4	2.9

The achieved possibility of employing a classifier trained on the data obtained in different conditions that do not match the conditions where it is used is very important since such a situation can easily arise during the actual operation. It should be noted that this is possible due to the abstract classification features.

7.4   Conclusions

The research conducted in the present work confirmed once again that automated landmine detection is a very challenging task. All stages of the development of such a system contain important unsolved problems, from low-level positioning of a mobile landmine detection platform in an unstructured environment, to the high-level recognition task complicated by non-selectivity of the landmine detection sensors.

The central problem addressed in this work is automatic landmine detection using the sensor data obtained by a mobile scanning platform. The landmine detection task was analyzed in detail, first, using the experimental data available from public databases of landmine signatures. This analysis revealed several specific properties of this task and allowed us to develop an optimal strategy consisting of a multi-stage approach. Such division of the detection process accounts for the fact that the landmine detection sensors considered are not selective for the landmines. Instead, each sensor distinguishes objects with differing values of some physical property that can also be present in a landmine. This may lead to a situation where the data quality is not enough to perform any recognition. The multi-stage strategy allows us to overcome this problem by terminating the detection process after an appropriate stage and safely considering all undistinguishable objects to be landmines. Thus, the detection process can be viewed as a step-by-step reduction of the false alarm rate where the next improvement is possible if the quality of the sensor data allows it.

The landmine detection stage provides the recognition of landmines among the detected suspicious objects. Analysis of large amounts of experimental data showed that this task should be divided into two stages in order to provide better separability of the classes at each stage: first, the suspicious objects are classified as man-made or natural, and then the landmines are identified among the detected man-made objects. A feature-based sensor fusion approach is used for the classification considering the Bayesian classifier and multivariate Gaussian model for representation of the feature parameters.

A number of new classification features, which reveal the nature of the object, are proposed, and several other features are adapted from other fields of pattern recognition. The main goal for the development of new classification features is the improvement of their independence from the nature of a particular sensor. Several of the proposed features meet this requirement and, thus, their values can be fused with the signatures of the same object obtained from different sensors by averaging, deriving a combined feature with better characteristics. The classification features considered were analyzed in detail with the help of mutual information and Hausdorff distance measures in order to investigate opportunities of improving the classification results. The main obstacle here is a very high overlap between the classes in the feature space. This problem naturally follows from the fact that the sensors utilized are not specific to landmine detection. The separation of the classification into two stages is one step toward this goal. However, the classification process was improved further as summarized below.

The modification of the classifier training process is proposed in this work, aimed at improving the separability of the analyzed classes in the case of landmine detection. Therefore, the concept of *selective training* was developed here. This approach benefits from the fact that some of the classification features have bimodal distributions for one of the classes where one of the maxima is better separated from the distribution of the other class. In this case ignoring the highly overlapped maximum is acceptable, since the second maximum confers better separability of classes. The model of the classifier is then changed to an additive Gaussian model accounting for the new relationship between the features where some parts of the distributions are ignored. However, only the presence of several classification features with this behavior can assure improvement in the classification results: if the better separated maxima characterize different sets of objects, a combination of several features can allow the classification of the whole set of objects even if large parts of the distributions are ignored. This result can be explained by the fact that the highly overlapped areas of the distributions do not provide any useful information for the classifier. Instead, they can even confuse the classification and reduce its performance.

Although the selective training markedly enhances the classification results in comparison to the naive approach, the concept of *dominant class* was additionally introduced here in order to account for one more specific in the landmine recognition task. The selective training reduces the classification error without any specific restrictions on the values of the resulting detection and false alarm rates, similar to most approaches of pattern recognition. However, the case of landmine detection is specific in this sense because it is essential to keep the highest possible level of detection rate (ideally 100 percent). In order to overcome this problem the concept of selective training was further modified to provide a possibility of specifying one of the classes to be a dominant class, modifying the distribution of the classification features before the training of the classifier.

In this case the distribution of the dominant class is unchanged, and the overlapping part of the other class distribution is ignored. The remaining parts of the non-dominant class distribution now represent the whole distribution, allowing the dominant class to prevail during the classification. This simple concept leads to a different level of performance consisting of higher detection rates and still bringing the reduction of the false alarm rate. The dominant class approach is especially useful if the quality of sensor data is low. The fine training of the classifier using the selective training can only be complicated in this case due to lack of information. The introduction of a dominant class makes the classification possible and in some cases reduces the false alarm rate.

The landmine detection algorithms developed were implemented on the mobile scanning platform LADERO after being tested on a large amount of experimental data from the public database. The specificities of the data acquired by the platform include low spatial resolution and deformations of the object signatures due to the positioning failures of the platform. Because of the low-quality data, the concept of dominant class was used for the classification. The important result is that the data acquisition problems do not affect the performance of the suspicious object detection algorithm. The data set obtained during the experiments was sufficient only for the evaluation of the classification and was not large enough to use a part of it for the classifier training. Thus, the classifier was trained using the data available from the public database, although the sensors installed on LADERO were not employed in that database. It is evident that such a process is not possible in the case of the classification features that are highly dependent on the sensor characteristics. However, there is a possibility of using the features that are relatively independent from the sensor as combined features. A special measure for estimating the degree of generality of the feature in respect to the dominant class was developed. Based on this measure, only the three best features were chosen to form the classifier. After being trained on the experimental data from the public database obtained from four different sensors, the classifier showed good performance on the experimental data obtained by LADERO, reducing the false alarm rate by the factor of 1.9. Such operation is only possible thanks to the special classification features, which include a high degree of abstraction from the nature of the sensor. This achievement proves that it is possible to train a classifier for landmine detection using experimental data obtained in different conditions that do not match the conditions where it will be used.

Summarizing the results obtained during this research it can be concluded that the application of a mobile scanning platform for automatic landmine detection is promising.

7.5   References

Acar, E. U., Choset, H., Rizzi, A. A., Atkar, P. N. & Hull, D. (2002). Morse decompositions for coverage tasks, *The International Journal of Robotics Research* 21(4): 331–344.

Acar, E. U., Choset, H., Zhang, Y. P. & Schervish, M. (2002). Sensor-based coverage of unknown environments: Incremental construction of Morse decompositions, *The International Journal of Robotics Research* 21(4): 345–366.

Acar, E. U., Choset, H., Zhang, Y. P. & Schervish, M. (2003). Path planning for robotic demining: Robust sensor-based coverage of unstructured environments and probabilistic methods, *The International Journal of Robotics Research* 22: 441–466.

Acar, E. U., Zhang, Y. P., Choset, H., Schervish, M., Costa, A. G., Melamud, R., Lean, D. C. & Graveline, A. (2001). Path planning for robotic demining and development of a test platform, Int. Conf. on Field and Service Robotics, Helsinki, Finland, 11–13 June, pp. 161–168.

Baertlein, B., Gunatilaka, A., Ho, K. C., Yarovoy, A. G. & Kovalenko, V. (1998). Improving detection of buried land mines through sensor fusion, *Proc. of SPIE, Detection and Remediation Technologies for Mines and Minelike Targets*, Orlando, Florida, April, Vol. 3392, pp. 1122–1133.

Breejen, E., Schutte, K., Cremer, F., Schutte, K. & Benoist, K. (1999). Sensor fusion for anti personnel landmine detection: a case study, *Proc. of SPIE, Detection and Remediation Technologies for Mines and Minelike Targets IV*, Orlando, USA, Apr., Vol. 3710, pp. 1235–1245.

Bruschini, C., Gros, B., Guerne, F., Piece, P.-Y. & Carmona, O. (1996). Ground penetrating radar and induction coil sensor imaging for antipersonnel mines detection, Int. Conf. on Ground Penetrating Radar, Sendai, Japan, 30 September–3 October, pp. 211–216.

Clark, G. A., Hernandez, J. E., Sengupta, S. K., Sherwood, R. J., Schaich, P. C., Buhl, M., Kane, R. J., Barth, M. J., DelGrande, N. K. & Carter, M. R. (1992). Computer vision and sensor fusion for detecting buried objects, Asilomar Conference on Signals, Systems, and Computers, Pacific Grove, USA, 1–3 October, pp. 466–471.

Clark, G. A., Sengupta, S. K., Aimonetti, D., Roeske, F. & Donetti, J. G. (2000). Multispectral image feature selection for land mine detection, *IEEE Trans. on Geoscience and Remote Sensing* 38(1): 304–311.

Clark, G. A., Sengupta, S. K., Buhl, M., Sherwood, R. J., Schaich, P. C., Bull, N., Kane, R. J., Barth, M. J., Fields, D. J. & Carter, M. R. (1993). Detecting buried objects by fusing dual-band infrared images, 27th Asilomar Conference on Signals, Systems, and Computers, Pacific Grove, USA, 1–3 November, Vol. 1, pp. 135–143.

Collins, L. M., Gao, P., Carin, L., Moulton, J. P., Makowsky, L. C., Reidy, D. M. & Weaver, R. C. (1999). An improved Bayesian decision theoretic approach for land mine detection, *IEEE Trans. on Geoscience and Remote Sensing* 37(2): 811–819.

Collins, L. M., Gao, P., Schofield, D., Moulton, J. P., Makowsky, L. C., Reidy, D. M. & Weaver, R. C. (2002). A statistical approach to landmine detection using broadband electromagnetic induction data, *IEEE Trans. on Geoscience and Remote Sensing* 40(4): 950–962.

Cremer, F., Jong, W., Schutte, K. & Bibaut, A. (2003). Fusion of polarimetric infrared features and GPR features for landmine detection, Int. Workshop on Advanced Ground Penetrating Radar, TU Delft, Delft, 14–16 May, pp. 222–227.

Cremer, F., Jong, W., Schutte, K., Yarovoy, A. G. & Kovalenko, V. (2003). Feature level fusion of polarimetric infrared and GPR data for landmine detection, Int. Conf. on Requirements and Technologies for the Detection, Removal and Neutralization of Landmines and UXO, Brussels, Belgium, Sep., pp. 638–642.

Cremer, F., Schavemaker, J., Jong, W. & Schutte, K. (2003). Comparison of vehicle-mounted forward-looking polarimetric infrared and downward-looking infrared sensors for landmine detection, *Proc. of SPIE, Detection and Remediation Technologies for Mines and Minelike Targets VIII*, Orlando, USA, 21–25 April, Vol. 5089, pp. 517–526.

Cremer, F., Schutte, K., Schavemaker, J. & Breejen, E. (2001). A comparison of decision-level sensor-fusion methods for anti-personnel landmine detection, *Information Fusion Journal* (2): 187.

Cremer, F., Schutte, K., Schavemaker, J., Breejen, E. & Benoist, K. (2000). Towards an operational sensor-fusion system for anti-personnel landmine detection, *Detection and Remediation Technologies for Mines and Minelike Targets V*, Orlando, USA, 24–28 April, Vol. 4038, pp. 792–803.

Dawson-Howe, K. M. & Williams, T. G. (1998). The detection of buried landmines using probing robots, *Robotics and Autonomous Systems* (4): 235–243.

de Santos, P. G., Galvez, J. A., Estremera, J., Garcia, E., Suganuma, S., Takita, K. & Kato, K. (2003). Silo4. A true walking robot for the comparative study of walking machine techniques, *IEEE Robotics and Automation Magazine* (12): 23–32.

Duda, R. O., Hart, P. E. & Stork, D. G. (2000). *Pattern Classification*, Wiley-Interscience Publication, New York.

Filippidis, A., Jain, L. C. & Lozo, P. (1999). Degree of familiarity art2 in knowledge-based landmine detection, *IEEE Trans. on Neural Networks* 10(1): 186–193.

Filippidis, A., Jain, L. C., Martin, N., Breejen, E. & Benoist, K. (2000). Multisensor data fusion for surface land-mine detection, *IEEE Trans. on Systems, Man and Cybernetics*, Part C 30(1): 145–150.

Frigui, H., Gader, P. D., Keller, J. M. & Schutte, K. (1998). Fuzzy clustering for land mine detection, Conf. of the North American Fuzzy Information Processing Society – NAFIPS, USA, 20–21 August, pp. 261–265.

Fritzsche, M., Lohlein, O., Acheroy, M., Breejen, E. & Benoist, K. (2000). Sensor fusion for the detection of landmines, *Subsurface Sensing Technologies and Applications* 1(2): 247–267.

Gao, P., Tantum, S. L., Collins, L. M., Weaver, D., Moulton, J. P. & Makowsky, L. C. (1999). Statistical signal processing techniques for the detection of low-metal landmines using EMI and GPR sensors, *Geoscience and Remote Sensing Symposium*, Hamburg, 28 June–2 July, 5: 2465–2467.

Gunatilaka, A., Baertlein, B., Ho, K. C., Yarovoy, A. G. & Kovalenko, V. (1999). Comparison of pre-detection and post-detection fusion for mine detection, *Proc. of SPIE, Detection and Remediation Technologies for Mines and Minelike Targets III*, Orlando, Florida, 5–9 April, Vol. 3710, pp. 1212–1223.

Gunatilaka, A., Baertlein, B., Ho, K. C., Yarovoy, A. G. & Kovalenko, V. (2001). Feature-level and decision-level fusion of noncoincidently sampled sensors for land mine detection, *IEEE Trans. on Pattern Analysis and Machine Intelligence*, Vol. 23, pp. 577–589.

Hirose, S., Yokota, S., Torii, A., Ogata, M., Suganuma, S., Takita, K. & Kato, K. (2005). Quadruped walking robot centered demining system – development of TITAN-IX and its operation, IEEE Int. Conf. on Robotics and Automation(ICRA), Barcelona, Spain, 18–22 April, pp. 1296–1302.

Hirose, S., Yoneda, K., Tsukagoshi, H., Guilberto, J., Suganuma, S., Takita, K. & Kato, K. (1997). TITAN VII: Quadruped walking and manipulating robot on a steep slope, IEEE Int. Conf. on Robotics and Automation (ICRA), Albuquerque, USA, April, pp. 494–500.

Ho, K. C., Collins, L. M., Huettel, L. G. & Gader, P. D. (2004). Discrimination mode processing for EMI and GPR sensors for hand-held land mine detection, *IEEE Trans. on Geoscience and Remote Sensing* 42(1): 249–263.

International Mine Action Standard (IMAS) 09.10. Clearance requirements (Online). United Nations Mine Action Service (UNMAS), http://www.mineactionstandards.org/IMAS_archive/Amended/Amended1/IMAS_0910_1.pdf (01.01.2003).

Ishikawa, J., Kiyota, M. & Furuta, K. (2005). Evaluation of test results of GPR-based anti-personnel landmine detection systems mounted on robotic vehicles, IARP Int. Workshop on Robotics and Mechanical Assistance in Humanitarian Demining, Tokyo, Japan, 21–23 June.

Joint Multi-Sensor Mine Signature Database MsMs (online). http://demining.jrc.it/msms/ (13.02.2007).

Kacelenga, R., Erickson, D., Palmer, D. & Gader, P. D. (2002). Voting fusion adaptation for landmine detection, IEEE Int. Conf. on Information Fusion, Annapolis, USA, 8–11 July, Vol. 1, pp. 333–340.

Kacelenga, R., Erickson, D., Palmer, D. & Gader, P. D. (2003). Voting fusion adaptation for landmine detection, *AES Magazine* 18(8): 13–19.

Kato, K., Hirose, S., Estremera, J., Garcia, E., Suganuma, S., Takita, K. & Kato, K. (2000). Proposition of the humanitarian demining system by the quadruped walking robot, IEEE Int. Conf on Intelligent Robots and Systems (IROS), Takamatsu, Japan, 30 October–5 November, pp. 769–774.

Kempen, L., Katartzis, A., Pizurica, V., Cornelis, J. & Sahli, H. (1999). Digital signal/image processing for mine detection. Part 1: Airborne approach, Euroconference on Sensor Systems and Signal Processing Techniques Applied to the Detection of Mines and Unexploded Ordance, Firenze, Italy, 1–3 October, pp. 48–53.

Kwak, N. & Choi, C.-H. (2002). Input feature selection for classification problems, *IEEE Trans. on Neural Networks* 13(1): 143–159.

Landmines must be stopped (1997). International Committee of the Red Cross, Geneva.

Larionova, S. (2007). Automated landmine detection by means of a mobile robot, PhD thesis, Faculty of Science and Technology, University of Coimbra.

Luo, R. & Kay, M. (1995). *Multisensor Integration and Fusion for Intelligent Machines and Systems*, Ablex Publishing Corporation, Norwood, CN.

Milisavljevi, N., Bloch, I., Acheroy, M., Breejen, E. & Benoist, K. (2000). *Characterization of Mine Detection Sensors in Terms of Belief Functions and their Fusion, First Results*, Vol. 2, pp. ThC3 – 15–22.

Milisavljevi, N., Bloch, I., Huettel, L. G. & Gader, P. D. (2003). Sensor fusion in anti-personnel mine detection using a two-level belief function model, *IEEE Trans. on Systems, Man and Cybernetics*, Part C 33(2): 269–283.

Neves, M. A., Gomes, R. R. & Costa, R. M. (2003). Robô com pernas para desminagem humanitária. Diploma thesis. University of Coimbra.

ORDATA database (online). http://www.maic.jmu.edu/ordata/ (13.02.2007).

Peichl, M., Dill, S., Suess, H. & Gader, P. D. (2003). Application of microwave radiometry for buried landmine detection, Int. Workshop on Advanced Ground Penetrating Radar, TU Delft, Delft, 14–16 May, pp. 172–176.

Perrin, S., Du A. (2004). Multisensor fusion in the frame of evidence theory for landmines detection, *IEEE Trans. on Systems, Man and Cybernetics*, Part C 34(4): 485–498.

Piramuthu, S. (1999). The Hausdorff distance measure for feature selection in learning applications, Int. Conf. on System Sciences, Maui, Hawaii, USA, 5–8 January .

Rachkov, M. Y., Marques, L. & de Almeida, A. T. (2005). Multisensor demining robot, *Autonomous Robots* 18(3): 275–291.

Roughan, M., McMichael, D. W., Ho, K. C., Yarovoy, A. G. & Kovalenko, V. (1997). A comparison of methods of data fusion for land-mine detection, Int. Workshop on Image Analysis and Information Fusion, Adelaide, Australia, 6–8 November.

Sarkar, N. & Chaudhuri, B. B. (1994). An efficient differential box-counting approach to compute fractal dimension of image, *IEEE Trans. on Systems, Man, and Cybernetics* 24(1): 115–120.

Schavemaker, J., Breejen, E., Cremer, F., Schutte, K. & Benoist, K. (2001). Depth fusion for anti-personnel landmine detection, *Detection and Remediation Technologies for Mines and Minelike Targets VI*, Orlando, USA, Apr., Vol. 4394, pp. 1071–1081.

Sheedvash, S. & Azimi-Sadjadi, M. R. (1997). Structural adaptation in neural networks with application to landmine detection, Int. Conf. on Neural Networks, 9–12 June, Vol. 3, pp. 1443–1447.

Stanley, R. J., Gader, P. D., Ho, K. C., Yarovoy, A. G. & Kovalenko, V. (2002). Feature and decision level sensor fusion of electromagnetic induction and ground penetrating radar sensors for landmine detection with hand-held units, *Information Fusion Journal* 3(3): 215–223.

Stiles, J. M., Apte, A. V. & Beh, B. (2002). A group-theoretic analysis of symmetric target scattering with application to landmine detection, *IEEE Trans. on Geoscience and Remote Sensing* 40(8): 1802–1814.

Test minefield at Meerdaal bomb disposal unit (online). http://www.itep.ws/facilities/Belgium/belgium.htm (13.02.2007).

Torrione, P. & Collins, L. M. (2004). Performance comparison of automated induction-based algorithms for landmine detection in a blind field test, *Subsurface Sensing Technologies and Applications* 5(3): 121–150.

Waltz, E. & Llinas, J. (1990). *Multisensor Data Fusion*, Artech House, Norwood.

Zhang, Y. P., Collins, L. M., Yu, H., Baum, C. E. & Carin, L. (2003). Sensing of unexploded ordance with magnetometer and induction data: Theory and signal processing, *IEEE Trans. on Geoscience and Remote Sensing* 41(5): 1005–1015.

Zhang, Y. P., Schervish, M., Acar, E. U. & Choset, H. (2001). Probabilistic methods for robotic landmine search, IEEE Int. Conf. on Intelligent Robots and Systems (IROS), Maui, Hawaii, Nov., pp. 1525–1532.

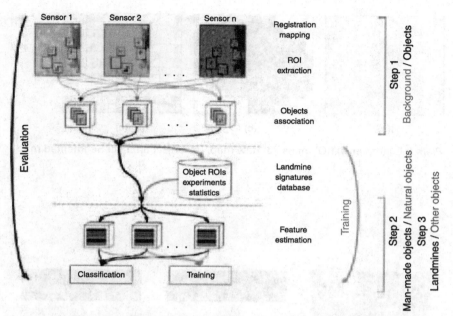

Plate I General structure of the proposed methodology for sensor fusion. One of the advantages of the proposed strategy is its ability to account for the quality of the sensor data as shown in Figure 7.4. The landmine detection process can be terminated at any stage if the data quality is not high enough to perform a more precise classification. Then, all objects of the appropriate class (shown with gray background in Figure 7.4) are considered to be landmines allowing the detection to be performed but with a higher false alarm rate. Thus, the developed landmine detection strategy naturally appears as a step-by-step reduction of the false alarm rate.

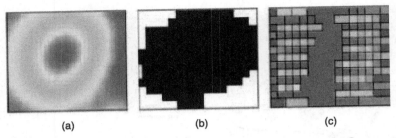

Plate II Examples of ROI maps: (a) Raw map. (b) Object area. (c) Segmented map.

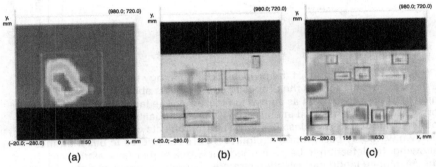

Plate III ATK mines: (a) Metal detector. (b) IR0. (c) IR1.

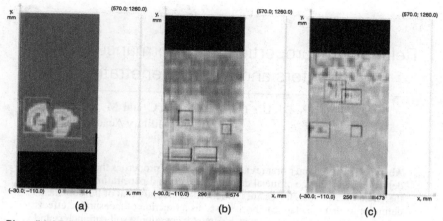

Plate IV AP PMN-2: (a) Metal detector. (b) IR0. (c) IR1.

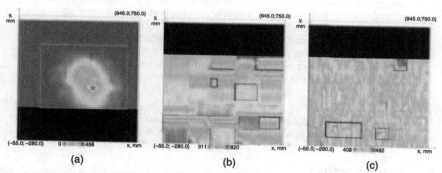

Plate V AP M35BG: (a) Metal detector. (b) IR0. (c) IR1.

8

Relating soil properties to performance of metal detectors and ground penetrating radars

P. DRUYTS, Y. YVINEC and M. ACHEROY,
Royal Military Academy, Belgium

Abstract: This chapter first reviews the basic soil properties that affect the detection performance of metal detectors and ground-penetrating radars, and then a number of soil classes are introduced. A crucial point of the soil class definition is that the class can be assessed as a function of measurable effects on the performance. We show how a database containing soil parameters and corresponding soil classes can be used to train a classifier. With such a classifier, making soil measurements is enough to assess the difficulty of a soil. It is not necessary to first select detectors and make specific performance tests with them. With the classifier, information on soil difficulty can thus be obtained with fewer measurements and the result is more general as it is not related to a single detector. The soil class can be used to plan de-mining operations rationally because it allows the planner, with few measurements, to define soil difficulty and choose the most suitable detector. No database is currently available. We provide guidelines on what the database should contain. Nevertheless, to illustrate the soil classification concept, a simple metal detector model is developed and used to simulate measurements, which are then used to train a classifier.

Key words: metal detector model, soil electromagnetic properties, metal detector and ground-penetrating radar performance, soil classification.

8.1   Introduction

Metal detectors are common tools for detecting mines, unexploded ordnance and other explosive remnants of war. One of their main drawbacks is the large number of small scraps of metal that they detect. This led to the design of dual sensors, which combine metal detectors and ground-penetrating radars to reduce the number of false alarms. Soils can affect the performance of both metal detectors and ground-penetrating radars. This may be mitigated to some extent by ground compensation (or ground cancellation). Studying the effect of soils on these sensors' performance is therefore of the utmost importance. A first attempt can be found in CEN (2008) but many questions are still unanswered.

To study the effect of soils on the performance of metal detectors and ground penetrating radars, a first step is to list the soil properties that are known, or expected, to affect performance. These properties include soil electromagnetic properties, the variation of these properties with space and time, soil surface roughness (or relief) and soil heterogeneity. These properties depend on factors,

189

called indirect parameters, such as soil water content or the signal picked up by the detector from the soil alone, which can also be used.

A critical point is to find ways to characterize the spatial and frequency variations of electromagnetic properties with a limited number of features that can be derived from a reasonable number of measurements.

It is also important to pay adequate attention to the methods used to measure sensor performance. Measuring the maximum detection depth is a good method but it assumes that the detection is deterministic. In practice detection will always be stochastic to some extent. The most appropriate method to measure performance may depend on the soil and the importance of the stochastic component of its effect. Estimating the probability of detection and the false alarm rate of the detector on a given soil can be used when detection is stochastic.

Careful measurements of relevant soil properties and adequate estimation of sensor performance could make it possible to study objectively and quantitatively the link between soil properties and performance.

This chapter reviews the basic parameters and describes a framework that can be used to relate some chosen soil parameters to detector performance in a soil classification. As not enough appropriate measurements are available the classification will be based on simulated data. The underlying model is presented and aimed as being as realistic as possible while remaining simple enough. In practice the relevance of the chosen parameters can be ascertained by the quality of the classification.

This chapter is organized as follows: the soil parameters relevant to metal detectors and ground-penetrating radars are introduced in 8.2. Classes of soils based on the effect on sensor performances are then defined in 8.3. A simple model for a metal detector is developed and used to classify a soil based on measurements of soil properties and sensor performance in 8.4. Section 8.5 discusses the usefulness of a soil database and 8.6 concludes this chapter. Conventions and notation are described in 8.8.

8.2   Soil parameters for metal detectors and ground-penetrating radars

Both metal detectors and ground-penetrating radars probe the soil with an electromagnetic field. Such a field is described by Maxwell's equations (Stratton, 1941) and the relevant parameters to characterize the medium (soil, mine, any inhomogeneities, etc.) are the magnetic permeability μ, the electrical conductivity σ and the electric permittivity ε. Since the magnetic permeability of a soil and the magnetic permeability in a vacuum are often very close to each other, their relative difference, called the magnetic susceptibility, is often used. It is defined as $\chi = \dfrac{\mu}{\mu_0} - 1$ where μ_0 is the magnetic permeability in a vacuum.

These parameters vary with spatial position and frequency. Furthermore, they are complex-valued, therefore two values are needed at each frequency. For example, the electric permittivity, $\varepsilon = \varepsilon' + j\varepsilon''$, is defined by its real (ε') and its imaginary (ε'') parts. It should also be noted that the electric permittivity and electrical conductivity always appear together, as a term $\sigma + j\omega\varepsilon$, in Maxwell's equation, in such a way that it is impossible to distinguish σ' from ε'', or σ'' from ε' by means of electromagnetic measurements alone. Therefore, effective parameters $\varepsilon_{eff} = \varepsilon' + \dfrac{\sigma''}{\omega}$ and $\sigma_{eff} = \sigma' - \omega\varepsilon''$ are sometimes defined.

Soil electromagnetic properties can vary with time particularly in response to prevailing weather conditions. Since electric permittivity depends mainly on soil water content it can increase after rain and decrease as the soil dries out. This change over time depends on the soil texture. For instance clay can retain more water and for a longer time than sand. Magnetic susceptibility is not significantly influenced by weather conditions. On the other hand it may change slightly with temperature.

8.2.1 Differences between metal detectors and ground-penetrating radars due to frequency range

Metal detectors typically work at frequencies below a few hundred kilohertz (Bruschini, 2002), whereas ground-penetrating radars work at centre frequencies around one gigahertz (Daniels, 2004). This difference in frequency causes a significant change in the field propagation mechanism and in the relevant electromagnetic parameters to consider.

For current sources composed of current loops, such as the coils of metal detectors, and in the frequency band used by metal detectors, the displacement current can be neglected in Maxwell's equations (Haus et al., 1998). The magnetic field $\mathbf{H}$ is then described by:

$$\nabla \times \left(\frac{\nabla \times \mathbf{H}}{\sigma} \right) = j\omega\mu\mathbf{H} + \nabla \times \frac{\mathbf{J_s}}{\sigma} \qquad [8.1]$$

$$\nabla.(\mu\mathbf{H}) = 0 \qquad [8.2]$$

In Equation [8.1] $\mathbf{J_s}$ is the current source in the coil that forms, together with the induced current $\sigma\mathbf{E}$, and the total current $\mathbf{J} = \mathbf{J_s} + \sigma\mathbf{E}$.

Equations [8.1] and [8.2] show that the electric permittivity has no influence on the magnetic field produced by the metal detector coil.

The following equations apply to the electric field $\mathbf{E}$:

$$\nabla \times \mathbf{E} = j\omega\mu\mathbf{H} \qquad [8.3]$$

$$\nabla.\left\{ \left(\varepsilon + \frac{\sigma}{j\omega} \right)\mathbf{E} \right\} = 0 \qquad [8.4]$$

Equations [8.3] and [8.4] show that the electric permittivity can influence the electric field if $\omega\varepsilon$ is significant when compared to σ. For a soil, the upper limit of the relative permittivity is around 40 (Daniels, 2004). For such a soil and a frequency of five kilohertz, the dielectric term is equal to the conduction term when $\sigma = 10^{-5}$ S/m. If the conductivity is sufficiently larger than this value, the permittivity has no significant effect.

For weakly conducting soils, the electric permittivity must be known or the electric field can only be computed up to a gradient term. Such a gradient electric field has, however, no influence on an ideal coil. It can nevertheless influence a real coil that deviates from an ideal one by the charge accumulation as a result of parasite capacitance. Coils of most modern metal detectors, however, are shielded, which should cancel the effect of a gradient electric field. Hence, for shielded coils, electric permittivity has no effect on the detector response, and for non-shielded coils it can have a second-order effect.

Another characteristic of Equations [8.1] and [8.2] is that they only include the first power of $j\omega$, which corresponds to a first order (parabolic) equation in the time domain, characteristic of diffusion mechanism. This is in contrast to the full-wave equation describing the fields at ground-penetrating radar frequencies, which is a second order (elliptic) equation, characteristic of wave propagation. In addition, for weakly conducting soils, the first power $j\omega$ becomes negligible when compared to the other terms. The magnetic field can then be computed as for static sources. This is the magneto-quasi-static approximation (Haus et al., 1998).

To assess whether soil conductivity can be neglected for a given soil, one can resort to the analytic solution of Maxwell's equations for a homogeneous medium. In that case, the magnetic field can be related to its source as follows:

$$\mathbf{H} = \int_{V_s} \nabla \times \{G(\mathbf{r}', \mathbf{r}) \mathbf{J}_s\} dV' \qquad [8.5]$$

where V_s is the volume containing the current sources, $\mathbf{r}'$ and $\mathbf{r}$ are respectively the source and field points and

$$G = \frac{e^{jkR}}{4\pi R} \qquad [8.6]$$

is the Green function in which R is the distance between $\mathbf{r}'$ and $\mathbf{r}$ and $k = \omega\sqrt{\varepsilon\mu}$ is the wave number. For a conducting soil, the wave number can be approximated by $k = \frac{1-j}{\delta}$ where $\delta = \sqrt{\frac{2}{\sigma\mu\omega}}$ is the skin depth.

If the skin depth is large when compared to the characteristic dimension of the problem, kR is small and the static Green function $G = \frac{1}{4\pi R}$ can be used. When this is the case, the conductivity can be neglected and the magneto-quasi-static approximation of the magnetic field can be used.

As an illustration, for a soil with $\sigma = 5$ S/m, which is 'likely higher than the conductivity of sea water saturated soil' (Das, 2004), the skin depth at five kilohertz is 3.2 m. For a detector coil of 20 cm diameter, which is typical for metal detectors, the soil at such a distance from the head may still have a significant influence on the response and the conductivity must be taken into account to compute the magnetic field in the soil. For most soils, however, the conductivity is much lower and the magnetic field may be computed as in free space. Even if the magnetic field can be computed as in the magneto-quasi-static state, the conductive soil could still produce a significant response when compared to the response of a mine. The magneto-quasi-static approximation may, however, be used to simplify the computation of the soil response.

8.2.2 Differences between metal detectors and ground-penetrating radars due to the magnitudes of soil properties

Another difference between the metal detector and the ground-penetrating radar can be highlighted by noting that the magnetic susceptibility is quite small for most soils. Many measurement campaigns (Van Dam et al., 2004; Guelle et al., 2005a; Guelle et al., 2005b; Van Dam et al., 2005a; Van Dam, 2005b; Lewis et al., 2006; Guelle et al., 2007) show that most soils have magnetic susceptibilities below 0.02. A noticeable exception is Playa Gorgana in Panama, which is rich in pure magnetite and for which magnetic susceptibilities close to one have been measured (Van Dam et al., 2004).

For such low magnetic susceptibilities, the magnetic field for metal detectors can be computed in the soil as in free space and the total response can be approximated by the sum of the target response and the soil response. Furthermore, the target response is not significantly affected by the soil and its free space response can be used.

It should be noted that the point we made about the electrical conductivity applies here too: even if the magnetic field can be computed in the soil as in free space, the soil could nevertheless produce a significant response when compared to the response of a mine because some mines contain only a very small amount of metal. The Born approximation can, however, be used to compute the soil response, which simplifies the problem (Druyts et al., 2009).

For the ground-penetrating radar, electric permittivity and electrical conductivity play a dominant role as they may exhibit a much larger variation than the magnetic susceptibility.

Typically the relative electric permittivity may vary from three for dry sand to 40 for wet clay (Daniels, 2004). With such high values, it is obvious that the fields in the soil may not be computed as in free space and the problem is much more complex than for the metal detector. The interactions between the mine and the soil must be considered and the mine response is significantly different in the soil than in free space (Bosch, 2006).

8.2.3 Relevant parameters for metal detectors

From a general analysis of Maxwell's equations in the frequency band used by metal detectors and from the order of magnitudes of the soil electromagnetic parameters, we have shown that the dominant parameter governing the response of a metal detector is the magnetic susceptibility. The electrical conductivity may have an effect for highly conducting soils and the electric permittivity may have a weak effect on unshielded heads. In addition, the total response can generally be approximated by the sum of the target response in free space and the soil response in the absence of a target.

We now concentrate on the dominant parameter, the magnetic susceptibility. We already mentioned that its frequency variation as well as its spatial variation (including soil relief) might affect metal detector performance. To understand how this happens and how the effect may vary with metal detectors, it is necessary to discuss the working principle of a metal detector.

A general presentation can be found in Guelle et al. (2003). A metal detector is composed of a transmitting coil, which produces a time-varying magnetic field, and a receiving coil, which measures the response. The detector will produce an alarm if the response is higher than a given threshold. This threshold is set up in a calibration phase by adjusting the detector sensitivity. In many cases, the sensitivity is set at the maximum value for which no alarm is generated when scanning above an area of soil that does not contain any metal fragment.

Spatial variation

From the above description of the alarm mechanism, a soil response that remains constant from point to point is not an issue. It is the variation of the soil response that is critical because the sensitivity is set according to the maximum soil response. If the soil response is lower, the target signal must be larger for the total response to reach the threshold in order to trigger an alarm.

Several factors may be at the origin of the response variation. The height of the head above the soil may vary during the sweeping. The higher the average soil response, the higher the variation of the response for a given height variation. The soil response may also vary due to soil relief or to spatial changes in the magnetic susceptibility of the soil. Hence some features characterizing the soil relief as well as the minimum magnetic susceptibility, the maximum magnetic susceptibility and the average value of the magnetic susceptibility are clearly important factors to consider.

Moreover, the detector head averages the soil response in the sensitivity volume (Druyts et al., 2009) and the detector internal processing may also include some spatial filtering. For these reasons, the way the magnetic susceptibility varies spatially is important. In general, the spatial correlation of the susceptibility must be considered. A feature often used to characterize the spatial variation is the correlation length.

What makes things more complicated is that the effect of soils on the performance of metal detectors is a combination of soil and detector characteristics. It might therefore be difficult to find parameters and a classification scheme that are valid for all detectors. A different classification may be required for each detector or for a number of detector groups.

Frequency variation

We have explained that the transmitting coil of a metal detector produces a time-varying magnetic field. Different time variations of the magnetic field may be used. The two main types of variations are a pulsed variation and a sinusoidal variation, which leads to the two principal types of metal detectors, the pulsed induction metal detectors and continuous wave metal detectors.

For *pulsed induction metal detectors*, the excitation includes a significant excitation in a large frequency band – typically from direct current (DC) to a few hundred kilohertz. The frequency variation of the magnetic susceptibility (also known as viscous remnant magnetization, magnetic viscosity, magnetic relaxation or magnetic after-effect) in that frequency band must be considered. Under some realistic assumptions the frequency variation is linear (on a log scale) in the frequency band of interest, and only the slope of the real part of the susceptibility ($\frac{\partial \chi'}{\partial \ln \omega}$) is needed to characterize the soil response (Das, 2006). Whether this result is valid for most soils is still an open question.

For *continuous wave detectors working at a single frequency*, the frequency variation of the magnetic susceptibility can obviously have no effect on the detector response as only the complex magnetic susceptibility at the frequency used by the detector matters. Note that under the assumptions mentioned above (Das, 2006, Equation (20)) we have:

$$\chi'' = \frac{\partial \chi'}{\partial \ln \omega} \qquad [8.7]$$

The frequency variation together with the real part of the susceptibility at a single frequency can then be used to characterize the magnetic susceptibility. These values can be measured, for instance by the Magnetic Susceptibility System MS2 from Bartington.

For *continuous wave detector using several frequencies*, the problem is more complex because the way the responses at the various frequencies are combined to generate the audio signal must be taken into account and this changes from detector to detector – and is often proprietary information. In general, the response may be a function of the susceptibility at the various frequencies used. If the variation of the real part of the magnetic susceptibility is linear for the frequency band used and if Equation [8.7] holds, the magnetic susceptibility at a single frequency and the slope of the magnetic susceptibility are again sufficient to fully characterize the magnetic susceptibility in this case.

8.2.4 Relevant parameters for ground-penetrating radars

We saw that the interaction between the ground-penetrating radar, the soil and the target depends on the electric permittivity and the electrical conductivity, and their spatial and frequency variations, including the variations due to the soil relief and roughness. In addition the interaction between the soil and the target must be taken into account. In order to describe in more detail how these parameters affect ground-penetrating radar performance, we now provide a general explanation of the ground-penetrating radar working principles.

A ground-penetrating radar is an instrument designed to detect electromagnetic contrasts in the soil and contains a transmitting antenna and a receiving antenna which allow it to send and detect electromagnetic waves at given frequencies. The transmitting antenna sends a wave that propagates into the soil. Whenever the wave encounters a variation of electromagnetic properties, part of it is reflected back to the surface (reflected wave) and the rest continues to propagate into the ground (transmitted wave). The receiving antenna detects the reflected wave.

A soil can have many effects on a ground-penetrating radar's ability to detect a mine, and these effects are related to the soil's electromagnetic properties.

The way the radar wave interacts with the targets, the soils and the inhomogeneities in the soils is complex. First, and unlike with metal detectors, the target response is significantly changed by the presence of soil. Therefore the target cannot be characterized by its free-space signature. Second, the far-field approximation may be inaccurate because the soil surface and even the target may be in the near field of the antennas. Therefore numerical codes are generally needed to accurately predict the radar response.

The following general discussion may, however, help to explain the principles behind this interaction.

First, the soil reflects a large part of the radar wave at the soil surface or at inhomogeneities, which provides additional responses that may be confused with responses of real mines and therefore produce false alarms.

These reflections also attenuate the transmitted wave that will reach the mine, and hence reduce the mine response.

Another cause of attenuation of the wave on its way to the mine and back is the absorption of energy in a lossy soil. Spreading loss is also a cause for attenuation.

Another important factor is the ratio of the incident wave to the reflected wave at the mine. If the soil electromagnetic properties are close to those of a mine, the wave may have reduced reflection when it reaches the mine, and detection at the receiving antenna might be difficult.

Figure 8.1 summarizes the principles of ground-penetrating radars and the wave propagation (indicated by figures). The transmitting antenna (Tx) is held at a certain distance about the ground (A) and sends a wave to the ground (1) that first reaches the soil surface (B). Since air and soil have very different electromagnetic properties a part of the wave is reflected back to the ground-

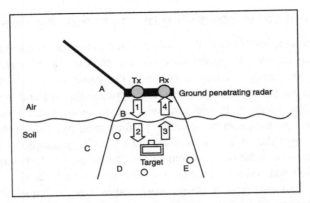

8.1 Simplified description of the basic principles of a ground penetrating radar: numbers represent wave propagation and letters sources of losses; see text for details. Source: CEN, 2008.

penetrating radar and a part goes through the ground (2). This reflection at the surface is heavily dependent on the soil roughness. The energy of the transmitted wave is not emitted equally in all directions but is concentrated inside a beam. When the beam enters the soil it is refracted and becomes narrower because of the difference of electromagnetic properties between air and soil (C). The wave energy per unit surface is then increased, which may lead to a better detection rate and a smaller footprint. When a wave propagates in a soil that exhibits some electrical conductivity, it is gradually attenuated (D). The presence of localized areas with electromagnetic properties different from the surrounding soil, such as stones, roots, rocks, or cracks, can create as many reflections (E). When reaching a mine, part of the wave is reflected. The reflected wave then travels back through the soil (3). Finally the surface is reached and part of the wave propagates through the air to reach the receiving antenna (Rx) of the radar (4).

The best detection is achieved when the signal that is reflected at the mine and reaches the receiving antenna is as strong as possible compared to noise. Therefore losses undergone by the wave from the transmitting antenna to the mine and back to the receiving antenna hamper detection. They increase with depth and this leads to a limit to the depths at which a mine can be detected.

In order to get a quantitative estimation of attenuation in the soil and reflection at the air–soil and soil–mine interfaces, we will use analytic solutions available for simple cases. Since these simple cases may differ significantly from reality, results should be used with caution.

From Equation [8.6] we can see that if k has an imaginary part then the argument of the exponential in the Green function has a real part αR with:

$$\alpha = \omega \sqrt{\frac{\mu\varepsilon'}{2}\left(\sqrt{1 + \left(\frac{\varepsilon''}{\varepsilon'}\right)^2} - 1\right)} \qquad [8.8]$$

For a plane wave (propagating in the z-direction), this gives an exponential attenuation $e^{-\alpha z}$ for the fields.

Considering that the attenuation of the radar wave is a key factor influencing the detection depth of ground-penetrating radars, Daniels has listed the expected attenuation range of various materials (Daniels, 2004).

The electrical size of an object gives a rough idea of how well an object can be detected. It is defined by the ratio of a characteristic dimension of the object to the wavelength in soil, $\lambda = \dfrac{2\pi}{\omega\sqrt{\mu\varepsilon}}$. It should be noted that the real parameter that describes how well an object would be detected is its radar cross-section in soil, which is a complex function of object geometry and the electromagnetic properties of both the object and the soil. There is no easy way to measure it, since measurement in free-space cannot be used.

The reflection coefficient between soil and mine describes how a radar wave propagating in the soil will be reflected by the mine. If the radar wave can be considered as a plane wave with a normal incidence on a flat homogeneous mine, then the reflection coefficient is computed from the characteristic impedances of the soil Z_s and the mine Z_m (Chew, 1990):

$$\Gamma = \frac{Z_m - Z_s}{Z_m + Z_s}$$ [8.9]

The characteristic impedance of an object or a medium is defined by (Daniels, 2004):

$$Z = \sqrt{\frac{\mu}{\varepsilon}} = \sqrt{\frac{\mu}{\varepsilon'\left(1 + j\,\dfrac{\varepsilon'}{\varepsilon''}\right)}}$$ [8.10]

To summarize, the following parameters should be considered:

- the characteristic impedance of the soil
- the attenuation of the soil
- the electromagnetic properties of the landmine (characteristic impedance)
- the electrical size of the landmine.

Features characterizing the soil inhomogeneities and soil surface are also required. As with metal detectors, a different classification may be needed to link these parameters to the performance of each ground-penetrating radar. But it should be noted that only a very small number of ground-penetrating radars and dual sensors are currently available for mine action.

8.2.5   Indirect parameters

Soil electromagnetic parameters fully characterize the effects of soil on metal detectors and ground-penetrating radars, but they may be difficult to measure. It may therefore be useful to link performance with indirect parameters such as soil water content or soil texture (Preetz et al., 2008). These indirect parameters might also provide indications on the expected variation of the electromagnetic soil properties, and therefore soil effects, with time.

8.2.6 Summary

For a metal detector

- The magnetic field is described by a diffusion equation in a conducting soil; in most soils with low or moderate conductivity, it can be computed as for a static source in air.
- The dominant parameter is the magnetic susceptibility.
- The electrical conductivity plays a secondary role, expected to have a significant effect for highly conducting soils only (Das, 2006; Billings et al., 2003).
- The electric permittivity has no effect on shielded heads and may have a weak effect on non-shielded heads.
- The output signal is the sum of the free-space signature of the mine and the soil signal.

For ground-penetrating radar

- The magnetic and electric fields are governed by a wave equation.
- The dominant parameters are the electric permittivity and the electrical conductivity.
- The magnetic permeability plays a secondary role due to the order of magnitude of the electromagnetic properties of natural soils.
- The output signal is more complex than just the sum of the signals from the mine and the soil; the interaction between the two must be considered.

Electric permittivity and electrical conductivity vary with environmental conditions, such as soil water content.

Magnetic permeability, and therefore magnetic susceptibility, remain stable but may be slightly affected by temperature.

For metal detectors and ground-penetrating radars, frequency variation, spatial variation of electromagnetic parameters and soil relief are important. Defining a limited set of parameters that efficiently characterize these variations and can easily be measured is a key issue.

8.3   Sensor performance and soil classes

8.3.1 Sensor performance

Since the goal is to link soil properties to sensor performance, it is imperative that performance is defined in a way that is measurable.

When assuming that detection is deterministic a good test to measure the performance of a detector is the fixed-depth detection tests, in which surrogate mines are buried at given locations and depths in a given soil; the detector is set up according to its user manual and is then swept over the locations. For each surrogate mine the operator records if it has been detected or not. This test provides indications on the effect of the soil on the detection capability of the detector.

If detection cannot be assumed to be deterministic the performance of a detector can be estimated by conducting blind tests in test lanes of given soils and with buried targets simulating mines and expected sources of clutter in order to estimate the detection probability and the false alarm rate.

More details on these test protocols can be found in CEN (2003, 2008).

8.3.2 Soil classes

Soil classes should be defined according to the effects that the soil has on sensor performance. Depending on metal detectors, a given soil may reduce more or less the detection capacity. Therefore a soil classification is dependent on a given sensor. For metal detectors, we propose the following classes.

A *neutral* soil is a soil that has no significant effect on a detector. Detection in such a soil is not significantly different from detection in air. In practice a neutral soil could be defined as a soil in which the maximum detection depth of a reference target by a given metal detector (that is, the maximum depth at which consistent detection of a reference target occurs) is more than 90 per cent of the detection distance in air of the same target by the same metal detector.

A *moderate* soil, or more precisely a soil having a moderate effect on the performance of a metal detector, is a soil where a metal detector can be used without soil compensation. It can be defined as a non-neutral soil in which a reference target, chosen by the operator, is always detected at the reference depth of detection by a given metal detector without soil compensation.

A *severe* soil, or a soil having a severe effect on the performance of a metal detector, is a soil that requires the use of a metal detector with soil compensation. It can be formally defined as a non-moderate soil in which a reference target is always detected at the required depth of detection but only with ground compensation.

A *very severe* soil, or a soil having a very severe effect on the performance of a metal detector, makes it impossible to use the metal detector even with soil compensation. It can be formally defined as a non-severe soil in which a reference target at a reference depth is not consistently detected by a given metal detector with soil compensation.

Since this classification depends on the metal detectors, it can be used to identify some groups of metal detectors for which the soils have the same effects on the performance. A different classification may be required for each metal detector or each group of metal detectors.

The classification also depends on the reference targets and the reference depth. The most difficult target expected to be found and the default depth of 13 cm can be chosen (UNMAS, 2003).

For ground-penetrating radars there is no such thing as a neutral soil. A 'simple' soil should be chosen as a reference. What 'simple' means may depend on the detector and the target electromagnetic properties (for instance its contrast with the soil's electromagnetic properties). A candidate might be dry uniform sand but it should first be checked if good performance is reached. Then a 'simple soil' can be defined as a soil where the maximum detection depth of a reference target is larger than 90 per cent of the maximum detection depth in the reference soil. 'Moderate,' 'severe' and 'very severe' soils could be defined by various ratios of the maximum detection depths in a 'simple soil.'

For metal detectors, CEN (2008) provides tables linking values of magnetic susceptibility and frequency dependence of magnetic susceptibility to the expected effect on detector performance. It also describes qualitative rules. These rules are empirical and have not been validated.

In the next section we describe a framework to link soil properties and sensor performance from a database of soil properties and sensor performance. This would lead to finding the parameters that have a real effect on performance based on physical considerations and learning from measurements. Ideally real measurements should be used. Some attempts made on data from trial campaigns for a systematic test and evaluation of metal detectors (Guelle et al., 2005a; Guelle et al., 2005b; Lewis et al., 2006; Guelle et al., 2007) showed that the number of soils on which data are available was not enough to link soil properties to performance-based soil classes. Therefore in what follows a model is used.

8.4   Simple model for a metal detector

As our objective is to highlight the principle of soil classification with a simple metal detector and not to discuss in detail the behaviour of a given detector, many simplifications have been introduced in the following model. We nevertheless made an effort not to use too simplistic a model in order to retain the major characteristics of a generic detector and we therefore expect the results to be realistic to some extent.

To illustrate soil classification, we consider the following simple scenario. A single frequency, continuous wave metal detector composed of a single circular coil is swept above a small metallic ball buried in a magnetic soil. We consider that the detector produces an alarm indication if the total response is above a threshold when the detector is located just above the ball. This location is chosen because it can be shown that, at some distance from the head, this is the location at which the maximum response occurs. The threshold is controlled by setting the detector sensitivity in a calibration phase. We assume that the detector is calibrated to reach the maximum sensitivity and therefore that the sensitivity is set at the maximum value that does not lead to false alarms due to the soil or other sources.

The total response can be approximated as the sum of the soil response V_s and the target response V_t (Druyts et al., 2006). An additional term V_n is introduced to take into account the noise generated inside the detector electronics and the noise induced at the coil terminals by the electromagnetic background. We assume that the noise can be modelled as a uniform distribution between V_n^{min} and V_n^{max}. The total response is thus:

$$V = V_t + V_s + V_n \qquad [8.11]$$

Let V_T be the metal detector threshold. There will be an alarm if

$$V > V_T \qquad [8.12]$$

8.4.1   Soil response

The soil response is a deterministic signal that may be computed once the soil magnetic susceptibility is known at each point. Due to soil inhomogeneities and relief, the soil response will vary from point to point. Since it is unrealistic to know the soil response at each location, we model it with a stochastic model. We assume that the soil response can be modelled as a noise uniformly distributed between V_s^{min} and V_s^{max}.

8.4.2   Reference target response

Soil classes are defined as functions of the detectability of a reference target. For simplicity we choose a small metal ball.

For a small target, the response can be expressed as (Baum, 1998, Equation (2.22)):

$$V_t = j\omega\mu_0 I_{TX} \breve{\mathbf{H}}_{RX}^{(fs)} \cdot \bar{\mathbf{M}} \cdot \breve{\mathbf{H}}_{TX}^{(fs)} \qquad [8.13]$$

where ω is the angular frequency, $\breve{\mathbf{H}}_{RX}^{(fs)}$ and $\breve{\mathbf{H}}_{TX}^{(fs)}$ are the magnetic field produced in the transmit (TX) and receive (RX) coils normalized by the corresponding current, $\bar{\mathbf{M}}$ is the magnetic polarizability dyadic and '.' indicates a dot product. For a metal ball, the magnetic polarizability dyadic is isotropic: $\bar{\mathbf{M}} = m\bar{\mathbf{I}}$ with $\bar{\mathbf{I}}$ the identity dyadic. m can be computed analytically as a function of the ball radius, electrical conductivity and magnetic permeability as well as the frequency (Baum, 1998, Equation (2.24)). For a small ball, the response can thus be expressed as:

$$V_t = \alpha \breve{\mathbf{H}}_{RX}^{(fs)} \cdot \breve{\mathbf{H}}_{TX}^{(fs)} \qquad [8.14]$$

where $\alpha = j\omega\mu_0 m I_{TX}$ is constant for a given detector and a given reference small metal ball.

In order to assess the response as a function of depth the magnetic fields $\breve{\mathbf{H}}_{RX}^{(fs)}$ and $\breve{\mathbf{H}}_{TX}^{(fs)}$, which are identical for the single-coil metal detector that we consider, are to be determined. For a circular coil of radius a carrying a current I, the magnetic field on the axis of the coil is given by (Stratton, 1941, Equation (16), p. 233):

$$\mathbf{H}_{coil} = \frac{I\hat{\mathbf{z}}}{2} \frac{a^2}{\left(a^2 + z^2\right)^{\frac{3}{2}}} = \frac{I\hat{\mathbf{z}}}{2a} \frac{1}{\left(1 + \zeta^2\right)^{\frac{3}{2}}} \qquad [8.15]$$

where x is directed along the coil axis, $\hat{\mathbf{z}}$ is the unitary vector along z and $\zeta = \frac{z}{a}$ is the dimensionless depth. Finally the response for a small sphere on the axis of a detector composed of a single circular coil of radius a is:

$$V_t = \frac{b}{\left(1 + \zeta^2\right)^3} \qquad [8.16]$$

with $b = \dfrac{j\omega\mu_0 m I_{TX} I_{RX}}{4a^2}$.

8.4.3 Calibration

During calibration, the threshold V_T is set to

$$V_T = V_s^{max} + V_n^{max} + V_{tol} \qquad [8.17]$$

where V_{tol} is a security factor that ensures that there will be no false alarm on the soil.

8.4.4 Normalization

The alarm condition in Equation [8.12] remains unchanged if both sides of the inequality are divided by the same constant. Hence, the number of parameters can be reduced by considering normalized response:

$$v = \frac{V}{V_n^{max} - V_n^{min}} \quad v_t = \frac{V_t}{V_n^{max} - V_n^{min}} \quad v_s = \frac{V_s - V_s^{min}}{V_n^{max} - V_n^{min}}$$

$$v_n = \frac{V_n - V_n^{min}}{V_n^{max} - V_n^{min}} \quad v_T = \frac{V_T - V_s^{min} - V_n^{min}}{V_n^{max} - V_n^{min}} \quad v_{tol} = \frac{V_{tol}}{V_n^{max} - V_n^{min}} \qquad [8.18]$$

Equations [8.11] and [8.12] can then be rewritten as $v_t + v_s + v_n > v_T$, which shows that the only effect of V_s^{min} and V_n^{min} is to move the threshold.

The normalized noise and the normalized soil response have a uniform distribution respectively between zero and one, and between zero and

$$v_s^{max} = \frac{V_s^{max} - V_s^{min}}{V_n^{max} - V_n^{min}}.$$

A reasonable choice is $V_{tol} = 0.1$ ($V_n^{max} - V_n^{min}$); then there remain only two parameters to characterize the problem: b and V_s. According to Equations [8.17] and [8.18] the dimensionless threshold is

$$v_T = 1.1 + v_s^{max} \qquad\qquad [8.19]$$

8.4.5 Probability of detection

The probability of an alarm is:

$$p_d = p(v_t + v_s + v_n > v_T) = p(v_s + v_n > v_T - v_t) \qquad [8.20]$$

In other words the probability of having an alarm indication is given by the cumulative distribution function of the sum of two uniform random variables. The density function can easily be computed and is illustrated in Fig. 8.2(a). By using the expression of the target response in Equation [8.16] with a metal detector coil radius of 10 cm the probability of detection as a function of depth can be computed as illustrated in Fig. 8.2(b). The solid line represents a soil with $v_s = 0$ with a target chosen to be detected at 20 cm in free space. The dashed line represents the case where v_s is equal to the threshold of moderate soil. See next section for more detail.

To understand the shape of the probability, let us consider Fig. 8.3(a): the threshold is defined as the sum of the maximum noise of the detector (in black), the maximum noise of the soil (in undulated lines) and a tolerance (in white) and is displayed as a horizontal line; (b) if the target response alone is higher than the

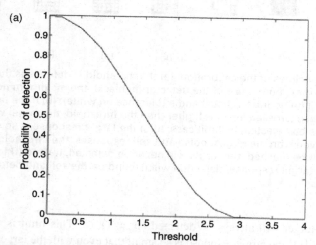

8.2 Example of probability of detection as a function of $v_T - v_t$ with $v_s^{max} = 2$ (a) and as a function of depth (b); the solid line represents the detection in air and the dashed line the detection with a soil at the threshold between moderate soil and severe soil.

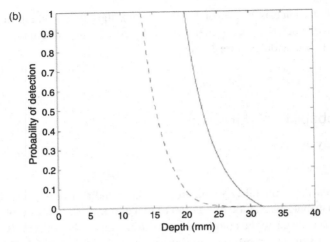

(b)

8.2 Continued

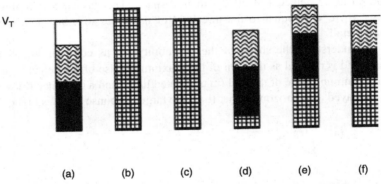

(a)   (b)   (c)   (d)   (e)   (f)

8.3 Principles of the calibration: (a) the threshold is defined as the sum of the maximum noise of the detector (in black), the maximum noise of the soil (in undulated lines) and a tolerance (in white); (b) if the target response (crossed lines) is higher than the threshold, the target will always be detected, (c) limit case; (d) If the target response is so small that even with the largest noise and soil responses, the threshold cannot be reached, the target will never be detected, (e) and (f) for a given target response, detection will depend on the soil and detector noises.

threshold, the target will always be detected ($p_d = 1$); (c) This limit is reached when $V_t = V_T$; (d) If the target response is so small that even with the largest noise and soil responses, the threshold cannot be reached, the target will never be detected ($p_d = 0$); (e) and (f) for a given target response, detection will depend on the soil and detector noises ($0 \leq p_d \leq 1$).

8.4.6 Neutral and moderate soil

As defined in 8.3.2 a neutral soil is such that the detection depth is 0.9 times its free space counterpart, which is chosen to be 20 cm for a metal detector coil radius of 10 cm. To get realistic results, we will choose b such that the free space detection depth is 20 cm for a coil radius of 10 cm ($\zeta_{20} = 2$). A certain number of reference targets are buried. A soil is then neutral if the detection depth for all buried targets is larger than 18 cm ($\zeta_{18} = 1.8$). Fig. 8.4(a) shows the probability

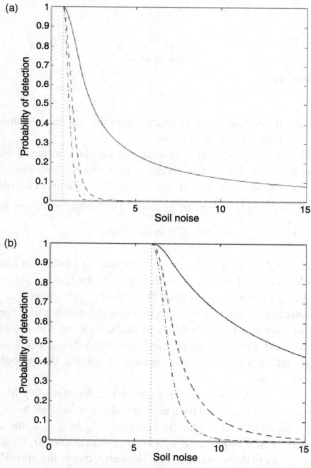

8.4 Probability of detecting all mines at 18 cm (a) and 13 cm (b) and 13 cm with ground compensation (c) as a function of soil noise (solid: one mine, dashed: five mines, dashdot: ten mines); the vertical dotted line shows the transition between neutral and moderate (a), moderate and severe (b) and severe and very severe (c).

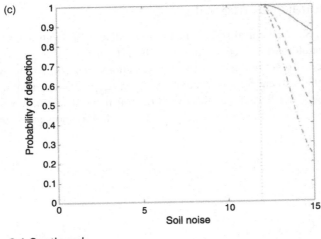

(c)

8.4 Continued

of detecting all buried mines at 18 cm as a function of the soil dimensionless noise V_s and for a various number of buried mines.

The limit value of soil noise for a neutral soil can be computed as follow. The target is chosen to be detected at 20 cm in free space. Therefore according to [8.19], its normalized response at 20 cm is 1.1. The normalized target response at 18 cm is then $\left(\frac{1+\zeta_{20}^2}{1+\zeta_{18}^2}\right)^3$ times as large. The soil is therefore neutral if $v_s = 1.1\left[\left(\frac{1+\zeta_{20}^2}{1+\zeta_{18}^2}\right)^3 - 1\right]$, which is around 0.70 in the example.

In practice a limited number of detection tests will be performed and a non-neutral soil can be classified as neutral if all targets in the test are detected. The probability of being classified as neutral is $P_d = (P_{18cm})^n$ where P_{18cm} is the probability of detection of the target at 18 cm and n the number of targets used. Figure 8.4(a) also indicates the probability of classifying the soil as neutral for various values of n. As expected the probability of classifying a soil as neutral tends to an optimal step function as n increases. Hence the probability of misclassification decreases.

A moderate soil is defined as a soil for which the detection depth is 13 cm ($\zeta_{13} = 1.3$). The reasoning presented for a neutral soil can be applied to a moderate soil by changing the detection depth. By replacing ζ_{18} by ζ_{13} in the equation above, we find that a soil is moderate if v_s is smaller than about 6.0. Figure 8.4(b) presents the probability of detection at 13 cm. Here also, due to the limited number of tests, a severe soil can be misclassified as moderate. Note that a neutral soil cannot be classified as moderate because by definition of a neutral soil, the reference target is always detectable at 18 cm and therefore the n test targets will be detected, whatever the value of n.

8.4.7   Severe soil

A soil has been defined as severe if the detection depth in the soil is less than 13 cm without soil compensation and more than 13 cm with soil compensation.

Many soil compensation algorithms may be implemented in a metal detector. Examples can be found in Candy (1996) and Bosnar (2001). In all cases, the objective is to process the signal in order to reduce the soil response without decreasing too much the target response. Here we will assume ideal soil compensation such that the soil response is halved and the noise and target response are unchanged. As a result the limit value for soil noise is twice as large as the limit value for moderate soil. See Fig. 8.4(c).

8.4.8   Relation to soil parameters

We have shown above that soil class can be related to the dimensionless maximum soil response $V_s^{\min}$. We would like to relate, however, the soil class to some parameters that can be measured independently of the detector. If the soil parameters are well chosen, the classification should be accurate with a crisp border between the classes. If the soil parameters are poorly chosen, for example if an important parameter is not taken into account, the resulting classification will be inaccurate with a blurred frontier between the classes.

Most natural soils have negligible electrical conductivity and low magnetic susceptibility. For such soils, the response can be expressed by (Druyts et al., 2009):

$$V_s = j\omega\mu_0 I_{TX} \int_{V_s} \chi S dV \qquad [8.21]$$

where $S = \check{\mathbf{H}}_{RX}^{(fs)} . \check{\mathbf{H}}_{TX}^{(fs)}$, which is the function of the spatial location, is called the sensitivity of the sensor head. As we consider a sensor head with a single coil, $\check{\mathbf{H}}_{RX}^{(fs)} = \check{\mathbf{H}}_{TX}^{(fs)}$ and therefore S is positive everywhere in space.

For a magnetic susceptibility varying between $\chi^{\min}$ and $\chi^{\max}$, the response will vary in the range

$$\left(V_s^{\min} = \alpha_{\text{head}}\chi^{\min}, V_s^{\max} = \alpha_{\text{head}}\chi^{\max}\right) \qquad [8.22]$$

where $\alpha_{\text{head}} = j\omega\mu_0 I_{TX} \int_{V_s} S dV$.

The above limits are reached only if the magnetic susceptibility remains constant, either at its minimum value $\chi^{\min}$ or its maximum value $\chi^{\max}$ in the whole volume of influence, which is defined as the volume of soil having a significant influence on the response (Druyts, 2009). This will only occur if the magnetic susceptibility varies slowly spatially. More precisely, the correlation length of the magnetic susceptibility should be large when compared to the characteristic dimension of the volume of influence.

The other extreme is when the correlation is much smaller than the volume of influence. The average magnetic susceptibility $\chi^{av} = \dfrac{\chi^{min} + \chi^{max}}{2}$ can then be used in Equation [8.21] leading to a constant response $V_{min} = V_{max} = \alpha_{head}\chi^{av}$.

To take the correlation length into account, we modify Equation [8.22] as follows:

$$\left(V_s^{min} = \alpha_{head}\left(\chi^{av} - \alpha_{corr}\frac{\Delta\chi}{2}\right),\ V_s^{max} = \alpha_{head}\left(\chi^{av} + \alpha_{corr}\frac{\Delta\chi}{2}\right)\right) \qquad [8.23]$$

where $\Delta\chi = \chi^{max} - \chi^{min}$ is the magnetic susceptibility range, and α_{corr} is a function of the correlation length. It varies from zero, when the correlation length is much smaller than the characteristic dimension of the volume of influence, to one, when the correlation length is much larger. One can easily verify that the more general Equation [8.23] includes the above-discussed limit case of the correlation.

The sensitivity α_{head} varies with the head geometry, orientation and height, and soil relief. To take into account the fact that the soil is never flat, that the head is never kept perfectly at a constant height and with constant orientation during the scanning, we assume that $0.9\ \alpha_{head}^{nom} < \alpha_{head} < 1.1\ \alpha_{head}^{nom}$ where α_{head}^{nom} is the nominal value of α_{head} obtained for a horizontal head at nominal height above a flat soil.

Introducing this variation in Equation [8.23] yields:

$$\left(V_s^{min} = 0.9\ \alpha_{head}^{nom}\left(\chi^{av} - \alpha_{corr}\frac{\Delta\chi}{2}\right),\ V_s^{max} = 1.1\ \alpha_{head}^{nom}\left(\chi^{av} + \alpha_{corr}\frac{\Delta\chi}{2}\right)\right) \qquad [8.24]$$

which leads for the normalized soil response:

$$v_s^{max} = \frac{\alpha_{head}^{nom}\left(0.2\chi^{av} + \alpha_{corr}\Delta\chi\right)}{V_n^{max} - V_n^{min}} \qquad [8.25]$$

Under the above assumptions, if the magnetic susceptibility is sampled adequately in a way that χ^{min}, χ^{max} and α_{corr} can be determined, then these features are linked unambiguously to the soil class. Indeed v_s^{max}, which, as explained above, is unambiguously related to the soil class, can be computed from them.

By normalizing the average magnetic susceptibility and the magnetic susceptibility range for a given metal detector, Equation [8.25] can be rewritten as follows:

$$v_s^{max} = 0.2\tilde{\chi}^{av} + \alpha_{corr}\Delta\tilde{\chi}$$

$$\tilde{\chi}^{av} = \frac{\alpha_{head}^{norm}}{V_n^{max} - V_n^{min}}\chi^{av} \qquad [8.26]$$

$$\Delta\tilde{\chi} = \frac{\alpha_{head}^{norm}}{V_n^{max} - V_n^{min}}\Delta\chi$$

The only misclassification should be due to the limited number of detection tests that are performed for each soil. This probability can be computed as a function of this number (n). The accuracy of the class map will then depend on the number of test soils available, their distribution in the parameter space, the number of detection tests for each soil and the classification algorithm used.

In practice, many parameters influence the response and only those having, or believed to have, a dominant effect will be measured. Furthermore, parameters will always be corrupted by some measurement noise. To illustrate this, we will assume that χ^{min} and χ^{max} are measured and that measurements include some measurement noise:

$$\hat{\chi}^{min} = \chi^{min} + \eta_\chi$$
$$\hat{\chi}^{max} = \chi^{max} + \eta_\chi$$

[8.27]

where η_χ is the measurement noise that we assume is uniformly distributed between $-\eta_\chi^{max}$ and η_χ^{max}.

α_{corr} is not measured and we assume that for the tested soils, it can be modelled as a random variable with uniform distribution between $\alpha_{corr}^{max} = 0.5 + \dfrac{\Delta\alpha_{corr}}{2}$ and $\alpha_{corr}^{min} = 0.5 - \dfrac{\Delta\alpha_{corr}}{2}$.

We can compute the probability that a soil belongs to a given class as a function of $\hat{\chi}^{min}$ and $\hat{\chi}^{max}$. This is illustrated in Fig. 8.5 and for various measurement noise and $\Delta\alpha_{corr}$. In Fig. 8.5(a) the soil is supposed to be completely known ($\Delta\alpha_{corr} = 0$). The normalized average and range of magnetic susceptibility completely define the soil and its probability of being neutral is either zero or one. When α_{corr} is equal to zero, the normalized magnetic susceptibility range has no

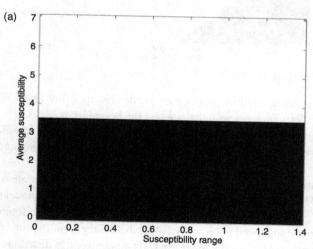

8.5 Probability of a soil belonging to class *Neutral* as a function of normalised magnetic susceptibility range $\Delta\tilde{\chi}$ (x-axis) and normalised magnetic susceptibility average $\tilde{\chi}^{av}$ (y-axis) from zero (white) to one (black) for α_{corr} equal to zero (a), 0.5 (b) and one (c).

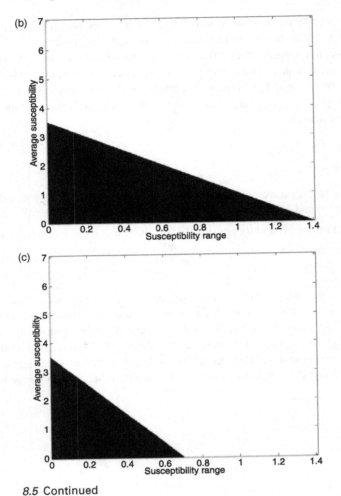

8.5 Continued

effect and only the normalized average magnetic susceptibility does. When it is non-zero, both features have an influence and the boundary is tilted.

The same procedure can be applied to the other soil classes: *moderate*, *severe* and *very severe* and boundaries for the different classes can be computed as seen in Figure 8.6.

In Fig. 8.7 the soil is not supposed to be fully known. α_{corr} is not measured and $\Delta\alpha_{corr} = 0.5$. The boundary becomes fuzzy and the probability that a soil is neutral can take any value between zero and one as indicated in the figure. When measurements are supposed to be noisy (b) the boundary is more blurred.

Finally, we present results obtained for 100 soils uniformly distributed in the $\hat{\chi}^{min} - \hat{\chi}^{max}$ space. Figures 8.5 and Fig. 8.7 showed that a linear boundary should provide accurate results. We therefore used logistic regression as a classification algorithm (Bishop, 2006, pp. 205–206).

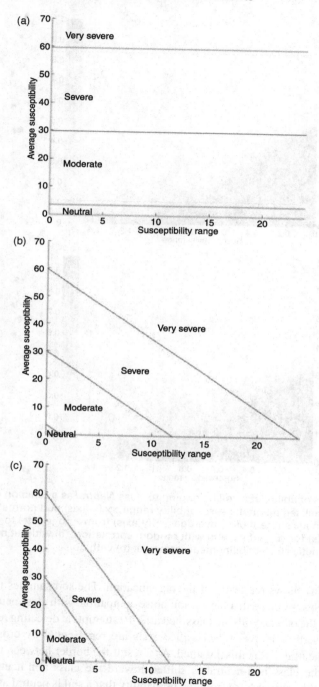

8.6 Boundaries of soil classes Neutral, Moderate, Severe and Very Severe for α_{corr} equal to zero (a), 0.5 (b) and one (c).

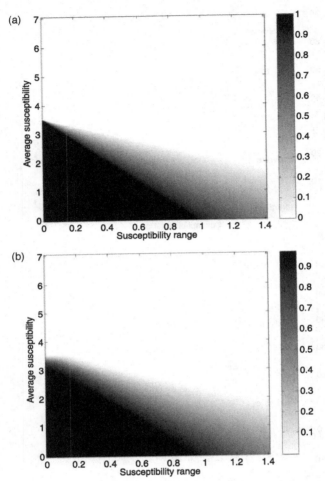

8.7 Probability of a soil belonging to class *Neutral* as a function of normalised magnetic susceptibility range $\Delta\chi$ (x-axis) and normalised magnetic susceptibility average χ^{av} (y-axis) from zero (white) to one (black) for α_{corr} of 0.5 and with random correlation; (a) without noise on magnetic susceptibility measurements; (b) with noise

Figure 8.8(a) shows the result of the classification. The soil samples used are shown as circles for those that have a soil noise compatible with class neutral and as squares for the others. Soils are classified after 10 attempts at detecting reference targets as described above. Misclassified soils are represented by crosses. As expected no neutral soil is misclassified. As a result the border between the class neutral and the class non-neutral is shifted toward the class non-neutral. The figure shows the contour lines from the probability that a soil is neutral as seen in Fig. 8.7(b), and the classification border.

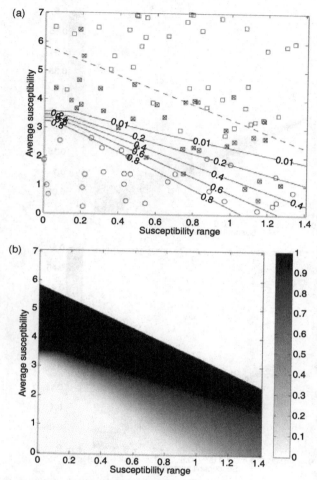

8.8 Results of classification in the same axis as the previous figures; (a) circles represent samples of neutral soils, and squares, samples of non-neutral soils; crosses indicate a misclassification; the line separating the two classes as given by the classification is drawn; contour lines from the probability that a soil is neutral with $\Delta\alpha_{corr} = 0.5$ are also indicated. (b) Probability of misclassification from zero (white) to one (black).

Figure 8.8(b) shows the probability of misclassification from zero (white) to one (black). Errors are located below the class boundary, as estimated by the logistic regression classification.

As a comparison an ideal clairvoyant classifier could classify a soil as neutral if its probability of belonging to the class (as seen in Fig. 8.7) is larger than 0.5. The corresponding error is illustrated in Fig. 8.9 without noise (a) and with noise (b) on the magnetic susceptibility measurements (corresponding to Fig. 8.7).

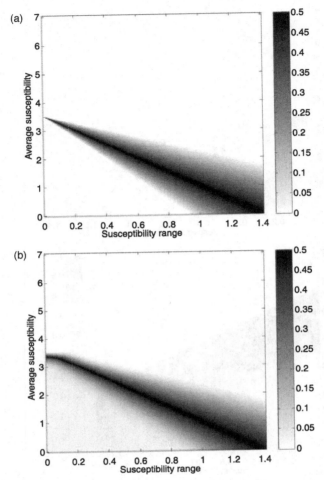

8.9 Probability of misclassification error of a clairvoyant classifier without errors of magnetic susceptibility measurement (a) and with such errors (b), from zero (white) to 0.5 (black); $\alpha_{corr} = 0.5$.

Simulations show that the results of the logistic regression classifier presented here tend to the results of the clairvoyant classifier when the number of trials of detection increases.

Errors are computed as the averages of errors of Fig. 8.8(b) and Fig. 8.9(a) and (b). They represent the probability of misclassification if the soils are uniformly distributed in the feature space, with the normalized average magnetic susceptibility from zero to seven and the normalized magnetic susceptibility range between zero and 1.4. The average error is 0.055 for the ideal classifier without noise (Fig. 8.9(a) and 0.065 with noise Fig. 8.9(b), compared with 0.30

for the classifier presented here. This shows that the dominant issue with our assumption is the number of detection trials (error of 0.30), then the fact that a feature is not measured (α_{corr}) and finally the noise (which is only responsible for the rise from 0.055 to 0.065).

The real probability of misclassification is a function of the true repartition of soils in the feature space. As explained above, the computed average errors are equal to the probability of misclassification for soils that are uniformly distributed in the feature space visualized on Fig. 8.7 and Fig. 8.8. If the soils of interest are equally distributed on a larger range of features, extending more towards the moderate soils, the average error should be computed on a larger area of the feature space, leading to a smaller average and hence the misclassification probability would be smaller.

Moreover, the number of soils considered for testing and their repartition in the feature space has an influence on the accuracy of the classifier. In the above-presented simulation, we have considered 100 soils uniformly distributed in the feature space around the neutral boundary. This may be seen as a rather ideal scenario and the error induced by the number and selection of test soils should be further analysed.

8.5   Usefulness of a soil database

Some very useful data to relate soil properties with metal detector performance are already available (Guelle, 2005a; Guelle 2005b; Lewis, 2006; Guelle, 2007), but they were collected during trial campaigns designed to test and systematically evaluate metal detectors, not link soil properties to performance-based soil classes. Their number of soils is usually too low to perform a quantitative analysis and some measurements for an accurate soil classification are missing. Moreover, only the average susceptibility (and frequency variation) is measured. It is therefore impossible to assess the effect of spatial variation.

In order to use the proposed method in practice, a database that would include well-distributed soils should therefore be created. For this, the set of parameters needed to characterize the soils must first be defined.

The dominant electromagnetic parameters are known (see section 8.2) but it may be difficult to ensure that a secondary electromagnetic parameter has no effect for a given soil. For example, the soil electrical conductivity may in some cases have an effect on a metal detector, even if it would have no effect in many cases. In addition, frequency and spatial variation of the electromagnetic parameters are known to be important but the required sampling step is not known. Therefore, if possible, all electromagnetic parameters should be measured with a small (conservative) step. It is then possible to analyse the data in order to objectively define the needed parameters and their optimal sampling. For this, several classifications based on several sub-sets of the measured electromagnetic parameters, including down sampling of the spatial and frequency variation, should be tested. The optimal parameters and sampling step may then be objectively defined based on the accuracy of classification. Indeed, provided enough well-distributed soils are used and enough detection tests are carried out per soil, the low accuracy of a classification based on the chosen parameters would indicate that some important electromagnetic parameter is missing or that the sampling step is too large.

Note that generally, it is not the raw data that will be used as input to the classifier. A limited number of features that efficiently characterize the spatial and frequency variation will first be computed from the measurements. The best features still need to be determined and for this the above-described method based on the classification accuracy can also be used.

Indirect parameters may also be considered for inclusion into the database and their relevance should be evaluated with the same methodology.

In summary, an ideal soil database would contain (a) enough well-distributed soils, (b) enough detection tests for each soil, (c) all electromagnetic parameters measured with (d) a small spatial and frequency sampling step.

In practice, such a high number of measurements will probably not be feasible and compromises will be needed. Additional research is required to perform the right choices for the chosen database measurements. From a practical point of view, such a database could be located in a website such as at James Madison University (JMU), the Geneva International Centre for Humanitarian Demining (GICHD) or the International Test and Evaluation Program for Humanitarian Demining (ITEP).

8.6   Conclusions

It is not currently possible to predict how difficult it would be to work in a specific soil with a given metal detector or ground-penetrating radar. Rules-of-thumb to link metal performance to some soil properties exist, but a lot of research is still needed (CEN, 2008). This chapter contributes to this research by describing a framework to define classes objectively from a database.

To our knowledge available soil data do not contain the data required for our analysis. This is the reason why we used a simple model for a metal detector and a classification based on simulated data and we suggested the creation of a soil database. The model we used aims to be as realistic as possible while remaining simple enough. It needs three parameters to fully determine a soil: the average of magnetic susceptibility, its range and a feature dependent on its correlation length. The consequences of not knowing or measuring an important soil parameter are analysed by considering a random non-measured value for this third parameter.

The following trend has been identified. The dominant source of errors was the limited number of targets that were used to measure the probabilities of detection used to define a soil class. The fact that the correlation length feature was not measured did not have a large effect on our simulation. This will not be the case for all parameters. For example, neglecting the effects of the magnetic susceptibility range would clearly lead to large errors. Finally the measurement errors of soil properties were less important.

8.7   Acknowledgements

The work presented here is partly based on CEN (2008) written by the European Committee for Standardization (CEN) Workshop 7. Besides the authors of this chapter, workshop contributors include Capineri Lorenzo (University of Florence), Cross Guy (Terrascan geophysics), Daniels David (ERA technology), Falorni Pier Luigi (University of Florence), Guelle Dieter, Hannam Jack (Cranfield University), Igel Jan (Leibniz Institute for Applied Geosciences), Lopera Olga (Royal Military Academy), Merz Armin (Vallon), Mulliner Noel (United Nations Mine Action Service), Preetz Holger (Leibniz Institute for Applied Geosciences), Sato Motoyuki (Tohoku University), Schoolderman Arnold (Netherlands Organisation for Applied Scientific Research) and Tollefsen Erik (Geneva International Centre for Humanitarian Demining).

8.8   Conventions and notation

A scalar is denoted x. A vector is denoted x and the corresponding unit vector $\hat{x}$. Unless otherwise specified, harmonic sources and fields are considered with an $e^{j\omega t}$ time variation assumed and suppressed, where j is such that $j^2 = -1$, ω is the angular frequency and t is time. Primes are used to express real parts of complex values, and double primes to express imaginary parts.

8.9   References

Baum, C. E. (1998), *Detection and identification of visually obscured targets*. Taylor and Francis.

Billings, S. (2004), in: Bloodworth, T.J. and Logreco, A. 2004: *Identifying and obtaining soil for metal-detector testing*. Technical Note No. I.04.117. JRC Ispra/IPSC/SERAC Unit, http://www.itep.ws/pdf/Technical_Note_I.04.117_JRC_Bloodworth.pdf (August 2004)

Billings, S. D., Pasion, L. R., Oldenburg, D. W. and Foley, J. (2003), The influence of magnetic viscosity on electromagnetic sensors: *Proceedings of EUDEM-SCOT2 2003, international conference on requirements and technologies for the detection, removal and neutralization of landmines and UXO*, Brussels, 15–18 September, 2003, http://www.geop.ubc.ca/~sbilling/papers/magsoils_eudem2003.pdf

Bishop, C. M. (2006), *Pattern recognition and machine learning*, Springer science + Business Media, LLC, ISBN-10: 0-387-31073-8, ISBN-13: 978-0387-31073-2, 2006 (corrected printing 2007)

Bosch, I. V. D. (2006), *Accurate modelling of ground-penetrating radar for detection and signature extraction of mine-like targets buried in stratified media*, Catholic University of Louvain, Louvain-la-Neuve, Belgium, Royal military academy, Brussels, Belgium, January. Ph.D. dissertation.

Bosnar, M. (2001), *Discrimination of metal targets in magnetically susceptible soil*, patent US6326791B1, December.

Bruschini, C. (2002), A multidisciplinary analysis of frequency domain metal detectors for humanitarian demining, PhD dissertation, VUB, Brussels, Belgium, http://www.eudem.vub.ac.be/files/PhDBruschiniFinalv2Lowres.pdf

Candy, B. H. (1996), *Pulse induction time domain metal detector*, Patent US5576624, November.

CEN (2003), CWA 14747-1:2003 *Mine action – Test and evaluation, part 1 Metal detectors*, CWA 14747:2003(E), European Committee for Standardization (CEN), Republished in 2009.

CEN (2008), CWA 14747-2:2008 *Mine action – Test and evaluation, part 2 Soil characterisation for metal detector and ground penetrating radar performance*, European Committee for Standardization (CEN), December.

Chew, W. C. (1990), *Waves and fields in inhomogeneous media*. Van Nostrand Reinhold.

Cross, G. (1999), *Soil properties and GPR detection of landmines: A basis for forecasting and evaluation of GPR performance*, Contract report DRES CR 2000-091 (October 1999)

Daniels, D. (2004), *Ground penetrating radar*, Daniels, David J. (Ed.), IEE-UK, ISBN: 0-86341-360-9 and 978-0-86341-360-5, UK 2nd edition.

Das, Y. (2004), A preliminary investigation of the effects of soil electromagnetic properties on metal detectors. In *Proc. SPIE conference on detection and remediation technologies for mines and mine-like targets IX*, 5415, pages 677–690, Orlando, FL, USA, April.

Das, Y. (2006), Effects of soil electromagnetic properties on metal detectors. *IEEE Transactions on geoscience and remote sensing*, Vol. 44, No. 6, (June) 10 pages (1644–1653), 0196-2892

Druyts, P., Das, Y., Craeye, C. and Acheroy, M. (2006), Effect of the soil on the metal detector signature of a buried mine. In *Proc. SPIE Defence and security symposium*, Orlando, FL, USA, April 2006.

Druyts, P., Das, Y., Craeye, C. and Acheroy M. (2009), Modeling the response of electromagnetic induction sensors to inhomogeneous magnetic soils with arbitrary relief. *IEEE Transactions on geoscience and remote sensing*. Accepted for publication, 2009.

Guelle, D., Smith, A., Lewis, A. and Bloodworth, T. (2003), *Metal detector handbook for humanitarian demining*, European communities, http://www.itep.ws/pdf/metal_detector_handbook.pdf

Guelle, D., Lewis A., Pike, M., Carruthers, A. and Bowen, S. M. (2005a), *Systematic test and evaluation of metal detectors (STEMD). Interim report field trial Lao, 27 September–5 November 2004*. Technical report S.P.I05.24, JRC, Joint research centre of the European Commission, March 2005. http://www.itep.ws/pdf/STEMD_Interim_Laos_final_Small.pdf

Guelle, D., Lewis, A. and Pike, M. (2005b), *Systematic test and evaluation of metal detectors (STEMD) Interim report field trial Mozambique*, European Commission Joint research centre Report EUR 21886 EN (November 2005) http://www.itep.ws/pdf/Interim_Final_Moz160108_web_optimized.pdf

Guelle, D., Gaal, M., Bertovic, M., Mueller, C., Scharmach, M. and Pavlovic, M. (2007), *South–East Europe interim report field trial Croatia (continuation of the ITEP-project systematic test and evaluation of metal detectors – STEMD) 25 September–18 October*

2006. Technical report, BAM, Bundesanstalt fuer Materialforschung und pruefung, March 2007. http://www.itep.ws/pdf/STEMD_Interim_Croatia_final.pdf

Haus, H. A. and Melcher, J. R. (1998), *Electromagnetic fields and energy*. MIT Hypermedia teaching facility.

Lambot, S., Weihermüller, L., Huisman, J. A., Vereecken, H., Vanclooster, M. and Slob, E. C. (2006), Analysis of air-launched ground-penetrating radar techniques to measure the soil surface water content. *Water resources research*, Vol. 42 (November), W11403, doi:10.1029/2006WR005097.

Lewis, A. M., Bloodworth, T. J., Guelle, D. M., Littmann, F. R., Logreco, A. and Pike, M. A. (2006), *Systematic test and evaluation of metal detectors (STEMD), interim report laboratory tests Italy*. Technical report EUR22536EN, JRC, Joint research centre of the European Commission, December, http://www.itep.ws/pdf/STEMD_Interim_Lab. pdf

Preetz, H., Altfelder, S. and Igel J. (2008), Tropical soils and landmine detection – An approach for a classification system. *Soil science society of America journal*, Vol. 72, No. 1 (January–February), 9 pages (151–159), ISSN 0361-5995 print; ISSN 1435-0661 online.

Stratton, J. A. (1941) *Electromagnetic theory*. McGraw-Hill Book Company.

UNMAS (2003), *IMAS 09.10 Clearance requirements*; 2nd Edition, UMAS, 1 January.

Van Dam, R. L., Hendrickx, J. M. H., Harrison, J. B. J., Borchers, B., Norman, D. I., Ndur, S., Jasper, C., Niemeyer, P., Nartey, R., Vega, D., Calvo, L. and Simms, J. E. (2004), *Spatial variability of magnetic soil properties. In Detection and remediation technologies for mines and minelike targets IX*, 5415, pages 665–676, Orlando, FL, USA.

Van Dam, R. L., Harrison, J. B. J., Hendrickx, J. M. H., Hirschfeld, D. A., North, R. E., Simms J. E. and Li, Y. (2005a), Mineralogy of magnetic soils at a UXO remediation site in Kaho'olawe Hawaii. In *18th Annual symposium on the application of geophysics to engineering and environmental problems*, Atlanta, USA, April.

Van Dam, R.L., Hendrickx, J.M.H., Harrison, J.B.J. and Borchers, B. (2005b), Conceptual model for prediction of magnetic properties in tropical soils. In *Detection and remediation technologies for mines and minelike targets X*, 5794, pages 177–187, Orlando, FL, USA.

9

The contribution of geophysics to landmine and unexploded ordnance (UXO) detection: Case study in the Egyptian environment

G. EL-QADY, A. MOHAMED, M. METWALY and M. ATYA,
National Research Institute of Astronomy and Geophysics, Egypt

Abstract: Landmines are essentially tactical and operational weapons that cause most widespread calamities, transcend humanitarian and sociological concerns, and bring severe environmental, economic and development problems. Worldwide, it is not known exactly how many landmines were planted and where these mines are located. Detecting mines with standard metal detectors is difficult as many are comprised primarily of plastic, with often only a firing pin as the sole metallic component. In this chapter, we describe the contribution of geophysical techniques specifically, electromagnetic metal detector (MD), ground penetrating radar (GPR) and electrical resistivity imaging scheme for detecting environmental clutters such as landmines, unexploded ordnance (UXO) and waste disposals. In addition, details about tests we carried out in Egypt using these techniques are discussed.

Key words: landmines, UXO, geophysics, metal detector, GPR, electrical resistivity, Egypt.

9.1   Introduction

Landmines are a type of inexpensive weapon widely used in conflict areas in many countries worldwide. They are one of the most widespread calamities, which transcend humanitarian and sociological concerns and bring severe environmental, economic and developmental problems. It is not known exactly how many landmines have been planted and where these mines are located. Nonetheless, it is estimated that about 80–120 million landmines have already been planted in many post-conflict areas, in about 90 countries (Berhe, 2007). However, the areas contaminated with mines directly and indirectly impact the surrounding community.

Landmines are essentially tactical and operational weapons, although on occasion they also have strategic implications. When used tactically, landmines are usually employed during battlefield engagements of relatively limited duration to disrupt an enemy's progress. Landmines are classified as the perfect soldiers. They are inexpensive; they do not eat; they do not fall asleep on duty; they do not require maintenance. Most field soldiers as well as civilians view landmines as an evil component of warfare. To them, landmines are undiscriminating weapons that kill and maim friends and enemies alike (Novakoff, 1992).

221

Landmines fall into two broad categories. Antipersonnel (AP) landmines are intended to kill or disable persons on foot. Antitank (AT) landmines are used against vehicles, such as tanks and armored personnel carriers. Mixed systems, which combine both AT and AP mines in the same munitions, are typically used against an enemy force that is mostly mounted but is accompanied by significant numbers of dismounted soldiers. AP in mixed systems are intended to prevent or discourage foot soldiers from penetrating or breaching an AT minefield.

Detecting these mines with standard metal detectors is difficult, as many are comprised primarily of plastic, with often only a firing pin as the sole metallic component. In most minefields, there is a plethora of benign sources that can trigger sensor detection. These sources may be man-made, such as metallic scrap, or natural, such as magnetic rocks. Without the ability to discriminate between these false alarms and the actual targets, every item detected must be treated as a landmine, resulting in a methodology that is both prohibitively expensive and time-consuming.

During the last decade, more modern search technologies were developed with automated arrays of sophisticated passive and active metal detectors. Digital geophysical mapping has become the goal for landmine and unexploded ordnance (UXO) searches (e.g. Gamey et al., 2000). Improved detection technologies have been coupled with data analysis systems of varying sophistication to allow either semi-automated or interactive analysis capability (e.g. Pasion and Oldenburg, 2001).

The scientific community has already invested considerably in technologies to find, remove, and dispose of land mines. Various approaches are being developed, each with advantages and disadvantages from a technical aspect, feasibility angle for implementation, or cost issue. Because of the variety of environments in which these must operate, the international community is moving towards an integrated suite of technologies as the most reliable and comprehensive method of detecting landmines, instead of relying on a single device (Donskoy, 1980). However, no single detecting technique can operate effectively in all environments and for all landmine types (MacDonald et al., 2003). The combination and integration between different techniques increases the rate of detection as well as giving some specific physical properties about the buried targets.

In this chapter, we describe the applicability of ground-penetrating radar (GPR) and the electrical resistivity imaging scheme for detecting environmental clutters, such as landmines, UXO and waste disposal. In addition there is a brief report on test results carried out in Egypt.

9.2 Landmine problem in Egypt

Due to its central geographical location between Africa, Asia and Europe, Egypt was the location for many battles. During the Second World War, the well-known El-Alamein battle, Western Desert (Fig. 9.1) was fought between British and German troops.

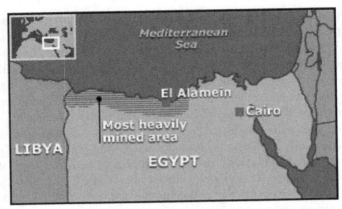

9.1 Location map of landmine field in the North Western Desert, during the Second World War.

As a result of this fighting, a large number of antitank and antipersonnel mines have been left. A total of 23 million landmines and UXO were emplaced in the country as a result of the many wars since the Second World War and this is considered to be about 21 percent of the total number of mines that are buried in the whole world (UNMAS, 2000). The presence of such active mines has caused many problems in Egypt.

Egyptian officials have stated that only 20–25 percent of these 'landmines' are really landmines, the remainder being other types of UXO. Antipersonnel mines believed to be in the Western Desert include German S-type bounding fragmentation mines and British Mk-2 mines. Anti-vehicle mines are thought to include German Riegalmine 43, Tellermine 35, Tellermine 42 and Tellermine 43 mines, Italian B-2 and V-3 mines, and British Mk.5 and Mk.7 mines (UNMAS, 2000; USAID and RONCO, 2002; and NCHR, 2005).

Despite UXO causing the loss of human limbs and even death, development proposals were hindered and prevented. Nowadays, the national trends of the country are focused on clearing the mines from the planted fields and adding them to reform projects. Therefore, modern scientific and technological methods have to achieve their aims by solving such problems.

Among several technical challenges faced by the experts in de-mining activities in areas such as Egypt are corrosion of the mines, which can cause them to explode, and the burial depth of the objects, which has changed due to the accumulation of sand dunes (Fig. 9.2).

In the Western Desert of Egypt, sand dunes are accumulated very fast. This makes the mines that were laid on the ground surface go down to depths of up to 3 meters sometimes. Such great depth makes it very difficult for them to be detected by conventional de-mining technologies. Hence, it is mandatory to find other suitable techniques to detect objects up to that range (2–3 m) of depth.

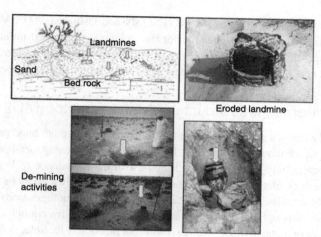

9.2 Landmines in Egypt, top left: cross section illustrates how deep the landmines are buried, top right: examples of eroded landmines, bottom: de-mining activities in the Western Desert of Egypt.

9.3   Geophysical techniques available for landmine and unexploded ordnance (UXO) detection

Ordnance is color-coded during manufacturing for identification purposes. However, color markings cannot be relied upon for identification. UXO markings can be altered or removed by weather or exposure to the environment. For geophysical work, the color code is not important but the physical parameters are, including for instance, magnetic permittivity and electrical conductivity.

Of more than 700 types of known landmines, the most common is the blast mine. It is buried at a shallow depth in the soil, rarely deeper than 10 cm, and triggered by the pressure of a footstep. Many of these mines are crudely constructed and deteriorate in the ground so the conditions under which they will explode cannot be predicted (Jefferson, 1993). These points require geophysical techniques to be used for landmine detection.

Several geophysical techniques have been proposed and utilized worldwide to achieve these objectives (Savelyev et al., 2007). Among these, ground penetrating radar (GPR) and metal detectors (MD) are considered the most effective, because they can locate both metallic and non-metallic landmines by non-invasive subsurface sensing (Gao et al., 2000; Chen et al., 2001; Daniels, 2004). However, it is well-known that the performance of GPR is influenced by the electromagnetic (EM) properties of the soil, particularly with an increase in the moisture and clay contents (Das et al., 2001; Lopera and Milisavljevic, 2007; Metwaly, 2007).

The two main types are metallic and non-metallic (mostly plastic) landmines. They are most commonly investigated by magnetic GPR, and MD techniques. These geophysical techniques, however, have significant limitations in locating the non-metallic landmines and wherever the host materials are conductive. Geoelectrical resistivity has been successfully addressed (El-Qady and Ushijima, 2005, Metwaly et al., 2008) and there is a high potential for the technique to be used in landmine detection through the contrast of electrical resistivity. Here we address those available techniques and how they could be adapted to work in the Egyptian environment.

9.3.1 Electromagnetic induction: Metal detector (MD)

Traditionally, landmines are located using metal detectors or through hand prodding of areas of soil which have been disturbed. Detecting landmines using metal detectors has become increasingly dangerous and difficult with the prevalence of landmines containing little or no metal content. The metal detectors currently used by de-mining teams cannot differentiate a mine from metallic debris, which sometimes leads to false alarm problems. Although the detectors can be tuned to be sensitive enough to detect the small amount of metal in modern mines, this is not practically feasible, as they will also be sensitive to ferrous soils, leading to the detection of smaller debris and augmenting the false alarm rate. The need for new, efficient and affordable de-mining technologies and sensor systems has become more obvious.

The detector operates on the principle of electromagnetic induction in which electric currents flow in the transmitter coil producing a primary magnetic field that penetrates the surrounding medium and the nearby metallic objects (Fig. 9.3).

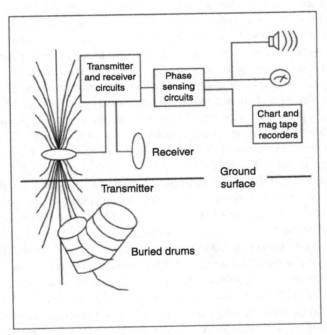

9.3 Block diagram of one MD coil arrangement and associated electronics. (Benson, Glaccum, and Noel, 1983)

The changes in the primary magnetic field strength with time produce eddy currents in the buried objects. These currents in turn radiate a secondary magnetic field, which can be sensed using the receiver coil. The MD system is able to detect even the small metallic component in the landmine; however, with the current innovation process in this technology, the MD technique has limitations. Currently there are some landmines that are made entirely of non-metallic components. Moreover, the sensitive MD cannot differentiate between landmine targets and changes in the ground conductivity, as well as other man-made clutter (MacDonald et al., 2003). Figure 9.4 illustrates the sensitivity of the MD in detecting such ordnance at limited depth.

9.3.2 Ground penetrating radar (GPR)

Ground penetrating radar (GPR) is the technique that employs radio waves, typically in the 15 MHz to 3 GHz frequency range, to map structures and features buried in the ground or in man-made structures (Annan, 2001). GPR has only recently been considered as an alternative technique for landmine detection. It provides a profile of subsurface features, which can be used to determine the

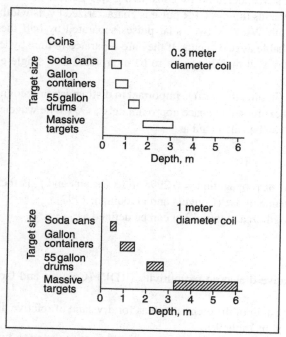

9.4 Approximate metal detector (MD) detection depths for various targets with two coil sizes. (Benson, Glaccum, and Noel, 1983)

location and depth of buried objects, including landmines. A GPR transmitter emits a specific band of electromagnetic pulses into the ground, which is reflected when changes in the characteristics of the medium occur. 'Reflectors' may be stones, shrapnel, landmines, or simply changes in the nature of the soil. Discriminating between landmines and other reflectors, within a complex background of variable soil characteristics, requires advanced processing of the GPR signal.

For landmines, a center frequency of 1 to 2 GHz seems to be adequate to locate mines with a diameter of 8–10 cm. Smaller mines might require correspondingly shorter wavelengths, which will shorten the investigated depth range too, but they are also buried closer to the surface.

GPR resolution

For landmine detection, the employed GPR unit must have a high frequency to achieve a good resolution, but since higher frequencies have a limited penetration power, the chosen range of frequencies is always a tradeoff between resolution and penetration depth. Resolution of the radar is defined as two components, which are the vertical resolution and the lateral resolution. These two resolutions are critical in order to be able to detect a specific target at a specific depth. Vertical resolution is most easily understood by considering two pulses, which occur as amplitude variation versus time. A GPR pulse is characterized by its width at half amplitude (pulse width W). Any two radar pulses separated by half their 'half width' are distinguishable as two events. If they are separated in time by less than this quantity then they will mainly expect to be interpreted as a single event as illustrated on Fig. 9.5.

So, the value W/2 (in time domain) is important to distinguish or recognize two targets. This value is expressed in space approximately as $\lambda/4$. The wavelength of radar wave λ_a in air can be calculated as:

$$\lambda_a = v / f_c \qquad [9.1]$$

where v is the speed of propagation = 0.2998 m/ns (in air) and f_c is the central frequency of the antenna in use (Conyers and Goodman, 1997).

Then, the wavelength in a material λm can be defined as:

$$\lambda_m = \frac{\lambda_a}{\sqrt{k_m}} \qquad [9.2]$$

where k_m is the relative dielectric permittivity (RDP) (Conyers and Goodman, 1997).

The quarter wavelength of different antennas for dry sand of relative dielectric permittivity 5 is listed in Table 9.1.

Figure 9.6 is a graphic presentation for the vertical resolution of 300 MHz, 500 MHz, 900 MHz and 1.5 GHz antennas.

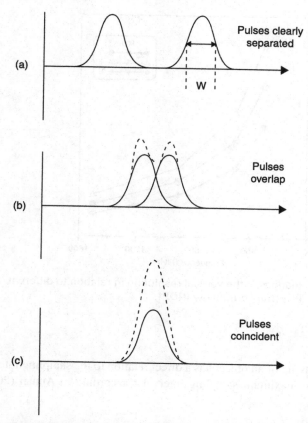

9.5 (a) Two radar return events are well resolved if the two transient signals are clearly separated in time. (b) When two events overlap each other in time, the question arises as to whether one or two events are present. (c) Two events are coincident completely (modified after Annan, 2001).

Table 9.1 The vertical resolution of 300 MHz, 500 MHz, 900 MHz and 1.5 GHz antennas

Frequency (MHz)	Dielectric permittivity	Velocity (m/ns)	Wavelength (m)	Quarter wavelength (m)
300			0.45	0.11
500		0.13	0.27	0.07
900	5		0.15	0.04
1500			0.089	0.02

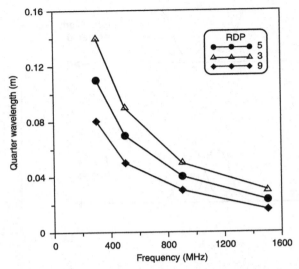

9.6 The changes in the vertical resolution in relation to different relative dielectric permittivity (RDP).

Also, the resolving of an object has a direct relation to the sampling rate of the radar traces. The maximum sampling interval t according to Annan (2001) is given by:

$$t = \frac{1000}{6f_c}$$ [9.3]

The maximum sampling interval of different antennas can be seen in Table 9.2.

Table 9.2 The maximum sampling interval of different frequencies

Antenna center frequency (MHz)	Maximum sampling interval (ns)
10	16.70
20	8.30
50	3.30
100	1.67
200	0.83
500	0.33
1000	0.17
1500	0.11

On the other hand, the lateral resolution is a function of the radar wave footprint. The footprint of the radar signal is defined as:

$$A = \frac{\lambda_m}{4} + \frac{d}{\sqrt{k_m - 1}}$$

[9.4]

where A is the approximate long dimension radius of the footprint, λ_m is the center frequency wavelength of the radar signal, d is the depth from the ground surface to the reflection surface and k_m is the average relative dielectric permittivity of the material from ground surface to depth d (Annan and Cosway, 1992). So, according to the previous equations, the wavelength of the 1.5 GHz antenna will be about 20 cm in air. Assuming a dry sand condition of the filling material of the site with relative dielectric permittivity of 5, the wavelength will diminish to 8.9 cm. The footprint of the radar wave was calculated at different depths and is shown in Fig. 9.7. Moreover, the attuning of the number of traces per meter during the survey work plays an important role in resolving the targets laterally.

9.3.3 Geoelectrical resistivity

Electrical resistivity can help identify buried objects, especially those non-metallic objects that are difficult to detect using metal detectors, magnetometers and GPR techniques, and have significant electrical conductivities in contrast relative to the host environment.

In highly conductive soil the electromagnetic waves diffuse quickly and therefore the GPR, which utilizes high frequency waves, is not able to see deeply into the ground. Similarly, the application of MD for landmine detection sometimes fails when mines are composed of non-metallic materials and/or the

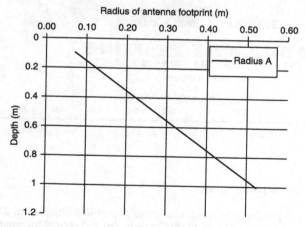

9.7 The footprint radius (A) of antenna 1.5 GHz as function of depth.

soil contains high concentrations of ferruginous minerals (Lopera and Milisavljevic, 2007). Therefore there is a strong need for applying another non-destructive surface technique, which is neither completely affected by the landmine materials nor by the EM properties of the soil. Such a proposed technique could be used either in combination with the GPR and MD techniques in routine landmine detection or as an independent confirmation tool for the assurance of landmine cleared areas. These requirements could be satisfied by using the electrical resistivity tomography (ERT) technique, particularly the capacitive resistivity (CR) dynamic system (Benderitter et al., 1994). The CR system is similar to the well-known conventional DC resistivity system with the main difference that the galvanic electrodes are replaced by capacitive sensors (Kuras et al., 2006). The ERT method generally provides a low cost and rapid tool for generating spatial models of subsurface physical properties (Chambers et al., 2006). Figure 9.8 illustrates the results of ERT forward modeling for UXO objects (El-Qady and Ushijma, 2005).

The classical mechanical installation of the steel electrodes is impractical and probably risky when used for landmine detection. Therefore, a need has arisen for an alternative resistivity imaging methodology like the capacitive electric resistivity system, which does not need direct coupling with the ground surface (Metwaly et al., 2008). The technique is based on a four-point electrode array that is capacitively coupled to the ground and acts as an oscillating non-grounded electric dipole (Kuras et al., 2007). The coupling mechanism between electrodes and the ground is then predominantly capacitive and the inductive effects are

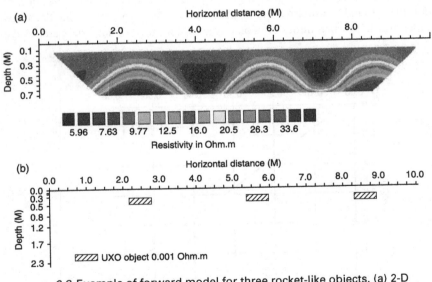

9.8 Example of forward model for three rocket-like objects. (a) 2-D response of the three buried UXO objects. (b) 2-D model for the buried UXO objects.

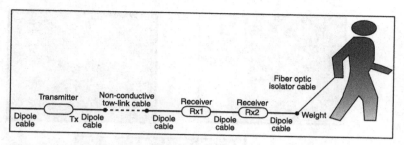

9.9 Sketch for the towed electrical resistivity system using capacitive electrodes.

negligible. The entire system is designed to be dragged or towed along the ground surface either manually or mechanically while resistivity can be measured continuously (see Fig. 9.9) (Milsom, 2003).

The chief problem with the images was low spatial resolution, which was of the order of 10 percent of the diameter of the electrode array. This was somewhat problematic since ER relies on imaging contrasts in conductivity to distinguish mines from rocks, soil inhomogeneities and other natural anomalies. Increasing the spatial resolution increases the probability of mine detection and reduces the false alarm rate (Church et al., 2006).

9.4 Experimental evaluation of metal detectors (MD) and ground penetrating radar (GPR) at test sites

9.4.1 Testing of MD

At the test area, metal detectors (MD) were used (Fig. 9.10) and it was not very easy to distinguish between false alarms and correct ones. This means that a lot of effort was required to dig and discover which were false and which were true.

(a)
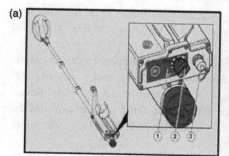

(b)

9.10 The Forester Minex 2FD. (a) Sketch of the Forester Minex 2FD showing the rotary switch (2) on the control panel (1) is used to set the sensitivity level and the headphones are connected to the socket (3). (b) The Forester Minex 2FD in operation.

9.11 False alarms due to insignificant objects.

Thousands of false alarms were triggered during the survey from objects lying on or just beneath the surface. Meanwhile, about 200 alarms were triggered from deeper objects buried up to 60 cm depth. These alarms were of special interest due to their burial depth. Therefore, they were manually excavated using small plastic shovels. However, almost all of them were false alarms from rusted cans, screws and other small metal objects (Fig. 9.11).

9.4.2   GPR test site

The test site has been dug out on a desert area with dimensions of 9.4 m × 7.5 m (Fig. 9.12). The mine-like bodies were selected to be very close to real mines in shape, weight and materials. The objects were distributed over the site in a scattered allocation, similar to the real situation in Egypt where the location of planted mines is so far unknown.

The site was surveyed using a SIR 20H GPR system connected to a monostatic antenna of 1.5 GHz central frequency. The device was adjusted to acquire 200 scans/m and 512 sample/trace. The range was attuned to 13 ns. The interval between the successive GPR profiles was 10 cm. The GPR profiles covered the survey site in two perpendicular directions (X and Y). A description of the mine-like objects used is shown in Table 9.3.

Preparing the simple test site with some landmine-like targets was essential to discover the different 'fingerprint' forms of various mine-like objects. The example test site has been constructed in the hot dry sand of the western Egyptian desert. The host sediments are friable sand of about one meter depth. A set of

Table 9.3 The parameters of the buried objects in the test site

No.	Object	Length (cm)	Width (cm)	Height (cm)	Depth (cm)
1	Vertical drum	87	60	87	14
2	Horizontal drum	87	60	87	34
3	Ceramic jar		38	55	26
4	Ceramic jar		46	85	0
5	Iron plate	181.5	155		71.5–77.5
6	Vertical can		10	15	25
7	Can N-S		7	12	25
8	Can 45 N-S		10	25	25
9	Can E-W		10	15	25
10	Can E-W		7	12	25
11	Vertical can		10	15	25
12	Can N-S		10	25	25
13	Triangle of six arranged in triangle shape		7	12	25
14	Aluminium pan cover				25
15	Vertical aluminium pan				25
16	Two iron disks		25–15	3	25
17	3 cans		10	15	25
18	Iron disk		25	3	25
19	Iron disk		15	3	25
20	Iron disk + computer mouse		25	3	25
21	Computer mouse E-W				25
22	Computer mouse N-S				25
23	Aluminium pan				25
24	Iron disk		15	3	25
25	Can E-W		10	25	25
26	Computer mouse N-S				25
27	Computer mouse E-W				25
28	Can N-S		10	25	25
29	Computer mouse N-S				25
30	Iron disk		25	3	25
31	Iron disk		15	3	25
32	Aluminium pan				25
33	Iron disk		20	3	25
34	Aluminium				25
35	Computer mouse E-W				25

Notes
1. The diameter of the cylindrical objects is shown in the width column.
2. Depth computed from the sand surface to the upper surface of the buried object.

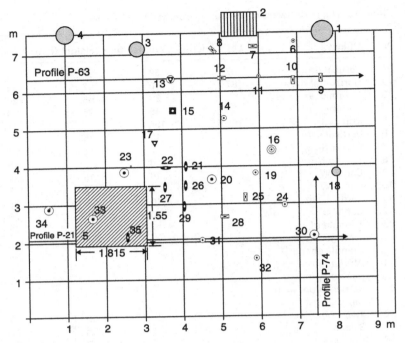

9.12 The distribution of objects in the test site.

Table 9.4 The physical and geometrical parameters used in the synthetic data modeling (ES unit denotes electrode separation)

	Quality	Dimension (in ES unit)	Depth (in ES unit)	Resistivity (Ohm.m)
Landmine	Metallic	1	1	0.01
	Non-metallic	1	1	100
Host soil	Dry	–		1000
	Wet	–		5

different metallic, plastic and partially metallic objects, similar to real landmines, has been buried regularly along two profiles at various depths (Fig. 9.12). The entire site was covered by a flat layer of clean sand. The metallic bodies comprise three rings with diameters 7.5, 14.5 and 25 cm (Fig. 9.12, No. 1, 3 and 4). The thicknesses of the inner solid part are 1, 10.5 and 10.5 cm respectively. The height of the three ring bodies is 2 cm. The solid round sheet (Fig. 9.12, No. 2) is 10 cm in diameter and 2 cm in height. The metallic bar body is 30 cm in length and 3 cm in diameter (Fig. 9.12, No. 5). See also Table 9.4.

The plastic bodies have cylindrical shapes with different dimensions (Fig. 9.13). The outer shell thickness is about 0.5 cm. Bodies number 7, 8 and 9

9.13 Models of landmines and cultural objects: (a) metallic landmine object, (b) real metallic antitank mine T-80, (c) different plastic landmine objects, (d) real plastic antipersonnel mine Ts-50, (e) low metallic landmine, (f) overview of the test site.

are partially filled with iron ash in order to simulate partially metallic bodies (Fig. 9.14). Generally all the metallic bodies are buried with their diameters oriented horizontally, while the plastic bodies are buried vertically except bodies no. 5 and 10, which are oriented horizontally (Fig. 9.13 and 9.14).

9.5   Results and discussion

The MD test was somewhat successful. However, the level of false alarms was still the biggest challenge when using this technique in wide area de-mining activities. For the GPR test, we tried to implement the cross-correlation analysis for a wider aspect in GPR applications. Using the cross-correlation approach to locate mine-like objects is much faster in operation than conventional interpretation procedures. Moreover, this procedure facilitates the ability to class the detection of a specific target and to map it successfully. That, in return, will lead to a minimization of the false alarm problem, which occurs due to the presence of some metallic objects that are not mines. By specifying the traces of a particular mine, the zone of work in that definite target will be limited. This reduces the consumption of time, money and effort.

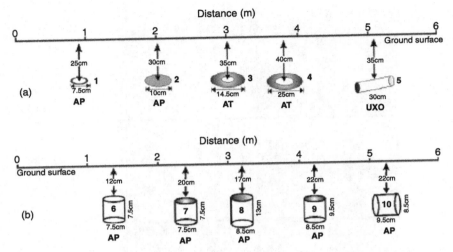

9.14 Shapes, geometries and buried depths for landmine-like objects. (a) Metallic objects, (b) low metallic content and plastic objects, AP = antipersonnel mine, AT = antitank mine, UXO = unexploded ordnance.

The cross-correlation process was achieved through two steps. Firstly, the correlation was done for a trace or group of traces of a specific target against the traces of the whole profile. Figure 9.15 shows the result of the correlation and the location. The selected target is well observed with a high correlation coefficient value, which was highest at the target trace.

Secondly, the correlation was done for a background trace or a number of traces (which will be stacked as one correlating trace). The background traces are the traces that did not suffer any effect due to the targets. This selected background trace was correlated against the rest of the traces in the profile. Figure 9.16 illustrates the minimum correlation coefficient amplitude at the location of the target.

The same two steps were executed using a trace (or correlating trace resulting from stacking a group of traces) of one profile to be correlated to a different profile. The result is presented in Fig. 9.17 and 9.18. It can be observed that the objects were more resolved when the background trace was used. Afterwards, the correlation procedure was conducted over the whole area of the site.

The cross-correlation approach is not affected by the target depth, while the time slice is. This gives the current procedure the power to map all the objects through the whole investigation depth. As it appears on Fig. 9.16, 9.17 and 9.18, the cross-correlation amplitude decays sharply. Therefore, each object covers a smaller area than it actually covers in the site. Consequently, the objects that are close together could be marked out with higher resolution. Also, this procedure could be progressed to a more complicated version to achieve interactive detection of mines. The approach had a satisfactory outcome and needs further efforts to gain more beneficial results (Abbas and Lethy, 2005).

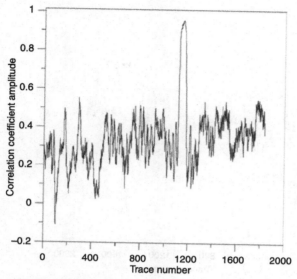

9.15 Cross-correlation using an over target trace against the same profile.

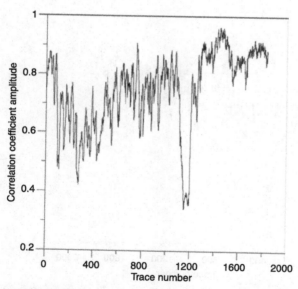

9.16 Cross-correlation using a background trace against the same profile.

It is noted that the slice maps include some anomalies which are not related to any objects. These anomalies may be due to external noise. The cross-correlation 2D presentation is free from such kinds of anomalies.

The forward calculations of DC resistivity are used for determining the apparent resistivity pseudosections applying seven different electrode configurations (see

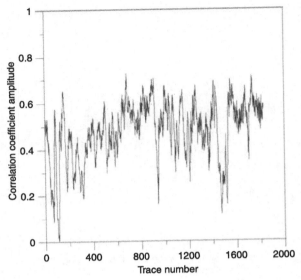

9.17 Cross-correlation using an over target trace against different profile.

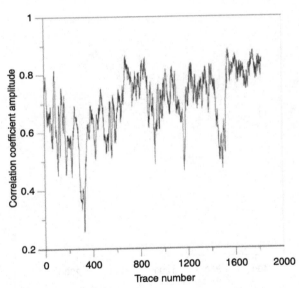

9.18 Cross-correlation using a background trace against different profile.

Plate VI in the colour section between pp 166–7), running over the center of the proposed buried landmine targets (Metwaly, 2007). The host sand resistivity is set to be 1000 and 10 Ohm.m for dry and wet host sand. The landmine resistivities were set to be 0.2 and 100 Ohm.m to represent metallic antitank and non-metallic antipersonnel landmines respectively placed horizontally at depths 0.1 and 0.2 cm

(Plate VI a and i). The electrode spacing (a) ranges from 0.05 to 1.5 m for all electrode configurations and (n) is set to one. The profile length is 4.5 m, which requires a maximum of 90 electrodes to cover the proposed profile. The two-dimensional model involves a finite element calculation mesh, which divides the subsurface into a number of rectangular blocks (Dey and Morrison, 1979). Then the inverse calculations have been carried out using least square sense using Marquardt's algorithm (Loke and Barker, 1996) to attain the physical models that are consistent with the forward calculations.

The pseudosection gives an approximate picture of the subsurface resistivity distributions because it depends principally on the type of electrode arrays rather than on the subsurface resistivity. For that reason, only the inverted resistivity sections will be considered in the following discussions.

Like the GPR and MD techniques, electrical resistivity imaging (ERI) can detect the metallic mine buried either in dry or wet environments. However the plastic mine shows significant electrical resistivity anomalies with varying accuracy that differs from one array to the other, based on the subsurface distributions of the current and potential lines. The metallic mine shows some evidence of distorted anomaly using the seven electrode configurations. The lower boundary of the mine cannot be detected effectively (Plate VI b–h). This is not a big landmine detection problem as the metallic mine buried in dry soil is the preferred target for the MD and GPR techniques. The interesting point is the detection of the plastic mine in the dry environment. All seven electrode configurations are effectively detecting this small (15cm × 20cm) plastic mine in dry conditions (Plate VI b–h). The dipole–dipole configuration shows the highest resistivity resolution and relatively undistorted signals (Plate VI g). The wet and conductive environments like beaches, rice paddy fields, marshes and swamps, which are considered harsh environments for using the GPR and MD techniques, are the best conditions for performing ERI. Such conductive environments ensure a good electrical coupling with the ground and consequently provide efficient resistivity data. The resolution of the inverted resistivity images for the metallic mine in the wet environment is better than it is in the dry environment. Almost all electrode arrays except the Wenner alpha successfully detect the mine body (Plate VI j–p).

However, the bottom of the mine is not defined well. It is likely that the plastic mine in the wet soil can be almost detected using the utilized electrode arrays (Plate VI j–p).

The landmines are 3D bodies. Therefore, it is essential to apply the 3D detection technique for getting the signature of these various small landmines buried at different depths in dry and wet environments using different electrode arrays (Metwaly et al., 2008). The landmines and the host soil parameters, which were used to calculate the synthetic data, are summarized in Table 9.4. The number of employed electrodes is set to be 20 in x and 10 in y directions. The dipole electrode array was applied with the dipole length equal to the unit electrode spacing (ES), while the dipole separation factor (n) is set to range from (1n) to (6n). Both the metallic and non-metallic landmines were modeled with a homogenous cube, whose side length is equal to the electrode separation (ES) unit and they are buried at a depth equal to the ES, as well as in homogenous soils. The RES3DMOD three-dimensional forward modeling program was used to calculate the synthetic apparent resistivity data for both electrode arrays.

The 3D inverted resistivity data using both L1 norm and L2 norm schemes for both wet and dry soil models are displayed as horizontal resistivity images at depths equal to the unit electrode separation (see Plate VII in the colour section) (Metwaly et al., 2008). Both inversion techniques detect the metallic and non-metallic landmines in wet and dry soils successfully, but with different spatial resolutions. The inversion results using L2 norm show a type of concentric, smeared out anomalies with gradational boundaries around the landmine targets (Plate VII a and c). The inversion results using L1 norm scheme (Plate VII b and d) give better landmine resolutions in wet and dry soils without such a high smearing effect. The inverted images for both metallic and non-metallic bodies exhibit relatively sharp boundaries between the landmines and the background soils, which are uniformly distributed around the landmines, compared with the L2 norm inversion results.

9.6   Conclusions

Since the available methods are strongly dependent on the response of the metal buried targets, the soil conditions in the survey area control the maximum detection depth and may cause a dramatic decrease in the detection sensitivity. We need the help of advanced technology to detect these unexploded mines and we think that there is a moral and legal responsibility upon the states that planted them. Even so, there is still a lack of success to date in developing a capability to characterize UXO, which is due, in part, to the reliance on magnetic sensors for the majority of UXO detection work, which has been only marginally successful. It seems clear that other sensors will be needed to distinguish UXO.

Research in the field of landmine detection is under way to develop new detection strategies that show a significant response to both the non-metallic mines and the buried landmines in very conductive environments. Thereby reducing the false alarm rate while maintaining a high probability of detection, and saving time and reducing the chance of injury to the de-mining crew. The current efforts are focused on producing an integrating multisensor system rather than concentrating on individual sensors. Another trend is concerned with applying the algorithmic fusion of data from individual sensors to develop the theory necessary to support an advanced multisensor system.

We propose that in the near future, humanitarian activities should focus on implementing new geophysical techniques and analysis for UXO detection, like electromagnetic ground mapping. At the same time, there should be joint assessments for different geophysical approaches to minimize the limitations of each geophysical technique. There should be diversity in the research and development projects from various scientific and engineering staff to facilitate the development of the present geophysical systems to simplify the survey and processing applications.

9.7   References

Abbas Mohamed Abbas and Ahmed Lethy (2005). Implementation of wave-let correlation to locate landmine-like objects using GPR Data as experimented on a test site, *Journal of Suez Canal University*, pp. 83–97.

Annan, A. P. (2001). *Ground penetrating radar, workshop notes.* Sensors & Software Inc., 1091 Brevik Place, Mississauga, Ontario, L4W 3R7, CANADA.

Annan, A. P. and Cosway, S. W. (1992). Ground penetrating radar survey design, *Proceedings of the Symposium on the Application of Geophysics to Engineering and Environmental Problems, SAGEEP'92*, April 26–29, 1992, Oakbrook, IL, pp. 329–351.

Benderitter, Y., Jolivet, A., Mounir, A., and Tabbagh, A. (1994). Application of the electrostatic quadripole to sounding in the hectometric depth range, *J. Appl. Geophys.*, 31, 1–6.

Benson, Richard, R.A. Glaccum, and M.R. Noel (1983). *Geophysical Techniques for Sensing Buried Wastes and Waste Migration*, 236 pp. 6375 Riverside Dr., Dublin OH 43107: National Ground Water Association.

Berhe, A. A. (2007). The contribution of landmines to land degradation, *Land Degrad. Dev.*, 18, 1–15.

Chambers, J. E., Kuras, O., Meldrum, P., Ogilvy, R. D., and Hollands, J. (2006). Electrical resistivity tomography applied to geologic, hydrologic, and engineering investigations at a former waste disposal site, *Geophysics*, 71, B231–B239.

Chen, C., Rao, K., and Lee, R. (2001). A tapered-permittivity rod antenna for ground penetrating radar applications, *J. Appl. Geophys.*, 47, 309–316.

Church, P., McFee, J. E., Gagnon, S., and Wort, P. (2006). Electrical impedance tomographic imaging of buried landmines, *IEEE T. Geosci. Remote*, 44, 2407–2420.

Conyers, L.B., Goodman, D. (1997). *Ground penetrating radar: an introduction for archaeologists.* AltaMira press.

Daniels, D. (2004). *Surface Penetrating Radar*, 2nd edition. The Inst. Electrical Eng., London.

Das, B., Hendrickx, J., and Borchers, B. (2001). Modeling transient water distributions around landmines in bare soils, *J. Soil Sci.*, 166, 163–173.

Dey, A. and Morrison, H. F. (1979). Resistivity modeling for arbitrarily shaped three dimensional structures, *Geophysics*, 44, 753–780.

Donskoy, D. M. (1980). Nonlinear vibro-acoustic technique for landmine detection, *Proc. SPIE 12th, Conference on Detection and Remediation Technologies for Mines and Minelike Targets IV*, ed. by Dubey, A. C. et al., pp. 211–217.

El-Qady, G. and Ushijima, K. (2005). Detection of UXO and landmines using 2-D modeling of geoelectrical resistivity data, *Proceedings of SAGEEP Meeting*, 1176–1182.

Gamey, J. Doll W. and Bell, D. (2000). Airborne UXO detection technology demonstration. *Proceeding of SAGEEP2000*, p.57–66.

Gao, P., Collins, L., Garber, P., Geng, N., and Carin, L. (2000). Classification of landmine-like metal targets using wideband electromagnetic induction, *IEEE T. Geosci. Remote*, 23, 35–46.

Jefferson, P. (1993). The Halo Trust, "An overview of demining, including mine detection equipment" in Symposium on Anti-Personnel Mines, Montreux, April 1993, ICRC, Geneva, pp. 125–132.

Kuras, O., Beamish, D., Melrum, P., and Ogivly, R. (2006). Fundamentals of the capacitive resistivity technique, *Geophysics*, 71, 135–152.

Kuras, O., Meldrum, P. I., Beamish, D., Ogilvy, R., and Lala, D. (2007). Capacitive resistivity imaging with towed arrays, *J. Environ. Eng. Geoph.*, 12, 267–279.

Loke, M. H. and Barker, R. D. (1996). Practical techniques for 3-D resistivity surveys and data inversion, *Geophysics*, 44, 499–523.

Lopera, O. and Milisavljevic, N. (2007). Prediction of the effects of soil and target properties on the antipersonnel landmine detection performance of ground-penetrating radar: A Colombian case study, *J. Appl. Geophys.*, 63, 13–23.

MacDonald, J., Lockwood, J., Altshuler, T., Broach, T., Carin, L., Harmon, R., Rappaport, C., Scott, W. and Weaver, R. (2003). *Alternatives for Landmine Detection.* RAND, USA.

Metwaly, M. (2007). Detection of metallic and plastic landmines using the GPR and 2-D resistivity techniques, *Nat. Hazards Earth Syst. Sci.*, 7, 755–763.

Metwaly, M., G. El-Qady, J. Matsushima, S. Szalai, N. S. N. Al-Arifi, and A. Taha (2008) Contribution of 3-D electrical resistivity tomography for landmines detection. *Nonlin. Processes Geophys.*, 15, 977–986.

Milsom, J. (2003). *Field Geophysics, 3rd Ed.* (The geological field guide series), John Wiley and Sons Ltd, p. 232.

National Council of Human Rights (NCHR) (2005). Egypt and the problem of landmines, Background paper, The International Conference for Development and Landmine Clearance in the North West Coast, Cairo 27–29 December, Abstr. No. 12.

Novakoff, A. K. (1992). FAA Bulk Technology Overview for Explosive Detection, Proceedings "Applications of Signal and Image Processing in Explosives Detection Systems", Boston, Massachusetts, 16–17 Nov. 1992 Volume 1824, pp. 2–12.

Pasion, L. R. and Oldenburg, D. W. (2001). A discrimination algorithm for UXO using time domain electromagnetics, *J. Env. And Eng. Geophys.* 6, 91–102.

Savelyev, T. G., Kempen, L., Sahli, H., Sachs, J., and Sato, M. (2007). Investigation of time-frequency features for GPR landmine discrimination, *IEEE T. Geosci. Remote*, 45, 118–128.

UNMAS (2000). *Mine Action Assessment Mission Report: Arab Republic of Egypt.* February, 2000.

USAID and RONCO Consulting Corporation (2002). *Arab Republic of Egypt, Mine Action Assessment Report and Proposed Organization*, April 2002.

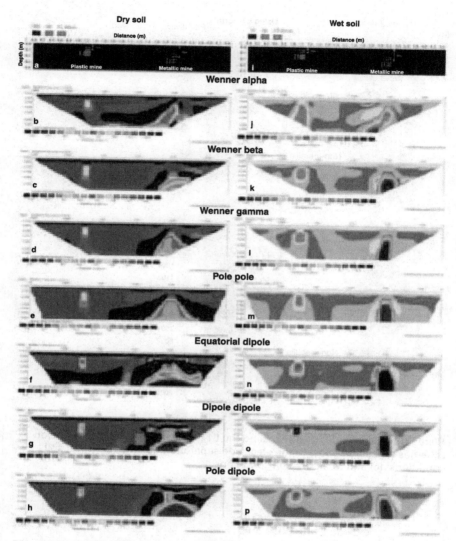

Plate VI 2D inverted ERI models using robust inversion for plastic and metallic landmine-like objects.

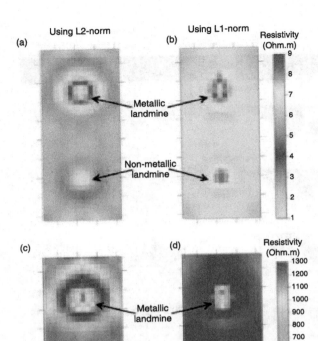

Plate VII Horizontal depth slices using the L1 norm and L2 norm 3D inversion schemes at a depth equal to the electrode separation. (a) and (b) in wet soil conditions, (c) and (d) in dry soil conditions.

10

Detecting landmine fields from low-resolution aerial infrared images

I. Y.-H. GU, Chalmers University of Technology,
Gothenburg, Sweden and T. TJAHJADI,
University of Warwick, UK

Abstract: This chapter primarily addresses the problems of landmine field detection through the use of low-resolution infrared (IR) images captured from airborne or vehicle-borne passive IR cameras. We describe a scale-space-based scheme for detecting landmine candidates, as indications of landmine fields. The scheme contains two parts: in the first part, a multi-scale detector, using a special type of isotropic bandpass filter, is employed. In the second part, refinement of landmine candidates is performed through a post-processing scheme that seeks maximum consensus of corresponding landmine candidates over image frames. The chapter also briefly addresses the problems of landmine detection from high-resolution IR images measured at close distances to ground surfaces. A detector based on the thermal contrast model is described. Experiments were conducted on several IR image sequences measured from airborne and vehicle-borne cameras, and on IR images measured at close distances to ground surfaces, where some results are included. Our experiments on these methods have shown that landmine signatures have been significantly enhanced after the processing, and automatic detection results are reasonably good. These methods may therefore be potentially employed for assisting humanitarian de-mining work.

Key words: landmine field detection, infrared images, multi-scale detector, isotropic bandpass filter, salient point feature detection, maximum consensus, infrared landmine modeling, airborne measurement, vehicle-borne measurement, hypothesis test.

10.1 Introduction

Humanitarian de-mining is concerned with the detection and subsequent removal of mines. The process consists of identifying mine fields and reducing the suspected area by discriminating individual landmine-like objects from clutter (e.g., bushes, rocks, petrified wood and animal burrows) in suspected regions, and of the actual landmine clearance. Several technologies based on different physical principles have been investigated for landmine detection. These include the use of metal detectors, ground penetrating radar, chemicals, and acoustic and optical sensors (Earp et al., 1996; Fillippis & Martin, 1999; Paik et al., 2002; Faust et al., 2005; Sahli et al., 2006). The various detection approaches examine different properties of surface-laid and shallowly buried landmines under various constraints, e.g., thermal, chemical, frequency, spatial and shape (Bruschini et al., 1999).

244

The metal detector (e.g., Bruschini, 2004), an electromagnetic-induction based landmine detector, measures the induced currents in buried conductors. Such a detector relies on the landmine casing to contain significant metal to trigger a response, and has several advantages over other detection approaches, including ease of use, system robustness, low cost, and high detection rates for subsurface landmines. However, metal detectors have limited discrimination capability and thus suffer from a high false alarm rate due to ubiquitous metal clutter. Furthermore, most currently developed landmines are either made of plastic or have very low metal content.

An ultra-wideband synthetic aperture radar (UWB SAR) has a ground penetrating capability to detect shallowly buried landmines and causes a metallic or plastic anti-tank landmine to generate two dominant scatters in the slant-range direction called the double hump signature (Jin & Zhou, 2007). An UWB SAR landmine detector operates on a low frequency and when mounted on an air- or vehicle-borne system the sensor provides landmine detection over wide areas from a safe standoff distance. A recent example of such an approach to landmine detection utilizes a support vector machine with a hyper-sphere classification boundary to distinguish landmines from clutter, and the hidden Markov model to represent the amplitude varying information of the double-hump signature of metallic and plastic landmines (Jin & Zhou, 2008).

A ground-penetrating radar (GPR) landmine detector measures the reflected electromagnetic waves from the changes in subsurface electrical soil properties, for example, the changes in dielectric, to detect plastic landmines with little metal content (e.g., Morrow & Genderen, 2002). The GPR signal from a landmine is dependent on the size, shape and composition of the mine as well as its burial depth and orientation. In addition to the properties of the mine, the electrical characteristics of the soil also affect the signature of landmines and clutter.

There are two main GPR-based approaches to landmine detection. In the first approach, a background model is defined and all objects that significantly deviate from this model are classified as targets. Such an approach involves techniques such as background subtraction (Wu et al., 2001) and background modeling (Ho & Gader, 2002). To address the randomness of the existence of objects, particle filters in association with the reversible jump Markov chain Monte Carlo method are proposed in Ng et al. (2008) to provide robust detection and accurate localization of landmines.

In the second approach, a target model is defined and then searched for in the GPR data. In Torrione and Collins (2007) the data tiling and 3D texture feature approach were applied to 3D time-domain GPR data sets (obtained from a GPR detector mounted on an unmanned ground vehicle). Landmine discrimination using the Wigner distribution as the target signature can be performed by singular value decomposition (SVD) and principal component analysis, and SVD-based features of the non-stationary target response that allow the discrimination of landmines from stones are proposed in Savelyev et al. (2007). In Kovalenko et al. (2007) the detection algorithm searches for a reference wavelet that represents a target response to the probing radar pulse (derived from a set of data acquired in controlled environment conditions) and suppresses responses with different waveforms, which are presumed to correspond to clutter. Noting that landmine targets and clutter objects often have different shapes and composition, which generates different amount of energy return at different frequencies, and thus different energy density spectra (EDS), the approach in Ho et al. (2008) uses the finite-difference time-domain numerical model to extract features from the spectral characteristics obtained from EDS. The features are then used to improve landmine detection and clutter discrimination.

The presence of buried objects including mines affects the heat conduction inside the soil under natural heating conditions such that the soil temperature on the ground above the buried objects is often different from that of unperturbed areas. For infrared (IR) images of mine fields measured from airborne sensors, landmines are indicated by spatial differences from their surroundings due to digging, or due to thermal and material signatures. However, the background in images usually consists of various types of noise and clutter, e.g., thermal noise, sand, gravel roads and vegetation, thus making the detection difficult. This chapter presents related theories and techniques by examining the spatial-frequency differences for detecting landmine candidates and locating landmine fields in low-resolution IR images. The chapter focuses on the following aspects:

- Describing the common characteristics of landmines and landmark points, and the characteristic differences in their surroundings.
- Formulating a single- and multi-scale isotropic feature detector that exploits the characteristic difference in the spatial-frequency domain, and is suitable for detecting/locating landmine candidates.
- Interpreting the detector using the 2D isotropic bandpass filter, matched filter, detection theory, and thermodynamic-based landmine models.
- Employing a multi-scale detector where the detectability of landmines is enhanced by automatically selecting an appropriate scale, and the localization of landmines is improved by inter-scale position tracing.
- Employing a post-processing scheme to refine the landmine detection.

IR images from vehicle-borne sensors as well as airborne sensors on a helicopter over the test bed scenarios provided by The Swedish Defence Research Agency (FOI) are used to demonstrate the effectiveness and robustness of the detector as a semi-automatic landmine field detection tool.

10.2   Related work

Airborne and vehicle-borne sensor-based techniques (Bruschini et al., 1999; Bishop et al., 1998, EU-IST, 2000) offer efficient and safe detection of mine fields. The techniques obtain measurements from passive sensors mounted on a vehicle or a helicopter at a height ranging from 10 to several hundred meters above the ground surface. For the latter case, sensors are applied in a few spectral regions, called atmospheric windows (Caniou, 1999). These measured images are typically of low spatial resolution where a landmine may be represented by only a few pixels. Thus, different detection strategies are utilized as compared to those (high resolution) images obtained by employing sensors at a close distance (e.g., 2–3m) to the landmines.

A common characteristic of landmines in low-resolution images and of features such as corners and landmarks is that they all appear to be point-like features. In Manjunath et al. (1996) feature points were detected by applying a scale-interaction model, where the difference of two directional Gabor filters of different spread is applied to an original image at various orientations, and the directions with the first several high outputs were used for extracting features. In Mokhtarian and Suomela (1998) a curvature scale-space based method was applied to multi-scale edge curves to extract corner features. Small targets or point features have been detected by spatial processing in IR search and track systems (Chan et al., 1990; Ferrara et al., 1998). However, these search and track systems are not directly applicable to landmine detection because they are only adequate for applications with little clutter and the targets are bright compared to the background. In Gu and Tjahjadi (2002b) a family of multi-scale isotropic bandpass filters for detecting landmarks and corners achieves good detectability and localization. The application of a special 2D isotropic bandpass filter from this family and involving an automatic scale level selection, a post-processing of peak picking and inter-scale position tracing on IR images measured from vehicle- and airborne sensors has demonstrated its potential as a semi-automatic tool for mine field detection (Gu & Tjahjadi, 2002a). Stochastic modeling may also be exploited, e.g., using the stochastic jump diffusion process in Markov random fields to detect junction points and using Markov chain Monte Carlo to model texture that could potentially be applied to remove clutter (Zhu et al., 2000).

There has also been much work on using local features that are invariant to full affine transformations for image matching and object recognition. The most relevant work is the scale invariant feature transform (SIFT) that transforms an image into a collection of local features identified using a cascade filtering approach (Lowe, 1999, 2004). The first stage identifies key positions over all scales by searching for locations that are maxima or minima of a difference-of-Gaussian function. Each point is used to generate a feature vector (called SIFT key) that describes the local image region sampled relative to its scale-space coordinate frame. The local image gradients are measured at the selected scale in the region around each keypoint and these are transformed into a representation that allows for a significant level of local shape distortion and change in illumination.

When considering the use of point features for landmine detection, it is worthwhile to consider models associated with landmines. In Lundberg and Gu (2000), Sjokvist et al. (1998) and Christiansen and Ringberg (1996), a thermal contrast-based model of landmines in IR images was proposed for high resolution images measured at 2–3 meters above a test bed. The landmines were assumed to be buried shallowly under the ground surface. A forward thermal contrast model aims to model the temporal difference of buried landmines to their surrounding soil under natural heating conditions. Most of the studies on IR-based landmine detection have focused on defining forward thermal models for buried objects. A thermal model has been applied on IR images of a soil area for landmine detection (Lopez et al., 2004) in which the depth of burial and the thermal diffusivity of buried objects were estimated, assuming their height and location in the soil volume are known. A 3D model for homogeneous soil containing a buried landmine that includes explosive (TNT) and a casing is proposed in Khanafer and Vafai (2002). The effect of a thin metal case and air gap on the simulated soil temperature at different depths, including the soil surface, and the top and bottom surfaces of the mines on IR images with buried mines were analyzed (Khanafer et al., 2003). However, a thermal model for landmine detection should take into account that IR signatures of soil surfaces depend on weather and environmental conditions and soil types. This led to a finite-difference method for estimating the soil and soil surface thermal properties from soil temperature measurements acquired using thermocouple to an estimation from acquired IR images in real mine fields (Thanh et al., 2007), which results in a realistic thermal model applicable to real mine fields.

Since there are significant differences between images measured from airborne cameras and cameras placed at close distances to ground surfaces, it is not clear as to whether the aforementioned models can be generalized to the former. However, the models could indicate what the patterns (features) of landmines look like and thus how to exploit the essential differences of landmines to their surroundings, including clutter, landmarks, and background noise. The above qualitative comparison of computer vision applications with the addressed landmine problem inspires us to consider it as detecting point-like features (yet conforming to some models) embedded in a special noisy and clutter background noise. This issue is investigated through examining two detectors respectively used for images measured from airborne cameras and from using cameras at close-distances to ground surfaces.

By examining the characteristic differences between landmines and their surroundings in an image in the spatial-frequency domain, a special type of isotropic bandpass filter is proposed in this chapter for a robust detector that is suitable for detecting/locating landmine candidates from airborne and vehicle-borne measured images. The performance of the detector is further enhanced by introducing its multi-scale version and a post-processing scheme that seeks the maximum consensus of landmine candidates through image frames.

10.3   Nature of infrared (IR) measured landmine fields

10.3.1 IR landmine field images

This section addresses the issue of landmine field detection by analyzing IR images. It is well known that with the passage of time, various surface-laid or shallowly buried landmines may be covered with soil and vegetation, and surrounded by metallic and non-metallic debris, making them more difficult to detect by the human eye. It is worth noting that for airborne or vehicle-borne IR images, a typical landmine-like object is very small, indicated by only a few pixels due to the height of measurements and the camera resolution. Furthermore, using IR sensors imposes limitations, for example:

- Landmines can only be detected if they are shallowly buried as the temperature differences between landmines and their surroundings become too weak to measure using IR sensors if mines are deeply buried.
- There should be no clouds between the IR sensors and the ground fields that are being measured as this can significantly impact on the quality of IR images.

Observing airborne-measured IR images, it can be seen that landmine candidates often differ spatially from their surroundings, e.g., structural changes in soil surfaces due to digging, intensity differences due to temperatures. Landmine candidates, among others, often look like affine invariant salient points (or landmarks) against their surroundings.

The basic idea of identifying landmine fields is through candidate landmine detection: if a field contains sufficient visual cues of landmine candidates, then the field is classified as a possible mine field requiring further analysis. Hence, the task of detecting landmine fields is equivalent to detecting landmine candidates from low-resolution images.

10.3.2 Characteristics of low-resolution mine field images

Before detecting candidates, let us examine some characteristics of landmines in low-resolution IR images from the spatial domain and the spatial-frequency domain. In the spatial domain, pixel points of landmines can be characterized by:

- insensitiveness to rotations and translations
- intensity differences to their surroundings.

In the spatial-frequency domain (e.g. obtained by applying 2D subband filters with different radial center frequency), landmine points are characterized by:

- having relatively large magnitude values when compared with their surroundings
- magnitudes that decrease monotonically with the increase of filter center frequency

- a bandpass magnitude spectrum profile of landmine points that has a central peak, and decreases monotonically when the spatial locations are away from the peak.

These phenomena indicate that a landmine candidate detector should be isotropic bandpass. Further, the spatial shape of landmines for a low resolution IR landmine is consistent with that (deterministic Gaussian shape (Lundberg & Gu, 2000)) in a high resolution IR image. Based on these characteristics, we formulate a detection filter that is rotational symmetric and bandpass in nature.

10.4   Multi-scale point detection for landmine candidates

10.4.1 Multi-scale landmine candidate point detector

In this section we first describe a multi-scale detector formed by a multi-stage anisotropic 2D bandpass filter (Gu & Tjahjadi 2001, 2002a), a special case of the detectors that can be formed from a family of isotropic bandpass filters (Gu & Tjahjadi, 2002b). It is worth noting that under this particular bandpass filter kernel, the detector is similar to the scale-invariant feature transform (SIFT) described in Lowe (1999, 2004). However, SIFT introduces a feature vector for each detected point which can be used for point matching or corresponding. We then describe a post-processing method that refines the detected candidate landmines by using their feature vectors as described in SIFT, followed by choosing RANdom SAmple Consensus (RANSAC) described in Fischler and Bolles (1981), which is designed to find consensus among landmine candidate points through image frames, while some spurious points due to clutter or noise can be further removed.

Specifications of a generic single-scale filter

The following specifications for a generic 2D bandpass filter suitable for detecting low-resolution landmines should be satisfied:

- The magnitude frequency response is rotationally symmetric (or isotropic).
- The radial center frequency is set at low frequency.
- The attenuation at the zero frequency is high (i.e., exclude the d.c.).
- The impulse response has a single peak, monotonically decreasing when away from the peak location.
- The filter has a linear-phase, or nearly linear-phase.
- The impulse response of the filter is real, symmetric with an effective finite support.

It is worth noting that the filter can be either FIR (finite impulse response) or IIR (infinite impulse response) with a small effective support in the spatial domain.

Furthermore, the (nearly) linear-phase is necessary when extending to a multi-scale detector.

A sample multi-scale detector

This section describes a sample multi-scale (e.g., L-scale) detector, which satisfies the above generic specifications. As shown in the block diagram of Fig. 10.1, a multi-scale detector is formed by repetitive cascading (L times) of the same single-scale detector followed by thresholding and peak picking. We focus on describing the fundamental building block of a single-scale.

Although many different types of bandpass filters satisfying the required specifications exist (Gu & Tjahjadi, 2002b), a straightforward way to formulate a bandpass filter in a single-scale detector is to use the difference of two 2D isotropic lowpass filters, being the same type but with different bandwidths (Lu & Antoniou, 1992):

$$g_k(x, y) = g_0^{(1)}(x, y) - g_0^{(2)}(x, y) \qquad [10.1]$$

where $g_0^{(1)}(x, y)$, $g_0^{(2)}(x, y)$ are the 2D lowpass filter kernels with a wider and narrower frequency bandwidth, respectively, the suffix g_k indicates the radial frequency f_k of the filter, and $g_0^{(i)}$ is a lowpass filter with an isotropic 2D Gaussian kernel,

$$g_0^{(i)}(x, y) = \frac{1}{2\pi\sigma_i^2} e^{-(x^2 + y^2)/2\sigma_i^2}, \quad i = 1,2 \qquad [10.2]$$

where $\sigma_2 = scale \cdot \sigma_1$, $scale > 1.0$. The 2D bandpass filter in [10.1] has an effective finite support despite its infinite support. To reduce the computation, the kernel is truncated after the absolute value of normalized kernel envelope becomes insignificant (i.e. below a small value). Figure 10.2 shows an example of the impulse response, magnitude spectrum and the 1D cross-section of such a 2D bandpass filter. It is worth mentioning that the parameter *scale* should be carefully selected so that the 2D filter magnitude spectrum attenuates all signal components at zero frequency (i.e., removes all d.c. elements). Failure to do so would lead to high false alarm in detection. Empirically, this requirement can be monitored by observing the 1D cross-section of the 2D magnitude response.

Introducing multi-scale into the detector is aimed at simultaneously introducing good detectability, localization, and noise robustness, i.e., all landmines should be

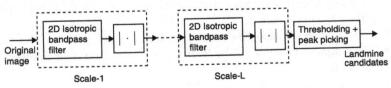

10.1 Block diagram of a multi-scale (L-scale) landmine detector, obtained by cascading a single-scale detector in L-times.

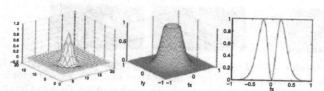

10.2 An isotropic 2D bandpass filter in Fig. 10.1. Left to right: 2D filter impulse response; 2D filter magnitude spectrum; the 1D cross-section of the 2D magnitude spectrum. The parameters used: $\sigma_1 = 1.4$, scale = 2.0, and the filter impulse response is truncated when the normalized amplitude envelop $\leq 10^{-5}$.

reliably detected; the spatial locations of detected mines should be as close as possible to their true locations; and the landmine detection should be resilient to noise.

Since the spread of measured mines could be different (e.g., due to variation in measuring height, different size of landmines) their contrast to backgrounds could be different due to variations in mine temperatures (e.g., due to sun radiation and heat transfer from/to different background materials such as soil, sand and gravel roads). It is difficult to set in advance a best suitable scale in the detector that matches the desired landmine spread pattern so as to result in the 'best' detectability. Further, good detectability and good localization cannot be simultaneously achieved because the product of spatial resolution and frequency resolution remains a constant governed by the uncertainty principle (Flandrin, 1999). To reliably detect landmine-like objects, the filter in the detector requires high attenuations at zero frequency and a very low passband center frequency, hence a very narrow transition bandwidth at the lower frequency. To satisfy this, the filter requires a long kernel length. However, a long filter kernel length implies a low spatial resolution, hence low accuracy for landmine position estimates. Such conflicting requirements are achievable by employing a multi-scale detector. In the multi-scale detector, detecting landmine candidates is done as follows: starting from the finest scale (level-1), if landmine candidates in the output image are not prominent against their surroundings, coarser scales are added by repetitively cascading one additional single-scale detector from the previous finer scale, until the 'best' scale level $L_{optimum}$ is reached, or a pre-determined maximum level L_{max} is reached. Since significant landmine candidates become more pronounced in a coarse scale when most spurious noise is removed, it is easier to detect candidates from local peaks at a coarser level. Once local peaks in the coarsest level are detected, locations of mine candidates are then determined by the inter-scale peak position tracing where better localizations are found in finer scales. Figure 10.3 shows an example of the detectability and localizations.

An automatic scale level selection method for determining the best scale levels $L_{optimun}$ was proposed (Gu & Tjahjadi, 2002a), which is based on the average local signal-to-noise ratio (SNR) estimates over all candidates in each scale. Let the

10.3 An example of multi-scale landmine detection showing that feature points are more pronounced in the coarser spatial scale detector but with more spread (or, less localization). From left to right: Original image, output image from a fine scale *l* = 2, output image from a coarse scale *l* = 3 (before peak-picking). The parameters used: σ_1 = 1.2 and σ_2 = $2\sigma_1$ (fine scale), σ_1 = 1.5 and σ_2 = $3\sigma_1$ (coarse scale) and threshold = 0.25.

noise power of a mine candidate be estimated from an annular region between the inner radius R_I and outer radius R_O centered at its peak position and the signal power be estimated from the output intensity at the peak position. The average local SNR at level *l* is computed by averaging local SNRs over all candidates as:

$$SNR_{avg}(l) = \frac{1}{N_{p_i(l)}} \sum_{i=1}^{N_{p_i(l)}} 10 \log \left\{ \frac{\tilde{I}_l^2(p_i(l))}{\sum_{m \in Region[R_I, R_O]} \tilde{I}_l^2(m)/N_i} \right\} \quad [10.3]$$

where $\tilde{I}_l(m)$ is the filtered image at scale level *l*, $p_i(l) = (x_i(l), y_i(l))$ the peak position of the *i*th detected candidate at level *l*, $N_{p_i}(l)$ the total number of detected candidates, N_i the total number of pixels in the region $[R_I, R_O]$ covered by the area between the inner and outer radius centered at $p_i(l)$. An optimal level $L_{optimun}$ is selected once $SNR_{avg}(l) \geq th_{SNR}$ is satisfied.

Relations to matched filters

It is shown in Gu and Tjahjadi (2002a) that such a detector can be interpreted by matched filters with the pattern associated with the thermodynamic landmine model. The impulse response of a 2D matched filter appears to be a spatially reversed replica of the landmine pattern being detected. Assuming $f(x,y), 0 \leq x \leq N_1$, $0 \leq y \leq N_2$, is a function of the pattern to be detected, the impulse response of the matched filter associated with this pattern is:

$$h(x, y) = f(N_1 - x, N_2 - y) \quad [10.4]$$

The matched filter output $I_O(x, y) = h(x, y)* I(x, y)$ results in a maximum local SNR value (Srinath & Rajasekaran, 1979), if a desired pattern is matched. This happens when the local region of the image $I(x,y)$ coincides or maximally correlates with the desired pattern, i.e. the bandpass filter that resembles a frequency-modulated and spatially truncated Gaussian function.

Interpreting from the detection theory

We apply the following hypothesis tests to a local image region R containing a maximum of one landmine $f_l(r)$,

$$H_0: I(r) = n(r)$$
$$H_1: I(r) = f_i(r) + n(r)$$

[10.5]

where $r = (x,y)$, $I(r)$ is the original image and $n(r)$ is the noise. Assuming Gaussian independent and identically distributed noise $N(0, \sigma^2)$ and deterministic landmine patterns $f_i(r)$, the probability density functions of $I(r)$ under H_0 and H_1 hypotheses respectively become:

$$p_I(r|H_0) = \prod_{r \in R} \frac{1}{\sqrt{2\pi}\sigma} e^{-\frac{I^2(r)}{2\sigma^2}}, p_I(r|H_1) = \prod_{r \in R} \frac{1}{\sqrt{2\pi}\sigma} e^{-\frac{(I(r)-f_i(r))^2}{2\sigma^2}}$$

[10.6]

Based on the Neyman–Pearson approach (Kay, 1998) that maximizes the detection probability P_D under the constraint of false alarm probability $P_{FA} \leq \alpha$, the log-likelihood ratio for detecting landmines is:

$$\Lambda(I(r)) = \ln \frac{\prod_{r \in R} p_I(r|H_1)}{\prod_{r \in R} p_I(r|H_0)}$$

[10.7]

Substituting [10.6] into [10.7] with some manipulations yields,

$$\Lambda(I(r)) = \frac{1}{\sigma^2} \sum_{r \in R} \left(I(r)f_i(r) - \frac{f_i^2(r)}{2} \right) \underset{H_0}{\overset{H_1}{\underset{<}{>}}} \eta$$

[10.8]

where η is dependent on the false alarm threshold α, $\sum_{r \in R} f_i^2(r)$ is the energy of the l feature. Since the landmine detection filter acts as a matched filter, and the output image $\tilde{I}(r)$ from the matched filter has an increased SNR (i.e., increased contrast between landmine points and background), $p_{\tilde{I}}(r|H_0)$ and $p_{\tilde{I}}(r|H_1)$ are more separable after applying the detector. From the above analysis, one may see the limitation of such a detector, i.e. it works when the noise is white and does not contain clutter. If the noise contains both clutter and white noise, $n(r) = n_c(r) + n_0(r)$, [10.6] does not hold since $I(r|H_0)$ and $I(r|H_1)$ in [10.5] are no longer Gaussian distributed. Explicit expressions of the likelihood could be difficult to obtain. Obviously, if the clutter and landmines have similar patterns, the clutter would also be enhanced and detected. To prevent such cases, a post-processing step is necessary to further analyze the detected candidates.

10.4.2 Post-processing for refining landmine candidates

This section describes a post-processing method aimed at refining the landmine candidates detected by removing some clutter from the candidates so as to reduce the false alarm. This can be described as jointly using SIFT descriptors of the detected candidates for point correspondences, and RANSAC for finding the subset of candidates that best fit for a given model. We first briefly describe SIFT descriptors and RANSAC, and then the joint post-processing scheme.

SIFT descriptors as landmine candidate features

The aim of employing SIFT features to landmine candidates is to find corresponding landmine candidates in two image frames. For each landmine candidate point obtained from Section 10.4.1, a SIFT feature vector is added according to the definition of SIFT keypoint descriptor (Lowe, 2004). This can be briefly described as follows. For each detected point (or keypoint), a feature vector (or descriptor) of size 128 is formed by first computing the gradient magnitude and orientation at each image sample point in a 16 × 16 pixel region centered in the keypoint and weighted by a Gaussian window with the variance of 1.5 times the keypoint scale. The region is then divided into 16 sub-regions, each of size 4 × 4. For each of the 16 sub-regions, samples are accumulated to form an orientation histogram, consisting of eight orientation bins starting from 0° where the interval between two neighbor bins is 45°. The value in each bin is obtained as the sum of the gradient magnitudes near that bin's orientation within the 4 × 4 sub-region. In this way, a feature vector of size 16 × 8 = 128 is formed for each detected peak point. An example of an 8 × 8 region case is shown in Fig. 10.4.

The correspondences between the detected landmine candidates in two different image frames are found by using the detected landmine candidates together with their SIFT features in the respective frames, where the images may undergo a transformation, e.g. affine transforms and projective transforms.

Consensus correspondence points

RANSAC (Fischler & Bolles, 1981) is an iterative method to estimate parameters of a mathematical model from a set of observed data points containing outliers. Lately, RANSAC has been increasingly used (Khan et al., 2009; Wang et al., 2009) and has become a standard for robust estimation of model parameters in images (Sonka et al., 1998). It produces a reasonable result with a certain probability that increases with more iterations. A basic assumption is that data consists of inliers and outliers. The distribution of inliers can be described by parameters, while outliers do not fit the model. The outliers are due to measurement

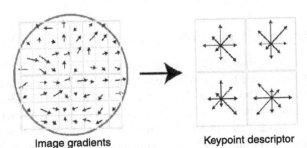

Image gradients   Keypoint descriptor

10.4 SIFT feature descriptor for a keypoint (from Lowe, 2004).

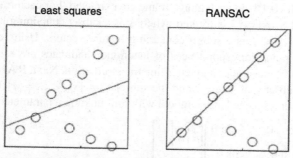

10.5 Using least squares and RANSAC for line estimation from a given set of points.

noise or incorrect hypotheses. RANSAC also assumes that, given a small set of inliers, it may estimate the parameters of a model that optimally explains or fits the hypothetical inliers. A simple example of RANSAC is the line estimation using a set of points containing inliers and outliers. As shown in Fig. 10.5, the least squares (LS) estimation performs poorly, while RANSAC, seeking maximum consensus points under a line model, results in a better line with more consensus points on it and better estimated parameters of the line.

For a given fitting problem and the model, RANSAC is performed as follows:

1. Randomly choose the minimum number of points needed to estimate the model.
2. Estimate the model parameters using these points.
3. Find the number of data points agreeing with the estimated model (within a pre-specified tolerance error)
4. Repeat Steps 1 to 3 for a predetermined number of times.
5. Choose the largest set of consensus points, and re-estimate the model parameters.

RANSAC is employed to find a subset of consensus points from the given set of landmine candidates in different image frames.

Maximum consensus of landmine candidate correspondences by jointly using SIFT features and RANSAC

To reduce the false alarm of the detected landmine candidates, we employ a post-processing scheme to refine the detection of landmine candidates. This scheme seeks a subset of maximum consensus landmine candidates by jointly using landmine candidates with their SIFT features and RANSAC. Most spurious candidates, due to noise or clutter, do not fit well with the given model, hence these would be removed.

The joint scheme can be described as follows. First, SIFT feature vectors associated with the detected landmine candidates (from 'A sample multi-scale detector' in Section 10.4.1) in each image frame are extracted. For each frame (e.g. frame t), a data set $\mathbf{S}_t = \{s_t, f_t\}$ is formed, where s_t is a subset containing landmine candidate positions, and f_t is a subset containing feature vectors. Using data sets from two frames, point correspondences of landmine candidates are established between these two frames under a given affine transformation. Next, RANSAC is employed to the subset s_t at frame t, and the subsets $S = \{s_{t-L}, \dots s_{t+L}\}$ from the nearby frames under a selected affine model with four unknown parameters,

$$\begin{bmatrix} x \\ y \end{bmatrix} = \beta \begin{bmatrix} \cos\theta & -\sin\theta \\ \sin\theta & \cos\theta \end{bmatrix} \begin{bmatrix} \tilde{x} \\ \tilde{y} \end{bmatrix} + \begin{bmatrix} dx \\ dy \end{bmatrix} \qquad [10.9]$$

where (x,y) and $(\tilde{x}, \tilde{y})$ are correspondences from two different frames, θ is the rotation angle, β is the scaling factor, and (dx, dy) are the translation between the two frames. The parameters can be estimated by the matrix inversion (for $k = 2$), or the least squares solution (for $k > 2$) of [10.5] using the selected correspondences,

$$\begin{bmatrix} \tilde{x}_1 & -\tilde{y}_1 & 1 & 0 \\ \tilde{y}_1 & \tilde{x}_1 & 0 & 1 \\ \vdots & & & \vdots \\ \tilde{x}_k & -\tilde{y}_k & 1 & 0 \\ \tilde{y}_k & \tilde{x}_k & 0 & 1 \end{bmatrix} \begin{bmatrix} \beta\cos\theta \\ \beta\sin\theta \\ dx \\ dy \end{bmatrix} = \begin{bmatrix} x_1 \\ y_1 \\ \vdots \\ x_k \\ y_k \end{bmatrix} \qquad [10.10]$$

where $\{(x_1, y_1), \dots, (x_k, y_k)\}$ and $\{(\tilde{x}_1, \tilde{y}_1), \dots, (\tilde{x}_k, \tilde{y}_k)\}$ are subsets of k correspondence points selected from the two frames. Choosing the affine model in [10.9] is based on the assumption of how images were measured. Under this model, we first use RANSAC to seek the maximum consensus landmine candidates between the current frame t and one neighboring frame t', $t' = t - L, \dots t - 1, t + 1, \dots t + L$. The total number for each landmine candidate being selected by RANSAC in 2L frames is then accumulated. Those consensus candidate points, the accumulated numbers of which exceed a given threshold, are then chosen as the refined landmine candidates. The remaining candidates are removed as being clutter points in order to reduce the false alarm.

10.5   Landmine detection from close distance measured IR images: thermal contrast-based models

Once a landmine field is located, each landmine candidate detected from low-resolution images needs to be zoomed in and further analyzed. This section briefly describes a thermal contrast-based method that aims to detect landmines from high-resolution IR images.

Studies were performed for shallowly buried landmines and cameras mounted at 2–3 meters above the test bed site. The thermal signature of buried mines is obtained from simulations of heat transfer in a finite volume containing a landmine and by incorporating boundary conditions, using the finite element method (FEM) (Sjokvist et al., 1998; Sjokvist, 1999). The model is used to determine whether there is sufficient contrast to detect landmines, but is not necessarily used for the detection. Based on this FEM, the thermal signatures related to buried landmines centered at the coordinate origin can be described by a deterministic 3D Gaussian shaped function in a normalized IR contrast image sequence (Lundberg & Gu, 2000, 2001):

$$q(r, t) = A\, e^{-r^2/B}\, e^{-(t-D)^2/C} *s(r) + n(r, t) \qquad [10.11]$$

where $r = (x,y)$, and $s(r)$ is a 2D signal based on physical landmine shape, $s(r,t) = A\, e^{-r^2/B}\, e^{-(t-D)^2/C} *s(r)$, A is a scaling factor, B is the spatial spread of landmine signature, C and D describe the temporal change of contrast, $n(r,t)$ is the background noise, * is linear convolution, and $q(r,t)$ is the modelled IR contrast image sequence. The actual contrast image sequence is computed from the IR image sequence $I(r,t)$ using:

$$q_{\text{actual}}(r, t) = \frac{I(r, t) - E(I(r, t; \text{background only}))}{\sqrt{E(I(r, t) - E(I(r, t; \text{background only}))^2}} \qquad [10.12]$$

where I(r, t; background only) is the original image sequence measured without the presence of landmines, and E(I) is the expectation of I over t. This removes the mean intensity and normalizes the standard deviation at a different time instant. Considering one IR image frame with a landmine in the center, [10.11] is degenerated to a 2D isotropic Gaussian-shaped function with additive background noise, $q(r, t) = Ae^{-r^2/B} *s(r) + n(r, t)$. Figure 10.6 shows an example of a real measured IR image and a synthetic one generated from the FEM model. Figure 10.7 compares the spatial temperature contrasts generated from the FEM and the Gaussian shape functions. These figures show that the Gaussian shape functions agree rather well with the FEM model. Furthermore, synthetic images from the Gaussian shape functions are close to the measured IR images (after noise removal).

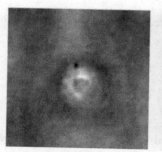 

10.6 Left: measured IR image for a shallowly buried plastic mine (<5mm); Right: synthetic image generated from the thermal contrast model based on the FEM (Lundberg & Gu, 2001).

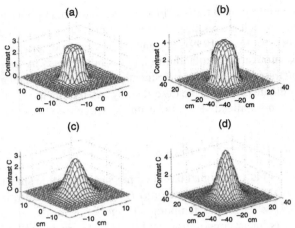

10.7 Comparisons between the spatial temperature contrasts of landmines modeled by the FEM and by the Gaussian shape function in (5.11). Row-1: from FEM; Row-2: from Gaussian function. Column-1: for a shallowly buried (<5mm) plastic AP-mine; Column-2: for an AT-mine buried at 5cm (from Lundberg & Gu, 2001).

It is also worth noting that the spatial shape pattern of this model agrees with the pattern used for landmine candidates in the low-resolution aerial IR images (Section 10.4). In fact, if the landmine is very small due to far-distance sensors and low image resolution, the spatial spread of Gaussian shape function is reduced (i.e., $s(x,y,t)$ in [10.11] is close to a δ function), hence the resulting landmine patterns agree with those of landmines in IR images measured from far distances. Similar to the low-resolution image case, this model does not distinguish landmines from clutter (for instance, stones) if they have a similar spatial shape pattern.

Hypotheses similar to [10.5] can be applied under this model for detecting landmines, i.e.,

$$H_0: q(r, t) = n(r, t)$$
$$H_1: q(r, t) = s(r, t) + n(r, t)$$

[10.13]

The log-likelihood ratio test (or, Neyman–Pearson detector) is shown to be

$$\Lambda(\mathbf{q}) = \mathbf{q}^T \mathbf{C}^{-1} \mathbf{s} = (\mathbf{C}^{-1/2} \mathbf{q})^T (\mathbf{C}^{-1/2} \mathbf{s}) \underset{\underset{H_0}{<}}{\overset{\overset{H_1}{>}}{}} \eta$$

[10.14]

where $\mathbf{q}, \mathbf{s}, \mathbf{n}$ are column-scanned vectors of $q(r,t)$, $s(r,t)$ and $n(r,t)$, respectively, and C is the covariance matrix of $\mathbf{n}$ (Lundberg & Gu, 2001).

More complex models for detecting landmines include, e.g. Lundberg (2003), detectors based on the generalized likelihood test (GLRT) where unknown parameters are introduced to landmines and noise; or, utilizing IR and visual band images for landmine detection where the visual band is used to reduce noise caused by ground surface clutter.

10.6   Experiments and results

10.6.1 Results from airborne or vehicle-borne IR images

IR cameras and measurements

IR image measurements were generated and kindly provided by the Swedish Defence Research Agency (FOI). The FOI used two passive sensor systems to obtain the IR measurements: one is an Agema 900 Thermovision camera system operating in the infrared wavelengths of 3–5 and 8–12 mm; the other is a multispectral vision system consisting of six narrow spectral bands in the wavelength ranging from 200 nm to 800 nm (Christiansen & Ringberg, 1996).

Test data sets and landmine and landmark settings

Test set (a): Vehicle-borne IR image sequences were measured by IR cameras mounted at 10 m height above the ground surface in the test site of FOI laboratory during a 24-hour period. The desired target feature points correspond to surface-laid and shallowly buried anti-personnel and anti-tank mines. One frame of IR images and the ground truth are shown in Fig. 10.8. The ground truth image in Fig. 10.8 contains several surface-laid anti-personnel mines (top row) and shallowly buried anti-tank mines (bottom row). Observations of measured IR images do not show clear visible contour information of the targets (landmines) due to the low spatial resolution. Landmines are associated with pixels with intensity differences to their surroundings due to different temperatures, or signs of structural changes in the soil surface due to digging.

Test set (b): Airborne measured IR images were captured by IR cameras on a helicopter at approximately 180 m above the ground surface when scanning one of the landmine test sites in Eksjo, Sweden. Each of the images contained more than a dozen surface-laid anti-tank mines between the two columns of man-made

10.8 Left: the ground truth estimated from visual observations through the IR image sequence (white circles: surface-laid landmines; black circles: buried anti-tank mines, two large dash line circles: large mines or stones). Middle and right: two IR landmine images (8–11 mm) of the same scene measured at different time of the day, using a stationary vehicle-borne IR camera at 10 m height.

10.9 Left: Ground truth marked on one image, containing man-made landmarks and landmines. White circles at the left column: square panel marks; black circles at the right column: astro turf marks; black rectangles at the bottom row: corn marks; small black rectangles in between the two columns: surface-laid anti-tank mines. Middle and right: two image frames from an IR image sequence (3–5 µm), were captured at slightly different scenes from a test bed site by an IR camera mounted on a helicopter scanning a large area of test bed landmine field at an altitude of 180 m.

10.10 Detected landmine candidates from an IR image sequence from vehicle-borne IR camera (3–5 mm) at 10 m height. First row: original image frames #1, 203, 493, 1582, 1723; and 2nd row: detected landmine candidates (marked with white rectangles) using the multi-scale detector. Parameters used: $\sigma_1 = [1.2, 2.0]$, *scale* = 2.0, threshold = [0.2, 0.35].

landmarks (i.e., the square panel, the astro turf and the corn marks). Figure 10.9 shows one frame of the IR images containing the above setting scenario.

Results

For test set (a): First, the multi-scale detector in Section 10.4.1 was applied to vehicle-borne IR images. Figure 10.10 shows some of the results from five image frames extracted from the vehicle-borne IR sequence in Fig. 10.8. One can see that the contrast of landmines and background is quite small in these frames. The cross digging signs are weakly visible, and the signatures of the landmines in the top row are very weak. One can observe from the results in the second row that the multi-scale detector has successfully picked up some of the landmines, however, some clutter points are also included as landmine candidates, probably due to their similar pattern to landmines.

Figure 10.11 shows another test example, where 10 frames of IR images in another time interval were extracted from the same sequence shown in Fig. 10.8, where the contrast between landmines and the background is somewhat improved. This time, the post-processing scheme in 'Maximum consensus of landmine candidate correspondences by jointly using SIFT features and RANSAC' in Section 10.4.2 was added after applying the multi-scale detector.

Observing the results in Fig. 10.11 shows that the multi-scale detector in this case has yielded reasonably good results in detecting anti-personnel mines on the

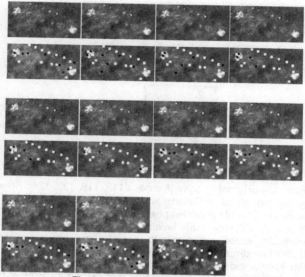

Final candidates based on 80% consensus

10.11 Rows 1, 3, 5: Ten original IR image frames in [#989, #998].
Rows 2, 4, 6: detected landmine candidates using the multi-scale
detector marked by black circles plus white rectangles. Parameters used:
$\sigma_1 = 1.4$, *scale* = 2.0, total scale levels $L = 4$, and threshold = 0.15. The
post-processing scheme results in the selected maximum consensus
candidates of landmarks and landmines marked by white rectangles. A
candidate is selected as a consensus point if it appears in ≥80% of the
scenes under consideration. The last image in the 6th row is the final
candidates based on consensus information in these 10 frames.

top row (with intensity difference), and also located several buried anti-tank
mines on the bottom row (mainly shown by cross digging signs). However, some
clutter points are also included. It is also observed that after the post-processing,
seeking consensus landmine candidates through frames, the final results have
been further improved (indicated by selected white rectangles).

For test set (b): The multi-scale detector in Section 10.4.1 was applied to
airborne IR images, where the desired targets correspond to man-made landmarks.
These landmarks were set for locating regions of interest. Figure 10.12 shows
some resulting frames extracted from the airborne IR sequence in Fig. 10.7. The
results from Fig. 10.12 show that the proposed multi-scale detector is effective in
detecting man-made landmark candidates (marked by while rectangles and black
circles). Similarly to the previous two examples, some clutter points are also
detected. Further, it is noticed that almost none of the anti-tank mines are detected,
mainly due to the too low spatial resolution of landmines using the given IR
cameras at the measurement altitude. Further, after applying the post-processing
scheme, the landmark and landmine candidates are chosen from the consensus

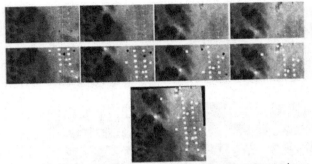

A panorama containing the final results after post-processing

10.12 Row-1: 4 original images (frames #115, 118, 123, 124); Row-2: detected candidates of landmarks and landmines using the multi-scale detector marked by black circles plus white rectangles. Parameters used: $\sigma_1 = 0.9$, *scale* = 2.0, total scale levels $L = 4$, and threshold = 0.24. The post-processing scheme results in the selected maximum consensus candidates of landmarks and landmines marked by white rectangles. A candidate is selected as consensus point if the candidate appears in more than 45% of the scenes containing that candidate. Row-3: the final results after the post-processing, all detected landmine and landmark candidates are marked by white rectangles in a panorama image (through image registration of the above four images).

using RANSAC, which further improves the detection results (marked by white rectangles only).

Performance evaluation

Automatic scale level selection criterion: To evaluate the effectiveness of the automatic scale level selection criterion in [10.3], 12 output images from a multi-scale detector were used to compute the average SNRs, with parameter settings $th_{SNR} = 20$ dB, $L_{max} = 7$, $R_I = 4$ and $R_O = 8$. Table 10.1 shows the results. A threshold SNR = 20 dB was empirically set for choosing the scale level L.

Detection and false alarm for the multi-scale detector: A rough evaluation was only performed on the multi-scale detector, in terms of the average detection rate and average false alarm rate over several image frames in the above two test sequences. The results are summarized in Table 10.2. From Table 10.2, one

Table 10.1 Estimated local SNRs in each level of a multi-scale detector

Spatial scale level	1 (finest)	2	3	4	5	6	7 (coarsest)
Average SNR (dB)	8.9581	11.4414	17.5021	19.8541	21.4127	21.7598	22.1895

Table 10.2 Performance of the multi-scale detector, averaged over several image frames from the two test sets

	Landmines	Landmarks
Average detection rate PD_D	0.7009	0.8190
Average false alarm rate P_{FA}	46.31	29.20

can observe that a multi-scale detector may provide a reasonably good detection rate in terms of the poor image quality. It can also be observed that the false alarm rate is relatively high. Although we have not performed any quantitative evaluation after the post-processing scheme, from the visual observations (see Fig. 10.11 and 10.12), the false alarm has been significantly reduced after the post-processing.

10.6.2 Results from close distance measured IR landmine images

The detector in Section 10.5 was applied to several measured IR images generated from the test box in FIO laboratory. Figure 10.13 shows two sets of results. The results in Fig. 10.13 show that the detector based on the thermal contrast model is very effective: buried landmines and objects in the IR images have been significantly enhanced, making the detection a much easier task. It is worth noting that these test bed examples are somewhat simplified versions of the real world landmines that had been buried for a long time. However, it is a positive step towards landmine detection and clearance.

10.7 Conclusions

For landmine field detection by detecting landmine candidates from low-resolution airborne or vehicle-borne measured IR images, the first part of this chapter is focused on describing a scheme that uses a multi-scale landmine detector followed by a refining process, seeking maximum consensus of corresponding landmine

10.13 Enhanced landmines from measured IR images. Two left images: a measured IR image containing a plastic land mine buried at 5mm depth, and the enhanced landmine of the corresponding image; two right images: a measured IR image containing a block of beeswax in the left and a plastic mine in the right buried at 3 cm depth, and the corresponding enhanced images (from Lundberg & Gu, 2001).

candidates through image frames. In the detector, the number of scale levels is automatically selected based on the average local SNRs. The detector significantly enhances the landmine signatures, making the detection feasible and more reliable. However, the number of selected landmine candidates is somewhat sensitive to the threshold setting for peak picking. Also, clutter with similar patterns to landmines can be picked up, causing false alarms. It is shown that these problems are mitigated by the post-processing of combining SIFT keypoint descriptors and RANSAC. This has led to an improved and refined selection of landmine candidates by removing inconsistent candidates to a given model, hence reducing false alarms. This has also led to a less sensitive threshold selection for the detector. Experimental results have shown the combined scheme is fairly robust for detecting landmine candidates in low-resolution images, which is key for locating regions of interest where landmine fields may exist. Furthermore, close zoomed-in examinations of the existence of landmines are required.

For landmine detection from high-resolution IR images measured from close distances to ground surfaces, the second part of the chapter (Section 10.5) contains a brief description of a thermal contrast model-based detector. It is interesting to note the associations between these two detectors, even though they are used for IR images from very different distance ranges of scenarios. As a matter of fact, the spatial patterns used for this detector are in agreement with the spatial patterns of the detector for low-resolution IR images. Furthermore, both detectors can be considered as matched filters, and interpreted by the detection theory. Experiment results have shown that the detector is fairly effective in enhancing mine signatures and hence for detecting mines. References for more sophisticated methods of landmine detection are also given.

10.8   Acknowledgement

The landmine data were kindly provided by the Swedish Defence Research Agency (FOI), Sweden.

10.9   References

Baertlein B & Gunatilaka A (1998), Improving detection of buried mines through sensor fusion. *Proceedings of SPIE Conference on Detection and Remediation Technologies for Mine and Minelike Targets III*, SPIE 3392, 1122–1133.

Bishop P K, Perry, K M. & Poulter M A (1998), Airborne minefield detection, *Proceedings of 2nd Int'l IEE Conf. on the Detection of Abandoned Land Mines*, Edinburgh, UK.

Bruschini C, De Bruyn K, Sahli H & Cornelis J (1999), Study on the state of the art in the EU related humanitarian de-mining technology, product and practice. *EUDEM report*, Vrije Univ. Brussel, Belgium.

Bruschini C (2004), On the low-frequency EMI response of coincident loops over a conductive and permeable soil and corresponding background reduction schemes. *IEEE Trans. Geoscience and Remote Sensing*, 42(8), 1706–1719.

Caniou J (1999), *Passive Infrared Detection, Theory and Application*, Kluwer Academic Publishers.

Casasent D & Ye A (1997), Detection filters and algorithm fusion for ATR, *IEEE Trans. Image Processing*, 6(1), 114–125.

Chan D S K, Langan D A. & Staver, D A (1990), Spatial processing techniques for the detection of small targets in IR clutter, *Proceedings of SPIE*, 1305, 53–62.

Christiansen A L & Ringberg S (1996), Optical mine reconnaissance at the National Defense Research Establishment Multispectral Imaging and Classification: Thermodynamic soil modeling, *Proc. IEE Conf. Detection of Abandoned Landmines*, 97–101.

Earp S, Hughes E, Elkins T & Vickers R (1996), Ultra-wideband ground-penetrating radar for the detection of buried metallic mines. *IEEE Aerospace and Electronics System Magazine*, 11(9), 14–17.

EU-IST Project Description (2000), ARC: Airborne minefield area reduction, GEOSPACE, GTD, SCHIEBEL, CROMAC, FOI, TNO, IMEC-ETRO.

Faust A A, Chesney R H, Das Y, McFee J E & Russell K L (2005), Canadian teleoperated landmine detection systems. Part I: The improved landmine detection project. Part II: Antipersonnel landmine detection. *International Journal of Systems Science*, 36(19), 511–543.

Ferrara C F, Fries R W, Rushnow W B & Mansur H H (1998), Adaptive signal processing for the detection of point targets in non-stationary clutter, *Proceedings of IRIS Specialist Group on Targets, Backgrounds and Discrimination*, II, 187–198.

Filippis A, Jain L & Martin N (1999), Using generic algorithms and neural networks for surface land mine detection. *IEEE Trans. Signal Processing*, 27, 176–186.

Fischler M A & Bolles R C (1981), Random sample consensus: a paradigm for model fitting with applications to image analysis and automated cartography, *Commun. of the ACM*, 24, 381–395.

Flandrin P (1999), *Time-Frequency/Time-Scale Analysis*, Academic Press, New York.

Gu I Y H & Tjahjadi T (2001), Detection of landmine candidates from airborne and vehicleborne IR images, in *Proc. of Detection and Remediation Technologies for Mines and Minelike Targets VI*, SPIE 4394, part one, 275–283, Orlando, USA.

Gu I Y H & Tjahjadi T (2002a), Detecting and locating landmine fields from vehicle- and air-borne measured IR images, *Pattern Recognition*, 35, 3001–3014.

Gu I Y H & Tjahjadi T (2002b), Multiresolution Feature Detection Using a Family of Isotropic Bandpass Filters, *IEEE trans. Systems, Man, and Cybernetics, part B*, 32(4), 443–454.

Ho K C & Gader P D (2002), A linear prediction land mine detection algorithm for hand held ground penetrating radar. *IEEE Trans. Geoscience and Remote Sensing*, 40(6), 1374–1384.

Ho K C, Carin L, Gader P D & Wilson J N (2008), An investigation of using the spectral characteristics from ground penetrating radar for landmine/clutter discrimination. *IEEE Trans. on Geoscience and Remote Sensing*, 46(4), 1177–1191.

Jin T & Zhou Z (2007), Ultrawideband synthetic aperture radar landmine detection. *IEEE Trans. on Geoscience and Remote Sensing*, 45(11), 3561–3573.

Jin T & Zhou Z (2008), Feature extraction and discriminator design for landmine detection on double-hump signature in ultrawideband SAR. *IEEE Trans. Geoscience and Remote Sensing*, 46(11), 3783–3791.

Kay S (1998), *Fundamentals of Statistical Signal Processing: Detection Theory*, Prentice Hall, New Jersey.

Khan Z H, Gu I Y H, Wang T & Backhouse A (2009), Joint Anisotropic Mean Shift and Consensus Point Feature Correspondences for Object tracking in Video, IEEE ICME'09, New York, USA.

Khanafer K & Vafai K (2002), Thermal analysis of buried land mines over a diurnal cycle. *IEEE Trans. Geoscience and Remote Sensing*, 40(2), 461–473.

Khanafer K, Vafai K & Baertlein B A (2003), Effects of thin metal outer case and top air gap on thermal IR images of buried antitank and antipersonnel land mines. *IEEE Trans. Geoscience and Remote Sensing*, 41(1), 123–135.

Kovalenko V, Yarovoy A G & Ligthart L P (2007), A novel clutter suppression algorithm for landmine detection with GPR. *IEEE Trans Geoscience and Remote Sensing*, 45(11), 3740–3751.

Lopez P, Van Kempen L, Sahli H & Ferrer D C (2004), Improved thermal analysis of buried landmines. *IEEE Trans. Geoscience and Remote Sensing*, 42(9), 1965–1975.

Lowe D G (1999), Object recognition from local scale-invariant features. *Proceedings of the International Conference on Computer Vision (ICCV 99)*, 2, 1150–1157.

Lowe D G (2004), Distinctive image features from scale-invariant keypoints, *International Journal of Computer Vision*, 20, 91–110.

Lu W S & Antoniou A (1992), *Two-Dimensional Digital Filters*, Marcel Dekker, Inc.

Lundberg M & Gu I Y H (2000), A 3D matched filter for detection of land mines based on spatio-temporal thermal model, *Proc. SPIE Conf. Detection and Remediation Technologies for Mines and Minelike Targets V*, SPIE 4038(1), 179–188.

Lundberg M & Gu I Y H (2001), Infrared detetion of buried land mines based on texture modeling, *Proc. of SPIE conf. Detection and Remediation Technologies for Mines and Minelike Targets VI*, SPIE 4394(1), 275–283.

Lundberg M (2003), Land mine detection using dual-band electro-optical sensing, PhD thesis, Dept. of Signals and Systems, Chalmers Univ. of Technology, Sweden.

Manjunath B S, Shekharm C & Chellappa R (1996), A new approach to image feature detection with applications, *Pattern Recognition*, 29(4), 627–640.

Mokhtarian F & Suomela R (1998), Robust corner detection through curvature scale space, *IEEE Trans. Pattern Analysis and Machine Intelligence*, 20(12), 1376–1381.

Morrow I L & Genderen P V (2002), Effective imaging of buried dielectric objects. *IEEE Trans. Geoscience and Remote Sensing*, 40(4), 943–949.

Ng W, Chan T C T, So H C & Ho K C (2008), Particle filtering based approach for landmine detection using ground penetrating radar. *IEEE Trans. Geoscience and Remote Sensing*, 46(11), 3739–3755.

Paik J, Lee C P & Abidi M (2002), Image processing-based mine detection techniques : A review. *Subsurface Sensing Technologies and Applications*, 3(3), 153–202.

Sahli M, Bruschini C & Crabble S (2006), Catalogue of advanced technologies and systems for humantarian demining. *Eudem 2 Technology Report*, Dept. Electron. Inf., Vrij Univ. Brussel, Belgium, v.1.3.

Savelyev T G, Kempen L V, Sahli H, Sachs J & Sato M (2007), Investigation of time-frequency features for GPR landmine discrimination, *IEEE Trans. Geoscience and Remote Sensing*, 45(1), 118–129.

Sjokvist S, Georgsonm M, Ringberg S, Uppsall M & Loyd D (1998), Thermal effects on solar radiated sand surfaces containing landmines – a heat transfer analysis, *Proceedings of International Conference on Computational Methods in Heat Transfer V*, Cracow, Poland, 177–187.

Sjokvist S (1999), Heat transfer modelling and simulation in order to predict thermal signatures – the case of buried land mines, Licentiate thesis (Thesis 796), Dept. of Mechanical Engineering, Linkoping University, Sweden.

Sonka M, Hlavac V & Boyle R (1998), *Image Processing, Analysis, and Machine Vision*, 2nd Ed. Chapman & Hall.

Srinath M D & Rajasekaran P K (1979), *An Introduction to Statistical Signal Processing with Application*, Wiley, New York.

Thanh N T, Sahli H & Hao D H (2007), *IEEE Trans. Geoscience and Remote Sensing*, 45(3), 656–674.

Torrione P & Collins L M (2007), Texture features for antitank landmine detection using ground penetrating radar. *IEEE Trans. Geoscience and Remote Sensing*, 45(7), 2374–2382.

Wang T, Gu I Y H, Khan Z H & Shi P (2009), Adaptive Particle Filters for Visual Object Tracking using Joint PCA Appearance Model and Consensus Point Correspondences, ICME 09, New York, USA.

Wu R, Clement A, Li J, Larsson E G, Bradley M, Habersat J & Maksymonko G (2001), Adaptive ground bounce removal. *Electronic Letters*, 37(20), 1250–1252.

Zhu S C, Liu X W & Wu Y N (2000), Exploring texture ensembles by efficient Markov chain Monte Carlo: toward a "trichromacy" theory of texture, *IEEE Trans. PAMI*, 22(6), 554–569.

11

GPS data correction using encoders and inertial navigation system (INS) sensors

S. A. BERRABAH and Y. BAUDOIN,
Royal Military School, Belgium

Abstract: This chapter introduces the increase in mobile robot positioning based on the global positioning system (GPS) using data from other sensors. Positioning using the GPS is determined, at any time, by measuring the time delay in a radio signal broadcast from several satellites, and using this and the speed of propagation to calculate the distance to the satellites. The position on earth is then calculated by triangulation of intersecting radio signals at the GPS receiver. Using the GPS for positioning is subject to several sources of errors: ionosphere and troposphere delays, signal multi-path, number of visible satellites, satellite geometry/shading, and so on. A typical civilian GPS receiver provides 6 to 12 meters accuracy, depending on the number of satellites available. This accuracy can be reduced to 1 m by using a differential GPS (DGPS) which employs a second receiver at a fixed location to compute corrections to the GPS satellite measurements. In order to increase the accuracy of the robot positioning, we use an extended Kalman Filter (EKF) to integrate data from the DGPS with data from an inertial navigation system (INS) and robot encoders. This will also allow kipping robot positioning even if no satellite is visible.

Key words: global positioning system, inertial navigation system, sensors, Kalman Filter, localization, data fusion.

11.1 Introduction

To reach a reasonable degree of autonomy, two basic requirements are needed: sensing and reasoning. Sensing is provided by an on board sensory system that gathers information about the robot itself and the surrounding environment. According to the environment state, the reasoning system must allow the robot to localize itself in the environment and to search for free paths.

The localization problem can be divided into two sub-tasks: global and local localization. In many applications an initial estimation of the robot pose (position and orientation) is known (supplied directly or indirectly from the user). During the execution of a task, the robot must update this estimate using measurements from its sensors. This is known as local localization or pose tracking.[3,10] Using only sensors that measure relative movements, the error in the pose estimate increases over time as errors are accumulated. Therefore external sensors are needed to provide information about the absolute pose of the robot. This is achieved by matching the sensors' measurements with a model of the environment. In pose tracking, if a good initial estimate is given, the correspondence or data

269

association problem is easier as it does not consider the entire space when looking for the correspondence, but rather only a relatively small region around the estimated pose.[4,10]

Global pose estimation[10,15] is the ability to determine the robot's pose in an a priori or previously learned map, given no other information than that the robot is somewhere on the map, i.e. it can handle the kidnapped robot problem, in which a robot is kidnapped and carried to some unknown location. Global localization is considerably more difficult than pose tracking because of the data association problem. The level of complexity of this task varies with the size of the environment, but also with the level of symmetry in the environment. It is only by integrating large amounts of data over time that these symmetries can be resolved.

The localization methods depend on the information used. In the localization methods using active beacons or landmarks, the absolute position of the robot is computed by triangulation or trilateration[11] from three or more active beacons or detected landmarks. Active beacons are transmitters, usually using light or radio frequencies, and they must be located at known sites in the environment. Landmarks are distinct natural or artificial features that the robot can recognize from its sensory input. The characteristics of the landmarks must be known and stored in the robot's memory before it starts using them for navigation.

Global vision for robot localization refers to the use of cameras placed at fixed locations in a work space of the robot to extend the local sensing available on board of it.[9,14] In global vision methods, characteristic points forming a pattern on the mobile robot are identified and localized from a single view. A probabilistic method is used to select the most probable matching according to geometrical characteristics of those percepts. One advantage of this approach is that it allows the operator to monitor robot operation at the same time.

Another well-known approach for robot localization is map-based localization. The robot uses its sensors to acquire information on the surrounding environment, and performs the localization task by comparing such data with the previously stored map of the environment.[15] Map-based positioning often includes improving global maps based on the new sensory observations in a dynamic environment and integrating local maps into the global map to cover previously unexplored areas.

In this chapter, we will introduce the data fusion from different sensors for robust robot localization. As an example, we describe an approach for outdoor robot localization (Figure 11.1) by merging data from the global positioning system (GPS), inertial navigation system (INS), and wheel encoders. Each of those sensors, when used separately for robot localization, is subject to a lot of error sources affecting the accuracy of the obtained robot positioning. Our work consists of combining data from those sensors for accurate position estimation. The following gives an overview of the functioning principles of the different sensors and the algorithm we propose for data integration.

11.1 Outdoor mobile robot ROBUDEM.

11.2   GPS based positioning

Positioning using global positioning systems (GPS) is determined, at any time, by measuring the time delay in a radio signal broadcast from several satellites, and using this and the speed of propagation to calculate the distance to the satellites. Position on earth is then calculated by triangulation of intersecting radio signals at the GPS receiver. Using the GPS for positioning is subject to several sources of errors, as listed below.[7,13]

Clock inaccuracies and rounding errors: Despite the synchronization of the receiver clock with the satellite time during the position determination, the remaining inaccuracy of the time still leads to an error of about 2 m in the position determination. Rounding and calculation errors of the receiver translate into approximately 1 m of error.

Multipath effect: The multipath effect is caused by reflection of satellite signals (radio waves) on objects. This effect mainly appears in the neighborhood of large buildings or other elevations. The reflected signal takes more time to reach the receiver than the direct signal. The resulting error typically lies in the range of a few meters.

Satellite orbits: Although the satellites are positioned in very precise orbits, slight shifts of the orbits are possible due to gravitation forces. The sun and the moon have a weak influence on the orbits. The orbit data are controlled and corrected regularly and are sent to the receivers in the package of ephemeris data.

Atmospheric effects: Another source of inaccuracy is the reduced speed of propagation in the troposphere and ionosphere (Fig. 11.2). While radio signals travel with the velocity of light in the outer space, their propagation in the ionosphere and troposphere is slower. These errors are mostly corrected by the receiver by calculations. The typical variations in velocity while passing the ionosphere for low and high frequencies are well known for standard conditions.

Other spatial and geometrical factors affect the accuracy of the GPS measurements. A typical civilian GPS receiver provides 6–12 meters accuracy,

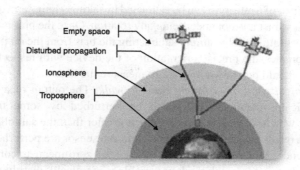

11.2 Atmospheric effects on GPS measurements.

depending on the number of satellites available. This accuracy can be reduced to 1 m by using a differential GPS (DGPS), which employs a second receiver at a fixed location to compute corrections to the GPS satellite measurements.

11.3   Inertial navigation system

The inertial navigation system (INS) is a self-contained navigation technique in which measurements provided by accelerometers and gyroscopes are used to track the position and orientation of an object relative to a known starting point, orientation and velocity.[1,2] INS typically contain three orthogonal rate-gyroscopes and three orthogonal accelerometers, measuring angular velocity and linear acceleration respectively. By processing signals from these devices it is possible to track the position and orientation of a robot on which the INS device is mounted.

Inertial navigation systems usually can only provide an accurate solution for a short period of time. As the acceleration is integrated twice to obtain the position, any error in the acceleration measurement will also be integrated and causes a bias on the estimated velocity and a continuous drift on the position estimate by the INS. Additionally, the INS software must use an estimate of the angular position of the accelerometers when conducting this integration. Typically, the angular position is tracked through an integration of the angular rate from the gyro sensors. These also produce unknown biases that affect the integration to get the position of the unit.

Constant bias: The bias of an INS is the average output from the INS when it is not undergoing any rotation (i.e: the offset of the output from the true value). For the gyro sensor, a constant bias error of ϱ, when integrated, causes an angular error which grows linearly with time $\theta(t) = \varrho \cdot t$. Whereas, for the accelerometer, a constant bias error of ϱ, when double integrated, causes an error in position which grows quadratically with time:

$$s(t) = \varrho \cdot \frac{t^2}{2} \qquad [11.1]$$

The constant bias error of a rate gyro can be estimated by taking a long-term average of the gyro's output whilst it is not undergoing any rotation. Once the bias is known it is trivial to compensate for it by simply subtracting the bias from the output.

For the accelerometer, the precise orientation of the device with respect to the gravitational field should be known in order to measure the bias.

Thermo-mechanical white noise/angle random walk: The output of a micro-electrical-mechanical system (MEMS) will be perturbed by some thermo-mechanical noise, which fluctuates at a rate much greater than the sampling rate of the sensor. As a result the samples obtained from the sensor are perturbed by a white noise sequence, which is simply a sequence of zero-mean uncorrelated random variables. In this case each random variable is identically distributed and has a finite variance $\sigma2$.

Let N_i be the ith random variable in the white noise sequence. Each N_i is identically distributed with mean $E(N_i) = E(N) = 0$ and finite variance $Var(N_i) = Var(N) = \sigma^2$. By the definition of a white sequence $Cov(N_i, N_j) = 0$ for all $i \neq j$.

Integrating the white noise signal $\varrho(t)$ over time $t = n \cdot \delta t$, using the rectangular rule:

$$\text{Gyro:} \int_0^t e(\tau)d\tau = \delta t \cdot \Sigma_{i=1}^n N i \qquad [11.2]$$

$$\text{Accelerometer:} \iint_0^t e(\tau)d\tau\, d\tau = \delta t \Sigma_{i=1}^n \delta t \Sigma_{j=1}^i N_j = \delta t^2 \Sigma_{i=1}^n (n-i+1)N_i \quad [11.3]$$

where n is the number of samples received from the device during the period and δt is the time between successive samples.

Flicker noise/bias stability: MEMS are subject to flicker noise. Flicker noise is noise with a 1/f spectrum, the effects of which are usually observed at low frequencies in electronic components. At high frequencies flicker noise tends to be overshadowed by white noise. Bias fluctuations which arise due to flicker noise are usually modeled as a random walk. Using this model, flicker noise creates a second order random walk in velocity where the uncertainty grows proportionally to $t^{3/2}$, and a third order random walk in position which grows proportionally to $t^{5/2}$.

Temperature: Temperature fluctuations due to changes in the environment and sensor self-heating induce movement in the bias. Such movements are not included in bias stability measurements, which are taken under fixed conditions. Any residual bias introduced due to a change in temperature will cause an error in orientation, which grows linearly with time. The relationship between bias and temperature is often highly nonlinear for MEMS sensors. Most inertial measurement units (IMUs) contain internal temperature sensors which make it possible to correct for temperature-induced bias effects.

11.4   Data fusion for robot localization

In order to increase the accuracy of the robot positioning, several methods have been proposed to combine data from different sensors. Those methods are often based on Kalman filtering. In this chapter, we present an approach using an extended Kalman filter to integrate data from a GPS with data from an INS and robot encoders. This will also allow kipping robot positioning even if no satellite is visible.[13]

The Kalman filter is an extremely effective and versatile procedure for *combining noisy sensor* outputs to estimate the *state of a system* with *uncertain dynamics*. In the GPS/INS/encoders integration case, noisy sensors include GPS receivers, INS and encoders components, and the system state includes the position, velocity, acceleration, attitude, and attitude rate of a vehicle. Uncertain dynamics include unpredictable disturbances of the host vehicle and unpredictable changes in the sensor parameters. Kalman filter optimally estimates position, velocity, and attitude errors, as well as errors in the inertial and GPS measurements.

The relatively low data output rate of GPS receivers (usually 1 Hz) might not meet the cm level accuracy requirements for robot positioning. This problem becomes more serious when the potential temporary loss of a GPS signal occurs or phase ambiguity resulting from cycle slips is considered. INS provides the dynamics of motion between GPS epochs at high temporal resolution and complements the discrete nature of GPS in the occurrence of cycle slips or signal loss.

In addition, positioning with INS requires integration with respect to time of accelerations and angular rates; the measurement noise accumulates and results in long wavelength errors. GPS errors do not accumulate, but in the short term, they are relatively larger and the measurements have poorer resolution.

Integrated systems will provide a system that has superior performance in comparison with either GPS or INS. The main strengths and weakness of INS and DGPS are summarized in Table 11.1.[1,2,8,6]

Table 11.1 INS/DGPS strengths and weaknesses

INS	DGPS
– high position velocity accuracy over the short term	– high position velocity accuracy over the long term
– accurate attitude information	– noisy attitude information (multiple antenna arrays)
– accuracy decreasing with time	– uniform accuracy, independent of time
– high measurement output rate	– low measurement output rate
– autonomous	– non-autonomous
– no signal outages	– cycle slip and loss of lock
– affected by gravity	– not sensitive to gravity

INS/DGPS
– high position and velocity accuracy
– precise attitude determination
– high data rate
– navigational output during GPS signal outages
– cycle slip detection and correction
– gravity vector determination

In the case of a long temporary loss of a GPS signal, positioning updating based only on INS can lead to an inaccurate solution. Therefore, a correction using the encoders' data is needed.

In the following section we describe in detail the procedure used for data fusion from GPS, INS and encoders sensors for the localization of the outdoor robot ROBUDEM.

11.4.1 Reference frames and transformations

The GPS, INS and encoder sensors work in different reference systems and in order to apply an integration process their data should be represented in the same reference system. In this paragraph we give an overview of the reference systems used to represent the data in the robotic navigation field.

Local geodetic or tangent plane

The local geodetic frame or tangent plane frame is the north, east, down (NED) or north, east, up (NEU) rectangular coordinate system we often consider in our everyday life (Fig. 11.3).

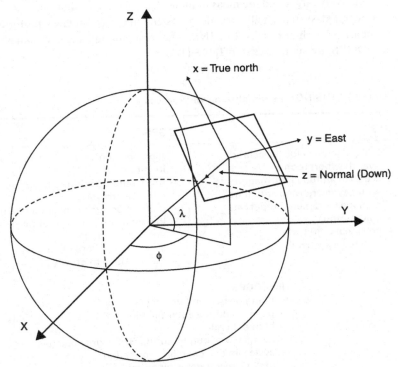

11.3 Local geodetic and ECEF coordinate systems.

It is determined by making a fictional tangent plane at the origin, just like presenting the globe as a map. The X-axis points north, the Y-axis points east, and the Z-axis points toward the interior of the earth, normal to the reference ellipsoid.

For a stationary system, located at the origin of the tangent frame, the geographic and local geodetic frames coincide. When a system is in motion, the tangent local geodetic is fixed, while the geographic frame origin is the projection of the platform origin on the reference ellipsoid of the earth.

Vehicle frame

This is the reference frame system carried by the vehicle. The origin is usually the center of gravity of the vehicle. Its X-axis points towards the defined front of the vehicle, the Z-axis points down, and the Y-axis points right to complete the right hand rule. This frame represents the vehicle states in six degrees of freedom (6DOF) with respect to a reference NEU-frame system, known as Cartesian coordinates (x,y,z) and roll, pitch and yaw (γ, θ, φ).

Earth centered-earth fixed frame

The earth centered-earth fixed (ECEF) frame is mostly used in the case of global-positioning based navigation. This frame has its center in the center of the earth, and the frame is stationary relative to the surface. Of all the possible combinations of ECEF coordinate systems, two are of particular importance.

The first representation frame gives its position in Cartesian coordinates, based on its distance from the center according to each axis. This is named the ECEF rectangular system but is usually just referred to as the ECEF system. Its x-axis points through the intersection of the prime median (0° longitude) and equator (0° latitude), its z-axis towards the true North Pole (parallel to the earth's rotational axes), and the y-axis to complete the right hand rule through the intersection of 90° longitude and equator (Fig. 11.3).

The other representation is called ECEF geodetic frame. This system expresses position in latitude, longitude and height, [Φ,λ,h] and is given in the spherical coordinates. The system takes its basis in the ECEF rectangular frame. The latitude is found by rotating around the z-axis until the x-axis crosses the projection from the position on to the x–y-plane. The longitude is then found by rotating around the y-axis until the x-axis coincides with the vector from the center of the earth to the position. The height is the distance from the nearest point normal on the assumed altitude.

WGS-84, World Geodetic System 1984

- Longitude λ: how far to the east or west we are from the Greenwich meridian. At the Greenwich meridian the longitude is equal zero ($\lambda = 0°$) and completes 360° in the east direction.

- Latitude Φ: how far we are from the equator. At the equator $\Phi = 0°$ and reaches $\Phi = 90°$ at the North Pole and $\Phi = -90°$ at the South Pole.
- Height h: height in meters above sea level.

Geographic frame

The geographic frame is defined locally, relative to the earth's geoid. Its origin moves with the system (vehicle) and is defined as the projection of the platform origin P onto the ellipsoid used to describe the surface of the earth. The x-axis points north, the y-axis points east and the z-axis points down, normal onto the ellipsoid. The latitude Φ and longitude λ define the position of the geographic frame origin on the reference ellipsoid.

Geocentric frame

The geocentric frame is equivalent to the geographic frame with the difference that the geocentric z-axis points from the system location towards the earth's center. The x-axis points towards the north in the plane orthogonal to the z-axis. The y-axis points east to complete the orthogonal.

ECEF geodetic to ECEF rectangular conversion

Using the WGS-84 ellipsoid model for the earth, the rectangular ECEF coordinates (x, y, z) are calculated based on the ECEF geodetic coordinates $(\emptyset, \lambda, h)$ as follows:

$$x = (N + h) \cos \emptyset \cos \lambda$$
$$y = (N + h) \cos \emptyset \sin \lambda \qquad [11.4]$$
$$z = [N(1 - e^2) + h] \sin \emptyset$$

Where
e is the eccentricity of the WGS-84 ellipsoid $e = 0.08181919$;
N is the prime vertical radius of curvature

$$N(\emptyset) = \frac{a}{(1 - e^2 sin^2(\emptyset))^{1/2}}$$

and a is the equatorial radius: $a = 6378137m$.

11.4.2 Kalman filtering modeling

In our application, the mobile car-like robot 'ROBUDEM' (Figure 11.1) travels through the environment using its sensors to localize itself. A world coordinate frame w is defined such that its x and y axes lie in the ground plane, and its z axis points vertically upwards. The system (ROBUDEM) state vector y_R in this case is defined with the 3D position vector $r = (y_1, y_2, y_3)$ of the center of gravity of the robot (taken to be in the middle of the axis between the rear wheels) in the world frame coordinates and the robot's orientations roll, pitch and yaw about the Z, X, and Y axes, respectively (γ, θ, φ).

$$y_R = \begin{bmatrix} y_1 \\ y_2 \\ y_3 \\ \gamma \\ \theta \\ \varphi \end{bmatrix}$$

The coordinates of a point (x^R, y^R, z^R) in the vehicle frame system are calculated in the world frame as follows:

$$\begin{bmatrix} x^W \\ y^W \\ z^W \end{bmatrix} = R_R^W(\gamma, \theta, \varphi) \begin{bmatrix} x^R \\ y^R \\ z^R \end{bmatrix}$$

[11.5]

where $R_R^W(\gamma, \theta, \varphi)$ is the rotation matrix of the vehicle frame R by γ, θ, and φ radians about the world frame system.

$$R_R^W(\gamma, \theta, \varphi) = \begin{bmatrix} c\varphi & s\varphi & 0 \\ -s\varphi & c\varphi & 0 \\ 0 & 0 & 1 \end{bmatrix} \begin{bmatrix} c\theta & 0 & -s\theta \\ 0 & 1 & 0 \\ s\theta & 0 & c\theta \end{bmatrix} \begin{bmatrix} 1 & 0 & 0 \\ 0 & c\gamma & s\gamma \\ 0 & -s\gamma & c\gamma \end{bmatrix}$$

$$= \begin{bmatrix} c\varphi c\theta & s\varphi c\theta & c\theta \\ -s\varphi c\theta + c\varphi s\theta s\gamma & c\varphi c\gamma + s\varphi s\theta s\gamma & c\theta s\gamma \\ s\varphi s\gamma + c\varphi s\theta c\gamma & -c\varphi s\gamma + s\varphi s\theta c\gamma & c\theta c\gamma \end{bmatrix}$$

with the notation $cx = \cos(x)$ and $sx = \sin(x)$.

The dynamic model or motion model is the relationship between the robot's past state y_R^{t-1}, and its current state y_R^t, given a control input u^t

$$y_R^t = f(y_R^{t-1}, u^t, v^t)$$

[11.6]

where f is a function representing the mobility, kinematics and dynamics of the robot (transition function) and v is a random vector describing the unmodeled aspects of the vehicle (process noise such as wheel sleep or odometry error).

The system dynamic model in our case, considering the control u as identity, is given by:

$$y_R^t = \begin{bmatrix} y_1^t \\ y_2^t \\ y_3^t \\ \gamma^t \\ \theta^t \\ \varphi^t \end{bmatrix} = \begin{bmatrix} [y_1^{t-1} + (v^{t-1} + V)\cos(\varphi^{t-1})\Delta t] \\ y^{t-1}{}_2 \\ y^{t-1}{}_3 + (v^{t-1} + V)\sin(\varphi^{t-1})\Delta t \\ \gamma^{t-1} \\ \theta^{t-1} \\ \varphi^{t-1} + (\omega^{t-1} + \Omega)\Delta t \end{bmatrix} \qquad [11.7]$$

Δt is the time interval between two sensor readings. v and ω are the linear and the angular velocities, respectively. At time step t, inputs v and ω are read from the wheel speed and steering angle encoders.

v is given by:

$$v = \sqrt{\dot{y}_1^2 + \dot{y}_3^2} \qquad [11.8]$$

and ω is calculated based on the steering angle of the front–rear and the linear velocity as follows:

$$\omega = \frac{\tan\varnothing}{l} v \qquad [11.9]$$

where l is distance between the front wheel and the rear wheel, $\varnothing$ is the steering angle between the front wheel and the body axis (Fig. 11.4), V and Ω are the Gaussian distributed perturbations to the robot's linear and angular velocity, respectively.

We have $[\dot{u},\ \dot{v},\ \dot{w}]$ as the true vehicle acceleration in the body frame. This acceleration is not the one measured by the INS accelerometer. The INS

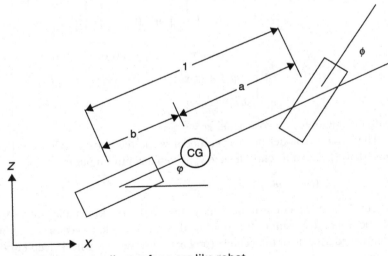

11.4 General coordinates for a car-like robot.

accelerometer sensors will measure the gravity vector, the vehicle acceleration, the centripetal acceleration, and some noise.

The acceleration due to angular velocity[5] is:

$$a_n = \frac{v^2}{r} = \frac{wrwr}{r} = wv \qquad\qquad [11.10]$$

In vector form this will be:

$$a_n = w \times v = w \times (w \times r)$$

where r is the radius of curvature (Fig. 11.5).

The vehicle accelerations are calculated by subtracting the gravity acceleration g_n from the measured accelerations a_n.

The gravity vector in the body frame does not depend on the yaw or heading of the vehicle:

$$\begin{bmatrix} g_x \\ g_y \\ g_z \end{bmatrix} = R_R^W(\gamma, \theta, \varphi)^{-1}[0,0,g_e]^T = \begin{bmatrix} -g_e\sin\theta \\ g_e\cos\theta\sin\gamma \\ g_e\cos\theta\sin\gamma \end{bmatrix}$$

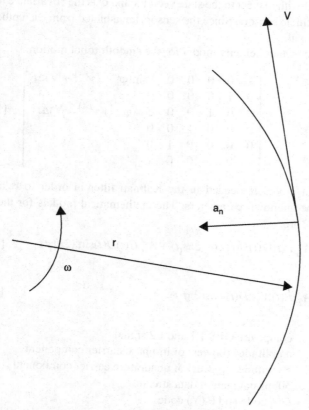

11.5 Acceleration related to curvature.

We get the θ and the γ angles from the INS and can calculate the gravity components in the body frame.

The GPS satellites transmit two carrier signals on the L-band frequency, a primary signal (L1) at 1575.42 MHz and a secondary signal (L2) at 1227.60 MHz[1–9].

The Kalman Filter maintains the state vector y_R based on sensors' measurements. It also maintains a covariance matrix P, which includes the uncertainties in the various states as well as correlations between the states.

At each time step of the filter we obtain the predicted state y_R and covariance P using the state transition function.

$$y_R^{t|t-1} = f(y_R^{t-1|t-1} u^t, v^t) \tag{11.11}$$

$$P^{t|t-1} = FP^{t-1|t-1}F^T + Q^{t-1} \tag{11.12}$$

where

$$F = \left.\frac{\partial f}{\partial y_R}\right|_{y_R^{t-1|r-1}}$$

is the Jacobian of f with respect to the state vector x and Q is the covariance matrix of noises in the vehicle sensors. Since the sensors are isolated from each other, we assume Q is diagonal.

Considering a constant velocity model for the smooth robot motion:

$$\left.\frac{\partial f}{\partial y_R}\right|_{y_R^{t-1|r-1}} = \begin{bmatrix} 1 & 0 & 0 & 0 & 0 & -\sin(\varphi^{t-1})(v^{t-1} + V)\Delta t \\ 0 & 1 & 0 & 0 & 0 & 0 \\ 0 & 0 & 1 & 0 & 0 & \cos(\varphi^{t-1})(v^{t-1} + V)\Delta t \\ 0 & 0 & 0 & 1 & 0 & 0 \\ 0 & 0 & 0 & 0 & 1 & 0 \\ 0 & 0 & 0 & 0 & 0 & 1 \end{bmatrix} \tag{11.13}$$

The measurement model is needed in the Kalman filter in order to relate the observation to the unknown parameters. The mathematical models for the GPS L1 and L2 signal[5] are:

$$S_1(t) = A_{1P}P(t)W(t)D(t)\cos(2\pi f_1 t) + A_{1C}C(t)D(t)\sin(2\pi f_1 t) \tag{11.14}$$

and

$$S_2(t) = A_{2P}P(t)W(t)D(t)\cos(2\pi f_2 t) \tag{11.15}$$

where

$S_1(t)$ and $S_2(t)$:	composite GPS L1 and L2 signal
A_{1P} and A_{2P}:	amplitudes (power) of in-phase carrier component
A_{1C}:	amplitudes (power) of quadrature carrier component
$D(t)$:	50bps navigation data stream
$C(t)$ and $P(t)$:	C/A code and P (Y) code
f_1 and f_2:	carrier frequency of L1 and L2

11.5   Conclusion

We presented an overview of the problem of mobile robot localization and introduced an approach for data fusion based on Kalman filtering for the data integration of sensors for more accurate mobile robot localization in a geo-referenced map. The sensors used are the global positioning system (GPS), inertial navigation system (INS) and wheel encoders.

11.6   References

1 Bar-Itzhack I.Y. and Berman N., Control Theoretic Approach to Inertial Navigation Systems, *Journal of Guidance*, Vol. 11, No. 3, 1988, pp. 237–245.

2 Barnea A., CoRoBa implementation for an INS sensor, Master Thesis, 2008, Royal Military Academy, Belgium.

3 Berrabah S. A. and Bedkowski J.: *Robot Localization based on Geo-referenced Images and Graphic Methods*, IARP WS RISE'2008, Benicassim, Spain.

4 Berrabah S. A. and Baudoin Y., *GPS data correction using encoders and INS sensors*, RISE2009, Brussels, Belgium.

5 Brown A. and Sullivan D. Precision Kinematic Alignment Using a low cost GPS/INS System, *Proceedings of ION GPS 2002*, Navsys Corporation, Oregon, 2002.

6 Gaylor D. and Lightsey E.G., *GPS/INS Kalman Filter design for Spacecraft operating in the proximity of the International Space Station*, University of Texas – Austin, Austin.

7 Grejner-Brzezinska D.A. and Wang J., Gravity Modelling for High-Accuracy GPS/INS Integration, *Navigation*, Vol. 45, No. 3, 1998, pp. 209–220.

8 Grewal M.S., Weill L. R. and Andrews A. P., *Global Positioning Systems, Inertial Navigation, and Integration*, John Wiley and Sons, New York, 2001.

9 Hayashi Y., Tohyama S. and Fujiyoshi H., Mosaic-Based Global Vision System for Small Size Robot League, RoboCup 2005, *Lecture Notes in Computer Science*, 2006, pp.593–601.

10 Jensfelt P., Approaches to mobile robot localization in indoor environment, PhD thesis, Department of Signals, Sensors and Systems, Royal Institute of Technology, Stockholm, Sweden, 2001.

11 Jimenez A. R., Seco F., Ceres R. and Calderon L., *Absolute Localization using Active Beacons: A survey and IAI-CSIC contributions*, Institute for Industrial Automation, CSIC Madrid.

12 Moon S.W., Kim J. H., Hwang D. H., Ra S. W. and Lee S. J., *Implementation of a Loosely Coupled GPS/INS integrated system*, Chungnam National University, Korea.

13 Navarro G., GPS Robot Positioning, Master Thesis, 2008, Royal Military Academy, Belgium.

14 Roque W. L. and Doering D., Trajectory planning for lab robots based on global vision and Voronoi roadmaps, *Robotica*, 2005, Vol.23, pp.467–477.

15 Se S., Lowe D. G. and Little J., Global localization using distinctive visual features, International Conference on Intelligent Robots and Systems, IROS, Lausanne, Switzerland, 2002, pp.226–231.

16 Wang J., Lee H. K. and Rizos, C., *GPS/INS Integration: A Performance Sensitivity Analysis*, University of New South Wales, Sydney.

17 Wolf R., Eissfeller B. and Hein G. W., *A Kalman Filter for the Integration of a Low Cost INS and an attitude GPS*, Institute of Geodesy and Navigation, Munich, Germany.

Part III

Autonomous and teleoperated robots for humanitarian de-mining

12

Environment-adaptive antipersonnel mine detection system: Advanced mine sweeper

T. FUKUDA, Nagoya University, Japan, Y. HASEGAWA, University of Tsukuba, Japan, K. KOSUGE, Tohoku University, Japan, K. KOMORIYA, National Institute of Advanced Industrial Science and Technology, Japan, F. KITAGAWA, Mitsui Engineering and Shipbuilding Co. Ltd., Singapore and T. IKEGAMI, TADANO Ltd., Japan

Abstract: In this chapter, an environment-adaptive antipersonnel mine detection system called Advanced Mine Sweeper is introduced. The Advanced Mine Sweeper is developed for safe and effective de-mining procedures after the Level II survey, based on sensing technologies, access-control technologies and system integration technologies. Advanced Mine Sweeper consists of a sensing vehicle/unit, an access vehicle and an assist vehicle. The sensing vehicle/unit is composed of an integrated sensor (metal detector and ground penetrating radar) and a small-reaction sensor head manipulator. The access vehicle is parked facing a minefield in order to control the sensing unit position in a global area using its boom. The assist vehicle is parked some distance from a minefield. It controls the sensing vehicle/unit and access vehicle and then displays the processed sensing information for landmine detection, receiving sensing information and sensing position. By using this system, experiments in the field with buried dummy landmines were carried out for utility and performance evaluations.

Key words: mine detection, metal detector, ground penetrating radar, small-reaction manipulator, sensing technology, Level II survey, information management system, IMSMA.

12.1 Introduction

It is reported that 120 million anti-personnel mines have been buried in more than 70 countries. These mines are buried not only on battlefields but also in residential areas. How to detect the landmines efficiently is one of the urgent issues for the de-mining of the enormous amount of landmines buried (GICHD, 2002; Moody and Lavasseur 2000; Trevelyan, 2000). It is expected that robot technologies are applied to hazardous fields such as space, disaster sites, nuclear power plants, and so on. Minefields are also dangerous areas, where we would expect to apply robot technologies. By applying robot technologies to the de-mining process, we can execute dangerous tasks safely in the landmine fields.

In this subsection, we introduce an environment-adaptive antipersonnel mine detection system called Advanced Mine Sweeper. This system is developed based

285

on robot technologies such as sensing technologies, access-control technologies, and system integration technologies for the safe and effective de-mining procedure after the Level II survey.

For sensing technology, we widen the bandwidth of a frequency modulated continuous wave (FMCW) ground-penetrating radar (GPR) up to 4GHz and integrate the GPR and a metal detector in order to minimize its size. A precise manipulator is also developed to realize high resolution of the ultra wide band (UWB) GPR. For the access-control technology, an access vehicle, which is developed based on a rough-terrain crane, is used to put the integrated sensor on the arbitrary place using its boom, such as a steep slope, a bank side, etc. For the system integration technology, an information management system is developed to control the integrated sensor and access vehicle. The information management system also outputs all kinds of information that is related to mine action in a compatible format with the Information Management System for Mine Action (IMSMA).

In the following sections, first, we introduce the system architecture of our mine detection system briefly. Next, we introduce the sensing technologies which are the integrated sensor system consisting of the metal detector, GPR and its signal processing methods, and the access-control technologies which consist of a small-reaction manipulator, sensing vehicle/unit, access vehicle, and assist vehicle. In addition, we introduce the information management system, which can support an operator by making a report that is submitted to some de-mining organizations. Finally, by using the developed advanced mine sweeper, we experiment using several kinds of minefields to evaluate the system performance.

12.2   Architecture of an antipersonnel mine detection system

An environment-adaptive antipersonnel mine detection system as shown in Fig. 12.1 has three sub-systems: a sensing vehicle/unit, an access vehicle, and an assist vehicle.

The sensing units (Fig. 12.2) are composed of an integrated sensor and a small reaction sensor head manipulator. The sensing vehicle/unit is put on a minefield with a small contact pressure without mine explosion and then the manipulator enables the sensor head to trace a ground surface with a small gap precisely in a local area. As a result, the system can detect a landmine even on rough terrain.

The access vehicle (Fig. 12.3) is parked facing a minefield in order to control the sensing vehicle/unit position in a global area, using its boom. This boom enables a wide area to be checked and it can even put the sensing unit in areas like a steep slope, which it would be almost impossible for a human or vehicle to access.

The assist vehicle is parked some distance from the minefield for safety. It controls the sensing vehicle/unit and access vehicle and displays the processed sensing information for landmine detection, receiving sensing information and sensing position.

12.1 Mine detection system: advanced mine sweeper.

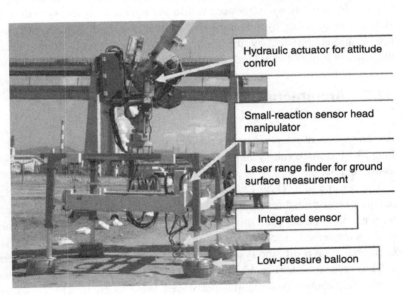

Hydraulic actuator for attitude control

Small-reaction sensor head manipulator

Laser range finder for ground surface measurement

Integrated sensor

Low-pressure balloon

12.2 Sensing unit.

All kinds of sensing information related to de-mining action go through the system so that they can be logged on it. An information management system is therefore developed and installed in the assist vehicle in order to control and display the information on demand. This management system also allows an

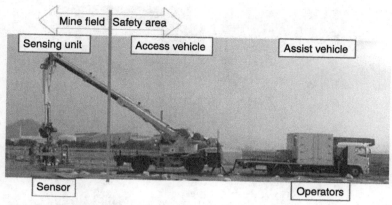

12.3 System location in de-mining action.

operator to make a report that can then be submitted to some de-mining organizations. Operators can stay in the assist vehicle during the whole de-mining procedure and they are never exposed to a minefield. The system provides a safe and effective de-mining procedure after the Level II survey. The following section explains each sub-system of the mine detection system.

12.3   Sensing technologies of advanced mine sweepers

12.3.1 Integrated sensor

A metal detector is one of the most reliable systems for the detection of landmines and is used for current de-mining actions. Although a metal detection system could detect a landmine with a metal effectively, the disadvantages of this system are that it is sensitive to any metal fragments and gives a lot of false signals for the mine detection procedure. In addition, it does not sense well for a small amount of metal buried deeper than 0.2 meters.

Currently, the ground penetrating radar (GPR) is also thought to be one of the most promising technologies for the efficient detection of landmines (van der Merwe and Gupta, 2000; Milisavljevic and Block, 2003). GPRs have the advantage of being able to detect plastic landmines and those buried deeply in the ground and they can also recognize the shape of buried objects. However, the GPR finds it difficult to detect landmines on the surface, because the GPR signal is influenced strongly by the refraction signal from the ground's surface.

To enable the effective detection of landmines in the real landmine fields, we integrated GPR and a metal detector, so that the integrated sensor could detect many objects including the landmines at any depth compared with the conventional metal detector. Figure 12.4 shows an overview of the developed integrated sensor and Figure 12.5 shows the antenna part of the integrated sensor. Plate VIII (in the colour section) also shows the experimental results using the integrated sensor. By

12.4 Overview of integrated sensor.

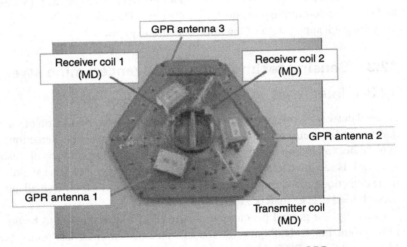

12.5 Integrated sensor: a transmission coil surrounds GPR antenna and two receiving coils that are located at the central area of the sensor.

using the GPR part of the integrated sensor, we can see the buried object clearly at any depth in the ground. The signal of the metal detector is shown in the red part of the experimental results. By analyzing signals from both the GPR and the metal detector, we can detect the position and depth of the landmine precisely in the real landmine field compared with the previous mine detection approach, which only utilizes the metal detector.

12.3.2 Signal processing for geography adaptive sensing

Image reconstruction is especially required for easier extraction of suspected mines, since a GPR response signal is reflection intensity versus time. The Kirchhoff migration is a well-known method for reconstruction of spatial (depth) image from time domain reflection signals in seismology. We reformed Kirchhoff migration to deal with signals acquired by the geography adaptive sensing. See Hasegawa et al. (2005) and Fukuda et al. (2007) for details.

12.4 Access-control technologies of an antipersonnel mine detection system

12.4.1 Sensor manipulation system

A small-reaction sensor head manipulator is developed to realize the high-speed and precise sensor head maneuver. The high-speed sensing will reduce the working time and improve the cost effectiveness. The small-reaction manipulator counteracts the reaction force of the sensor head motion by using counter weights. Our sensing unit/vehicle is connected with the ground through low-ground-pressure tires/balloons. They are extremely flexible and elastic; therefore the fast sensor head motion without the counter weights causes oscillations on the system.

The oscillation of the robot impairs the accuracy of the sensor position and makes resolution of the integrated sensor low. By using the small-reaction manipulator, we can execute the high-speed and precise motion of the sensor head without oscillations.

We also developed the ground adaptive manipulation method of a sensor head. The purpose of the ground adaptive manipulation is to reduce an effect of the reflection signal of GPR from the ground surface. This reflection signal has a large proportion of the received signal and the complicated reflection signal hides the signal from a buried object. The ground adaptive manipulation keeps a distance between the integrated sensor and the ground surface as a constant in order to keep reflection signals from the ground surface constant, so that we could detect the landmine clearly by removing the refraction signal from the ground surface.

To realize the ground adaptive manipulation, first, we measure the rough terrain of the ground by using the laser range finder. Next, from the data of the laser range finder, we generate a path for the sensor head of the integrated sensor for moving its sensor head along the rough terrain as shown in Fig. 12.6.

The small-reaction manipulator has six degrees of freedom of motion, so that the sensor head attached to the end effector of the small-reaction manipulator could move while keeping the small distance between the integrated sensor head and the ground surface constant as shown in Fig. 12.7. See Yabushita et al. (2005) for details.

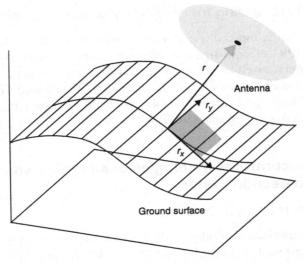

12.6 Coordination system for ground adaptive manipulation.

12.7 Ground adaptive manipulation.

12.4.2 Sensing vehicle

An unmanned vehicle has an advantage over a sensing unit hanging from the access vehicle of surveying a minefield from the viewpoint of sensing speed and sensing area when a minefield is relatively flat. The vehicle must move around the minefield without mines exploding. In order to prevent mines from exploding, flexible low-pressure tires are employed on the vehicle. The tire deforms and increases contact area with the ground when load is added. The wide contact area decreases the contact pressure, and the force that acts on a mine also decreases.

Since there is no commercially available low-pressure tire suitable for the sensor vehicle especially, a tire tube was tested for the contact pressure with the ground. The tube was set around a center wheel, and load acted at the hub. Pressure at a depth of 5 cm in the ground was measured as the result of the preliminary experiments showed that pressure in the ground peaks at a depth of about 5 cm. Figure 12.9 shows the result of the experiment. The diameter of the tube is about 70 cm. The air pressure inside the tube is 0 hPa (no load), and 30 hPa (25 kgf load). Explosive pressure of the PMN-S mine (one of the most explosive antipersonnel mines) is 0.064 kgf/cm^2. The result shows that the tube tire is able to support about 25 kgf without explosion. Since estimated weight of the sensing vehicle was 250 kgf, a 16-wheeled vehicle was specified for the sensing vehicle (Fig. 12.8).

Weight and keeping all tires/balloon in ground contact are key points for designing the sensor vehicle/unit. The light tire weight contributes to the low contact pressure with the ground. The pressure exerted on the ground by the contacting tires increases when some tires out of the 16 wheels lose contact with the ground. In the sensing vehicle especially, the basic concept of the locomotion mechanism is similar to the four-wheeled automobile. The 16 wheels are distributed among four units. Each unit has four wheels, and it works as one wheel. The front two units have steering function and the rear two units have traction. Steering and traction are driven independently by an actuator mounted in each unit. In order to ensure contact with the ground, each unit is connected to the body frame by a passive joint around the roll axis.

The small-reaction manipulator described in the previous section is set in the body frame. A photograph of the developed sensing vehicle is shown in Fig. 12.9. It is 3.8 m long, 2.0 m wide, and weighs 270 kg. For navigation, GPS and an inertial sensor are fitted. A prism for optically measuring the position from outside

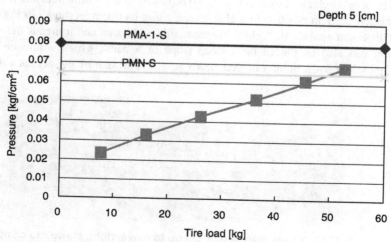

12.8 Relative pressure from tire at a depth of 5 cm.

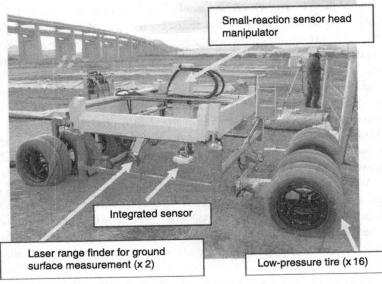

Small-reaction sensor head manipulator

Integrated sensor

Laser range finder for ground surface measurement (x 2)

Low-pressure tire (x 16)

12.9 Unmanned sensing vehicle.

the vehicle is also attached. In the field test for mine detection, path following control was applied. The course was flat and 15 m long, 19 sensing areas were specified. The sensor vehicle moved at a speed of 0.3 m/sec, and the standard deviation of error from the planned path was 3 cm.

12.4.3 Access vehicle

The access vehicle shown in Fig. 12.10 is parked facing a minefield and it controls the sensing unit position in a global area, using its boom as shown in Fig. 12.11. This boom enables the vehicle to sense in a wide area and it means the sensing unit can also be placed on a steep slope or a bank, which would be almost impossible for a human or vehicle to access. The size of the access vehicle is

12.10 Access vehicle (left: set up to move, right: crawler to climb steep slope).

12.11 Sensor manipulation system manipulated by access vehicle.

3,100 mm height, 9,100 mm length, and 2,100 mm width. The total vehicle weight is 14,000 kg. The maximum speed is also 50 km/h, and the grade ability is 25 degrees with a crawler shown in Fig. 12.10 (right).

12.4.4 Assist vehicle

The assist vehicle is parked some distance away from the minefield as shown in Fig. 12.3. The cockpit of the assist vehicle is shown in Fig. 12.12. We can manually control the small-reaction manipulator, the sensing vehicle/unit and access vehicle by using the joystick or switches intuitively, based on the display images captured by charge-coupled device (CCD) cameras. An information management system also automatically controls the small-reaction manipulator, the sensing vehicle/unit and the access vehicle and then displays the processed sensing information for landmine detection, receiving sensing information and sensing position from them.

The absolute position of the assist vehicle is measured by a real time kinematic global positioning system (RTK-GPS), and the relative position between the assist vehicle and the sensing unit/vehicle is measured by a total station system that uses a laser beam and a homing device.

12.5 Information management system

An information management system makes the planning for de-mining procedure and controls the sensing vehicle/unit and the access vehicle based on the plan. It also provides information on the current status or past de-mining results in order to share the information with operators and other de-mining organizations.

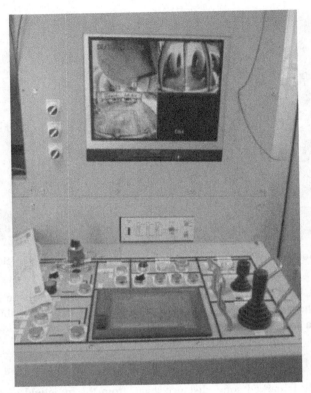

12.12 Cockpit in assist vehicle.

The information management system is composed of three sub-systems as follows:

- Controller for sensing unit/vehicle and access unit (MV-CPU): All kinds of information related to the de-mining action comes from the information management system. The operator knows the current status of the de-mining action and makes a decision to start de-mining tasks. Once the operator sends a start command, the controller (MV-CPU) automatically controls the sensing unit position, sending operational commands to the access vehicle. Next the controller senses the ground configuration using the laser range finder and then starts scanning the ground surface. After sensing one area, the controller moves the sensing unit to the next area by controlling the boom of the access vehicle.

- Mine detection support system (PC-GPR): The sensing information is also sent to the information management system. A mine detection support system, which is part of the information management system, processes the sensing information so that the operator can detect landmines by processing the

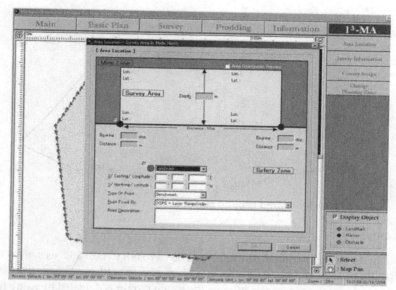

12.13 Information management system: 'Integrated information interface for mine action (I3MA)'.

sensing information intuitively. This system displays the sensing results integrating the sensing information from the metal detector and GPR. In addition, the landmine candidates are marked by a mine detection algorithm in advance in order to reduce operator tasks and avoid operator errors.

- Integrated information interface for mine action (I3MA): The I3MA is also installed in the information management system in order to control and to display the information on demand, as shown in Fig. 12.13. It also supports an operator making a report that is submitted to de-mining organizations. The report format is based on the international standard Mine Action Extensible Mark-up Language (maXML).

12.6 Experiments

To evaluate the validity of the developed advanced mine sweeper, we experiment in a test field where a lot of dummy landmines are buried. This field was prepared in the Kagawa prefecture, Japan and several kinds of experiments for evaluating the developed advanced mine sweeper were carried out in February 2005 for about a month. One example of the experiment results is shown in Plate IX (in the colour section between pp 166–7). In this experimental field, the ground surface is flat. Several pairs of Type 72-S landmines are buried at a depth of 10 cm, and the distances between each pair of landmines are 10 cm, 15 cm, 20 cm, and 25 cm. To illustrate the resolutions of the developed GPR, we generate the underground image based on GPR signal. We can see each landmine clearly.

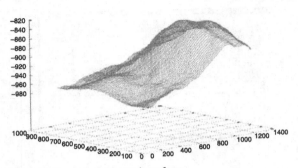

12.14 Measured ground surface.

In the next experiment, the minefield has a rough terrain and, as well as dummy landmines, several kinds of objects such as plastic bottles and nails are buried. In this experiment, the information on the rough terrain is detected by using the laser range finder, which generates a path for the integrated sensor to move along the rough terrain. A three-dimensional image of the ground surface is shown in Fig. 12.14. The experiment results using an integrated sensor are shown in Plate X (in the colour section). Even though this was a difficult field for mine detection, the developed advanced main sweeper was still able to detect the dummy landmines.

12.7   Summary

In this chapter, we have introduced the environment-adaptive antipersonnel mine detection system called Advanced Mine Sweeper. Using sensing technologies, access-control technologies and system integration technologies, an advanced mine sweeper for a safe and effective de-mining procedure after the Level II survey has been developed.

The Advanced Mine Sweeper consists of an integrated sensor, which includes a metal detector and the UWB-GPR, a small-reaction manipulator, sensing vehicle/unit with low-pressure tires/balloon, an access vehicle, and an assist vehicle. The information management system is used for planning the de-mining procedure and controlling the sensing vehicle/unit and the access vehicle based on the plan. It also stores and displays information about the current status or past de-mining results and can share this data with operators and other de-mining organizations.

The developed advanced mine sweeper was tested using a landmine field in which dummy landmines were buried, and its performance evaluated through several experiments. The experiment results show the validity of the developed advanced mine sweeper.

12.8 Acknowledgements

This project was successfully carried out with technical cooperation from Kazunori Yokoe, Yasuhiro Kawai, Yasuhisa Hirata, Hidenori Yabushita, Mitsuhiko Kanehama, Hironori Adachi, Takanori Shibata, Koichi Sugita, Chihiro Jyomuta, Toru Kenmizaki, Fujio Oka, Koichi Sato, Shyoji Sakai, Naoto Aomori, Yoshiyuki Sakamoto, Takahiro Yoshida, and Kanji Hara, in addition to the authors.

The authors would also like to thank Japan Science and Technology Agency (JST) who provided us with the experimental field and financially supported our research and development.

12.9 References

Fukuda, Toshio, Hasegawa, Yasuhisa, Kawai, Yasuhiro, Sato, Shinsuke, Zyada, Zakarya and Matsuno, Takayuki, GPR Signal Processing with Geography Adaptive Scanning using Vector Radar for Antipersonal Landmine Detection, *Int J of Advanced Robotic Systems*, 2007, 4(2), 199–206.

GICHD (Geneva International Centre for Humanitarian Demining), http://www.gichd.org/

GICHD, *Mine Action Equipment: Study of Global Operational Needs*, 2002, http://www.gichd.ch/22.0.html

Hasegawa, Y. et al., Automatic Extraction for Mine Suspects from GPR, *Proc. of the IARP Int workshop on Robotics and Mechanical Assistance in Humanitarian Demining*, 2005, 27–32.

James P. Trevelyan, *Technology Needs for Humanitarian Demining*, Department of Mechanical and Materials Engineering, University of Western Australia, 2000.

van der Merwe, Andria and Gupta, Inder J., A Novel Signal Processing Technique for Clutter Reduction in GPR Measurements of Small, Shallow Land Mines, *IEEE Transaction on Geoscience and Remote Sensing*, 2000, 38(6), 2627–2637.

Milisavljevic, Nada and Bloch, Isabelle, Sensor Fusion in Anti-Personnel Mine Detection Using a Two-Level Belief Function Model, *IEEE Transaction on Systems, Man, and Cybernetics-Part C: Applications and Reviews*, 2003, 33(2), 269–283.

Moody, K. A. and Lavasseur, J. P., *Current and Emerging Technologies for Use in a Hand-Held Mine Detector*, Land Force Technical Staff Course V, Department of Applied Military Science, The Royal Military College of Canada, 2000.

Yabushita, Hidenori, Kanehama, Mitsuhiko, Hirata, Yasuhisa, Kosuge, Kazuhiro, 3D Ground Adaptive Synthetic Aperture Radar for Landmine Detection, *Proceedings of the 2005 IEEE/RSJ Int Conf on Intelligent Robots and Systems*, 2005, 1861–1866.

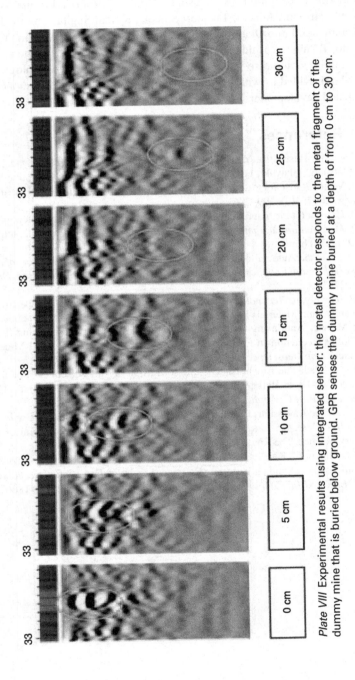

Plate VIII Experimental results using integrated sensor: the metal detector responds to the metal fragment of the dummy mine that is buried below ground. GPR senses the dummy mine buried at a depth of from 0 cm to 30 cm.

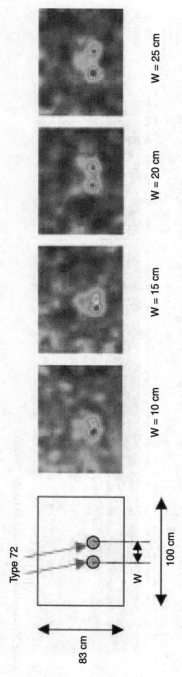

Plate IX GPR resolution test: the distance (W) between the centers of two Type 72 dummy mines changes from 10 cm to 25 cm with five-centimeter intervals.

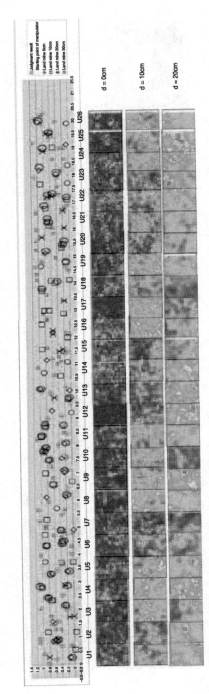

Plate X Experimental results in rough terrain.

13

Mechanical mine clearance: Development, applicability and difficulties

M. K. HABIB, The American University in Cairo, Egypt and
Y. BAUDOIN, Royal Military Academy, Brussels, Belgium

Abstract: Landmines are prominent weapons and they are very effective, yet cheap, and easy to make. They lie on or just under the ground surface. The removal and destruction of all forms of dangerous battlefield debris, particularly landmines and explosive remnants of war (ERW) are vital prerequisites for any region in order to recover from the aftermath of war. Although de-mining has been given top priority, mine clearing is currently a labor-intensive, slow, very dangerous, expensive, and low technology operation. Hence the development of new machines, technologies, sensors, techniques, and inexpensive field-oriented tools that are efficient and fast, which is important in humanitarian mine clearance. This chapter focuses on the problems associated with landmines, the trend of mechanical de-mining, the development of new machines for improved performance and adaptability to local circumstances and needs.

Key words: landmines, mine clearance, de-mining, humanitarian de-mining, manual de-mining, mine detection, mine clearing machines, mechanically assisted de-mining, de-mining robot, clearance accuracy.

13.1   Introduction

Although de-mining has been given top priority, mine clearing is currently a labor-intensive, slow, very dangerous, expensive, and low technology operation. The development of new machines, technologies, sensors, and techniques is important in humanitarian mine clearance. In addition, efforts must be made to develop new procedures, standards and techniques as well as better, faster, simpler and inexpensive field-oriented tools and equipment that can enhance cost efficiency and productivity, while maintaining safe practices and enhancing operational safety.

Technology has become the solution to many long-standing problems, and while current mine detection and clearance technologies may be effective, they are far too limited to fully address the hugely complex and difficult landmine problem facing the world. The challenge is in finding creative, reliable and applicable technical solutions in such a highly constrained environment. Applying technology to humanitarian de-mining is a stimulating objective. Greater resources need to be devoted to de-mining both for immediate clearance and for the development of innovative detection and clearance equipment and technologies. Mine clearance itself can be accomplished through different methods with varying levels of technology and accuracy. Many methods, technologies and techniques

299

have been developed for mine detection and clearance (Habib, 2001a), such as, manual mine clearance, mechanical equipment and tools for mine clearance (mechanization approach), mine detection and sensing technologies, robotic solutions for mine detection and clearance, etc.

Mechanical approaches rely on the use of motorized mine-clearers, the design of which is initiated and influenced by military de-mining requirements. A number of mechanical mine clearing machines have been constructed or adapted from military vehicles or armored vehicles of the same or similar type and of varying sizes. In addition, commercial and agricultural machines/vehicles have been developed, modified, adapted, tested or used to suit humanitarian de-mining purposes. Some machines can carry out several applications with appropriate modifications. While they are not yet an answer to the problem, some of them are more applicable to humanitarian mine clearance, and they have the potential to provide a useful contribution until better technologies are developed. The general trend is moving from 'mechanical de-mining' towards 'mechanically assisted de-mining', adaptable to local circumstances and needs.

The fundamental goals of humanitarian de-mining are to detect and destroy/remove landmines and other explosive remnants of war (ERW) from infected areas efficiently, economically, reliably, rapidly and as safely as possible in order to make these areas economically viable and useful for development. The removal and destruction of all forms of dangerous battlefield debris, particularly landmines, are vital prerequisites for any region in order to recover from their impact. Many varied fundamentals have to be observed, such as soil and type of topology and terrain, as well as the type of contamination, climate, vegetation, etc. (Nicoud and Habib, 1995; Hewish and Less, 1995; Nicoud, 1996; King, 1997; Habib, 1998; US Department of Defense, 1998; Green, 1999; Habib, 2001a, 2002b; Baudoin and Acheroy, 2002; Habib, 2007a, 2008a, 2008b).

Landmines are prominent weapons because they are simple devices, very effective, yet inexpensive, readily manufactured anywhere, easy to place and yet so difficult and dangerous to find and destroy. A landmine is a type of self-contained explosive device, which is placed on to or into the ground to constitute a minefield, and it is designed to destroy or damage equipment or personnel. A mine detonates by the action of its target (a vehicle, a person, an animal, etc.), the passage of time, or controlled means. A number of fuse activation mechanisms may activate a landmine, such as pressure (step on or drive over), pressure release, movement, sound, magnetic influence (change of magnetic field around the mine), vibration, electronic, and command detonation (remote control). Minefields may be laid by several means: manually, by mine-laying launchers on vehicles, fired by artillery from a distance, dropped from both rotary and fixed-wing aircraft, ejected from cruise missiles, etc. The most labor-intensive way of laying mines is to have assigned personnel bury the mines.

Landmines can be categorized into two groups: anti-personnel (AP) and anti-tank (AT) mines (Habib, 2007a, 2008b, 2008c).

AP mines are widely considered to be ethically problematic weapons with the ability to kill or incapacitate their victims and they can damage unarmored vehicles. Pressure, tripwires, tension or pressure release, electro-magnetic influence, and seismic signals can detonate AP mines. Some landmines are 'hardened' against neutralization by explosives and other landmines have anti-disturbance mechanisms. About 2000 types of landmines exist around the world; among these, there are more than 650 types of AP mines. AP mines are harmful because they are hidden, and for the variety in explosive load, activation means, action range, and the effect on human bodies. Most AP mines can be classified into one of the following four categories: blast, fragmentation, directional, and bounding devices. These mines range from very simple devices to high technology (O'Malley, 1993; US Department of State, 1994; Habib 2002b, 2007a, 2008b, 2008c). Landmines may have been in place for many years and they may become corroded, waterlogged, impregnated with mud or dirt, and can behave quite unpredictably. Some mines were buried very deeply to stop more organized forces finding them with metal detectors. Deeper mines may not detonate when the ground is hard, but later rain may soften the ground to the point where even a child's footstep will set them off. Modern landmines are fabricated from sophisticated non-metallic materials and incorporate advanced electronics. This new mine technology makes it impossible for metal detectors to discriminate between the mine's metal and other types of metal.

The major side effect of landmines is to deny access to land and its resources, causing deprivation and social problems among the affected populations. In addition the medical, social, economic, and environmental consequences are immense. The International Committee of the Red Cross estimates that the casualty rate from mines currently exceeds 26,000 people every year. It is estimated that 800 people are killed and 1200 maimed each month by landmines (US Department of Defense, 1998; International Committee of the Red Cross, 1993 and 1996; Habib, 2002b, 2008b). The primary victims are unarmed civilians; children are particularly affected.

The production costs of anti-personnel mines are between 3 and 30 US$. New smart mines have been developed or are under development at a slightly higher cost. In contrast, the current cost rate of clearing one mine is 300–1000 US$ (depending on the mine infected area and the associated false alarm rate). The United Nations Department of Human Affairs (UNDHA) estimates that there are between 60 and 100 million mines that pose significant hazards in more than 68 countries around the world. New mines continue to be manufactured and laid every year. The current rate of mine clearance is far slower. Humanitarian de-mining requires that each individual mine in an area is located, uncovered and removed or destroyed. According to UNDHA, there is a need to have at least a 99.6 percent successful detection and removal rate and a 100 percent success rate to a certain depth according to International Mine Action Standards (IMAS) (Blagdan, 1993; Habib, 2002b). Different de-mining methods have been adopted, such as, manual (usually associated with the use of metal detector, dogs, etc.) (Busuladzic and Trevelyan, 1999; GICHD, 2005; Habib, 2007a), mechanical (Habib 2001b, 2002a; GICHD, 2004, 2006a), mine detection techniques (van Westen, 1993; Hewish and Ness, 1995; Cain and Meidinger, 1996; McFee and Carruthers, 1996; Habib, 2001a; GICHD, 2006b; Habib, 2007b), robotization (Nicoud and Habib, 1995; Gage, 1995; Habib, 1998; Baudoin and Acheroy, 2002; Havlik, 2005; Habib 2007a, 2008c), etc.

However, current clearance technology and techniques are typically labor-intensive, expensive, slow, dangerous, low technology operations, with a range of clearance reliability.

13.2   Core components of the humanitarian mine action plan

The objective of the humanitarian mine action plan is to reduce the risk from landmines to a level where people can live safely, where economic, social and health development can occur free from the constraints imposed by landmine contamination, and in which the victims' needs can be properly addressed.

The process of landmine clearance comprises five stages (Habib, 2002b):

1. Locate, identify and mark any of the recognized minefields. This includes: surveying, assessment and planning, mapping, prioritization of marked minefields and resources, etc. This should be associated with mine risk education, human skill development and management, the public awareness process, information management, safety and benchmark consideration, etc.
2. Prepare the marked minefields for the clearance operation by cutting vegetation and clearing the fields, collecting metal fragments, etc. Area reduction is considered during this stage too.
3. Apply suitable mine clearance techniques that suit the relevant minefield, to locate and mark individual landmines within the identified area.
4. Remove the threat of the detected mines by neutralization: removal, or detonation.
5. Apply quality control measures (post-clearance inspection). There is a need to verify and assure with a high level of confidence that the cleared area is free from mines.

In parallel to the above, healthcare, rehabilitation, and medical support should be provided to affected persons. In addition, implementing continuous educational and awareness programs, infrastructure building, job creation and initiating economic support should be established.

13.3   The mechanical approach to mine clearance

A good deal of research and development has gone into motorized mechanical mine clearance, the early design of which was influenced by military de-mining requirements. The use of such machines aims to unearth mines or force them to explode under the pressure of heavy machinery and associated tools and to avoid the need for de-miners to make physical contact with the mines. A number of mechanical mine clearing machines have been constructed or adapted from military vehicles, armored vehicles, or modified commercially available agriculture vehicles of the same or similar type, of varying sizes (Habib, 2001b, 2002a, 2004, 2007a; GICHD, 2006a).

Mechanical methods have emerged with their own strengths and weaknesses. A single mechanical mine clearance machine can work faster than 1000 de-miners over flat fields. They are most appropriate and cost effective in large and wide areas without dense vegetation or steep gradients. In small paths, thick bush, or soft or extremely hard soil such machines simply cannot maneuver. Mechanical clearance equipment is expensive and it cannot be used on roadsides, steep hills, around large trees, inside a residential area, on soft terrain, heavy vegetation or rocky terrain. Mobility and maneuverability also limit vehicles: wheeled vehicles cannot travel efficiently on anything other than flat surfaces, tracked vehicles cannot travel in areas with steep vertical walls, and machines in general cannot climb undefined obstacles or squeeze through narrow entrances. Using machines often does not destroy all mines in a contaminated area. In some cases, AP mines may be pushed to one side or buried deeper or partially damaged, making them more dangerous.

In addition, mechanical clearance has its own environmental impact on land, such as, erosion and soil pollution. The logistical problems associated with transporting heavy machinery to remote areas are critical in countries with little infrastructure and resources.

The aim of using machines is typically not to fully clear land of mines, but to prepare the ground for post-machine full clearance using manual methods and mine detection dog teams (GICHD, 2004), along with other possible technologies. Hence, none of the equipment within this category has been developed specifically to completely fulfill humanitarian mine clearance objectives. There is no available mechanical mine clearance technology (with the exception of flat terrain) that can give the required clearance ratio to help achieve humanitarian mine clearance standards effectively while minimizing the environmental impact. It has been suggested that few AP blast mines are left behind in a functional condition after treatment by certain machines in suitable terrain, and in order to achieve better clearance rates, manual de-miners and mine detection dog teams should follow up to compensate for the likely residual mine threat left by those machines.

Mechanical mine clearance systems are divided into three basic categories: (1) vegetation cutters and removers, (2) manned mechanical systems, and (3) remotely controlled mechanical systems. A manned system is controlled from a protected position within the de-mining vehicle. A remotely controlled mechanical system is managed from a safe distance using a hand-held radio-frequency module. All categories may contain single or multifunctional mechanisms.

Cost effective and efficient clearance techniques and mechanisms (flexible and modularized) for clearing both landmines and vegetation have been identified as a significant need by the de-mining community. Hence, it is important to highlight the importance of determining the clearance potential of current and future mechanical machines in order to use their speed and potential cost-efficiency.

De-mining operations conducted by some mechanical machines have shown promising results that need to be enhanced further given suitable conditions against an appropriate target (GICHD, 2004). In order to enhance the possibility of a successful usage of de-mining machines, it is important to understand the physical limits imposed upon a de-mining machine by its operational environment and ecological needs. This includes factors of topography, soil, ordnance type and machine. Furthermore, there is an urgent need to standardize the method of recording mechanical clearance data (GICHD, 2004) and set up proper benchmarks for evaluations, testing and risk assessment.

13.4   Mechanical equipment and tools for mine clearance

A number of mechanical mine clearing machines have been tested in the past. The general trend has moved from 'mechanical de-mining' towards 'mechanically assisted de-mining', which is adaptable to local circumstances. Some examples of mechanical clearance equipment include, but are not limited to, vegetation cutters, flails and light-flails, the Panther mine clearing vehicle, armored bulldozers, plows and the rake plows, the M2 Surface 'V' mine plow, earth tillers, mine sifters, mechanical excavation, armored wheel shovels, mine clearing cultivators, floating mine blades, mine rolling, mine-proof vehicles, the Swedish mine fighter (SMF), armored road graders, etc. (US Department of Defense, 1998; *Humanitarian Mine Action Equipment Catalogue*, 1999; Department of Defense, 2002; Habib, 2002a, 2004; GICHD, 2006a; Habib, 2007a, 2008c).

13.5   Vegetation cutters and removal

Dense vegetation is a common phenomenon in minefields. In many countries and not only in tropical ones, vegetation is a large problem facing de-mining and often poses major difficulties to the de-mining efforts. Lush vegetation spreads over uncultivated areas very quickly and such vegetation impedes the work of manual de-miners. Because of the extremely high risk of danger from fragmentation and tripwire landmines amid the undergrowth, it can be very risky to cut the undergrowth by hand. In addition, vegetation removal can take up a substantial amount of time and properly mechanized vegetation cutting and removal mechanisms are needed.

Men and dogs cannot enter a mined area if it is overgrown with vegetation, even if it is just tall grass. Hence, cutting the vegetation ahead of manual de-mining increases the speed of manual clearance by enabling easier access for de-miners and enhancing their safety. Therefore, there is an urgent need for effective vegetation clearance technologies and techniques that avoid detonating mines. These machines should be designed to cut down on the time required for de-mining.

In their simplest form, vegetation cutters consist of adequately modified commercial devices (e.g. agricultural tractors with hedge cutters or excavators).

Mulching down vegetation is an important role in mechanically assisted manual de-mining actions (*Humanitarian Mine Action Equipment Catalogue*, 1999; Habib, 2001a, 2004; GICHD, 2004, 2006a; Habib, 2007a).

The MgM Mulcher vegetation cutter was developed by the German non-governmental organization (NGO) Menschen gegen Minen (MgM) [People against Landmines] in Angola (Fig. 13.1a, b) and is one of the first machines to be used specifically for vegetation removal to solve a big problem for de-miners, especially in tropical and sub-tropical areas. It is mounted on a South African mine-proof vehicle (a Wolf). MgM's special armored vehicle with its hydraulically powered vegetation mulching tools that are attached to a long crane arm speeds up the process and enhances safety.

Similar concepts have been tested by the HALO Trust in Cambodia (Fig. 13.1f HALO Trust vegetation cutter) using a standard four-wheel drive agricultural tractor unit with armored cab and fitted with the Bomford Turner (UK) hydraulically operated vegetation cutter. HALO has vegetation cutters rear mounted on different units including MTZ 82 R, MTZ 1507 (Belarus), Muir Hill A5000 (UK) and New Holland 8340 (UK). Front mounted cutters are deployed on a Volvo 4400 medium wheel shovel (Nicoud, 1996; US Department of Defense, 1998; Humanitarian Mine Action Equipment Catalogue, 1999; Habib, 2004; GICHD, 2006a; Habib, 2008c). The system works by cutting vegetation in swathes with the 1250 mm cutting head attached to a hydraulically operated arm with an 8000 mm reach. Figure 13.1c shows the Bell buncher/cutter from South Africa. It is a commercial forestry machine fitted with lightweight armor to protect against fragmentation. It could be very useful for vegetation removal and greatly increase operation speeds. Flails have been used but they have demonstrated limited ability to reduce vegetation problems. Tempest is a multipurpose teleoperated machine (Fig. 13.1e introduces Tempest in Cambodia) being developed by Warwick University (UK) and a flail can be attached to it for vegetation clearing. Also, the teleoperated ordnance disposal system (TODS) (Fig 13.1d) has been used for vegetation cutting purposes.

HALO has developed the standard European tractor-mounted hedge and verge cutter and deployed units to cut years' worth of vegetation growth in minefields in Cambodia, Africa and the Caucasus. Medium-wheeled loaders have been adapted and armored for clearing mine-contaminated rubble in Kabul and the Caucasus, and also for sifting laterite soils in Africa. The Belarus 1507 tractor has been armored and adapted for vegetation cutting and area reduction in Cambodia and Chechnya. The UNO Corporation of Japan has also demonstrated its ground milling machine performing vegetation removal in Cambodia.

Another technology that might be of help to assist with several aspects of vegetation removal is abrasive water jets. Water jets, which entrain sand and grit, can cut through steel, concrete and stone with ease. They will also cut through mines and fuses without detonating them. Research and testing is needed to confirm that this concept will work in practice.

13.1(a)–(f) Examples of vegetation cutting machines.

13.6 Manned and hybrid mechanical systems

13.6.1 Large flail systems

Flails have been used primarily by the military to clear lanes through minefields, and several versions are deployed in humanitarian de-mining operations as shown in Fig. 13.2. They consist of a large number of chains with clearing elements similar to hammers attached to them and connected to a rapidly rotating drum, which beat and mill the ground. The flails hit the ground and either detonate or destroy all types of landmines. They are resistant to anti-personnel mines. Each piece of ground is struck at least twice. The slower the forward speed of such machines, the greater the clearance depth, thus enabling depth control depending on the speed of the flail, direction of travel, and weight of the hammers or knives secured to the end of the chains. Flails are also very good at removing vegetation, saplings, and even small diameter trees. Most surface laid and near surface mines are generally detonated or sometimes moderately torn apart. However, unless they come in direct contact with the flail, some blast resistant mines are able to survive these systems. The detonation of AT mines typically causes one or two chains to be lost. In dry conditions, the

Aardvark MK4   UOS-155 Belarty

Armtrac-325   Hydrema 910 MCV

13.2 Different examples of large flail systems.

chains create dense clouds of dust. Recent designs incorporate airflow control to keep dust clouds away from the driver's view. These machines are large, expensive, and difficult to maintain, and can damage the terrain by removing the topsoil. Maintenance costs are high, since chains destroyed by landmines or wear must be replaced. A number of units have also been produced and tested during recent years.

13.6.2 Earth tillers and rollers

Large and bulky clearance machines employing one or more rotating horizontal drums, with special metal teeth similar to a rock crusher mounted on their circumference, are capable of tilling the soil to a variable depth. Some tillers are able to reach landmines as deep as 50 centimeters. The machines use speed, impact and mass to destroy mines as they travel round the field. They can be mounted on a prime mover such as a mine-hardened vehicle; however, these machines are large and heavy. This limits their effectiveness in certain type of terrains, and their maintenance costs are high. Many examples of such machines have been built and used, such as MaK (Rhino system), Mine-breaker 2000 in Germany, and Bofors Mine-Guzzler in Sweden.

Rollers are usually pushed or pulled over terrain by another vehicle in the hope that the pressure exerted by their weight will either crush or detonate landmines. Rollers are particularly effective for proofing roads that are suspected of mine contamination. Rollers can be most effective in the early stages of humanitarian operations to allow the establishment of supply routes. Smaller rollers can easily be manufactured to achieve low costs and easy repairs. Numerous roller systems exist, but they tend to be heavy and require a powerful prime mover. They are fairly effective except on undulating or stony ground, or heavily vegetated areas.

The terrain and environment can limit their effectiveness. If heavy rain falls within a few months of use, sloping ground could become unstable and erode away causing environmental problems. In southern Africa, a machine based on similar principles has been used to assist clearance of dense perimeter minefields around townships. Manual clearance can be easier after the vehicle has passed because the ground is softened by the machine's action.

The Mine Breaker 2000 (Fig. 13.3) is a German-built mechanical mine clearing machine that carries out the first stage of a three-stage clearing program. The developmental version of Mine Breaker 2000 utilizes a spinning 'rotor-tiller' action through the use of its metal teeth, which quickly smash mines and ordnance. The second stage is to cover the same ground with a sifting machine to ensure that what remains of the ordnance is brought to the surface, and then teams using dogs carry out the third stage.

The machine spins, by hydraulic action, a large spiked cylinder on the front of the chassis. It works at a maximum rate of nine meters per minute and can bite a swath of soil with its metal teeth at a rate of up to 1000 square meters per hour under optimum conditions The hydraulic action is generated by a hydraulic pump unit mounted on the back of the Leopard chassis for protection of the hydraulic unit and for counterbalance of the front nose of the machine. The tilling action destroys mines faster than they can detonate. The operator sits safely mid-center in an armored cab. This machine is used in Croatia, Bosnia (Sarajevo) and currently in Kosovo and Angola.

The RHINO mine clearing system (Fig. 13.4) represents a generation of large area mine clearing systems that are remotely controlled and equipped with devices to crush both AP and AT mines as well as other debris, and can survive a maximum AT blast with minimal repair. The RHINO earth tiller de-mining system is designed to force all materials to pass through a small gap, with a maximum width of 3 cm, between two counter-rotating drums. This mine clearing system is based on construction machine technology, RHINO.

13.3 Mine Breaker 2000.

13.4 Mine clearing system RHINO.

The earth tiller consists of:

* the RHINO vehicle equipped with a grader blade
* a de-mining unit with automatic depth control
* a dozer blade
* a remote control system
* a set of tools
* a data collection and transmission subsystem
* protective shields for the driver cabin and other sensitive components.

The video monitored remote control unit allows operation from up to 1000 meters away from the vehicle. The total weight of the machine is 56 metric tons and it has a crawler chassis and is powered by a commercial A 656 kW Caterpillar diesel engine. Even after the detonation of AT mines, the vehicle itself survives with minimal damage. One RHINO earth tiller system is in operation in Croatia; a second system was purchased by Korea. The German government loaned a third RHINO for use in Cambodia. The US tested its system in Jordan. The test of this system shows that it may not be effective against small AP mines such as the

PFM1 butterfly mine. After crushing mines, the RHINO may leave potentially hazardous landmine components in its wake, necessitating a follow up inspection of the cleared path. It requires a substantial logistics and maintenance infrastructure to support successful de-mining operations.

The Bofors Mine-Guzzler (Fig. 13.5) is based on a commercial double track arrangement. A de-mining roller is carried on hydraulic supports at the front of the vehicle and powered by a 650 kW engine with hydrostatic drive. The vehicle drives forward into the suspected area, revolving the tiller unit. The de-mining roller, which is tillable to follow ground undulations, is adjustable for depth and has automatic deep holding. It is made up of a series of plates fitted with tungsten carbide teeth, which either cause the mines (anti-personnel and anti-tank) to detonate or chew them into small pieces. Damaged plates can easily be replaced by oxyacetylene cutting and welding in the field. The Mine-Guzzler can clear AP and AT mines to a depth of 50 cm and over a width of 320 cm. Maximum de-mining speed is 4 km/h. The Mine-Guzzler, which weighs approximately 45,000kg, may be operated remotely, using onboard television cameras, or from

13.5 Bofors Mine-Guzzler.

the protection of the driver's cabin, which is further protected against fragments. The Mine-Guzzler is built to stand detonations from 12 kg of explosives and is a large, tracked vehicle that needs special logistical consideration.

13.6.3 The Spitfire ORACLE

This is a technique integrated within an armored prime mover to form a total de-mining system called ORACLE (Obstacle Removal And Clear Land Equipment) for clearing mines placed both in and on the ground (Fig. 13.6). The tool has a rotating roller mounted on a rebuilt and armored caterpillar. Specially shaped metal teeth are strategically placed on the roller. They tear up the ground and simultaneously destroy the mine before it detonates. The system is built for

13.6 The ORACLE.

full performance in ambient temperatures ranging from +45°C to −30°C. It can be equipped with nine different track configurations to ensure enough traction for virtually any type of soil or ground conditions. The ORACLE can clear up to 5000 square meters per hour.

13.7   Swedish Mine Fighter (SMF)

The Swedish Mine Fighter (SMF) is mine clearance equipment consisting of modules of spikes. These spikes are especially hardened steel and are closely fitted to each other. This means that no mine can escape the spikes, not even if the mine is lying on end. The SMF mine clearance equipment consists of a system of modules that can be fitted together or work as separate units. The modules can be fitted onto suitable prime movers and they can be produced in small or large sizes. The small size version is 1 × 0.3m and the full size version is 3 × 1m. The weight varies from 400kg to 4000 kg, depending on the size of the module. Figure 13.7 shows the de-mining equipment fitted to an excavator. The excavator can rotate around its axle to make it possible to pass through areas with fluctuating pass-ability.

13.8   MgM Rotar on Cat 916

MgM mine clearance group is a German NGO that has been developing and testing several types of mechanical mine clearing machines. The Rotar is a drum, which can easily be mounted/fitted to all types of front-end loaders, wheel loaders, skid steer loaders, hydraulic excavators, and backhoe loaders to replace the vehicle's usual scoop. The drum has a large opening for loading large quantities of material at a time with the objective of providing the capability to sift mines

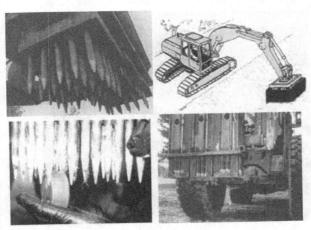

13.7 The Swedish Mine Fighter (SMF).

13.8 MgM Rotar on Caterpillar 916 MK-I.

from the soil. The first step of Rotar action is scooping soil from a suspected mine area into an open bucket with its rotating sieving drum. Then, the drum closes, rotates and sifts out soil and debris smaller than mines in relation to the selected grid size, i.e., to allow smaller objects to fall through the grid, out of the modified bucket. After sifting, suspected items are retained in the drum and are displayed in the inspection area so they can be dealt with safely later in a specially developed disposal system. After dumping, the Rotar will automatically return to position to process another load.

The MgM Rotar is an armored vehicle system that is constructed from a commercially available rubble sifter, modified and mounted on a mine protected front-end loader. Strong armored steel and armored glass has been used to protect the operator and enable the vehicle to withstand explosions from both blast and fragmentation AP mines with minimal damage. The Rotar is available in 10 different models from HPL 400M for skid loaders up to V5400H for heavy-duty applications. For various separation requirements the grid is constructed of spring steel with meshes from 6 mm up to 45 mm. The attachment size is 290 cm × 222 cm diameter, attachment weight is 1700 kg and the base vehicle for de-mining is the 916 Caterpillar (Fig.13.8). The prototype system has undergone extensive testingfor picking up landmines in Namibia and has demonstrated a good clearance capability. The Rotar can be used in a wide range of applications. But, in relation to de-mining the use of MgM Rotar is limited to use in areas where the front loader can operate. Also, it may not sift effectively in highly cohesive and vegetated soils. In addition, large blast mines, in particular AT, may damage the Rotar.

13.9 Minelifta

The Minelifta was developed by Lodlifta Ltd. It is a humanitarian anti-personnel landmine clearance machine (Fig. 13.9). It is designed to clear large areas of AP mines and return cleared agricultural land back to the local population. The machine is designed to be entirely simple in operation to enable the indigenous population to maintain and repair the machine in the field. The machine has a front scraper to lift the soil onto the machine to assist processing and to unearth buried mines. Armored cowling represents the unique part of the Minelifta and this contains the blasts from AP landmines and prevents secondary fragmentation. The debris is then channeled into a central furrow for further processing to enable secondary processing by other machines or by manual de-miners. In addition, the machine has a high inertia flail that detonates and pulverizes all types of mines. It has a rear plow/furrow system to channel the ground into a furrow at the rear of the machine, which enables secondary processing by other machines and safer quality assurance verification to be carried out by manual de-miners. A magnetic separator removes magnetic particles including metallic mine parts from the furrow to make manual verification safer.

13.9 The Minelifta.

13.10 Remotely controlled mechanical systems

13.10.1 Light and remotely controlled flail

Smaller and cheaper versions of the flail systems have been developed with chains attached to a spinning rotor to beat the ground and they are integrated with remotely controlled, line-of-sight, skid loader chases. The use of light-flails aims to safely clear light to medium vegetation, neutralize AP-mines and UXOs from footpaths and off-road areas, and assist in the area reduction of a minefield (see Fig. 13.10). These machines are developed to provide a capability to remotely clear AP mines and proof areas that have been cleared (US Department of Defense, 1998; GICHD, 2006a). The machines were designed for dealing with vegetation clearance and tripwires as a precursor to accelerate manual clearance. These flail systems are not designed for heavily vegetated or extremely rough terrain. Some systems can clear AP mines from off-road locations and areas that are not accessible by larger mechanical mine clearing equipment. The light-flail can defeat bounding, tripwire, fused, and simple pressure AP mines. In addition, these machines have a flail clearance depth of between 150 mm and 200 mm and a working width ranging between 1.4 m and 2.22 m. These machines are designed to withstand blasts of up to 9 kg of TNT. They are remotely controlled up to a range of 5000 m through feedback sensors and up to 500m away (line-of-sight distance) if they are operating in an open space. An armored hood is available to protect the machines against AP mine blasts. Furthermore, there is a set of tracks that can be installed over the tires when working in soft soil conditions to improve traction.

Different machines made by different manufacturers with almost similar concepts are available and have been used in real minefields (Likco and Havlik, 1997; US Department of Defense, 1998; Green, 1999; GICHD, 2004; GICHD,

(a) Armtrac 25.
(Armtrac Ltd., United
Kingdom)

(b) Bozena 4
(Way industry J.S. Co,
Slovak Republic)

(c) Mini-flail.
(US Department of
defense)

(d) Diana 44T
(Hontstav S.R.O.,
Slovak Republic)

(e) Minecat 140.
(Norwegian demining
consortium)

(f) The MV-4.
(DOK-ING d.o.o., Croatia)

13.10 Different types of light flails in action.

2006a; Danielsson & Blachford, 2003; Danielsson & Coley, 2004; Leach, 2004). Some of these are mentioned below:

- Two Armtrac 25 machines are in service with the UK Ministry of Defence with no information for actual usage in a real minefield.
- More than 110 Bozena machines have been produced. These machines have been, or are currently, in service in Afghanistan, Albania, Angola, Azerbaijan, Bosnia and Herzegovina, Cambodia, Czech Republic, Eritrea, Ethiopia, Iraq, Kenya, Kosovo, Lebanon, The Netherlands, Poland, Slovakia, Sri Lanka, and Thailand.
- The Compact Minecat 140 was developed in 2001 as a direct follow-up improvement of the MineCat 230 and has not yet been used in real minefields.
- There are 62 MV-4 light flails that have been purchased by various organizations/de-mining companies. Some of the organizations are the US Army (21 units), Swedish Army (5 units), Croatian Army (2 units), Irish Army (2 units), International Mine Action Training Centre (IMATC) Kenya (1 unit), Croatian Mine Action Centre (CROMAC) (4 units), Iraqi National Mine Action Authority (4 units), Norwegian People's Aid (NPA) (3 units), Swiss Foundation for Mine Action (FSD) (5 units), etc.
- Mini-flails have been tested extensively in Kuwait, Bosnia, Kosovo, and Jordan. Currently, six mini-flails are deployed in the Balkans, and four systems are deployed in Afghanistan. The new version 'Mini-Deminer' incorporates improvements to the problems associated with the US Army's original mini-flail that were identified during field evaluations. Development testing of the Mini-Deminer took place during the spring and summer of 1999.
- There is no information available from the manufacturer on the actual usage of Diana 44T machine in real minefields.

All light flail machines are small and compact in size, easy to transport on a light trailer, remotely controlled, easy to maintain and repair, and have a powerful engine with efficient cooling system.

Light flail machines have difficulty in operating with precision from a long distance (this applies to all remotely controlled machines), as they require line-of-sight operation with suitable feedback. The ground flailing systems create large dust clouds and tall vegetation will restrict the view of the operator on the machine. They also exhibit difficulty in flailing in soft soil, and can inadvertently scatter mines into previously cleared areas. These machines are not intended for use in areas where AT mines are present, and they may not be usable in steep or rocky terrain.

13.10.2 Enhanced teleoperated ordnance disposal system (ETODS)

The enhanced teleoperated ordnance disposal system (ETODS) is a remotely controlled teleoperated system that is based on a modified commercial skid loader with a modular tooling interface which can be field configured to provide the abilities to remotely clear light vegetation, detect buried unexploded ordnance (UXO) and landmines, excavate, manipulate, and neutralize UXO and landmines, to address the need of various mechanical clearance activities associated with humanitarian de-mining (Eisenhauer et al., 1999). ETODS has an integrated blast shield and solid tires.

ETODS includes a heavy vegetation cutter and a rapidly interchangeable arm with specialized attachments for landmine excavation. Attachments include an air knife for excavation of landmines, a bucket for soil removal, and a gripper arm to manipulate certain targets. Remote control capability combined with a differential GPS subsystem and onboard cameras enable the system to navigate within a minefield to locations of previously marked mines. Mines or suspicious objects already marked or identified with GPS coordinates can be checked and confirmed with an on-board commercial detector, and then excavated with a modified commercial backhoe, an air knife, excavation bucket, or gripper attachment. ETODS was developed and configured for the US Department of Defense (DoD) humanitarian de-mining research and development program starting in 1995. It has been through many field test activities, and was found to be suitable for use in humanitarian de-mining (HD) operations.

The HD issues that have been evaluated include accuracy, repeatability, and feasibility of usage in remote environments. In relation to vegetation cutting, three attachments have been tested: one front-mounted bush hog and two side-mounted boom mowers. In this case, the HD issues that have been evaluated include the ability to cut dense undergrowth, the proper preparation of the ground for ensuing detection activities, and the ability of the operator to effectively and efficiently clear an area under remote control. For the commercial backhoe, which can be field mounted to the ETODS, the HD issues that have been evaluated include the effectiveness and efficiency of locating and excavating mines, operator training requirements, inadvertent detonation rates, techniques for deeper excavations, techniques to identify mines and their status (e.g. booby trapped), and blast survivability/repair. A chain flail attachment converts the ETODS into a system capable of clearing AP mines through detonation, and in this case the HD issues that have been evaluated include the minimum sized mine cleared, depth of clearance, effectiveness of clearance, speed of clearance, and blast survivability/repair. During testing, ETODS was subjected to a 12 lb TNT blast replicating an AT mine detonation. ETODS drove away with field repairable damage. ETODS has proven effective in detonating M14 AP mines and is survivable through repeated 1.0 lb TNT detonations (OAO-Robotics, website). ETODS provides the safe, effective delivery of tools necessary for the clearance of landmines and UXO. ETODS is simple, rugged, and can provide a high technology indigenous de-mining capability in remote environments.

The ETODS has completed operational field evaluations in Jordan and Egypt, where it was found to have several significant limitations that make it less than

13.11 The ETODS in action.

suitable for humanitarian de-mining operations (Fig. 13.11 shows the ETODS in action). These include the tendency to become mired in mud or desert sand conditions, as well as the requirement for significant training to develop teleoperation skills (Department of Defense, 2002).

13.10.3 TEMPEST

TEMPEST is designed to safely clear light to medium vegetation, clear tripwire-fused mines, and assist in area reduction as a precursor to accelerated manual clearance. DTW began production of the TEMPEST Mk I in 1998–99, when it was designed purely as a vegetation-cutting device, and currently, the TEMPEST Mk V is in production. The TEMPEST Mk V is a remotely controlled, lightweight multi-tool system with vegetation cutting and trip wire clearing abilities (see Fig. 13.12).

TEMPEST is a low cost, small size and lightweight radio controlled AP mine blast-protected multi-purpose ground-based system. These features aim to give the system ease of transport and agility over difficult terrain. It can support a variety of interchangeable clearance heads to clear vegetation, uses large and small magnets to remove metal fragmentations, can engage the ground with a flail head, and neutralize tripwires, etc. It is designed to clear AP mines from off-road areas inaccessible to large-area mine clearers. The TEMPEST system consists of a diesel-powered hydraulically driven chassis, a radio control subsystem, and each of its four hydrostatic wheels is driven by an independent motor to improve maneuverability. The wheels are easy to remove, repair and replace. The TEMPEST also has a 1.2 meter wide horizontal chain flail with vegetation cutting

13.12 Tempest during operational field evaluation.

tips, and an adaptable flail head with hydraulic feedback system that can sense the load on the flail, i.e., the operator can set the speed control to maximum and the TEMPEST will automatically control its cutting rate and drive speed, and progress accordingly. The TEMPEST's ground engagement flail is designed to dig into the soil in order to destroy or expose mines by cutting 10 cm deep into the ground to initiate surface and subsurface mines at that level. Its V-shaped chassis and sacrificial wheels minimize damage from anti-personnel mine or UXO detonation and provide some protection against anti-tank mines. TEMPEST's vertical axis 'slasher' is capable of cutting through difficult vegetation such as bamboo and vines and its large magnetic array is capable of extracting ferrous material from the ground. The TEMPEST has demonstrated clearing capability of up to 1000 m^2/hour of light vegetation (75 cm of high grass, occasional small bushes), and up to 630 m^2/hour of medium vegetation (100 cm high grass, 2 m high bushes at a density of two per m^2). It also can clear up to 450 m^2/hour of moderate to heavy vegetation (100 cm high grass, 2–3 m high bushes at a density of three per m^2, and occasional 2–5 m tall trees). In addition, it can cut 20 cm diameter, 5 m tall conifers to 2 m above ground and mulch foliage in less than 4 minutes (Department of Defense, 2002). The latest versions of the TEMPEST systems have better and enhanced functions and performance. TEMPEST features ease of operation, maintenance, and repair.

TEMPEST is inexpensive to purchase and operate relative to other vegetation clearance systems. Currently, TEMPEST is produced in Cambodia as well as the United Kingdom, thus representing a regional capability in Southeast Asia (Department of Defense, 2002).

TEMPEST is an excellent example of how an operational evaluation can lead to improvements that realize the potential of a prototype design. The early prototype of TEMPEST underwent extensive tests in Cambodia for AP and AT mines. TEMPEST began an operational evaluation in Thailand in January 2001. Although it was effective at clearing vegetation in mined areas, Thai operators identified overheating problems. The unit's promising performance warranted the investment of funds to improve the system. TEMPEST Mk IV was tested in Mozambique during 2003. The actual use of TEMPEST systems and the continuous evaluation resulted in TEMPEST Mk V being a reliable system with more speed and engine power capacity compared to the previous versions.

As evaluated by the manufacturer, the hydraulic hoses are vulnerable to fragmentation attacks, and the machine is not intended to be used in areas where AT mines are present. As evaluated by de-miners, the TEMPEST requires the operator to maintain direct line of sight with the system from a minimum of 50 meters and the operator can only be this close if behind the system's portable shield. This poses a problem in dense vegetation or rolling terrain. The TEMPEST has limited traction on wet muddy terrain due to the steel wheels clogging with mud. The machine has the ability to clear both mines and vegetation, but with limitations: the ground flailing system creates large dust clouds, the view of the operator on the machine can be restricted and the air filters can become clogged (Leach et al., 2005).

Currently, there are 25 machines operating in Angola, Bosnia, Cambodia, DR Congo, Mozambique, Sri Lanka and Thailand. The TEMPEST is currently used by seven de-mining organizations around the world (GICHD, 2006a). The new TEMPEST Mk VI will mitigate the highlighted problems by use of a new remote control system and the integration of tracks in place of the steel wheels to enable the vehicle to operate in most soil conditions and terrains.

13.10.4 The armored combat engineer robot (ACER)

Mesa Robotics has developed a series of teleoperated mobile platforms targeting a range of applications. Among these are MARV, MATILDA and the armored combat engineer robot (ACER) Robotic Platform.

The mobile base platform of ACER is armored with ballistic steel, measures 83" W × 62" H × 56" L and weight 4500 lb. It is powered by a 12 VDC NiMH battery with a possible operating time of between 1 and 2 hours. It has a hydraulic driven system with maximum speed of 6.3 mph and its payload capacity is 2500 lb. A driving color camera with IR is integrated into the ACER. The vehicle can negotiate obstacle up to 10 inches and can move up or down slopes of 60 degrees. ACER accepts a range of custom and standard attachments such as flail, blades, buckets, etc. and it has a towing capacity of 25,000 lb and arm lift capacity of 1000 lb. The vehicle's fording depth is 2 inches with zero turning radiuses (see Fig. 13.13).

ACER can be remotely controlled by one person through a belly-box operator control unit (OCU) with a control range of about 500 meters (see Fig. 13.13). The OCU has a 900 MHz digital control, 1.8 or 2.2 GHz analog video system, 6.4" display and two control joysticks: one for the vehicle and the other for arm control.

13.13a The mobile base unit of ACER with some possible attachments.

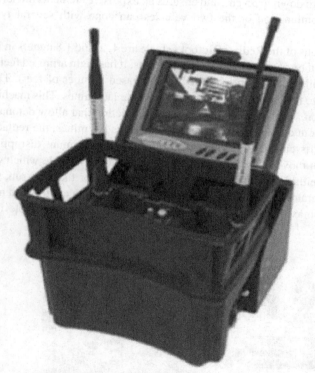

13.13b Belly-box operator control unit (OCU).

ACER weighs 6 lb and is powered by a 12 VDC NiMH battery with a 120 VAC adapter.aaaa

ACER provides a variety of capabilities for remote operations: UXO handling and removal, clearing and breaching, combat engineer support, hazardous material handling, logistics support, decontamination, and fire fighting.

ACER is still new and no testing for de-mining has been reported yet.

13.10.5 Redbus land mine disposal system (LMDL)

Redbus Limited has developed a combination of two machines (the Redbus Bigfoot and the Redbus Mineworm), which are dedicated to destroying/detonating mines and removing unexploded ordnance (Fig. 13.14). Both vehicles are remotely controlled. The clearing operation starts first with the Redbus Bigfoot applying pressure to detonate live mines, then the Redbus Mineworm excavates the topsoil to remove fuses, live ammunition, faulty mines and other ordnance. The system detonates all mines that can be exploded and brings unexploded ordnance to the surface. Topsoil covered by the machines is filtered and fragmented to a chosen depth down to 55 cm, and ensures no explosive elements are left in the ground. The combination of the two vehicles can cope with several types of terrain.

The dimensions of the Redbus Bigfoot vehicle are 4, 2, and 1.8 meters in length, width and height respectively and it weighs 5 tons. It has eight armored feet (other models are available with an increased or decreased number of feet). The feet apply pressure to the ground in sequence to detonate landmines. This machine has shock absorption, blast deflection and energy dissipation that allow detonation of AP mines without damage. For larger mines, such as AT mines, the replacement of low cost parts can be performed in the field with minimum disruption. As Redbus Bigfoot moves along the path, an attachment is available which can cut vegetation on either side of the vehicle in preparation for the next run. CCTV enables the operator to maneuver Redbus Bigfoot by remote control and monitor progress. The maximum rate of ground covering in a de-mining role is 1000

13.14 The Redbus Bigfoot and the Redbus Mineworm.

square meters per hour. The area covered or cleared will vary depending upon terrain, ground type and weather conditions.

The dimensions of the Redbus Mineworm vehicle are 3 m in length for the main unit and 2 m length for the fragmentor, 2 m in width and 2 m in height. The main unit weighs 4 tons and the fragmentor weighs 1.75 tons. The Mineworm follows behind Redbus Bigfoot to remove ordnance and ferrous metal to leave the area in a safe condition. It excavates the soil to chosen depths, up to 55cm, removing all unexploded ordnance for destruction in its crushing mechanism or for safe disposal. In addition, it is remotely controlled with cameras to monitor progress and inspect the excavated material. At the front of the machine are a soil breaker and root cropper. This prepares the ground for the rotating excavator that lifts the soil and passes it through the Redbus Mineworm. The maximum rate of ground clearance at 25 cm depth is 1000 square meters per hour.

13.11 Conclusions

Most of the available mechanical mine clearing equipment contributes to solving the mine problem. This contribution is bounded by their limitations as there is no one mechanical mine clearance technology that can clear all mines under the diverse environment, climate and terrain conditions. Despite the technological advances it is unlikely that mechanical clearance alone will, at least in the near term, meet the criteria required for humanitarian de-mining.

Mechanical techniques are only appropriate in large areas without dense vegetation or steep gradients. On small paths or in thick bush the machines simply cannot maneuver. Maneuverability on wet ground is limited, and combined with inclined terrain can be very difficult. It is difficult to imagine mechanical methods being applicable, for example, in defensive ditches, around large trees, in a residential area, on soft terrain, etc. Wheeled vehicles cannot travel efficiently on anything other than flat surfaces, tracked vehicles cannot travel in areas with steep vertical walls, machines in general cannot climb undefined obstacles, and machines cannot in general change shape to get through narrow entrances. The high cost of advanced technologies often makes them unaffordable to developing countries, along with the logistical problems associated with transporting heavy machinery to remote areas in countries with little infrastructure. Also with this technique, machines often do not destroy all mines in a contaminated area and AP mines may be pushed on their sides or buried deeper or partly damaged making them more dangerous. In addition, mechanized solutions to mine clearance have environmental effects, such as erosion and soil pollution due to exploded mine residuals, which have not always been duly studied.

To improve the applicability and effectiveness of mechanical mine clearance systems, they should be designed taking into consideration modularity, the safety of operators, transportability/minimum on-site logistic support and they should be less complicated, more cost effective, be easy to maintain and operate by local staff, cost effective and employ the use of local materials. They should have efficient and flexible locomotion structures to facilitate maneuverability under different circumstances. Such capabilities should enable these machines to have higher climbing angles and operate under different terrain. The design of such machines should be considered in association with multi functions that can deal for example with mines, vegetation, tripwires, etc. These machines should be able to work under different environmental and operating conditions, such as temperature, humidity and soft terrain. They should be capable of withstanding explosive blasts without suffering major damage and they should have dust filters for engine and operators. Mechanical vehicles and associated mechanisms should be equipped with well-protected sensing capabilities and should be versatile to perform tasks beyond mine action alone.

13.12 References

Baudoin, Y. and Acheroy, M. (2002). Robotics systems for humanitarian demining: modular and generic approach and cooperation under IARP/ITEP/ERA Networks. IARP Hudem'02 WS, Vienna.

Blagden, P. M. (1993). Summary of UN Demining. *Proceedings of the International Symposium on Anti-Personnel Mines*, Montreux, April 1993, CICR/ICRC, pp 117–123.

Busuladzic, D. and Trevelyan, J. (1999). An Ergonomic Aspect of Humanitarian Demining. http://www.mech.uwa.edu.au/~jamest/demining/tech/dino/ergonomics.html.

Cain, B. and Meidinger, T (1996). *The Improved Landmine Detection System*. EUREL, pp. 188–192.

Danielsson, G. & Blachford, P. (2003). DIANA 44T Test and Evaluation – August 2003. Ref.: 13 345:60629, Swedish Armed Forces, Swedish EOD and Demining Centre, 2003.

Danielsson, G. & Coley, G. (2004). Minecat 140 Test and Evaluation – Sept. 2003. Ref.: 13 345:60099, Swedish Armed Forces, Swedish EOD and Demining Centre, 2004.

Department of Defense, Humanitarian Demining Research and Development (R&D) Program (2002). Other Completed Mine/Vegetation Clearance Equipment. *Development Technologies Catalog 2001–2002*.

Eisenhauer, D. J., Norman, C. O., Kochanski, F. K., and Foley, J. W. (1999). Enhanced Teleoperated Ordnance Disposal System (ETODS) for Humanitarian Demining. *Proceedings of the 8th International Meeting of the American Nuclear Society (ANS)*, Pittsburgh, PA.

Gage, D. W. (1995). Many-robot MCM Search Systems. Proceedings of the Autonomous Vehicles in Mine Countermeasures Symposium, Monterey, CA, April 1995, pp.9.56–9.64.

Geneva International Centre for Humanitarian Demining (GICHD) (2004). *A Study of Mechanical Application in Demining*. Geneva International Centre for Humanitarian Demining, Geneva, Switzerland.

Geneva International Centre for Humanitarian Demining (GICHD) (2005). *A Study of Manual Mine Clearance (Books 1–5)*. Geneva International Centre for Humanitarian Demining, Geneva, Switzerland.

Geneva International Centre for Humanitarian Demining (GICHD) (2006a). *Mechanical Demining Equipment Catalogue*. Geneva International Centre for Humanitarian Demining, Geneva, Switzerland.

Geneva International Centre for Humanitarian Demining (GICHD) (2006b). *Guidebook on Detection Technologies and Systems for Humanitarian Demining*. Geneva International Centre for Humanitarian Demining, Geneva, Switzerland.

Green, W. E, (1999). The Case for the Flail Mechanical Landmine Clearance for the Humanitarian Application: A Manufacturer's View. *Mine Action Information Center Journal*, Vol. 3, No. 2, 1999.

Habib, M. K. (1998). Multi Robotics System for Land Mine Clearance. *Proceedings of the International Conference on Robotics and Computer Vision (ICRACV'98)*, Singapore, Dec. 1998.

Habib, M. K. (2001a). Mine Detection and Sensing Technologies: New Development Potentials in the Context of Humanitarian Demining. *Proceedings of the IEEE International Conference of Industrial Electronics, Control and Instrumentation (IECON'2001)*, USA, 2001, pp. 1612–1621.

Habib, M. K. (2001b). Machine Assisted Humanitarian Demining Mechanization and Robotization, *Proceedings of the International Field and Service Robots 2001*, Finland, pp. 153–160.

Habib, M. K. (2002a). Mechanical Mine Clearance Technologies and Humanitarian Demining: Applicability and Effectiveness. *Proceedings of the 5th International Symposium on Technology and Mine Problem*, California, USA, April 2002.

Habib, M. K. (2002b). Mine Clearance Techniques and Technologies for Effective Humanitarian Demining. *International Journal of Mine Action*, Vol.6, No.1.

Habib, M. K. (2004). Mechanization Technology of Mine Clearing Operations and Humanitarian Demining. The Eleventh World Congress in Mechanism and Machine Science. Tianjin, China, pp. 973–978.

Habib, M. K. (2007a). Humanitarian Demining: Reality and the Challenge of Technology – The State of the Arts. *International Journal of Advance Robotic Systems*, Vol. 4. No.2, June 2007, pp. 151–172.

Habib, M. K. (2007b). Controlled Biological and Biomimetic Systems for Landmine Detection. *Journal of Biosensors and Bioelectronics*, Elsevier Publisher, Vol. 23, pp. 1–18.

Habib, M. K., (2008a). Humanitarian Demining and the Challenge of Technology. *Proceedings of the 7th IARP International Workshop on Robotics and Mechanical Assistance in Humanitarian De-mining and Similar risky interventions (HUDEM 2008)*, Cairo, pp. 90–96.

Habib, M. K. (2008b). Humanitarian Demining: Difficulties, Needs and the Prospect of Technology. *IEEE International Conference on Mechatronics and Automation 'ICMA2008'*, Takamatsu-Kagawa, Japan, Paper Reference WC2-2.

Habib, M. K. (2008c). Humanitarian Demining: The Problem, Difficulties, Technologies and the Role of Robotics. Chapter 1, In *Humanitarian Demining, Innovative Solutions and the Challenges of Technology*, Maki K. Habib (ed.), ARS-pro literature Verlag Publishers, pp. 1–56.

Havlík, Š. (2005). A modular concept of robotic vehicle for demining operations. *Autonomous Robots*, Vol. 18, 2005, pp. 253–262.

Hewish, M. & Ness, L. (1995). Mine-detection Technologies. International Defense Review, October 1995, pp. 40–46.

Humanitarian Mine Action Equipment Catalogue (1999), German Federal Foreign Office.

International Committee of Red Cross (1993). *Antipersonnel Mines: An Overview*. Geneva, September 1996. (See also: http://www.icrc.org/).

International Committee of Red Cross (1996). *Antipersonnel Mines – Friends or Foe?* ICRC Publication, Ref. 0654, Geneva, 1996.

King, C. (1997). Mine Clearance in the Real World. *Proceedings of the International Workshop on Sustainable Humanitarian Demining*, Zagreb (SusDem'97), pp. S2.1–8.

Leach, C. (2004). Bozena 4 Mini Mine Clearance System Assessment Phase 1: QinetiQ/ FST/LDS/CR044502/1.0. Farnborough, 2004.

Leach, C., Blatchford, P., Coley, G. & Mah, J. (2005). *TEMPEST V system with Ground Engaging Flail Cambodia Trials Report*. Farnborough: NETIQ/FST/LDS/ TRD052379, 2005. p. 3.

Likco, P. and Havlik, S. (1997). The demining flail and system BOZENA. International Workshop on Sustainable Humanitarian Demining.

McFee, J. E. and Carruthers, A. (1996). Multisensor Mine Detector for Peacekeeping: Improved Landmine Detector Concept. *Proceedings of SPIE Technical Conference on Detection and Remediation Technologies for Mines and Minelike Targets*, Abinash C. Dubey; Robert L. Barnard; Colin J. Lowe; John E. McFee (eds.), Volume 2765, May 1996, pp. 233–248.

Nicoud, J.-D. and Habib, M. K. (1995). PEmex-B Autonomous Demining Robots: Perception and Navigation Strategies. *Proceedings of the IEEE/RSJ International Conference on Intelligent Robots and Systems (IROS'95)*, Pittsburgh, August 1995, pp. 419–424.

Nicoud, J.-D. (1996). A Demining Technology Project. *Proceedings of the International Conference on Detection of Abandoned Land Mines (MD'96)*, Edinburgh UK, October, 1996, pp.37–41.

OAO-Robotics website, http://www.manitgroup.com/oao.htm

O'Malley, T. J. (1993). *Seek and Destroy – Clearing Mined Land. Armada International*, Vol. 17, No. 1, February–March 1993, pp. 6–15.

US Department of Defense (1998). *Humanitarian Demining Development Technologies*. Catalogue, USA, 1998.

US Department of State (1994). Hidden Killers: The Global Landmine Crisis. Report to Congress, Washington D. C., Publication 10225, December 1994 (See also: http:// www.state.gov/).

Van Westen, C. J. (1993). Remote Sensing and Geographic Information Systems for Geological Hazard Mitigation. *ITC-Journal*, No. 4, 1993, pp. 393–399.

14

Robotic tools for de-mining and risky operations

Š. HAVLÍK, Slovak Academy of Sciences, Slovakia

Abstract: The chapter deals with applications of mobile robotic technology for performing risky tasks, mainly de-mining operations. The robotic approach and problems related to the de-mining process are analyzed in more detail and some specific performance features that an unmanned robotic agent should exhibit are discussed. The modular concept of the remotely controlled robotic vehicle, which consists of the general mobility system and a set of exchangeable task oriented tools, is also discussed. As an illustrative example, the brief history and current state of development of the remotely controlled vehicle 'Božena', with flailing activation mechanism and other available tools, are presented.

Key words: de-mining vehicle, robotic approach, modular concept, mine detection, remote control.

14.1 Introduction

The use of robotic technology to perform risky tasks to prevent threats to human life or large economic losses, as well as ecological damage, has became a major challenge for researchers in several domains, mainly in robotics. Dangerous operations include clearing terrain of landmines, fire fighting in uneven terrain, aiding after nuclear accidents or physical catastrophes (earthquakes, sea or river disasters, etc). As human safety is the highest priority, the best option is to remove the human operator from the scene and/or totally substitute him with an 'intelligent' agent, which is expected to provide the operator with the means that would enable him to perform the same mission safely. These application fields require specific robotic systems that exhibit specific features and display some limited level of autonomy for mobility and functions.

There are several task-oriented vehicles or mobile systems for inspection, searching dangerous terrain or fighting forest fires. But the actual need for humanitarian de-mining is one of the most challenging world issues for the development of new de-mining technologies, which should be safe, fast and reliable.

New research achievements that have been reached in robotics and other related domains could provide a new qualitative base for further development of more sophisticated approaches that can be applied for solving problems related to all risky operations. There has been progress in the following disciplines:

- Microelectronics and micro engineering. New materials and technologies mean robotic devices are smaller, lighter and more reliable in operation.
- New sensing principles and sensors.
- Computer and information processing technologies.

327

- Control and possibility of GPS localization/navigation.
- Communication technologies.

Naturally these results go together with achievements in the theoretical elaboration of particular problems. Examples include artificial intelligence (AI) approaches and the development of methods based on neural networks (NN) for target recognition or navigation/control.

This chapter deals with some common features of robotic devices for performing operations in dangerous terrain. The main focus of the chapter is the robotic tools for de-mining operations, as they are the most common and complex activity at the forefront of research interest.

14.2   Features and requirements of robotic tools for de-mining and risky operations

Performing risky operations in dangerous environments requires three principal features to be satisfied by an intervention robotic agent: self-recovery capability, minimal risk assessment and maximal reliability in all actions.

Self-recovery

This is an important and specific feature directly related to particular tasks. Its main purpose is to prevent/avoid losses or self-destruction of the agent and to finish a specified action in a risky environment without sustaining serious damage. The self-recovery strategies should occur in unwanted situations as follows:

- The failure of a system (communication, engine, control system, sensory system, etc). The problem is to remove the agent from the dangerous terrain without any risk to persons.
- In the case of erroneous decisions made by operators.
- There is no or not enough information for further action and it seems to be risky for the agent to stay/to continue action, as planned.

A general requirement is to build the agent and all its functional parts so that they are as reliable as possible. For instance, a typical requirement is that communication should involve two independent systems.

Consider an agent performing a de-mining operation in the minefield. Each motion of the vehicle in this dangerous terrain should be carefully judged, otherwise any wrong movement could result in its destruction. The principal strategy should be based on the following rule: any motion of the vehicle should take place in a direction where no accident can arise (mines have been reliably detected or all the mines have been destroyed/removed).

Another problematic situation arises in the event of any failure that means the vehicle cannot perform its desired activities. The difficulty is in removing the vehicle from the dangerous terrain without any risk to persons. One of the simplest ways to solve this problem is to use a cable and pull the vehicle out with a winch mechanism. Another possibility involves using another vehicle to remove the first one from the minefield.

Minimal risk

Solving any situation creates a decision problem for the operator/operation system: how to decide on the next course of action if an unexpected situation arises. The general rule is that the operator decides on the next step for the agent to take in order to minimize any risk of damage to the agent itself. This procedure represents the standard decision algorithms according to the risk assessment scheme in Fig. 14.1.

It is obvious that some decisions can be represented by relatively simple routines working over a given set of options: action: <STOP/GO BACK/ ... > if < CONDITION: SENSOR xx >. On the other hand, the operator's decisions require

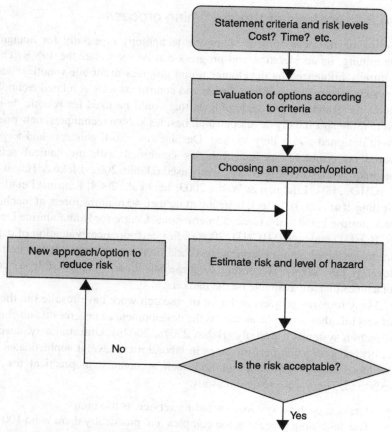

14.1 The general risk assessment and decision procedure.

a much more complex assessment of the possible risks with respect to the given criteria. Such a typical situation arises during the automatic de-mining operation when the mine detection systems do not give reliable information about the presence of mines and there is only a suspicion of whether 'something is inside'. Then, the operator should decide whether to continue the same task, or if some other techniques of searching using different detection systems and fusion are needed.

Reliability

The reliability of performing a task should be considered with respect to the criteria given for a particular operation. To compare different tasks, the actions and criteria that are used for performing different risky operations are listed with examples in Table 14.1.

14.3   Mine clearance

14.3.1 Analysis of the de-mining process

Following the development of robotic technology especially for humanitarian de-mining, its acceleration and progress can be seen after the 1980s. This was naturally influenced by development and progress in mobile robotics, detection systems, computing, communication and control, as well as related technological domains. Some existing technologies that could be used for robotic de-mining were more specifically developed and, besides known techniques, new principles were designed and widely studied. Despite this effort, concepts and techniques based on remotely operated vehicles combined with mechanical activation techniques are still the most frequently used (Habib, 2002; Licko & Havlík, 1997; GICHD, 2004; Lindman & Watts, 2003; Ide et al., 2004; Kaminski et al., 2003; Stilling et al., 2003). In fact, there are more then 40 manufacturers of machines for this purpose listed in the Geneva International Centre for Humanitarian Demining (GICHD) catalogue (GICHD, 2008a). For performance evaluation of particular machines the common criteria and test conditions were accepted at the international level (CEN, 2004). According to these standards it is possible to compare machines that are available on the market.

Many research projects and a lot of research work have resulted in the design of several other concepts, as well as the development of new robotic machines and detection systems, especially (Habib 2007a, 2007b). Unfortunately, despite this effort and promising performances in laboratories, several sophisticated robotic solutions and systems did not find such acceptance in practical use as was expected. There were several reasons:

- The cost of these systems, including service, is too high.
- The de-mining process is too complex and practically there is no 100 percent repeatability in performing particular tasks.

Table 14.1 Description and criteria of some risky operations

Operation	Task description (example)	Description of the action	Criteria
Humanitarian demining	Steps in demining: – Clearing vegetation, if any exists – Detection of mines and localization – Destruction of mines in place or removing	Robotic agents for performing particular tasks are operated from the control center. Agents are equipped with detection systems and tools for destruction/neutralization.	• High reliability of clearing (up to 100%) • Safety • Cost/speed
Fire fighting	Active intervention: (water/foam/sand gun, placement explosives) to extinguish the fire, removing objects/ obstacles, etc.	Mobile robot carries extinguishing material and tools to put materials into fire, or other actions. Poor/no visibility, unknown obstacles, toxic gases, high temperature, navigation problems, recovery. Sensory equipment: GPS, vision and thermal imaging.	• Robustness to work in harsh conditions
Nuclear power plant actions	Surveillance and manipulation with nuclear materials.	The robot hand executes manipulations with dangerous materials. The detection– measuring system verifies level of radiation. The operation space is visible and quite well defined. Equipment: Remotely-operated transport platform with on-board robot-manipulator, tools, vision system, gamma-detector (gamma locator)	• Reliability
Antiterrorist action	Airport action: Remove an object as potential/ suspected explosives or other dangerous materials inside. The agent should remove dangerous object from the space as soon as possible.	The remotely operated mobile robotic agent will approach the object, takes it by an arm and inserts the object into a container on its platform. The object is then carried out into a safe space for further recognition. The environment is quite well defined as to terrain, visibility and weight of objects. Equipment: Mobility system, robotic arm, vision, hand held camera, remote control (~200m), other sensors/ detectors	• Minimum time • Prevention of accidental explosion

- The danger of explosion is very high. For this reason some robust techniques, as for instance flailing, are preferred.
- To achieve desired performance (safety, reliability, cost, robustness, etc.), there are still problems that need to be solved.
- Economy: the market in the domain of robotic technology for de-mining, comparing to manufacturing robotics, is relatively small. On the other hand, the problems that need to be solved are very complex.
- Practically, there is no unification or standardization of parts. For these reasons large manufacturers of industrial robotic technology have very limited interest in engaging in this domain.

The overviews of existing research projects, techniques and equipment that have been developed for performing particular tasks are listed in several databases: www.gichd.ch, www.eudem.vub.ac.be, www.hdic.jmu.edu, www.state.gov/t/pm/rls/rpt/, www.demining.brtrc.com/r_d, and in proceedings from HUDEM conferences dealing with humanitarian de-mining organized by the International Advanced Robotics Programme (IARP) community.

When we talk about robotic technology for de-mining we have in mind robotic devices that have equipment with the following features:

- An unmanned vehicle/platform that is able to move and work in the minefield automatically, semi-automatically, or remotely controlled.
- A vehicle that is equipped with on board manipulation systems (robot arm, manipulator) with some desired motion and sensing capabilities for performing prescribed operations.
- A vehicle equipped with detection systems and sets of special task-related tools for performing given tasks from the de-mining process.

In principle, advanced robotic technology can be applied to performing specific tasks in all phases of the de-mining process:

- Searching large areas and localization of mines and any explosives (UXO) using fast and reliable methods. Scanning terrain and the detection and localization of mines is crucially important in the whole de-mining operation. Considering the great number of types of mines produced (different forms, plastic materials, colors, etc.), the variety of terrains, as well as the possibility of hiding mines in various terrains, reliable detection equipment is highly desirable. A complex detection system should work with more sensing principles and fusing sensory information. This naturally includes the development of fast and reliable recognition algorithms. Special robotic devices/vehicles should be used as carriers of detection systems for performing searching/scanning motions in the minefields and precise localization of detected targets.
- Preparing infected terrain for reliable detection, as well as for neutralization procedures, which involves removing vegetation and any obstacles that could prevent detection or safe neutralization. When the activity of the human

de-miner is analyzed, preparing the terrain represents a very large portion of the time in which he is under great psychological pressure. The principal problem is that after years of mine deployment, the terrain is frequently covered by soil and vegetation or other mechanical objects. This is one of the most critical and dangerous tasks and requires the application of special tools and procedures. This fact naturally results in difficult decisions about which technique will be used in a particular case. Applying robotic techniques for this kind of operation creates serious problems due to the real possibility of unknown objects and unforeseen situations. Performing such complex tasks require a multi-degrees of freedom (d.o.f.) robotic arm and sets of exchangeable tools together with additional sensory equipment (such as haptic–tactile and force/torque sensors, proximity and position sensors, cameras in the hand, etc). Further development of techniques together with exchangeable task/oriented tools for removing different objects and clearing vegetation are highly desirable. Naturally, a much more sophisticated interface and feedback control system is needed.

- Neutralization/destruction of mines. In principle, there are three methods of neutralization. The first method results in the activation of an explosion of the mines directly in the place of their occurrence. For this purpose the flailing technology is the most frequently used. The second approach involves the neutralization/destruction of mines in situ without explosions. Several techniques including burning, abrasive water jet cutting, etc. have been and are under development. The third method involves removing mines from the terrain and their safe transport to another location for further neutralization. For this purpose special grippers are needed.

As to the time needed for performing particular tasks, about 20–30 percent of it is spent on detection, 50–60 percent on clearing the vegetation and preparing the terrain and only 10–30 percent of the total time is spent on neutralization. Statistical data from MacDonald et al. (2003) states that of 200 million items excavated within a specific period only about 500,000 items (less than 0.3 percent) were anti-personnel (AP) mines or other explosive devices. Thus more than 99 percent of the total working time of the de-mining crew was spent on confirmation and excavation of scrap items that were a result of false signals from metal detectors. Combining the above data one can deduce an important fact that the most significant time savings in de-mining can be expected if the rate of false alarms is reduced. Due to safety rules, any false alarm should be considered as a potential mine or an explosive object/UXO and the same procedure should be followed as in the case of a mine occurrence. Thus, reliable detection and verification is one of the most time consuming procedures. It can be said that if a mine or any explosive on the field is reliably detected/recognized and localized the problem of de-mining is practically only 90–95 percent solved. For this reason, performing reliable detection and recognition of hidden explosive objects automatically is the most important task of the whole de-mining process.

14.3.2 The robotic approach

The classic definition of the mobile robotic system describes it is as 'a machine', able to move in a more or less structured environment and perform given operations automatically or according to a given plan. The functionality of such a robotic system during any step of control should include three principal performance features in the cognitive process: perception, recognition and decision making. It is obvious that the autonomy of the whole robotic system directly corresponds to sensory equipment, processing sensory information and decision algorithms as well. As to robotic de-mining there is one principal question: what level of autonomy should such a robotic vehicle exhibit?

To answer this question let us compare the de-mining process with other technological or service processes to which advanced robotic systems have been successfully applied. Consider some principal criteria and working conditions as follows:

Safety

During all activities, including preparing the terrain and neutralization, there is always a serious danger of explosion and the destruction of the robotic equipment; therefore all action has to be aimed at preventing explosions. Although there is a similar risk during the operation of robots for the liquidation of dangerous objects (as for instance in the case of explosives hidden in public places), de-mining in an unknown environment is much more complex (dimensions, terrain, unforeseen obstacles, vegetation, etc.).

Reliability

The humanitarian de-mining process should guarantee practically 100 percent reliability that all mines have been neutralized.

Complexity

The de-mining process is not very well defined in all three principal phases: detection, preparing terrain and neutralization. This includes environment, target objects, positions, and the sequence of particular operations. There are problems especially with:

- The large diversity of objects to be manipulated, varying in form, weight and positions (mines, stones, vegetation, etc.).
- The variety of terrain and environment (unknown; cannot be specified in advance).
- The severe outdoor working conditions.

It can be said that practically there is no second chance in performing particular tasks. Naturally, the system cannot exhibit a higher level of autonomy. The development of robotic machines and detection systems should focus on technologies that are able to acquire as much information as possible about the minefield prior to clearance. These machines need a variety of information-gathering tools to investigate a range of explosive risks.

Similarity

There are numerous common features and similarities between robotic systems for de-mining and other outdoor mobile robots that could be adopted, or partially integrated to form a common solution. These include mobility mechanisms, manipulation arms, communication, navigation/control, general sensory equipment, cognitive features, etc. Beside these general purpose systems there are several specific de-mining tools directly oriented for particular tasks, such as prodders, grippers, shovels, sand suckers, cutters, detectors, etc.

It is obvious that any robotic system can work effectively under standard and expected conditions/environment. When performing de-mining activities, there are some limits to capabilities, with respect to maneuvering of the mobility system, reliable detection of mines and desired confidence level of neutralization equipment. It is preferable to use automatic de-mining technology for clearing large homogenous terrain without complex obstacles (vegetation, trenches, etc.). Besides such complex automatic equipment, several task-oriented semi-automatic or remotely controlled devices can also work effectively.

Thus, as to the autonomy of robotic systems and considering the 'state of the art' technology in robotics, there is one relatively simple rule which can be adopted: 'The operator monitoring the whole de-mining process should be consulted about any activity of the robot that could result in the risk of an explosion'. This naturally requires transmission and monitoring of all relevant information from the de-mining scene. Practically, this means that all of the most dangerous actions will be performed in the 'master/slave' control mode. This can be seen on the information–control scheme in Fig. 14.2.

These specific working conditions for vehicles and robotic tools and security reasons require that the control system should work in two independent modes:

- Automatic/programmable control mode through communication with the operation center. A communication system for automatic modes to transmit control and sensory data: way-points/trajectory, control statements for vehicle and motor, images from camera (remote vision), vehicle and motor states, warning error situations, etc.
- Manual remote or master/slave control mode using a joystick/control panel enables operations to be performed in cases when the process requires the decisions of an operator on how dangerous situations will be solved. Dangerous

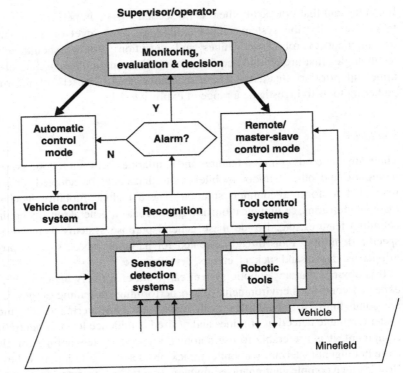

14.2 Information–control scheme of a semi-autonomous de-mining system.

situations are announced by an alarm activated by detection systems. After evaluation the operator decides on further procedures.

Manual control is used in cases where the vehicle has to be removed from the minefield and recovered as a result of the failure of any other system (programs, communication, etc.). It is also used when loading/unloading the vehicles during transport and while testing. This control mode directly operates steering and motor control loops. Communication is limited and corresponds to main statements for limited maneuvering motions.

Based on the above discussion one can state the priorities for solving problems and developing devices as part of the robotic de-mining technology:

Priority 1

Tools for detection and localization of mines/targets. Besides detection systems there are robotic arms with various sensory systems that enable dangerous terrain to be scanned/searched. In principle, the sensory systems can be situated on a special platform of the mobile vehicle performing the scanning of dangerous terrain or they can be located on the end of a robotic arm. Obviously, some principles are more suited for searching large areas and detecting the existence of minefields (infrared, chemical, bacteria) and others enable precise location of particular targets (metal detectors (MD), ground penetrating radar (GPR), camera, tactile prodder, and others). The positional data of detected targets can be directly set into minefield geographic information system (GIS) maps. Besides the tools used together with detection systems, other tools for marking positions of targets are also needed. The most frequently used are color sprayers.

Priority 2

Tools for preparing terrain. These tools represent a relatively broad class of devices that can be used in a large variety of situations according to terrain conditions. These include tools for removing obstacles or clearing vegetation. Several years after they have been deployed, mines become covered by sand (in desert conditions), ground, vegetation, masking means, etc. To remove these obstacles different remotely operated tools with sensory feedback should be developed, for example, saws, sand suckers, cutters, shovels, special grippers, diggers and probes, etc.

Priority 3

Devices and technologies for neutralization/removing mines. Besides mechanical systems including rollers, ploughs, flails, rakes, hammers, etc., several other tools for activating mines can be used. These include explosive hoses, fuel–air mixture burners, air–sand cutting jets, directed energy systems, lasers, microwave sources or sniper rifles. For mine removal tasks there are special end-effectors in the form of double shovels, diggers, soil separators, etc. The input data for these systems includes positions/coordinates of mine targets as well as actual sensory information (vision, proximity, tactile, force).

When considering the development of robotic technology a key idea is to construct a universal all terrain robotic system, which is highly mobile and lightweight, and that could be immediately set to carry out de-mining work. Although such an idea could possibly be realizable, there is no reason to spend so much money and enormous human effort on developing such a complicated, high-tech and enormously expensive system. In comparison with military devices any robotic system for humanitarian de-mining should respect some specific aspects and rules that should be taken into account and directly influence the choice of adequate technology. These are listed below.

- Minefields are not laboratories. Construction and control techniques should be robust and reliable to correspond to the harsh working conditions and environment. This includes having so called 'self-recovery strategies' for critical situations (occasional explosions, errors in systems/operators, loss of communication, etc.). This means that the robotic system can be removed from the minefield without access or intervention of humans.

- The reliable detection and localization of mines (UXO) as targets is the task of primary importance. If mines are reliably detected and localized exactly, then the neutralization procedure can be directly carried out at the place of their occurrence.
- Any new de-mining technology should be easily accepted by local authorities and people. The robotic system should satisfy specific conditions related to its local application demands (local people and their education and experiences, infected terrain, climatic conditions, type of mines, maintenance, etc).
- There are no universal solutions. Robotic technology cannot totally replace humans in all phases of the de-mining process. Some robotic approaches should replace some of the most dangerous searching/neutralization methods. Automatic technology is especially suited for primary detection and clearing large areas under some homogenous conditions (obstacles, mines, vegetation, etc.).
- Economy. Mines are deployed in post-battle regions where local materials, local manufacturing and local manpower should mainly be used to carry out the de-mining operation and to maintain all technology. The technical knowledge of the people is very limited and access to high-technology components is almost nonexistent. Usually the economy of such regions does not work or it has been totally destroyed. It is obvious that under such conditions using low-cost de-mining equipment, including standard hand searching and neutralization technologies, is preferred. Compared to standard de-mining technology, sophisticated robotic equipment is mostly complicated and much more expensive.
- Psychological aspects. Humanitarian de-mining requires a high level of confidence that all mines have been detected and neutralized. This naturally results in the fact that any technology should guarantee practically 100 percent reliability of clearing. De-miners (professionals or locally engaged) carrying out this dangerous task are always under intense psychological pressure and, usually, they are not able to master the complex robotic system including its operation, maintenance and possible repairs. Specialists need to be trained and this considerably increases the cost of de-mining work. It should be mentioned that the confidence of de-mining personnel in the technique also plays an important role.
- Any new solution should minimize risks for people and also minimize the risk of damage to the relatively expensive technology. This risk of damage or the expected lifetime for the new technology should be calculated in the expected comparable total cost for de-mining the surface area.

It is obvious that all parts of the system should be adequately robust to sustain not only harsh working and climatic conditions but also possible accidental explosions as well. The possibility of carrying out complex repairs on the minefields is very limited.

The most effective way is if the modularity approach for the construction of robotic tools is adopted. Typical examples are electronic systems, computer technology or bicycle solutions. A number of parts, mainly mechanisms, can be used as 'general purpose' mobility or manipulation systems. These systems can be combined with other de-mining task related tools. The modularity approach should be applied to all functional parts: the mobile platforms, robotic arms, tool changing systems, tools for grasping objects, etc. The majority of these parts are manufactured in large quantities and they are available on the general market. On the other hand, many specific devices for performing de-mining operations should be newly developed, such as 'task-oriented' tools. The modular concept is then based on the separation of these functional parts as shown in Fig. 14.3. Such an approach could minimize the cost of general purpose parts and the whole system itself.

14.3.3 Modular concept

The principal feature of a multi-purpose machine is that besides de-mining tasks it is able to perform other activities. These include, for instance, civil engineering

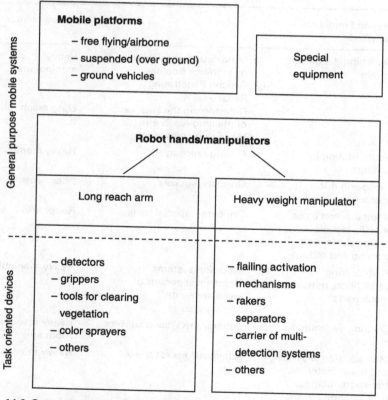

14.3 General purpose and task oriented parts of the de-mining system.

works (transport of materials, drilling, raking, manipulation of soils and loose materials, etc.) The general purpose parts can be used for other risky operations including fire fighting, disaster rescue and anti-terrorist actions, etc.

When analyzing aspects of de-mining, including available technologies and technical possibilities, the functional requirements can be related to on-board vehicle equipment and tools, as shown in Table 14.2.

Following this concept and considering the actual state and availability of robotic and de-mining technologies, the most general solution includes the following functional elements:

- a mobility system represented by the land vehicle
- a heavy manipulator
- a platform carrying various detection or destruction systems
- a long reach robotic arm

Table 14.2 Tasks for robotic tools

Task and minefield activities	Tools	Robotic arm
Searching dangerous terrain	Mine detection systems on platform (mainly Ground Penetrating Radar – GPR)	Heavy manipulator
	Detectors on the end flange of the long reach arm Marking system of targets	Long reach robot arm
Neutralization by flailing	Flailing mechanism	Heavy manipulator
Removing mines, obstacles ...	Grippers, suckers	Robot arm
Neutralization by using specific techniques: posing explosives, burning and others.	Grippers – special tools	Robot arm
Post-clearing verification, removing metal parts ...	Detection systems Magnetic separator of metal elements Soil separators	Heavy manipulator
Clearing vegetation	Cutters, saws, sand suckers ...	Heavy manipulator, robot arm
Manipulation with soil and loose materials: transport, loading, digging, drilling, etc.	Additional accessories/ tools	Heavy manipulator

- sets of exchangeable tools for removing obstacles/vegetation and marking positions of targets
- neutralization/destruction tools
- additional tools and attachments
- sophisticated sensing, control and communication systems.

Taking into account the possible situations that can arise in the de-mining process, a most general version of the ground vehicle, which enables the combination of various functional equipment, can be adopted, as depicted in the design study in Fig. 14.4.

Here we describe and specify requirements of particular functional parts.

The vehicle and its mobility system

The vehicle and its mobility system should provide good maneuvering capability on various terrains. Numerous mobile robotic devices moving on wheels, belts or legs have been developed. Besides these 'classic' mobility principles, several new locomotion principles are under study. For example, snake-like motion systems, a combination of legs and wheels, and others. With respect to the general requirements given, as discussed before, mobility that combines wheels and belts seems to be one of the best and most widely used solutions (Havlík 2003a, 2005). As a carrier of multi-sensor system and other robotic tools, the vehicle should exhibit some autonomy features, such as collision avoidance, automatic stop after the detection of a mine, and self-recovery capabilities. For security reasons the vehicle has to be protected against mine explosions (not only anti-personnel but anti-tank mines too). From the point of view of mechanical performance there are several specific requirements that vehicles should satisfy: maximal pressure on the ground, velocity related to speed of detection systems, noise and temperature limitation, reliable power and communication systems, etc.

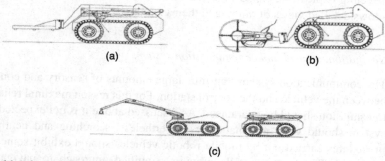

14.4 A modular concept of robotic vehicles: (a) the vehicle with sensory platform; (b) the vehicle with flailing activation mechanisms; (c) the long reach robot arm.

A heavy three to four degrees of freedom manipulator

The heavy weight manipulator in front of the vehicle with about 1000–1200 kg payload capacity is the main carrier of heavy sensory systems, as well as other mechanisms (flail with protective cover, tools for cutting vegetation, tools for manipulation of soils, etc.). It is equipped with a distance/proximity sensor, which enables the tracking of terrain at a given vertical distance, as well as range detectors for collision protection.

A long reach robotic arm

A long reach robotic arm as depicted in Fig. 14.4(c) performs some specific tasks especially in situations as follows:

- Targets are not localized exactly and, as a result, more precise searching/detection procedures using hand held detectors should be carried out.
- Targets are hidden by vegetation/stones or targets are in inaccessible positions for removing or neutralization. In these cases special de-mining procedures and tools have to be applied.

In general, a six d.o.f. remotely controlled robot hand can exhibit the payload capacity of about 20 kg with a reach of 3 m. It can be controlled in Cartesian hand references as well as with vehicle reference coordinates related to camera systems. It is assumed that the vehicle is equipped with a set of exchangeable tools for performing fine operations. One of the desired tasks involves laying additional explosives beside mines in situations where other neutralization procedures are not reliable or could be too dangerous.

Sets of exchangeable tools and attachments

The set of tools consists of probes, cutters, various grippers, additional sensors for detecting explosives, etc. On the end of the arm there is a small camera, which allows detailed views of mines and their vicinity.

Navigation, control and communication systems

The communication system transmits large amounts of sensory and control data between the vehicle and the control station. For this reason maximal reliability of transmission should be guaranteed. As discussed above it is not expected that the system should work automatically. Nevertheless, searching and neutralization procedures carried out by mobile robotic vehicles should exhibit some level of autonomy. This fact naturally requires a unified approach to navigation and control. The general scheme of the control system, in Fig. 14.5, shows some main components arranged in four control loops: global positioning, steering control

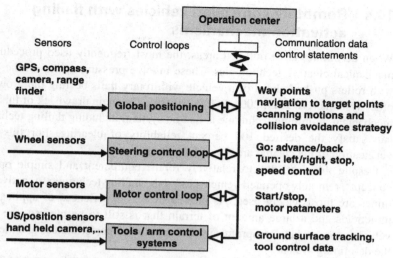

14.5 Components of the vehicle control system.

loop, motor control loop and loops for control of various on board equipment (robot arm, manipulator, tools, and others).

14.3.4 Expectations and reality

After more than 15 years of development of robotic tools for de-mining one can briefly evaluate the state of application of robotic technology in this field. Much research effort has been devoted to the development of new mobility/locomotion principles for vehicles of varying degrees of complexity, which are able to move in partially unknown/uneven terrain and perform neutralization (Ide, 2004; MacDonald et al., 2003; Santana, 2007; Wetzel, 2003; Tojo, 2004). Mechanics using sophisticated sensory and control systems have also been studied (Wetzel, 2006). Naturally, the main focus has been on searching and detection (Mori, 2005; Faust et al., 2005; Baudoin, 2008). As result of this originally academic research, several promising solutions of robotic vehicles with detection systems have been tested in real minefield conditions and are still under development (Nonami, 2007; Masunaga, 2007). An overview of these robotic devices is shown in Baudoin (2008) and Habib (2008). Besides the development of unmanned mobile vehicles with robotic mechanisms for scanning terrain and localization of mines, some special tools for clearing/cutting vegetation, neutralization or removing mines have also been developed (Wojtara, 2005; Denier et al., 2000).

Although all these systems have had a very good impact and represent progress in de-mining technologies, until now, they have not had more extended use for reasons previously discussed.

14.4   Remotely controlled vehicles with flailing activation mechanisms

When de-mining large polluted areas, the most frequently used procedures are mechanical clearing technologies. These involve pressure acting on the ground with rollers pushed ahead of a vehicle with rotary flails beating the ground and other systems (GICHD, 2004, 2008a, 2008b). The main drawback of this purely mechanical de-mining technique is that no system, including flailing technology, can satisfy the desired 100 percent reliability of clearing. For this reason verification of the cleared area is still required.

Despite this fact, due to relatively robust equipment and simple operation concepts, remotely operated vehicles using the flailing technique for activation of mines are most widely used. As proved by real experiences of application on minefields and a large amount of terrain that is still to be cleared, the flailing vehicle seems to be a compromise solution between performances and the cost of the de-mining system.

To keep up to date and in order to show the course of development of this kind of technology, the 'Božena' family (www.wayindustry.sk) of machines is detailed below, as one representative example.

14.4.1 History and experience

The primary concept of the vehicle was based on modifications of the small loader i.e. the machine for the manipulation of loose materials, terrain works, etc. The flailing activation mechanism (i.e. the rotating shaft with chains and hammers on their ends) is fixed on the end flange of the hydraulically powered manipulator at the front of the vehicle. The radio communication system enables remote mobility control, the manipulation of active mechanisms, as well as the monitoring of relevant parameters of the whole system. The next generations of these mini-flail machines 'Božena 2 and 3', were constructed as task-oriented machines and were produced between 2000 and 2002. They have a more powerful driving unit, better maneuvering and control capabilities, and much more reliable activation was reached, due to the greater amount of power available for driving the flailing mechanisms. (Havlík, 2007a, 2007b)

The further development of these machines represents important design changes, and improvement of their performance to reach greater applicability. New machines, except the flailing mechanism, enable the use of various additional attachments. The latest generations of these machines and more detailed descriptions are given next. The total number of machines actually in active operation is more than 130, in practically all the countries infected by landmines.

As confirmed through experience, the remotely operated flail using vehicles exhibit some important advantages. These include:

- Fast speed and high productivity when performing the clearing operation. Compared to classical hand de-mining procedures, systems based on the mechanical flailing technology are at least 10 times faster.
- The system is low cost and highly efficient, especially when the infected terrain is covered by grass or low vegetation. The flailing technology is especially suited for clearing large areas where it is difficult to detect mines due to terrain and vegetation.
- There is a universal technical solution of the system based on using multi-purpose soil machines as small loaders. The loader and maneuvering unit can be combined with several additional attachments for use in the de-mining process. Such a concept guarantees the availability of spare parts, verified reliability and good maintenance.
- Relatively low weight, fast and low-cost transport to the place of use is highly advantageous.

Results from more than 15 years of experience of using these machines can be summarized as follows:

- The application of purely mechanical destruction techniques, as well as the flailing technique cannot guarantee 100 percent reliability when clearing terrain of mines and other explosives. Despite this fact the vehicle with flail mechanisms is an effective tool especially in all cases when the exact positions of mines are not known, for example, if the terrain is covered by vegetation or if there is some uncertainty about the identity of the object. To satisfy maximal reliability the post verification of the clearing process can be realized using vehicles that are equipped with mine detection systems.
- Using remotely operated vehicles minimizes the psychological pressure and improves the safety of persons. The vehicles can be a useful aide for the operator if some functions are performed automatically, for instance, the flailing process with respect to advance speed or straight-line control routines.
- The efficiency of the whole de-mining process will be improved if mines have been previously detected and localized. Then the destruction vehicle can be directly navigated to the positions where mines are expected.
- In line with extensive user experience, the vehicle should be constructed as a 'multi-purpose' machine, which is able to perform various activities. It should be equipped with additional attachments.

The 'Božena 4' in Fig. 14.6 is the fourth generation of the mini-flail vehicles mainly oriented at clearing large areas of AP mines as well as anti-tank (AT) mines equivalent to up to 9 kg of TNT.

The last generation machine of this family, 'Božena 5', belongs to the category of midi-flail systems. This much more powerful machine is about twice as productive when clearing comparable terrain. In order to have good maneuvering capability in various terrains, a combination of wheels and belts was adopted.

14.6 'Božena 4' (left) and 'Božena 5' (right) in de-mining action.

14.7 Robust camera and visual monitoring–control box.

The control of this vehicle and all its mechanisms is carried out from the cabin where all data and information about the process, machine and its environment are transmitted. The operator can use the special portable control box with keyboard and joystick. Some principal control routines can be pre-programmed.

To improve controllability of de-mining actions an on-board remote vision system has been developed and installed. In the most complex configuration it consists of two stable cameras for observing the environment at the front and rear of the vehicle and one adjustable camera system, as shown in Fig. 14.7. This robust camera is fixed on the two d.o.f. mechanism which enables adjustable observation of the whole 360° area around the vehicle with +/– 20° of tilting. Pictures from the cameras are digitally transmitted to screens in the operation center. Thus, by combining the visual images with GPS data it is possible to identify the actual situation on the minefield (terrain, obstacles, trenches, trees, etc.) and to make correct decisions.

14.4.2 Additional tools and attachments

The concept of the multi-purpose machine includes two categories of tools and attachments. These are:

- Equipment directly related to the de-mining process: the platform for detection systems, flailing mechanism, target marking system, vegetation saw/cutter, system for removing metal parts, grippers, etc.

- Equipment for engineering works such as digging, drilling, loading and transportation of soil or loose materials, removing obstacles, etc.

The principal tools for the set of attachments that have been developed for Božena machines are detailed below.

Flailing mechanism

The well-known flailing principle consists of a rotating shaft with a set of chains and hammers on their ends. The crucial problem is to design a flailing system that maintains maximal efficiency and quality together with high productivity during the clearing process. To achieve this performance many parameters and characteristics need to be studied and experimentally verified. Among these are length of chains, shape and material of the hammers, positions of the chains on the shaft, speed of rotation, impact energy of the hammers, advance speed with respect to the depth of penetration, soil, etc. Besides technical criteria, the mechanism should be very robust to resist explosions of AP mines and possible AT mines too.

The flailing mechanism in Fig. 14.8 is designed as an independent system powered by two hydro-motors capable of reverse rotation. The flailing process, including advance speed, shaft rotation speed, depth, and following the terrain, is fully controlled by pre-programmed routines.

Collector of magnetic parts

After the clearing process on a minefield, great numbers of metal parts, such as shells, ammunition cartridges, mine fragments, or other ferromagnetic parts such as wires, screws, etc. are usually spread over the field. Obviously, these spread parts result in false signals of metal detectors when the verification procedure is carried out. To pick up all small ferromagnetic parts, a special attachment, a magnetic collector, as shown in Fig. 14.9, has been designed.

14.8 Flailing mechanism.

14.9 Magnetic collector.

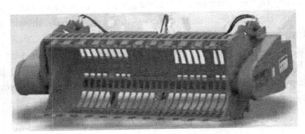

14.10 Separator for sifting and recycling soil.

Soil separator

Another useful attachment is the mechanism for sifting and recycling soils where AP mines and UXO are expected. This attachment picks up the material (soil, waste) and, after closing the drum, the content is sifted with a turning motion. The objects, such as AP mines, remain inside the drum and may be dumped afterwards after opening the jaw. The grated design of jaws is the best solution as it enables the blast waves to be spread if explosions occur inside the drum. As the procedure is remotely controlled, the safety of the operator is assured.

Other attachments

Besides the direct de-mining process, there are many dangerous procedures that should be carried out in remote operation mode only. The main reason for this is to protect people if any threat of explosion or another possible hazard situation should arise. There are several useful accessories that can be directly attached to the end flange of the heavy load manipulator. Some of those most frequently used for major works are shown in Fig. 14.11.

14.11 Accessories for remotely operated machines.

14.5   Conclusions

One of the principal lessons to be learned from history is: one may expect that people in war conflicts will use any available methods or items that could give them an advantage over the other side. This fact naturally includes using mines too. The sole solution lies in the development of fast, effective and reliable de-mining technologies, which will result in the removal of these weapons. The advantages that their deployment gives for one side of the conflict should be eliminated by the availability of fast and reliable detection and neutralization technologies. Then the reasons for their use will be reduced.

As can be seen, the problem of de-mining has crossed the borders of infected countries and research is being undertaken in many laboratories in cooperation at the international level. The sole way and contemporary way to reduce the risks for people carrying out this dangerous operation and improve the productivity of clearing operations is the application of advanced robotic technologies. As discussed in this chapter there are several specific requirements that a robotic system for performing risky operations, especially de-mining, should satisfy. Beside technical parameters, operational performance and cost criteria, better applicability can be reached if all functional parts are relatively simple.

Ce qui est simple est toujours faux.
Ce qui ne l'est pas est inutilisable

From Paul Valéry, *Mauvaises pensées.*

14.6   Acknowledgements

The author appreciates the help of WAY Industry Inc. and thanks them for photographic material used in this chapter.

14.7   References

Baudoin Y (2008), 'Mobile robotic systems facing the humanitarian demining problem state of the art (SOTA)', *Proc. 7th IARP Int. workshop HUDEM'2008*, Cairo, March 28–30.

CEN (2004), 'Workshop agreement: test and evaluation of demining machines', CWA 150 44, July 2004.

Denier, R. et al. (2000), 'The use of waterjets in the location and exposure of landmines for humanitarian demining', Rep. grant DAAG55-97-1-0014, Univ. of Missouri – Rolla.

Faust A. et al. (2005), 'Canadian tele-operated landmine detection systems. Part I: The improved landmine detection project'. *International Journal of Systems Science*, Vol. 36, No. 9, July 2005, 511–528.

GICHD (2004), *A Study of Mechanical Application in Demining*, Geneva International Centre for Humanitarian Demining, Geneva, Switzerland.

GICHD (2005), *A Study of Manual Mine Clearance (Books 1–5)*, Geneva International Centre for Humanitarian Demining, Geneva, Switzerland.

GICHD (2006), *Guidebook on Detection Technologies and Systems for Humanitarian Demining*, Geneva International Centre for Humanitarian Demining, Geneva, Switzerland.

GICHD (2008a), *Mechanical Demining Equipment Catalogue*, Geneva International Centre for Humanitarian Demining, Geneva, Switzerland.

GICHD (2008b), *Proc. Second Mine Action Technology Workshop*, Geneva International Centre for Humanitarian Demining, Sept. 8–10, 2008, Geneva

Habib M K (2002), 'Mechanical mine clearance technologies and humanitarian demining. applicability and effectiveness'. Proceedings of 5th International Symposium on Technology and Mine Problem. Monterey, CA, USA Apr. 22–25.

Habib M K (2007a), 'Humanitarian demining: reality and the challenge of technology – the state of the arts'. *International Journal of Advance Robotic Systems*, Vol. 4. No.2, June 2007, 151–172.

Habib M K (2007b). 'Humanitarian demining: the problem, difficulties, priorities, demining technology and the challenge for robotics'. In M K Habib (ed.) *Humanitarian*

Demining: Innovative Solutions and the Challenges of Technology, Vienna, Austria, I-Tech Education and Publishing, 1–56.

Habib K H (2008). 'Humanitarian demining: the problem, difficulties, priorities, demining technology and the challenge for robotics'. In M K Habib (ed.) *Humanitarian Demining Innovative Solutions and the Challenges of Technology*, Vienna, Austria, I-Tech Education and Publishing, 1–56.

Havlík S (1993). 'A reconfigurable cable crane robot for large workspace operations', *Proc. Int. Symp. on Industrial Robots*, Nov. 4–6, 1993, Tokyo, 529–536.

Havlík S (1998), 'Humanitarian demining: the challenge for robotic research', *Journal of Humanitarian Demining*, Issue 2.2, USA, May 1998.

Havlík S. (2002). 'Mine clearance robots'. *Proc. IARP Int. Workshop on Robots For Humanitarian Demining, HUDEM'02*, Nov. 3–5,2002, Vienna, Austria, 33–38,

Havlík S (2003a). 'A concept of robotic vehicle for demining', *Proc. EUDEM2-SCOT – 2003 Int. Conf. on Requirements and Technologies for Detection, Removal and Neutralization of Landmines and UXO*, Brussels, Belgium, Sept. 15–18, pp. 371–376.

Havlík S (2003b). 'Some concepts and design consideration in building robots for humanitarian demining', *Proc. IEEE ICRA 03, Int. Conf. on Robotics and Automation, Workshop. The State of the Art of Robotics in Humanitarian Demining*. Thai-Pei, Taiwan, Sept. 14–19.

Havlík S (2004), 'Robotic agents for dangerous tasks. Features and performances', *Proc. Int. Workshop on Robotics and Mechanical assistance in Humanitarian Demining and Similar Risky Interventions*, IARP, Brussels-Leuven, Belgium, June 16–18.

Havlík S (2005), 'A modular concept of robotic vehicle for demining operations. *Autonomous Robots*, 18, 253–262.

Havlík S (2007a), 'Some robotic approaches and technologies for humanitarian demining'. In M K Habib (ed.) *Humanitarian Demining: Innovative Solutions and the Challenges of Technology*, Vienna, Austria, I-Tech Education and Publishing, 289–314.

Havlík, S. (2007b). 'Land robotic vehicles for demining'. In M K Habib (ed.) *Humanitarian Demining: Innovative Solutions and the Challenges of Technology*, Vienna, Austria, I-Tech Education and Publishing, pp.315–326.

Ide K et al. (2004). 'Towards a semi-autonomous vehicle for mine neutralization'. *Proc. Int. Workshop Robotics and Mechanical assistance in Humanitarian Demining and Similar Risky Interventions*, IARP, Brussels, Belgium, June 16–18.

Kaminski L et al. (2003), 'The GICHD Mechanical Application in Mine Clearance Study'. *Proc. EUDEM2-SCOT –2003 Int. Conf. on Requirements and Technologies for Detection, Removal and Neutralization of Landmines and UXO*. Brussels, Belgium, Sept. 15–18, 335–341.

Licko P and Havlik S (1997). 'The demining flail and system BOZENA'. *Proc. International Workshop on Sustainable Humanitarian Demining, SUSDEM 97*, Zagreb, Croatia, Sept. 29–Oct. 1, S.4.8–S.4.11.

Lindman A R and Watts K A (2003), 'Inexpensive mine clearance flails for clearance of anti-personnel mines'. *Proc. EUDEM2-SCOT–2003 Int. Conf. on Requirements and Technologies for Detection, Removal and Neutralization of Landmines and UXO*, Brussels, Belgium, Sept. 15–18, 356–359.

MacDonald J A, J R Lockwood and J McFee (2003) 'Research plan for a multi-sensor landmine detector'. In *Proc. Int. Conf. On Requirements and Technologies for the Detection, Removal and Neutralization of Landmines and UXO*, Sept. 15–18, 2003, Brussels, Belgium, 625–632.

Masunaga S (2007), 'Controlled metal detector mounted on mine detection robot', *Int. J. of Advanced Robotic Systems*, 4(2), 237–245.

Mori Y (2005), 'Feasibility study on an excavation-type demining robot'. *Autonomous Robots*, 18, 263–274.

Nonami K (2007). 'Mine detection robot and related technologies for humanitarian demining'. In Habib M K, Humanitarian Demining Innovative Solutions and the Challenges of Technology, Vienna, Austria, I-Tech Education and Publishing, 235–262.

Santana P (2007). 'Developments on an Affordable Robotic System for Humanitarian Demining'. In M K Habib (ed.) *Humanitarian Demining: Innovative Solutions and the Challenges of Technology*, Vienna, Austria, I-Tech Education and Publishing, 263–288.

Stilling D S D et al. (2003), 'Performance of chain flails and related soil interaction'. *Proc. EUDEM2-SCOT – 2003 Int. Conf. on Requirements and Technologies for Detection, Removal and Neutralization of Landmines and UXO*. Brussels, Belgium, Sept. 15–18, 349–355.

Tojo Y (2004). 'Robotic system for humanitarian demining – development of weight-compensated pantograph manipulator'. *Proc. IEEE Int. Conf. on Robotics and Automation (ICRA 2004)*, New Orleans, LA, 2004, 2025–2030

Wetzel J P and Smith B O (2003). 'Landmine detection and neutralization from a robotic platform'. *Proc. EUDEM2-SCOT –2003 Int. Conf. on Requirements and Technologies for Detection, Removal and Neutralization of Landmines and UXO*, Brussels, Belgium, Sept. 15–18, 365–370.

Wetzel J P (2006). 'Modular robotic control system for landmine detection'. *Proc. 6th Annual Intelligent Vehicle Systems Symposium & Exhibition*, Michigan, June 2006.

Wojtara T (2005), 'Hydraulic master–slave land mine clearance robot hand controlled by pulse modulation'. *Mechatronics* 15, pp. 589–609.

15

RAVON: The robust autonomous vehicle for off-road navigation

C. ARMBRUST, T. BRAUN, T. FÖHST, M. PROETZSCH,
A. RENNER, B. H. SCHÄFER and K. BERNS,
University of Kaiserslautern, Germany

Abstract: This chapter describes the work of the Robotics Research Lab at the University of Kaiserslautern in the field of autonomous off-road robotics. It introduces concepts developed for hazard detection, terrain classification, and collision-free autonomous navigation. As an example of a system implementing the described techniques, the mobile off-road robot RAVON is presented. Experiments have been carried out to prove the effectiveness of the approaches.

Key words: autonomous off-road robotics, topological off-road robot navigation, 3D obstacle detection, behaviour-based control architecture, sensor data representation.

15.1 Introduction

Using a robotic vehicle for risky interventions poses high demands on the vehicle's hardware as well as its software: the vehicle must be robust enough to cope with hostile environmental conditions. Its kinematics must allow sophisticated manoeuvres on different types of terrain. The vehicle must be equipped with an entire range of sensor systems to detect a rich set of obstacle types. Finally, its control system must be able to deal with complex situations while being flexible, extensible, and easy to maintain.

With these demands in mind and with the desire to solve fundamental scientific problems in the field of off-road robot navigation, the Robotics Research Lab at the University of Kaiserslautern started the development of an entirely autonomous vehicle that is able to operate safely in rough and highly vegetated terrain. The experimental platform used is RAVON, the Robust Autonomous Vehicle for Off-road Navigation (see Fig. 15.1). A sophisticated hazard detection mechanism which is based on different sensor systems, reliable obstacle avoidance, and a flexible and versatile high-level navigation system in complex environments have been identified as key features.

The purpose of this chapter is to provide information about the platform RAVON and to present the concepts underlying the work in detail.

15.1.1 The need for autonomy

In recent years, the demand for robotic vehicles that can autonomously operate on difficult terrain has strongly increased in various domains. These include unmanned

353

15.1 The autonomous off-road robot RAVON.

space travel (Sherwood et al., 2001; Tunstel et al., 2001; Schenker et al., 2003), automatic gathering of measurements (Ray et al., 2005), archaeological exploration (Gantenbrink and Belless, 1999), agricultural automation (Thuilot et al., 2001; Lenain et al., 2003; Wellington and Stentz, 2004), and the evolution of military devices (Hong et al., 2002; Zhang et al., 2001; Debenest et al., 2003).

Unmanned vehicles could patrol borders, guard industrial estates, fulfil reconnaissance tasks in hostile environments, or assist in clearance duties in cases of severe accidents or natural disasters. In all of these scenarios, the risk to people's lives could be reduced by providing them with information about a previously unclear situation, or even by keeping them completely out of potentially dangerous situations. However, despite the strong need for autonomous vehicles, most of the implementations still appear more like enhanced remote-controlled cars (Kunii et al., 2001) than autonomously acting and decision-taking, thus 'intelligent', robots.

While the task of going from one place to another is usually simple for a human, it can be far from trivial for an autonomous robot. Even in highly structured environments (cp. section 15.2.1), there are still many problems to be solved before robotic vehicles can operate fully autonomously without posing a threat to their environment. Therefore, it is generally regarded as safer to let humans make most of the decisions necessary for a robot to accomplish its mission.

One crucial drawback of this approach (which is referred to as 'semi-autonomous') is the fact that radio communication often suffers from limited bandwidth and connection interruptions due to obstacles between a robot and its operator. Apart from this problem, high bandwidth communication with the control headquarters can consume a considerable amount of the energy the robot carries around. Furthermore, remotely deciding which actions are safe is not always easy from the sensory data provided by a mobile robot. Most of the time this is based on video images, which are normally 2D and therefore limit the depth perception. In general, a lot of training is necessary in order to master vehicle control in difficult situations. Introducing more sophisticated perceptional systems and preprocessing units would probably render remote-controlling even more complex.

Recapitulating, current remote-controlled vehicles suffer from the typical drawbacks of radio systems and result in long operation times as well as high costs for qualified personnel. Both issues will be addressed with self-dependent navigation systems which enable mobile robots to find their way to a given target area fully autonomously.

15.1.2 Structure of this work

The following section gives an overview of the state of the art technology in the wide field of autonomous outdoor navigation, with the focus lying on the work dealing with the problems of off-road navigation. After an overview in section 15.3, secton 15.4 deals with the different hardware components and the sensor data preprocessing of the autonomous off-road robot RAVON. The main sensor systems are presented as well as the robot's major sensing capabilities. Sections 15.5 to 15.7 are dedicated to the different layers of RAVON's navigation system, ranging from short-range over mid-range to long-range navigation. The purpose of section 15.8 is to discuss the current state of the project on the basis of expressive experiments. It leads to the concluding remarks and an outlook on future work, which are presented in section 15.9.

15.2 State of the art technology

Worldwide, various research projects deal with autonomous navigation of a robot in outdoor terrain. However, the target environments of these approaches differ significantly in terms of structure and types of obstacles. Many of the projects can be separated roughly into those targeting urban environments and those dealing with non-urban terrain. The latter can be further divided into projects aimed at on-road and off-road environments, respectively.

This section gives an overview of the state of the art technology in the field of outdoor robot navigation by presenting relevant projects.

15.2.1 Navigation in urban terrain

Many of the projects in the field of outdoor robotics focus on urban environments. Though fully autonomous cars are far from being available, a number of systems known as advanced driver assistance systems have been developed and integrated into standard cars during the last few years. Such systems include adaptive cruise control and automatic parking. They increase safety or comfort by providing partial autonomy that can help the driver in certain situations.

The success of these systems and the attention drawn to the field by the DARPA Urban Challenge[1] (see Fig. 15.2) motivated researchers to put effort into the work on robots that autonomously navigate in urban environments. Research topics include road sign detection (Fang et al., 2003), road following (Baker and Dolan, 2008), and pedestrian detection (see (Zhao and Thorpe, 2000) and Fig. 15.3 for

15.2 Boss, the winning vehicle of the DARPA Urban Challenge (www.tartanracing.org/hires/01.jpg).

15.3 Navlab 11 (http://www.cs.cmu.edu/afs/cs/project/abu/www/img/NL 11park.jpg).

[1] http://www.darpa.mil/grandchallenge/

work of the Navlab group, and Navarro-Serment and Hebert (2008) for work that is actually not focused on urban environments).

However, the differences between urban and rural or even off-road terrain are large and so the results of this research area can only be partly applied to the field of off-road robotics.

15.2.2 Off-road navigation in terrain with little to no vegetation

The projects that belong to this category often focus on autonomous navigation in desert-like areas (see Urmson et al., 2004). This has been fostered by the DARPA Grand Challenges held in 2004 and 2005, in which vehicles had to autonomously follow a route defined by GPS points (see (Singh, 2006a, 2006b) and Fig. 15.4). The focus of these challenges lay in high-speed (over 50 km/h) navigation in a terrain with minor jaggedness and only sparse vegetation. Furthermore, the route followed a path that could be recognised using vision systems in order to assist the navigation.

Another huge field of application for autonomous navigation is the space exploration sector. Needless to say, vegetation is no problem in this application. However, the terrain is often dominated by jagged rocky formations and there are also hazards like holes in the ground and chasms. As a consequence, the approaches published by the National Aeronautics and Space Administration (NASA)[2] and

15.4 Stanley, the winning vehicle of the DARPA Grand Challenge (www.braunschweig.de/politik_verwaltung/fb_institutionen/ staedtische_gesellschaften/bsmportal/presseinfos/Stanley_Image2.jpg).

[2] http://www.nasa.gov/.

15.5 Artwork showing a rover of the Mars Exploration Rover Mission (http://www.nasa.gov/centers/jpl/images/content/136489main–PIA 4413-feature-browse.jpg).

the Jet Propulsion Laboratory (JPL)[3] (see Olson (2000) and Fig. 15.5) are aimed at this type of terrain. The approach for visual terrain mapping described in Olson et al. (2007) is even more specialised: it uses not only surface images from a rover, but also images taken by a spacecraft as it approached the planetary surface, and images from spacecraft orbiting a planet.

A robot system that features elaborate facilities for local obstacle detection, localisation, and navigation is introduced in Lacroix et al. (2002), yet the computational effort only allows for very low velocities (about 10 cm/s).

Due to the high signal propagation delay between a control station on the earth and a spacecraft, telecommanding is difficult and solely relying on it increases the risk of accidents that can damage or even destroy a vehicle. Some approaches try to overcome these problems by providing a rover with a set of waypoints, which it follows autonomously between communication cycles (see Ravine, 2007). However, the current state of the art technology in this field uses little autonomy.

15.2.3 Off-road navigation in highly vegetated terrain

Only a few research projects deal with the problem of autonomous robot navigation on highly vegetated terrain. The different types of hazards in such environments

[3] http://www.jpl.nasa.gov/

15.6 The Experimental Unmanned Vehicle (XUV) of the DEMO III project (http://www.globalsecurity.org/military/systems/ground/images/xuv-pic3.jpg).

put high demands not only on a robot's navigation strategies, but also on the algorithms used for sensor processing.

Of the projects in this category, a large number have a military background. Among them is the DEMO III project, the goal of which was to advance and demonstrate the technology required to develop future unmanned ground combat vehicles that can operate autonomously in cross-country vegetated terrain (see Shoemaker and Bornstein, 1998 and Fig. 15.6). In the context of this project, many approaches for obstacle detection and terrain classification have been developed and investigated. Several contributions about such approaches have been made by researchers of the JPL (see Bellutta et al., 2000; Talukder et al., 2002; Manduchi et al., 2005). Other work dealt with architectural concepts of the control system (see Albus, 2002). Discriminatory for the DEMO III project (and other projects with military background) is the large variety of expensive technology like laser detection and ranging (LADAR), high-resolution stereo systems and powerful computational units as well as the semi-automated system architecture. The system does not operate fully autonomously, but is supported by a human operator when exploring, navigating, as well as tackling difficult situations.

More recent work about obstacle detection in highly vegetated terrain includes processing point clouds generated from LADAR data (see Vandapel et al., 2004), building up a probabilistic terrain model in order to improve ground estimates and obstacle detection (see Wellington et al., 2006), and the use of red and near-infrared reflectance of obstacles to discriminate between vegetation and other obstacles (see Bradley et al., 2007). Still, the problem of hazard detection in off-road terrain is far from being solved.

15.3 Overview of RAVON

The purpose of this secton is to provide information about the general ideas and concepts behind RAVON's control software. A high-level overview of its components is depicted in Fig. 15.7. As can be seen, RAVON's control system can be separated into four layers. The lowest layer contains the hardware platform together with the software that serves as the interface for the higher layers. The sensor data preprocessing, which creates short-term memories of the robot's environment, is also part of this layer. The components of the second-lowest layer realise a collision-free target approach. Abstract views on the environment are extensively used here. The second-highest layer contains components operating on a mid-range view. Finally, the highest layer contains the global navigation system that mainly operates on a topological map of the environment.

The following sections describe the four layers in detail. Section 15.4 presents the mechatronics and the sensor data preprocessing. The subsequent sections deal with the three layers of RAVON's navigation system, providing information about the short-range (section 15.5), the mid-range (section 15.6), and the long-range (section 15.7) navigation.

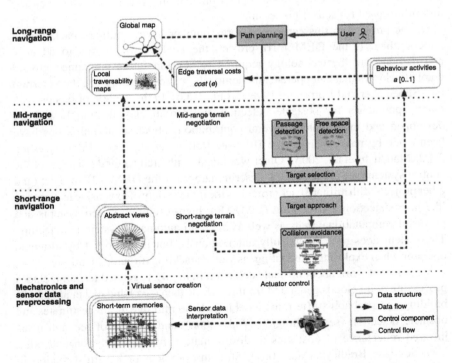

15.7 A high-level overview of the control system's components.

15.3.1 Software environment

All four of RAVON's industrial PCs run Gentoo Linux. The robot control software has been implemented using the Kaiserslautern branch of the C++ robot control framework Modular Controller Architecture (MCA). This framework was originally developed at the Research Center for Information Technology (FZI)[4] in Karlsruhe. It is now being developed under the name MCA-KL at the University of Kaiserslautern's Robotics Research Lab. The KL branch is a strongly extended version with a large number of additional features and user libraries. It is used on all of the robots developed at the Robotics Research Lab. There are also MCA components that do not run on a PC, but on the DSPs used on custom-design DSP boards (e.g. the board for RAVON's IMU), which allows for the preprocessing of sensor data before it is sent to a PC. MCA-KL, further libraries, and tools are published under the GPL.[5]

15.3.2 Behaviour-based framework

The behaviour-based architecture upon which RAVON's control system is based is called integrated Behaviour-Based Control (iB2C). It is an extension of the architecture introduced in Albiez et al. (2003). The modifications made are partly exemplified in Proetzsch et al. (2005). Performance and flexibility of the approach have been demonstrated in numerous case studies, one of which is presented in Schäfer et al. (2005a). The iB2C architecture has been implemented using C++ and has been seamlessly integrated into MCA-KL.

The main concept of iB2C is the decomposition of a given task into behaviours providing a standardised interface such that complex networks of behaviours can be built up. Each behaviour is realised using the fundamental unit as depicted in Fig. 15.8. Here, data at the input vector $\vec{e}$ is processed by a transfer function $\mathbf{F}$ in

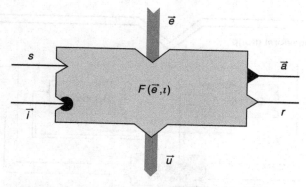

15.8 The interface of a behaviour module, including activity vector $\vec{a}$, target rating r, stimulation s, and inhibition $\vec{i}$.

[4] FZI: Research Center for Information Technology (http://www.fzi.de/)
[5] http://rrlib.cs.uni-kl.de/

order to generate an output vector $\vec{\mathbf{u}}$. The possible set of transfer functions is not limited; therefore arbitrary algorithms can be used here.

The most important output of a behaviour is its activity $\mathbf{a} \in \vec{\mathbf{a}}$ which represents a measure of the relevance for each behaviour. It is used for stimulating other behaviours via their input $\mathbf{s}$ (i.e. allow a high activity) and for inhibiting them via their input $\vec{\mathbf{i}}$ (i.e. reduce their possible activity). In combination, stimulation and inhibition result in the activation $\iota = \mathbf{s} \cdot (1 - \mathbf{i})$ of a behaviour, with $\vec{i}$ being the maximum component of $\vec{\mathbf{i}}$. Furthermore, the coordination of competing behaviours is based on the assumption that highly active behaviours gain more influence than less active ones. It is realised by so-called fusion behaviours, which again provide the standardised interface like any other behaviour.

Besides the activity signal, each behaviour generates an output called target rating $\mathbf{r}$. It is a measure of how satisfied a behaviour is in the current situation. This allows a generic evaluation of situations based on local conditions and is used, for example, for cost evaluation in the navigation component, see section 15.7.

The presented components are used to build up large behaviour networks for the implementation of complex systems by applying standardised interaction principles. The coordination via fusion behaviours allows several mechanisms like priority-based arbitration, superposition, or voting. The modular structure of the resulting network supports the extension of the control system without touching existing behaviours and interconnections. In order to cope with networks of increasing size, a hierarchical abstraction can be realised via behavioural groups (see Fig. 15.9), which comply with the same interface as behaviour modules. More information about iB2C can be found in Proetzsch et al., 2010.

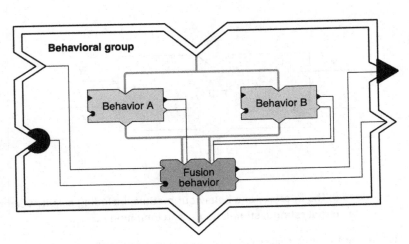

15.9 An example of a behavioral group.

15.4 RAVON mechatronics and sensor data preprocessing

This section presents the hardware of the autonomous mobile off-road robot RAVON. The first part provides information about the basic platform, while the second part presents the robot's various sensor systems.

15.4.1 Hardware platform

The basis of RAVON is a robot platform manufactured by Robosoft, called robuCAR TT (see Fig. 15.10). As the original platform did not meet the high demands on an off-road platform's stability and endurance, many parts have been reinforced or replaced by the members of the Robotics Research Lab, among them the mountings for the shock absorbers and the transverse links. Several sensor systems, computers, and other components have been added, resulting in the current version of the robot RAVON, which is depicted in Fig. 15.11.

Including all components, RAVON measures 2.4 m by 1.4 m by 1.8 m (length × width × height) and weighs 750 kg, which makes it comparable to a small car in terms of size and weight (see Fig. 15.12).

RAVON's two axles can be steered independently, which supports the robot's agility and allows for the execution of advanced manoeuvres like double Ackermann steering. Thus, RAVON can make sharp turns and increase the distance to lateral obstacles without changing its orientation. Many tasks like clearance duties after a

15.10 robuCAR TT, the platform manufactured by Robosoft.

15.11 RAVON during the ELROB 2008.

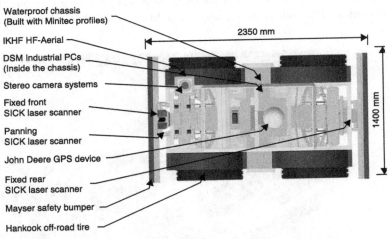

Waterproof chassis
(Built with Minitec profiles)

IKHF HF-Aerial

DSM Industrial PCs
(Inside the chassis)

Stereo camera systems

Fixed front
SICK laser scanner

Panning
SICK laser scanner

John Deere GPS device

Fixed rear
SICK laser scanner

Mayser safety bumper

Hankook off-road tire

2350 mm

1400 mm

15.12 A technical drawing of RAVON (top view).

severe accident or a natural disaster require such a high agility as there are typically many obstacles around which the vehicle has to manoeuvre.

The robot is powered by four independent electric motors that have kindly been provided by Johannes Hübner Giessen.[6] Each of them has a power of 1.9 kW and

[6] http://www.huebner-giessen.com/

is controlled by a high performance motor controller provided by Unitek Industrie Elektronik[7]. These controllers are connected to the main PC via controller area network (CAN) bus. With this equipment, RAVON can reach a maximum velocity of 3 m/s, and in combination with its Hankook off-road tires it can climb slopes of up to 100 per cent.

RAVON is equipped with four industrial PCs assembled and provided by DSM Computer[8]. They are responsible for running the high-level control systems for local navigation (Intel Core2Duo T7400), global navigation (Intel Core2Duo E4300), as well as low-res image processing and data collecting (Intel Pentium M 780 and Intel Core2Duo E6700). Five DSP boards developed at the Robotics Research Lab are attached to the computers using two CAN buses. They run low-level programmes that realise, for example, the motor control.

The power for all of RAVON's systems is delivered by eight OPTIMA YT S 4.2 accumulators made available by OPTIMA Batteries[9]. Each of them has a capacity of 55 Ah. This suffices for an operation time of 3 to 4 hours, depending on factors like driving speed and steepness of the terrain. For longer operating times, a power generator (Honda EU10i) can be attached to RAVON's rear part.

A custom casing has been built up of MiniTec[10] profile elements to protect the electronic components from dirt and humidity. The upper part can be opened at the back so that RAVON's interior can be accessed easily.

Although RAVON is designed to operate fully autonomously, for testing purposes a wireless connection is needed in order to be able to send commands to the robot and receive telemetry data. Therefore, a standard wireless local area network (WLAN) connection is available. In order to establish connections over longer distances, RAVON is equipped with a universal data transceiver produced by IK Elektronik[11]. In combination with a 500 mW amplifier distances of up to 10 km can be bridged at a data rate of 115 kbit/s. If this transceiver is used, an appropriate omnidirectional HF aerial can be connected. It is mounted onto a gimbal to ensure vertical alignment of the antenna on severe slopes (see Fig. 15.13).

15.4.2 Sensor systems

In order to be well prepared for the large variety of situations RAVON can face, it is equipped with a number of different sensor systems (see Fig. 15.12). They can be separated into two groups: sensor systems that estimate the robot's pose (i.e. position and orientation) and sensor systems designed for gathering information about the robot's environment.

[7] http://www.unitek-online.de/
[8] http://www.dsm-computer.de/
[9] http://www.optimabatteries.com/
[10] http://www.minitec.de/
[11] http://www.ik-elektronik.com/

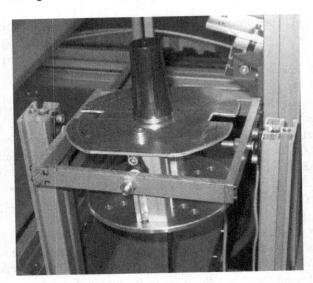

15.13 The HF aerial.

An encoder is attached to each of RAVON's four motors. They are used as a basis for a pose calculation using wheel odometry. As the robot often operates on slippery or uneven terrain, this only yields a rough estimate of the true pose, so other components are integrated into the estimation process.

A custom inertial measurement unit (IMU) that measures movements along and rotations around all three axes (see Koch et al., 2005) and a magnetic field sensor are mounted at the upper rear part of the casing (see Fig. 15.14). Furthermore, a John Deere[12] StarFire iTC and a u-blox[13] AEK-4H are employed as absolute position sensors. The estimates delivered by these sensors are combined using a Kalman filter (see section 15.4.3).

RAVON is equipped with three SICK[14] laser range finders. Two LMS 291s are attached to the lower front and rear of the robot, respectively (see Fig. 15.15). An LMS 291 has a field of vision of 180°, an angular resolution of 0.5°, and a distance resolution of 0.5 cm. The disadvantage of a fixedly mounted laser scanner is that it can only detect obstacles that intersect its scan plane. Relying solely on such sensors may be sufficient in very simple, structured environments. However, the highly unstructured environments in which RAVON operates demand more sophisticated obstacle detection capabilities. Therefore, a third SICK laser scanner (an S300) is attached in a custom-made mounting bracket that can pan the scanner to the left and to the right. As can be seen in Fig. 15.15, this scanner is attached to a sensor tower built up of MiniTec profile elements. By

[12] http://www.deere.com/
[13] http://www.u-blox.com/
[14] http://www.sick.de/

15.14 Cube of inertial measurement unit and magnetic field sensor.

Long range color stereo system
Large scale terrain traversability estimation

Short range color stereo system
Obstacle detection

Rotating 2D laser scanner
Obstacle detection / 3D local memory

Planar 2D laser scanner
Obstacle detection / safety system

Spring-mounted bumper
Tactile vegetation discrimination

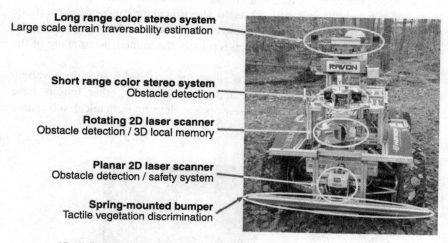

15.15 RAVON's sensor tower.

mounting the scanner with its scan plane upright and panning it continuously, 3D information about the robot's environment can be gathered. Using this data, RAVON's sensor processing system is able to detect hazards like water, overhanging obstacles and holes in the ground. How this is done is explained in section 15.4.4. In the current state of the system, the panning laser range finder is the main sensor system delivering information about the terrain on which the robot moves.

Two stereo vision systems are also mounted on the sensor tower. The lower of the two consists of two Point Grey Research Dragonfly2. It can be used for obstacle detection or visual odometry. The cameras are mounted on a pan/tilt unit so that they can monitor a large area in front of the robot. The higher of the two stereo vision systems consists of two high resolution cameras (Point Grey Research SCOR-20SOC-CS CCD). It is used by the global navigation system to gather detailed terrain information at certain navigation points. These two cameras are also mounted on a pan/tilt unit, which is placed on the highest position of the sensor tower. Hence the cameras have a good view of a large part of the area around the robot.

A MOBOTIX Q22 panorama camera provided by MOBOTIX[15] has recently been attached to the front of RAVON. It has been used for detecting ERICards (Emergency Response Intervention Cards) during the European Land Robot Trial (ELROB[16]) in 2009 and will be used for place recognition in the future.

Two spring-mounted safety bumpers manufactured by Mayser[17] are attached to the front and rear, respectively (see Fig. 15.16). In a special operation mode called 'tactile creep', the spring system is used to separate soft obstacles from rigid ones in situations where the geometric obstacle detection alone cannot be used. If one of the bumpers collides with an obstacle that cannot be pushed away easily, the robot's safety chain is triggered, which results in the immediate stopping of the vehicle.

The sensor systems described here allow for hazard detection in a variety of environments. Depending on the application scenario, other sensors have to be added. For example, special sensors for detecting chemical substances

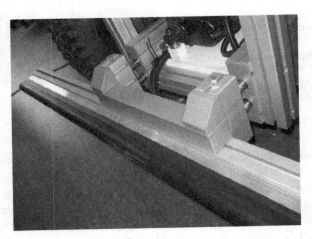

15.16 One of RAVON's bumpers.

[15] http://www.mobotix.com/
[16] http://elrob.org/
[17] http://www.mayser.de/

would be needed if the robot was used after an accident in a chemical plant. For patrolling or guarding tasks, additional sensors for detecting humans may be needed.

15.4.3 Localisation

For many applications of robots, a good pose estimation is necessary. In indoor environments, SLAM approaches are often used (see, for example, Eliazar and Parr, 2005). However, indoor environments are typically highly structured, which reduces the noise in the data delivered by distance sensors. Furthermore, indoor robots usually do not need a three-dimensional pose – their position on the ground and their orientation along the vertical axis are sufficient. On off-road terrain or in areas where a severe accident has occurred, by contrast, a full three-dimensional pose is often needed. The roll and pitch angles are especially important in order to detect critical situations. Although there are approaches for dealing with the application of SLAM in outdoor environments (see Montemerlo et al. 2003), this is difficult in extremely noisy environments or on terrain with few features like grassland and clearings.

Therefore, a fault-tolerant system architecture has been chosen for RAVON's control system, which can cope with dynamic changes in sensor availability and fulfils the requirements for a robust and adaptable solution (see Schmitz, Koch, Proetzsch and Berns, 2006; Schmitz, Proetzsch and Berns, 2006). The system is based on a linear Kalman filter and a flexible model of the system and sensors. It integrates the data provided by the vehicle's odometry, the IMU including the magnetic field sensor, and one of the GPS devices. A switching between the two devices is realised to account for the different characteristics of the devices (high robustness and imprecise position estimation vs. lower robustness, but much more precise position estimation). The visual odometry is not integrated in the current version of the localisation system and is left to future work.

15.4.4 Hazard detection

RAVON features various sensor systems that provide different specialised views of its environment. Depending on a specific sensor's capabilities different environment properties can be detected and stored in a grid map that forms a sensor-specific local short-term memory. The bumper system, for example, yields information about traversable ground or solid obstacles by sweeping through the environment along with the robot's movements. This manifests in the property of 'traversable space' for every position that was covered during the operation time of the robot. Another example is the panning laser scanner, which gathers three-dimensional information about the robot's frontal environment. Using the measured sizes and altitudes of scanned objects allows for their classifying and for distinguishing between rough

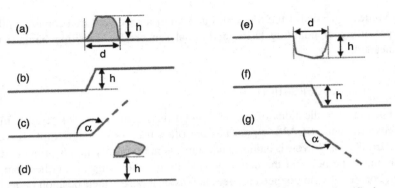

15.17 Different types of obstacles that can be detected and classified using the panning laser range finder: (a) positive obstacle, (b) positive step, (c) positive slope, (d) overhanging obstacle, (e) negative obstacle, (f) negative step, (g) negative slope.

ground and severe obstacles like big stones or overhanging branches (see Fig. 15.17 and Schäfer, Hach, Proetzsch and Berns, 2008).

The three-dimensional data from the panning laser scanner can also be used for guessing if detected obstacles are rigid or soft vegetation, using statistical analyses along with a voxel penetration technique (ray-tracing) (see Fig. 15.18, left). This enables proceeding at low speed and pushing through high grass, etc.

Another algorithm to extract properties from the raw three-dimensional data of the panning laser range finder on RAVON is used to detect water. It searches for void readings surrounded by ground points (see Fig. 15.18, right) and evaluates the sizes and densities of found regions. Numerous experiments have shown that the vehicle is able to recognise water hazards at distances up to 9 m. This detection range allows for sufficient foresight in control of the vehicle concerning obstacle avoidance.

For each sensor-processing algorithm a robot-local, orientation-fixed grid map implements a short-term memory. Each grid map has a size of 20 by 20 metres. In such a grid map, the bumper system mentioned above stores information about traversed space. The 3D obstacle detection stores the extracted properties in a second grid map. A filtered version, which is cleansed of readings that probably belong to vegetation, is stored in a third one. Two further grid maps are used for the data on water detection and the laser-based 2D proximity detection, respectively.

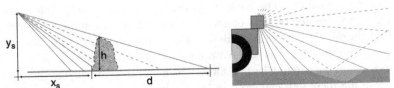

15.18 The basic principles of the voxel penetration technique (left) and of the water detection algorithm (right).

An obstacle detection based on the low-res stereo vision system was implemented on RAVON and described in Schäfer et al. (2005b), but it is not used at the moment due to research on laser-based obstacle detection.

The unified representation in the form of grid maps is used to decouple the development of the various sensor systems from problems in higher levels of the control system. The capabilities of the sensors and the direct processing of their data, i.e. filtering or feature extraction, define a set of properties that can be assigned to the observed environment by this particular sensor or its processing system. These properties are stored in the grid map cells of the specific short-term memory for each sensor. Figure 15.19 depicts a schematic drawing of such a grid map, while Fig. 15.20 shows several of the properties that can be stored in a grid map cell. For an example of a grid map created from real sensor data see Fig. 15.22. Figure 15.21 shows RAVON in a typical application environment, while Fig. 15.22 depicts a grid map filled with data from the panning laser range finder. The different brightness of data points represents different types of terrain.

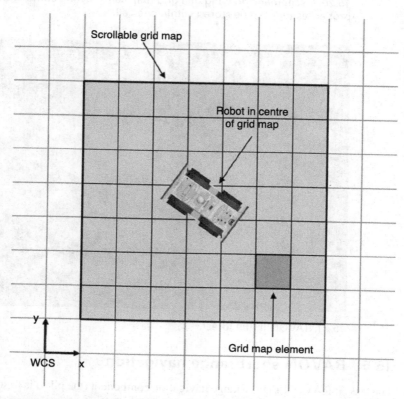

15.19 A schematic drawing of a scrolling, local, orientation-fixed grid map used as a short-term obstacle memory.

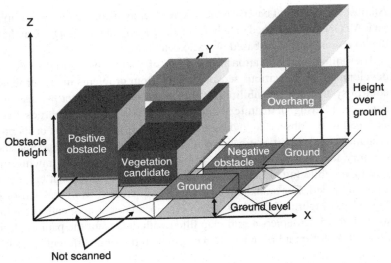

15.20 A schematic drawing of a grid map cell showing some of the properties that can be stored within one cell.

15.21 RAVON in the forest.

15.5   RAVON short-range navigation

The task of RAVON's short-range navigation component (the pilot) is to solve the problem of local navigation, i.e. realising collision-free operation in any situation and driving the robot more or less directly to a target provided by higher layers.

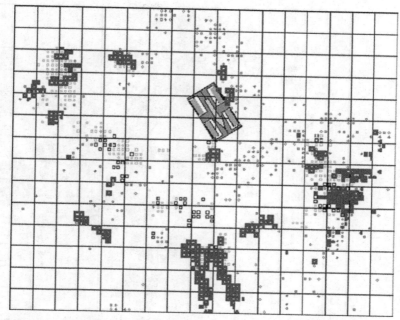

15.22 A grid map containing information about the environment.

15.5.1 Control-driven sensor processing

As the short-term memory introduced in section 15.4.4 is strictly predetermined by the development of the sensor systems, an abstract view for further processing is needed. Thus, a generic data structure that facilitates access to data of various sensors by providing a uniform interface called a sector map (Armbrust et al., 2007) is used.

The underlying grid maps provide fast access to areas defined by simple geometric objects like lines or convex polygons. Using them, it is easy to define areas and relevant properties within a grid map of a specific sensor. Area and property definitions are used to generate sector maps that divide the regions they cover into several sectors. Then, the sector maps for each sector store whether it contains one of the given properties and, if so, the distance from the sector's origin to the most relevant representative for the current configuration.

The mapping from the sensors' grid maps and properties to the abstract view in the sector maps is formulated in terms of aspects and can be used to break down the information needed for different control aspects into a set of relevant properties of appropriate sensors, hence the term 'control-driven sensor processing'. Using this virtual sensor layer, an action/perception-based design of RAVON's control software is realised (Schäfer, Proetzsch and Berns, 2008). Figures 15.23 and 15.24 show polar and Cartesian sector maps, respectively, created from the grid map depicted in Fig. 15.22.

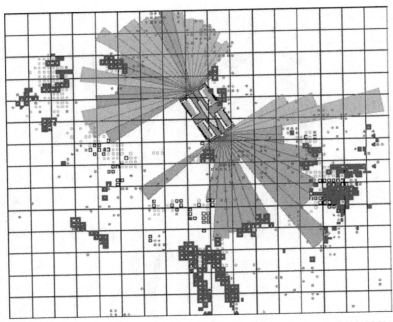

15.23 Two of the polar sector maps monitoring the area around RAVON.

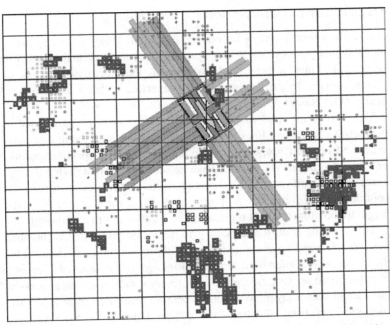

15.24 Some of the Cartesian sector maps monitoring the area around RAVON.

15.5.2 Collision-free navigation

One of the pilot's tasks is to assure the basic safety requirements in respect to collisions. The pilot directly utilises the virtual sensor layer to benefit from the uniform and abstract representation, which allows for behaviour module reuse as well as straightforward integration of new sensors or sensor data processing algorithms. Figure 15.25 shows some of the many sector maps that monitor RAVON's environment and are used by the anti-collision system.

The following characteristics can be differentiated:

- Arbitrary sensor coverage can be realised by extracting the corresponding region from the local grid map, e.g. front, rear, and side regions, with obstacles at different height levels.
- Several sensor data processing algorithms can be provided: raw laser scanner data evaluation, voxel penetration methods (for vegetation discrimination), water detection, etc.
- Different properties can be represented: rigid, soft, overhanging obstacles, holes (i.e. negative obstacles), water.

The behaviours of RAVON's safety system are connected in a network realising collision avoidance. This network is provided with drive commands originating from operator input behaviours, point access behaviours, as well as behaviours for preferring open space, driving along structures, etc. The safety behaviours gradually overwrite higher-level commands by inhibition and by providing corrective motion both for velocity and steering commands.

In order to meet the basic safety requirements, the first level (in a bottom-up design process) consists of iB2C behaviours that can trigger an emergency stop based on bumper events, overhanging or negative obstacles, as well as rigid

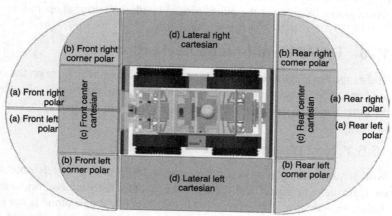

15.25 Some of the virtual sensors that are used by RAVON's anti-collision system.

obstacles at the sensor height. As these obstacle types might harm the robot hardware, all drive commands issuing from higher layers are blocked.

The next layer consists of reactive behaviours for slowing down, rotating away from obstacles, and sideward motion for obstacle avoidance. As RAVON traverses scenarios with different characteristics, three drive modes ('control-aspects') are defined. Each of them is represented by a behavioural group (see section 15.3.2). These groups are instances of the same class, provided with sensor information suitable for the respective drive mode. They are stimulated by corresponding drive mode behaviours, which implement the state switching using the concept of iB2C inhibitions. The following list describes the three drive modes:

- Fast driving requires long-range obstacle detection and the evaluation of all detected obstacles.
- Moderate drive through vegetation is provided with data from the voxel penetration evaluation for vegetation density determination. The resolution of the sensor systems only allows for the detection of close-range areas (up to 3 m). Therefore, the maximum velocity is limited accordingly.
- Dense vegetation requires additional sensors, as optical sensing cannot discriminate between solid and flexible objects. Therefore, the tactile creep mode issues a very slow drive command and supervises the deflection of the bumper system. If rigid obstacles are hidden in the vegetation, the vehicle stops and backs off again. Furthermore, overhanging obstacles that might damage the other sensor systems result in the same reaction.

Following a behaviour-based approach has several advantages during the development of a robot for risky interventions. The reactive aspect facilitates fast reactions to changing sensor data, while the modular structure makes a behaviour-based anti-collision system like the one implemented on RAVON tolerant to failures of a single sensor system. The behaviours operating on data of a failing sensor stop working, but the others proceed with their normal operation.

15.6   RAVON mid-range navigation

Mid-range navigation targets environmental structures which are on the one hand too large to be properly dealt with by the short-sighted collision avoidance, but on the other hand too small for the coarse-grained global navigation.

15.6.1 Complex situations

With its anti-collision system (see section 15.5.2), RAVON is able to safely navigate the typically difficult terrain that a robot for risky interventions encounters during its operation. However, reliable collision avoidance alone is not sufficient in complex situations like the one shown in Fig. 15.26. The robot would probably not collide with the various obstacles, but as the collision avoidance works locally

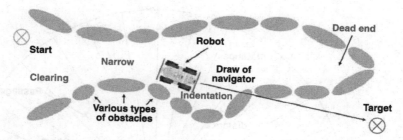

15.26 RAVON faces a complex situation while driving towards its target.

and misses the 'big picture', the robot may get stuck somewhere. A typical long-range navigation system (see section 15.7) uses a coarse-grained world model and thus also cannot deal with structures like narrows, indentations, and dead ends. A human, by contrast, would detect these structures and use his knowledge about them to decide where to go next.

15.6.2 Using passages to support navigation

As a first step towards improving the robot's navigation by using structures in its environment, the detection, evaluation, and use of so-called passages have been integrated. Passages are defined here as paths leading through obstacles. They are usually not represented by road-like structures, but by consecutive areas of traversable terrain surrounded by different types of obstacles. In the following, the concept of using passages is illustrated on the basis of an example (see Fig. 15.27).

When driving straight towards the target, the long-range navigation leads the robot into the indentation to its right, until the collision avoidance becomes active. The robot then has to back off or it might even get stuck between the obstacles. A human would normally decide to stay more to the left and follow the passage leading through the obstacles.

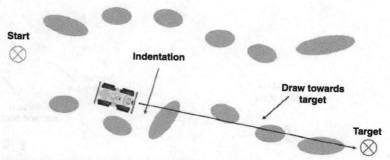

15.27 RAVON drives into the indentation instead of following the passage.

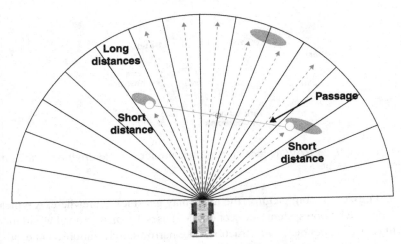

15.28 The polar sector map that is used for detecting passages.

The data source for the detection of such passages are virtual sensors represented by sector maps (cp. section 15.5.1) as they provide an easy and uniform access on the sensor data. The search algorithm loops over all sectors and compares the distances to the obstacles contained in the sectors. Gaps in the distances are interpreted as passage entries. The basic idea of the algorithm is illustrated for a polar sector map in Fig. 15.28.

A special type of virtual sensor is used to gather information about a passage's length and orientation. The pose of such a sensor is not fixed in terms of the robot coordinate system, but attached to some point of interest. As this resembles placing a sensor somewhere in the environment, this special type of virtual sensor is referred to as a virtual sensor probe (see Fig. 15.29). By placing virtual sensor probes at the entry of a passage, information about several properties of the passage, including its width, length, and orientation, can be gathered. This

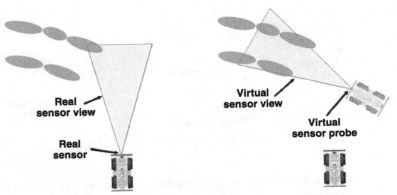

15.29 The views of a real sensor and a virtual sensor probe.

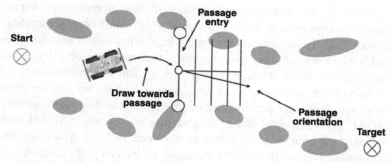

15.30 RAVON avoids driving into the indentation by following the passage.

information is then evaluated in order to decide whether a particular passage will be used or not.

If a passage has been classified as relevant for navigation, a behaviour-based system guides the robot towards the passage in a way that the robot is oriented properly when reaching the entry. Like the short-range navigation components described in section 15.5.2, this system has been designed based on the concepts of the iB2C (see section 15.3.2). Using passages in the described way has several advantages: the robot is kept away from indentations and reaches a passage properly oriented, thus it can smoothly enter it. Figure 15.30 takes up the situation depicted in Fig. 15.27 and illustrates how the mid-range navigation system uses the passage to improve the robot's navigation.

Detailed information about the detection and evaluation of passages and their use to support off-road robot navigation can be found in Armbrust et al. (2009). In the context of future work, further components will be added to the mid-range navigation layer.

15.7   RAVON long-range navigation

The global navigation layer of RAVON (the navigator) is responsible for navigation decisions that have a spatial extent significantly above the sizes of the short-range navigation system's grid maps and the structures detected by the mid-range navigation components. Thus, the global navigation focuses on robot movements starting at the scale of about 10 m, typically directed to goal positions up to several kilometres away.

Many established long-range robot navigation systems approach such a task by scaling the ideas underlying local obstacle maps up to the entire working range of the robot. Over time, they construct a single, large traversability grid map, which stores the obtained world information. Long-range path planning is then performed using D*-type algorithms (see Ferguson and Stentz, 2004 and Howard et al., 2005 for examples). This approach has been demonstrated to yield good results for semi-rugged terrain such as sparsely vegetated desert or dryland environments.

However, the construction and maintenance of a large global map is computationally expensive. Furthermore, successful mutual registration of new sensor data with the existing map also requires accurate robot localization at all times. Unfortunately, a highly vegetated off-road environment makes accurate robot localisation even more difficult than semi-rugged desert terrain. This is partially due to a strongly degraded GPS signal caused by overhanging foliage, but also an effect of the viewpoint dependent penetrability of vegetation that introduces much noise in LIDAR generated point clouds used for map registration in SLAM methods. In summary, the creation and registration of a global metric map in RAVON's target terrain would be computationally extremely taxing and prone to error.

In order to avoid the drawbacks associated with the use of a single, large traversability grid map for global navigation, the navigator developed for the RAVON system is based on a topological map, instead of relying on exact metric information. Topological maps focus on describing single waypoints (topological 'nodes'), which are relevant for navigation and the connections between them ('edges') (see Fig. 15.33 and 15.41 for examples of topological maps). This compact and abstract representation significantly reduces memory and computational demands. Furthermore, temporarily inaccurate robot localization is less problematic since global data registration is not required and navigation waypoints are only loosely coupled via topological links. Using the topological representation, the navigator deliberately abstracts from local aspects of the environment or the robot's exact trajectory which emerges upon edge traversal. This is only possible because the navigator can offload these issues to the already presented short- and mid-range piloting subsystems which maintain a metrically more detailed, but spatially limited world representation.

The abstraction from metrical detail is a mixed blessing. As already mentioned, the topological model is more compact and better maintainable in cluttered environments. On the negative side, the lacking knowledge greatly complicates the computation of a cost measure that actually reflects the characteristics of the driven path. But without such a measure, cost-optimal path planning across easily traversable terrain becomes effectively impossible on the global scale.

In order to solve this problem and provide physically plausible edge traversal costs to the topological navigator, two novel techniques have been developed. For one, an a posteriori cost learning method has been proposed. The main underlying idea of this method is to learn a consistent edge traversal cost measure each time a topological edge has been traversed by retrospectively observing the short- and mid-range behaviour performance during movement. A second methodology has been developed to allow the cost prediction for topological edges that have not been traversed yet, which is a common situation for exploration scenarios. This method re-uses information already contained in the local obstacle maps by conserving it in abstracted form in so-called local traversability maps (LTMs), which are attached to topological nodes. Further information for these LTMs is extracted from a long-range stereo vision image system specifically tailored to this task. In combination, cost learning and cost prediction approaches add a physically plausible cost measure to the abstract topological representation and allow efficient high-level navigation path planning in unstructured and cluttered outdoor terrain.

In the next section, the topological cost measure and the a posteriori cost learning scheme will be presented. After this, the main ideas underlying the cost prediction technique will be introduced.

15.7.1 A posteriori topological edge cost estimation

As a basis for cost-efficient topological path planning, a multi-dimensional cost measure has been added to the topological edges (Braun, 2009). This cost measure records the 'risk', 'effort' and 'familiarity' cost aspects of each connection. In short, the risk cost factor quantifies the amount of evasive actions required by the pilot whenever the edge is traversed. Likewise, the effort factor records the energy needed for edge traversal. Finally, the familiarity value is a virtual cost which quantifies the amount of cost knowledge already accumulated for each edge. Its purpose is to allow explicit influence on the robot's explorative behaviour, e.g. its desire to traverse either well-known or new paths.

In order to determine cost values that are consistent with the real terrain properties without the need to analyse the terrain extensively and construct a highly detailed world model, a technique to learn such consistent cost values from scratch based on feedback from the robot's pilot during operation has been developed (Braun and Berns, 2008). Its overall structure is presented in Fig. 15.31.

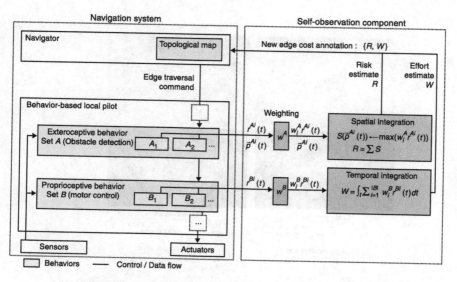

15.31 Topological edge cost learning scheme based on self-observation.

The feedback that is used to build the risk cost estimate is the target ratings $\mathbf{r}$ of all obstacle avoidance fusion behaviours ($\mathbf{r}^{Ai}$) located in the collision avoidance subsystem of the short-range navigation component (see section 15.5.2). Each time an avoidance behaviour becomes uncontent (target rating > 0) due to the proximity of a local obstacle, the numerical value of the increased target rating (weighted with a user-defined importance factor $\mathbf{w}_i^A$) is recorded in a coarse local grid map $\mathbf{S}$ at the spatial position $\vec{\mathbf{p}}^{Ai}$ which corresponds to the behaviour stimulus location, e.g. the obstacle position. For each cell of this grid map, the maximum target rating is retained. As elaborated upon in Braun and Berns (2008), this spatial integration scheme avoids double counting obstacles which are detected over more than one sensor cycle. Once a topological edge has been traversed, the target ratings of all grid cells are summed, resulting in a scalar estimate $\mathbf{R}$ of the amount of obstacle avoidance reactions performed by the local navigation layer. This estimate has been experimentally validated as a suitable measure of an edge's 'risk'. Similar to this, the energetical 'effort' estimate $\mathbf{W}$ can be deduced by integrating the motor control and driving behaviour target ratings $\mathbf{r}^{Bi}$ (corresponding to the electrical current powering the actuators).

Figure 15.32 exemplarily shows two traces of the pilot's spatially accumulated driving and obstacle avoidance behaviours (shown as green/light grey and red/dark grey dots), which are observed by the navigator and ultimately integrated to form risk and effort costs.

It can be observed that the overall increase in the estimated risk and effort cost (μ) for the edge $\mathbf{n}_0 \rightarrow \mathbf{n}_1$ corresponds well with the subjective increase in the amount of obstacles along the emerging trajectory. It has to be pointed out that

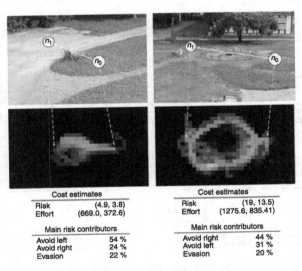

Cost estimates	
Risk	(4.9, 3.8)
Effort	(669.0, 372.6)

Main risk contributors	
Avoid left	54 %
Avoid right	24 %
Evasion	22 %

Cost estimates	
Risk	(19, 13.5)
Effort	(1275.6, 835.41)

Main risk contributors	
Avoid right	44 %
Avoid left	31 %
Evasion	20 %

15.32 Cost learning based on pilot behaviour traces.

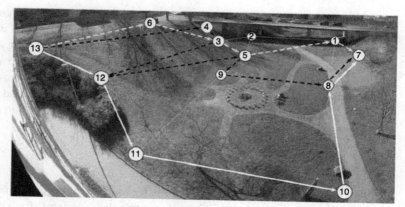

15.33 Fish-eye view of testing area and overlaid topological map.

this physically plausible measure has at no point dealt with actual sensor data directly. By basing the high-level global cost estimate solely on the reactions and judgement of the lower driving layer, the consistent estimate of the navigational difficulty of every driving situation is guaranteed.

Figure 15.33 shows the result of a further experiment that has been conducted to evaluate the effect of cost learning on path planning in a large real world scenario. The figure shows a panoramic image of the test area, overlaid with the topological map that was provided to the robot prior to cost learning. The testing ground covers approximately 100 by 100 metres and exhibits a maximum height difference of about 7 metres. The steepest part of the testing grounds (slope more than 20°) is located around node 3, while the most problematic obstacle configurations are below the bridge around nodes 1 and 2.

To obtain the path with shortest metrical distance, the path planner was commanded to plan a path from node 13 to node 7 using the unannotated, initial map. The result is marked in the figure with thick, white arrows (Path 13-12-11-10-8-7). Then the robot was issued several dozens of random edge traversal commands in order to build up cost estimates. After cost learning, the path planner was requested to generate the connection between nodes 13 and 7 that minimises either the risk or effort cost sum. The resulting paths are indicated in the picture with dashed black (minimal risk, Path 13-4-3-12-5-9-8-7) and dashed light grey (minimal effort, Path 13-6-3-5-1-7) edges. Both paths differ substantially from the purely distance-based path. The minimal-effort path saves energy by exploiting the steepest slopes around node 3 and using a relatively direct connection. This comes at a price of traversing difficult terrain around node 1. In contrast to this, the minimal-risk path contains a lot of lengthy detours in order to avoid this area and the vicinity of the hedge (11–13). Both paths are intuitively plausible in the context set by the cost measure.

This experiment and further validation in simulation prove the claim that the layered navigation design and the proposed extensions of the topological map allow the navigator to build a minimal world model which is compact enough to be easily scalable up to large environments. By refining the initially rough cost estimates continuously based on self-observation of the pilot, a more and more consistent measure of all cost-relevant aspects of the complex rugged off-road environment is obtained. With this approach, the additional knowledge invested into the global navigation layer can be kept at a minimum, and a high degree of robustness against sensor noise or inaccurate self-localisation can be achieved.

15.7.2 Topological cost prediction

To allow the prediction of traversability costs for topological edges that have not been traversed yet, an entire set of methods to extrapolate edge costs from existing cost information has been developed (Braun, 2009). Three of these methods re-use information stored in the a posteriori cost estimates of the topological map edges, but incorporate data on different levels of locality. The roughest prediction technique relies on a global cost model constructed from all available cost annotations using an outlier-robust linear regression of risk and effort cost factors. This model correlates the estimated travel length of the hypothetical edge with its probable cost by extrapolating global, overall terrain cost characteristics. A local cost model is built by the second prediction method, which restricts the spatial extent of edges eligible for cost transfer. This models local fluctuations of terrain properties better than the global cost model, improving risk and especially effort predictions. The third approach predicts costs based solely on an edge's direct inverse twin, which is the spatially closest source of information available.

In order to account for terrain properties that are not captured by the extrapolation of topological map information, a fourth cost prediction algorithm was proposed which uses local traversability maps (LTMs) attached to the topological nodes as the information source. These metrical maps store cost modifiers for possible exploration directions in the vicinity of topological nodes in a compact form. A LTM is composed of radially arranged 'seclets' (mini segments), each covering an area of a given angular extent (e.g. 10°) and depth (1 m). One LTM is visualised at the bottom of Fig. 15.34.

As shown in Braun (2009), the cost modifiers are the key to predicting the traversal costs of topological edges that lead into up-to-now unknown terrain. By combining the learned costs for traversed edges with the modifiers stored in the local metrical map, the accuracy of cost prediction can be greatly improved. This is especially relevant for the risk cost factor, which depends on the amount of obstacle evasions that will take place during edge traversal.

In total, three strategies to fill local traversability maps have been proposed. Most importantly, the local obstacle memory of the pilot is re-used by constructing two abstract views as described in section 15.5.1. These two abstract views both consist of polar sectors centred at the robot and covering a 360° field of

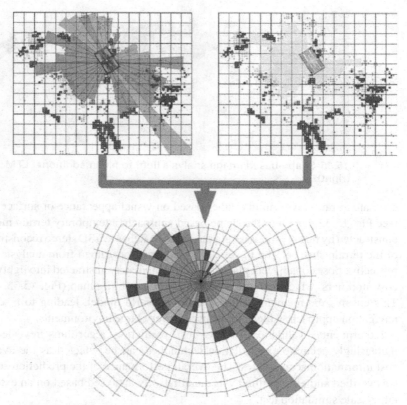

15.34 Generation of a traversability map: two abstract views (top row) are combined into a local traversability map (LTM). The cost modifier of each 'seclet' is visualised using colours: light grey for low, medium grey for medium, and dark grey for high traversal cost modifiers.

view. The first view (Fig. 15.34, top left) contains all terrain which is traversable in the corresponding direction in at least one pilot drive mode (either fast driving, moderate driving or tactile creep). It thus models the maximal operational space of RAVON at that location. The second view (Fig. 15.34, top right) only contains space that can be traversed in fast driving mode, and thus focuses on the spatial subset, that is very easy to traverse. If the robot reaches a topological node, both abstract views are generated, segmented into seclets and attached to the node as a persistent estimate of the overall traversability in different directions leading away from this node.

In addition to the re-use of the local obstacle map in an abstracted form for global navigation, two dedicated long-range image analysis techniques have been proposed in Braun, Bitsch and Berns (2008), Braun, Seidler and Berns (2008) and Zolynski et al. (2008) to fill two additional LTMs with information covering up to 30 m around a topological node. They use a long-range stereo camera system to

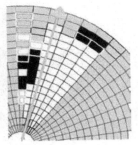

15.35 Shape-based image analysis (left) to fill an additional LTM (right).

estimate terrain traversability either based on visual appearance or surface shape (see Fig. 15.35, left). For the shape based approach, a temporary terrain model is constructed by fitting piecewise planar surface patches to a 3D stereo reconstruction of the terrain surface. The traversability information gained from analysis of the respective poses of adjacent planes is then immediately abstracted into lightweight cost modifiers, which are stored in a local traversability map (Fig. 15.35, right). This ensures the maintenance of a minimal world model, leading to a scalable navigation approach, which can be used in very large environments.

Experimental validation of the cost prediction algorithms revealed that increasingly accurate extrapolation techniques can be selected as the available cost information accumulates. The overall performance of the prediction strategy was verified and each method was quantitatively analysed based on an extensive large-scale simulation test.

15.8   Experiments

Numerous experiments have been conducted with the RAVON robot in different environments in order to test the system and prove the effectiveness of the concepts presented here. Three test runs are presented in the following sections.

15.8.1 Experiment 1: collision avoidance

The main purpose of this test run was to validate the correct operation of the components dealing with hazard detection and collision avoidance. It was conducted in the Palatinate Forest around Kaiserslautern. RAVON was given a target position in terms of a GPS coordinate that was located at approximately 1 km air-line distance from the starting point. Intermediate waypoints were neither provided nor generated by the high-level navigation system (see section 15.7). Thus the robot was continuously drawn in the direction of the target, while the low-level behaviour-based navigation system (see section 15.5.2) avoided collisions based on the obstacle data provided by the sensor processing system (see section 15.4.4). The result was autonomous movement towards the target.

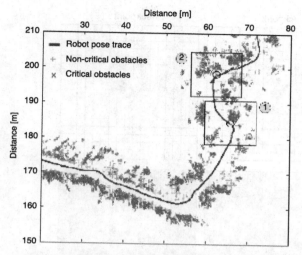

15.36 The pose trace of experiment 1: the two checkpoints are marked with numbers.

Figure 15.36 shows part of the complete test run. The black line indicates the path of the robot estimated by the localisation system (see section 15.4.3). Here, objects displayed with + symbols are potentially negotiable, i.e. the robot may be able to pass over them. Obstacles marked with × symbols, by contrast, are critical, i.e. a collision has to be avoided by any means. Note that RAVON's control system does not generate a global metric map of its environment. Therefore, the pose trace has been created offline on the basis of sensor and pose data recorded during the test run.

The situation at two checkpoints (marked with numbers in Fig. 15.36) is presented in detail. Figure 15.37 depicts the contents of the local obstacle memory at these checkpoints. The rays going to obstacles symbolise the contents of some relevant sector maps used by the anti-collision system.

15.8.2 Experiment 2: passage detection

In this experiment, the operation of the component that uses passages to support navigation (see section 15.6) was tested. It has also been conducted in the Palatinate Forest. Again, RAVON was given a target several hundred metres away from the starting point and had to get there fully autonomously. Figure 15.38 shows a short pose trace of the run. Obstacles are marked in a similar way as on Fig. 15.36. Passages that have been detected are indicated by three circles connected by a line. The colour is determined by the estimated quality of a passage (black: best; light grey: suitable to support the navigation). The lines starting at the middle of the three circles provide information about a passage's orientation. The dark grey line starting at the robot and going to the right indicates the direction in which the robot was drawn by the high-level navigation, i.e. the direction to the target.

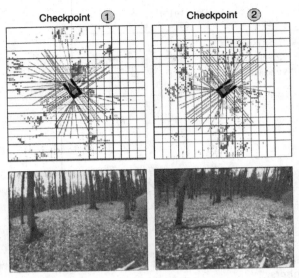

15.37 The contents of the local obstacle memory and images taken at two checkpoints.

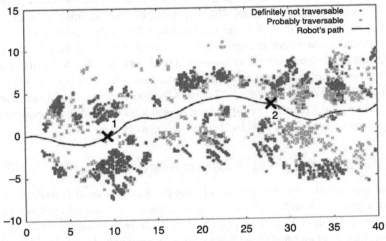

15.38 The pose trace of experiment 2 with the checkpoints marked.

At checkpoint 1 (see Fig. 15.39), the direction of the path leading through obstacles deviated a lot from the direction to the target. The draw of the high-level navigation would have led the robot directly into the undergrowth to its right. However, as two suitable passages were detected, the robot did not follow this draw, but stayed on the path in order to pass through the passages.

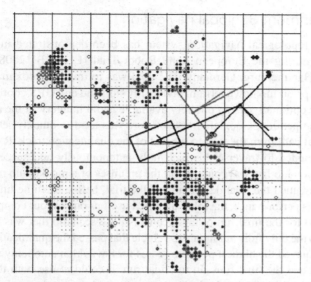

15.39 The grid map and the passages at checkpoint 1.

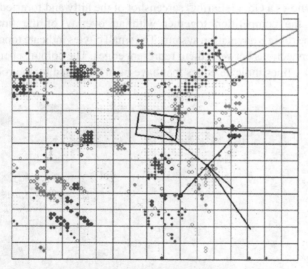

15.40 The grid map and the passages at checkpoint 2.

A similar situation is depicted in Fig. 15.40 (checkpoint 2). Again, following the drag of the high-level navigation would have resulted in the robot driving into the undergrowth. As a result, the collision avoidance behaviours would have stopped the robot if it had got too close to the obstacles to its left. Then backing off would have been necessary. However, by detecting and using passages, these time-consuming manoeuvres could be avoided.

15.8.3 Experiment 3: global navigation

The large-scale topological navigation system has been demonstrated in a real world scenario during the European Land Robot Trial (ELROB) 2008 robot competition in Hammelburg, Germany. The goal was to travel approximately 1 km through an a priori unknown terrain, given only start and end GPS coordinates of the track, and optionally, using publicly available aerial imagery of the area.

During the competition, GoogleEarth® software was used to quickly set up an initial topological path from the start to the end of the qualification track. RAVON was then commanded to traverse this path fully autonomously and successfully arrived at the second to last waypoint within the premises of a mocked-up military camp, which contained the final destination of the trial, before the time limit was reached. With this performance, the robot qualified itself as one of four competitors (out of 11 teams) for the final run of the reconnaissance scenario that was held at night. It was also the only fully autonomous vehicle that passed the qualification and consequently achieved the highest possible scores for navigation autonomy in both the qualification and the final runs. In the final run, it could re-use both the cost information and the adapted topological structure learned during the qualification run and was able to arrive at the target location well within the assigned time frame.

Figure 15.41 shows a GoogleEarth® screenshot of the hilly and forested terrain covered during the competition together with an overlay of the final map constructed by the robot's navigation system. The small circles indicate topological nodes and the connections between them represent edges. The colour scheme is as follows: light grey lines indicate edges that have been traversed by the robot and

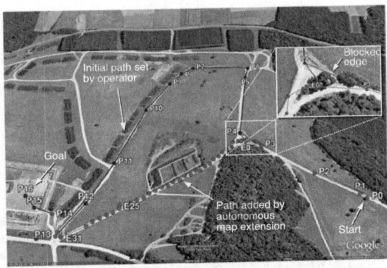

15.41 Autonomous map extension during European Land Robot Trial 2008.

nodes with attached LTMs, while black and dark grey elements signify not yet traversed and not traversable (blocked) elements of the topological map.

Initially, the robot was commanded to travel from node P0 on the extreme right of the image to the target node P16 visible on the left, along the route specified by the intermediate P nodes. The local navigation layer successfully followed this route until the connection of P3 and P4 had to be traversed. Here, a mixture of debris and tall vegetation, both not visible in the aerial image, blocked the path. Consequently, the edge traversal failed. In response to that, the topological navigator marked the edge as untraversable and inserted a new node E0 at the current robot position. As this broke the only connection to the goal node P16, a map extension algorithm was executed. This algorithm has not been presented in detail in this chapter due to space restrictions, but a precise description can be found in Braun (2009). This algorithm evaluated the cost modifiers stored in the LTM of E0 and decided that the cheapest alternative topological connection towards the goal P16 should start off in a southwestern direction from E0, following an easily traversable dirt road.[18] Upon initiating the traversal of the new connection, further topological nodes were successively placed at intervals of 20 metres each (the range of the map extension algorithm lookahead). Using the information obtained from each node LTMs, the new nodes were all placed accurately along the easily traversable dirt road. The map extension terminated once one of the inserted nodes coincided with the P13 node. At this point, the path planner 'jumped back on track' and followed the originally set waypoints to enter the military compound. At this time, the qualification run was ended by the jury because the time limit had been reached. Nevertheless, this successful demonstration of global navigation skills led to the qualification of the team for the final scenario of the competition. RAVON's performance during ELROB 2008 is described in more detail in Braun et al. (2009).

15.9   Conclusions and future work

This chapter gave a high-level overview of the autonomous off-road robot RAVON. The declared goal of the RAVON project – to develop a vehicle which is capable of driving fully autonomously through highly vegetated, difficult terrain – poses a number of requirements of the hardware as well as the software. Possible target applications of such a vehicle are various types of risky interventions, e.g. humanitarian de-mining or reconnaissance missions after natural disasters.

Due to the large variety of hazards it may encounter, different sensor systems had to be installed on RAVON and they have different characteristics and monitor different areas around the robot. Local obstacle memories are employed to keep track of obstacles when they leave the areas monitored by the sensors.

[18] Although this road is the best track to the goal, it had not been used for the original route set-up because the other, more difficult road would have provided a higher score for the competition.

In order to accomplish fast reactions to changing sensor data and to allow easy modifications and enhancements of the system's functionality, a behaviour-based approach has been chosen for control system development. The short-range navigation deals with tasks concerning collision avoidance and point access manoeuvres, while the long-range navigation provides a coarse driving direction and keeps track of the overall progress in reaching the target. In between, the mid-range navigation tries to detect structures in the robot's environment and use them for navigation. Together, the three layers build a powerful navigation system.

In the current state, RAVON is able to navigate autonomously in a large variety of environments. Numerous experiments have been conducted on grassland and in forested areas to prove the effectiveness of the proposed concepts. These experiments confirm that RAVON could be used as a platform for carrying various devices needed in the context of risky interventions.

However, tests have also shown that the control system cannot cope with complex situations and certain types of hazards. Therefore, the sensors and the sensor processing system will need to be enhanced in the future. For example, the perception abilities of RAVON should be extended by additional sensors or processing algorithms to detect muddy terrain or debris. For use in risky environments, special sensors will have to be added in order to detect hazards like toxic chemical substances.

At the moment there is no fusion on the sensor level. Only control signals are fusioned. In the future, the fusion of several short-term memories may provide higher control levels with a more combined view of the environment. Improvements in the virtual sensor layer will provide abstract sensor fusion on the level of sector maps, allowing already existing components to operate on a combined view without any change in the control algorithms. Furthermore, the mid-range navigation layer will have to be extended. The detection and usage of passages to improve navigation are a first step, but there are lots of complex situations for which concepts have to be developed in order to deal with them. These situations include RAVON reaching a dead end on its path, which requires turning around and finding another way.

Additional research will go into large-scale terrain classification to improve the navigator's abilities to plan in the 'big picture'. Points of interest are object modelling/fitting for place-recognition and extracting hints from the large-scale view of the surrounding terrain in order to use them for path planning.

15.10 Acknowledgements

Team RAVON thanks the following companies for their technical and financial support: IK elektronik, Mayser, Hankook, MiniTec, SICK, DSM Computer, Hübner Giessen, John Deere, Optima, ITT Cannon, MOBOTIX, Unitek Industrie Elektronik, and Werkzeug Schmidt GmbH.

15.11 References

Albiez, J., Luksch, T., Berns, K. and Dillmann, R. (2003). Reactive reflex-based control for a four-legged walking machine, *Robotics and Autonomous Systems*, Elsevier .

Albus, J. (2002). 4D/RCS: A reference model architecture for intelligent unmanned ground vehicles, *Proceedings of the SPIE 16th Annual International Symposium on Aerospace/ Defense Sensing, Simulation and Controls*, Orlando, FL, USA.

Armbrust, C., Koch, J., Stocker, U. and Berns, K. (2007). *Mobile robot navigation support in living environments, 20.* Fachgespräch Autonome Mobile Systeme (AMS), Kaiserslautern, Germany, pp. 341–346.

Armbrust, C., Schäfer, H. and Berns, K. (2009). Using passages to support off-road robot navigation, *Proceedings of the 6th International Conference on Informatics in Control, Automation and Robotics 2009 (ICINCO 2009)*, Institute for Systems and Technologies of Information, Control and Communication (INSTICC), Milan, Italy, pp. 189–194.

Baker, C. and Dolan, J. (2008). Traffic interaction in the urban challenge: Putting Boss on its best behavior, *Proceedings of the 2008 IEEE/RSJ International Conference on Intelligent Robots and Systems*, Nice, France, pp. 1752–1758.

Bellutta, P., Manduchi, R., Matthies, L., Owens, K. and Rankin, A. (2000). Terrain perception for DEMO III, *Proceedings of the IEEE Intelligent Vehicle Symposium 2000*, pp. 326–331.

Bradley, D., Unnikrishnan, R. and Bagnell, J. D. (2007). Vegetation detection for driving in complex environments, IEEE International Conference on Robotics and Automation.

Braun, T. (2009). *Cost-Efficient Global Robot Navigation in Rugged Off-Road Terrain*, RRLab Dissertations, Verlag Dr. Hut. ISBN 978-3-86853-135-0.

Braun, T. and Berns, K. (2008). Topological edge cost estimation through spatio-temporal integration of low-level behaviour assessments, in W. Burgard et al. (eds), *Intelligent Autonomous Systems (IAS-10)*, IOS Press, pp. 84–91.

Braun, T., Bitsch, H. and Berns, K. (2008). Visual terrain traversability estimation using a combined slope/elevation model, *Proceedings of the 2008 KI Conference*, pp. 177–184.

Braun, T., Schaefer, H. and Berns, K. (2009). Topological large-scale off-road navigation and exploration – RAVON at the European Land Robot Trial 2008, IEEE International Conference on Intelligent Robots and Systems (IROS), St. Louis, MO, USA.

Braun, T., Seidler, B. and Berns, K. (2008). Adaptive visual terrain traversability estimation using behavior observation, *Proceedings of IARP Workshop on Environmental Maintenance and Protection*.

Debenest, P., Fukushima, E. F. and Hirose, S. (2003). Proposal for automation of humanitarian demining with buggy robots, IEEE/RSJ International Conference on Intelligent Robots and Systems.

Eliazar, A. and Parr, R. (2005). Hierarchical linear/constant time SLAM using particle filters for dense maps, *Advances in Neural Information Processing Systems (NIPS 2005)*, Vol. 18, pp. 339–346.

Fang, C.-Y., Chen, S.-W. and Fuh, C.-S. (2003). Road-sign detection and tracking, *IEEE Transactions on Vehicular Technology* 52(5): 1329–1341.

Ferguson, D. and Stentz, A. (2004). Planning with imperfect information, *Proceedings of the IEEE International Conference on Intelligent Robots and Systems (IROS)*, Vol. 2, pp. 1926–1931.

Gantenbrink, R. and Belless, S. (1999). The UPUAUT project – a robot for pyramid exploration, http://www.cheops.org.

Hong, T.-H., Rasmussen, C., Chang, T. and Shneier, M. (2002). Fusing ladar and color image information for mobile robot feature detection and tracking, in M. Gini, W.-M. Shen, C. Torras and H. Yuasa (eds), *7th International Conference on Intelligent Autonomous Systems*, IOS Press, pp. 124–133.

Howard, A., Seraji, H. and Werger, B. (2005). Global and regional path planners for integrated planning and navigation: Research articles, *Journal of Robotic Systems* 22(12): 767–778.

Koch, J., Hillenbrand, C. and Berns, K. (2005). Inertial navigation for wheeled robots in outdoor terrain, 5th IEEE Workshop on Robot Motion and Control (RoMoCo), Dymaczewo, Poland, pp. 169–174.

Kunii, Y., Tada, K., Kuroda, Y. and Kubota, T. (2001). Tele-driving system with command data compensation for long-range and wide-area planetary surface explore mission, IEEE/RSJ International Conference on Intelligent Robots and Systems.

Lacroix, S., Mallet, A., Bonnafous, D., Bauzil, G., Fleury, S., Herrb, M. and Chatila, R. (2002). Autonomous rover navigation on unknown terrains: Functions and integration, *International Journal of Robotics Research* 21(10–11): 917–942.

Lenain, R., Thuilot, B., Cariou, C. and Martinet, P. (2003). Adaptive control for car like vehicles guidance relying in RTK GPS: rejection of sliding effects in agricultural applications, IEEE International Conference on Robotics and Automation.

Manduchi, R., Castano, A., Talukder, A. and Matthies, L. (2005). Obstacle detection and terrain classification for autonomous off-road navigation, *Autonomous Robots* 18(1): 81–102.

Montemerlo, M., Thrun, S., Koller, D. and Wegbreit, B. (2003). FastSLAM 2.0: An improved particle filtering algorithm for simultaneous localization and mapping that provably converges, *Proceedings of the Sixteenth International Joint Conference on Artificial Intelligence (IJCAI)*, Acapulco, Mexico, pp. 1151–1156.

Navarro-Serment, L. E. and Hebert, M. (2008). LADAR-based pedestrian detection and tracking, *Proceedings of the 1st Workshop on Human Detection from Mobile Robot Platforms*, IEEE ICRA 2008, IEEE.

Olson, C. F. (2000). Probabilistic self-localization for mobile robots, *IEEE Transactions on Robotics and Automation* 16(1): 55–66.

Olson, C. F., Matthies, L. H., Wright, J. R., Li, R. and Di, K. (2007). Visual terrain mapping for mars exploration, *Computer Vision and Image Understanding* 105(1): 73–85.

Proetzsch, M., Luksch, T. and Berns, K. (2005). Fault-tolerant behavior-based motion control for offroad navigation, 20th IEEE International Conference on Robotics and Automation (ICRA), Barcelona, Spain, pp. 4697–4702.

Proetzsch, M., Luksch, T. and Berns, K. (2010). Development of complex robotic systems using the behavior-based control architecture iB2C, *Robotics and Autonomous Systems* 58(1).

Ravine, D. P. M. M. (2007). Semi-autonomous rover operations: An integrated hardware and software approach for more capable mars rover missions, *Proceedings of the NASA Science Technology Conference 2007 (NSTC 2007)*.

Ray, L., Price, A., Streeter, A., Denton, D. and Lever, J. (2005). The design of a mobile robot for instrument network deployment in antarctica, *Proceedings of the 2005 IEEE International Conference on Robotics and Automation 2005 (ICRA 2005)*, Barcelona, Spain, pp. 2111–2116.

Schäfer, H., Hach, A., Proetzsch, M. and Berns, K. (2008). 3D obstacle detection and avoidance in vegetated off-road terrain, IEEE International Conference on Robotics and Automation (ICRA), Pasadena, USA, pp. 923–928.

Schäfer, H., Proetzsch, M. and Berns, K. (2005a). Extension approach for the behaviour-based control system of the outdoor robot RAVON, *Autonome Mobile Systeme*, Stuttgart, Germany, pp. 123–129.

Schäfer, H., Proetzsch, M. and Berns, K. (2005b). Stereo-vision-based obstacle avoidance in rough outdoor terrain, International Symposium on Motor Control and Robotics (ISMCR), Brussels, Belgium.

Schäfer, H., Proetzsch, M. and Berns, K. (2008). Action/perception-oriented robot software design: An application in off-road terrain, IEEE 10th International Conference on Control, Automation, Robotics and Vision (ICARCV), Hanoi, Vietnam.

Schenker, P., Huntsberger, T., Pirjanian, P., Dubowsky, S., Iagnemma, K. and Sujan, V. (2003). Rovers for intelligent, agile traverse of challenging terrain, International Conference on Advanced Robotics.

Schmitz, N., Koch, J., Proetzsch, M. and Berns, K. (2006). Fault-tolerant 3D localization for outdoor vehicles, IEEE/RSJ International Conference on Intelligent Robots and Systems (IROS), Beijing, China, pp. 941–946.

Schmitz, N., Proetzsch, M. and Berns, K. (2006). Pose estimation in rough terrain for the outdoor vehicle RAVON, 37th International Symposium on Robotics (ISR), Munich, Germany.

Sherwood, R., Mishkin, A., Estlin, T., Chien, S., Backes, P., Norris, J., Cooper, B., Maxwell, S. and Rabideau, G. (2001). Autonomously generating operations sequences for a mars rover using AI-based planning, IEEE/RSJ International Conference on Intelligent Robots and Systems.

Shoemaker, C. and Bornstein, J. (1998). The Demo III UGV program: A testbed for autonomous navigation research, *Proceedings of the IEEE International Symposium on Intelligent Control*, Gaitersburg, MD.

Singh, S. (ed.) (2006a). *Journal of Field Robotics – Special Issue on the Darpa Grand Challenge (Part I)*, Vol. 23, Wiley Blackwell.

Singh, S. (ed.) (2006b). *Journal of Field Robotics – Special Issue on the Darpa Grand Challenge (Part II)*, Vol. 23, Wiley Blackwell.

Talukder, A., Manduchi, R., Rankin, A. and Matthies, I. (2002). Fast and reliable obstacle detection and segmentation for cross-country navigation, IEEE Intelligent Vehicles Symposium, pp. 610–618.

Thuilot, B., Cariou, C., Cordesses, L. and Martinet, P. (2001). Automatic guidance of a farm tractor along curved paths, using a unique CP-DGPS, IEEE/RSJ International Conference on Robots and Systems.

Tunstel, E., Howard, A. M. and Seraji, H. (2001). Fuzzy rule-based reasoning for rover safety and survivability, *Proceedings of the 2001 IEEE International Conference on Robotics and Automation (ICRA 2001)*, Seoul, South Korea, pp. 1413–1420.

Urmson, C., Anhalt, J., Clark, M., Galatali, T., Gonzalez, J. P., Gowdy, J., Gutierrez, A., Harbaugh, S., Johnson-Roberson, M., Kato, H., Koon, P. L., Peterson, K. and Smith, B. K. (2004). High speed navigation of unrehearsed terrain: Red team technology for grand challenge 2004, *Technical Report CMU-RI-TR-04-37*, Robotics Institute, Carnegie Mellon University, Pittsburgh, PA.

Vandapel, N., Huber, D., Kapuria, A. and Hebert, M. (2004). Natural terrain classification using 3-D ladar data, IEEE International Conference on Robotics and Automation (ICRA), Vol. 5, pp. 5117– 5122.

Wellington, C., Courville, A. and Stentz, A. (2006). A generative model of terrain for autonomous navigation in vegetation, *The International Journal of Robotics Research* 25(12): 1287–1304.

Wellington, C. and Stentz, A. (2004). Online adaptive rough-terrain navigation in vegetation, *Proceedings of the IEEE International Conference on Robotics and Automation (ICRA)*, Vol. 1, pp. 96–101.

Zhang, Y., Schervish, M., Acar, E. U. and Choset, H. (2001). Probabilistic methods for robotic landmine search, IEEE/RSJ International Conference on Intelligent Robots and Systems.

Zhao, L. and Thorpe, C. (2000). Stereo and neural network-based pedestrian detection, *IEEE Transactions on Intelligent Transportation Systems* 1(1): 148–154.

Zolynski, G., Braun, T. and Berns, K. (2008). Local binary pattern based texture analysis in real time using a graphics processing unit, *Proceedings of Robotik 2008*, Vol. 2012 of VDI-Berichte, VDI, VDI Wissensforum GmbH, pp. 321–325. ISBN 978-3-18-092012-2.

16
Computer training with ground teleoperated robots for de-mining

G. KOWALSKI, J. BĘDKOWSKI, P. KOWALSKI and A. MASŁOWSKI, Warsaw University of Technology, Poland

Abstract: This chapter presents a study on the simulation platform and training of a mobile robot's operator. First, simulation and training of general machine operators is shown. Then problems of teleoperated robots operator's training are presented and user's needs are highlighted. Subsequently a multilevel training concept is described and extended by a virtual and augmented reality approach. Finally a platform for mobile robot operator training based on computer simulation using NVIDIA PhysX Engine is presented.

Key words: teleoperated robot simulation, e-training, mobile robot operator training.

16.1 Introduction

Landmines left in battlefields are a huge danger for civilization, spreading and extending into non-urbanized areas. The number of antipersonnel mines left after wars is estimated at more than 110 million, lying in the territories of more than 80 countries on non-marked ground. This situation brings a lot of problems for citizens as they would like to occupy the ground, build new roads and industrial facilities. The number of injuries and deaths is estimated to be 20,000 civilians per year. All these problems were described when the European Commission raised four priority levels concerning de-mining.[2,36]

The first problem is finding the mines and other potential threats. There are many means of finding the explosives based on electromagnetic, electro-optic, biosensors or nuclear and chemical technologies, for example, nuclear quadrupole resonance (NQR),[6] special sensors,[4] ground penetrating radar (GPR),[1,9] ultra-wide-band radar (UWB), infrared cameras and others. Some of these methods are applied in sensors, which are carried on semi-autonomous or autonomous unmanned vehicles (for example: Robudem,[2,5] Silo-6,[34] Pedipulator[2]) or teleoperated ones (for example: AMS[2]).

The second problem arises when it comes to de-mining the area. Apart from dangerous manual clearance, teleoperated de-mining vehicles are used. Some of them are used to cause the explosion of mines (e.g. MineWolf[35] – see Fig. 16.1[37]), and others are used to pick up the bombs and put them into a disposal box (e.g. Inspector,[23] see Fig. 16.2, Expert,[19] Ibis,[20] see Fig. 16.3, Scout,[22] see Fig. 16.4, Grizzly-1,[24] and Knight[50]).

397

16.1 MineWolf teleoperated vehicle near mine field.

16.2 Inspector transporting an old bomb.

Teleoperated robots have a complex control system. They usually come with single or multi-display consoles, complete with many buttons, levers and joysticks, light and sound status indicators (e.g. expert robot control console, see Fig. 16.5). It is obvious that when a new machine is bought, its future operators must become familiar with the controlling device.

16.3 Ibis robot transporting a package.

16.4 Scout robot reaching for a suitcase.

Unaware of the robot functionality and limitations, operators expose their newly bought machines to danger, because training processes must be carried out using real machines in various simulated situations. While the training is carried out, the robot can be, in best cases, scratched, and in the worst cases it can break down as a result of serious mechanical damage, which can halt the use of the robot for a period of time. Then the robot must be repaired, which brings additional costs for the user.

16.5 Expert robot control console.

Controlling this kind of machine is very complex, as the operators must not only worry about steering, but must also plan missions in order to achieve their goals. When it comes to the use of additional equipment, such as X-ray devices or firearms, the situation is even more complicated. In this case, the whole mission becomes a management task, where many operator skills, such as mission planning, orientation, steering abilities, etc., are involved. Thus, several trials must be carried out before allowing the operator to carry out the task during a stressful mission (handling explosives) where a lot of precision and patience is necessary.

A similar kind of situation can be seen when designing other new products, for example cars or planes. Building a prototype for initial testing is an expensive and time-consuming task. Computer-aided engineering (CAE) tools help the processes of designing and testing of early phase prototypes. The special abilities of such software enable the simulation, for example, of the behavior of the vehicle in exemplary static, dynamic and kinematic conditions (e.g. Adams[38]). These activities carried out in the real world would be expensive for the developer, so the aid of computer tools is important. Thus, computer software, simulating the work of the teleoperated robot in the virtual environment, can decrease the cost of training for the operators of these robots and lower the risk of damaging the machines.

16.2   Simulation and training in handling teleoperated robots

Virtual trainers are commonly used in the army to train soldiers in handling a variety of military equipment, starting with light firearms,[26,28] rocket launchers,[47,48] through to helicopters,[14,33] planes,[12,13,15,16,32] combat vehicles,[27,49] and finishing with tanks[17,25,45,46] or even ships[31] and submarines.[18] Preliminary training is necessary for military purposes, because there is a strong emphasis on the safety of weapon users and their surroundings as well as on decreasing the cost of the training. Although used by the military, which drives the market, many of the trainers were developed for civilian applications. Searching the Internet, one can find many trainers for different vehicles, like cars,[30] trucks,[29] locomotives[11] and building machines, for example cranes.[43,44] Use of this system is, as above, motivated by decreasing the cost of the training and increasing the safety for users and surroundings.

No matter whether the trainer is for military or civilian use, all trainers are very similar in structure and principles of operation. The most important aspect of these systems is the presence of a control console, which is the same or almost the same as one connected to the real device/vehicle. This enables the trainee to switch from the trainer to the real machine faster and decrease the time of training in effect. All trainers are equipped with a multi-display system, consisting of simple computer displays or big screens, which often enable the operator to have almost a 180° view in the cabin, built into the trainer's construction. Trainers mainly provide full-input support (e.g.[49]), a full view from the 'cabin' (e.g.[43]) or six degrees of freedom motion system, complete with a replica of the aircraft's cockpit and electric controls loading system (e.g.[15]). Sometimes trainee stations are set together with instructor stations (e.g.[45]). All stations are designed according to the rules of ergonomics.

Vehicle trainers provide the following hardware functionality:

- sets of real input imitators
- multiple display systems
- seclusion of trainees in containers, helping them to focus on the mission only
- acoustic background simulation system
- multi-DOF motion systems to move the trainer platform
- pilot seat vibration generators
- controls loading systems
- cannon kick imitators
- modular structure, which enables new equipment to be added quickly.

The instructor's station is usually less well equipped than the trainee's. The reason for this is that the instructor only controls and aids the trainee. Functionality of the instructor's station includes:

- distribution of training task for trainees
- regulation of the training process by introducing task parameters and result evaluation
- observation of the task realization course
- aiding the trainee using audio channel.

Good quality input and visualization devices are only half of the story. Software also plays a very important role here. The producers are concerned with making the simulation as realistic as possible. PC-based hardware is a host for real-time operational systems, which provide full control of both trainee and instructor stations. Advanced physics engines, which introduce main functionality to the trainers, are used to produce realistic 3D worlds with buildings, roads, urban infrastructure, vegetation, water tanks, different weather and light conditions. Such a variety of additions helps the simulation to be more realistic and to present many possible situations to cope with. The more complications a trainee faces during training, the better he will be at solving problems faced during the real mission. The main simulation features are:

- 2D and 3D still or moving targets, like tanks, transporters and other vehicles
- support for fog, lightning, dependent on time of day, season of the year, precipitation, shadows
- customizable scenarios and missions
- indicators reporting vehicle's status
- collision detection
- vehicle rollover
- vehicle parts snag and break
- mistake loggers
- global overview of mission field (instructor's ability).

16.3   Teleoperated robot operator's training

Teleoperated robots, which are used for explosives disposal, are constructions that move by means of caterpillar tracks or wheels. Wheeled robots are usually faster, but have limited mobile functionalities. Caterpillar-based vehicles are able to cross rough terrain with tall obstacles or drive up and down stairs. Although there are several differences between these kinds of robots, the controlling principles are the same, using buttons and joysticks to move the manipulator arm and mobile platform. Explosive ordnance disposal (EOD) robots can be steered via a control stand (see Fig. 16.6[40]), sometimes with a detachable console,[41] or case,[50] or totally commercial means, like PlayStation's joystick.[24]

Robots are usually equipped with additional devices, which enable them to carry out non-disposable tasks like detonating explosives, clearing the robot's path and seeking for danger. The main extensions are:[21,39,42]

- different grippers
- firearms
- tow hooks
- barbed wire cutters
- cable cutters
- saws
- glass breakers
- drills
- X-ray systems
- laser sights
- omni-directional microphones
- chemical sensors
- additional cameras and light sources.

Some of these tools are mounted on the robot before missions; others can be exchanged with those stored in an on-robot tool repository during missions.[42] All this equipment aids the robot to achieve its goal of disposing of or destroying explosives in a variety of dangerous situations, like clearing minefields or carrying out antiterrorist actions, and in different locations, for example, open fields, buildings and public transport vehicles.

The complex control of robots and their accessories is a task for which operators should be prepared. Searching the Internet, one cannot find open courses for teleoperated robot operators. This is connected with the fact that robot producers offer training only for customers, i.e. robot buyers. Therefore, there is no information about the procedures or teaching aids used during the training, as the companies keep this to themselves. Additionally, users of robots do not share any information about their training methods either.

Recently, in Poland, research based on surveys was carried out.[8] It showed that training provided by a producer of the robots, PIAP (Industrial Research Institute

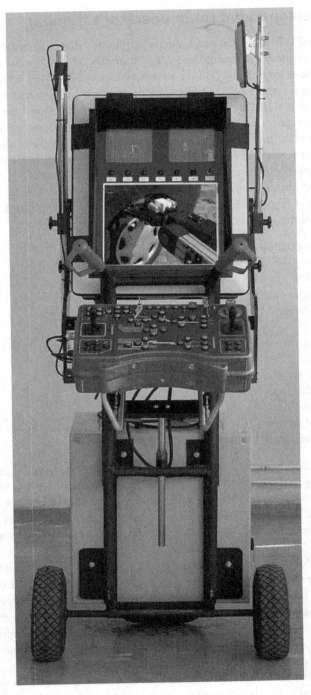

16.6 Control stand.

for Automation and Measurements), is estimated to take around 24 hours. During this time trainees are taught the basics of the robot: how to steer, service and maintain it. The workload of the robot during training is limited to between five and ten people at the same time, and so far around 135 people have been trained. The methods used do not involve any virtual teaching aids and trainees work only with a real robot and its accessories. Moreover, training is carried out using only the robot's manual, which contains step-by-step instructions about servicing, but without any extended chapters or procedures about the steering itself. Training is cost-free when buying a new robot.

Training among users of the robots – the Polish army and police – is also carried out without details on how to steer it. Basic training (using a real robot) takes around 20 hours, but it was pointed that this is not a precise time approximation and it depends on the trainee's skills. Getting familiar with service and maintenance takes several hours, the same as for the basics of the robot. The total time listed for the training is around 30 hours for three to four people training on one robot at the same time. Trainees continue learning during individual courses that they have to organize themselves at convenient times. The courses describe the goals, limitations, and time boundaries and try to achieve proficiency.

Interviewees admit that virtual training would enable them to take part in more effective training without any risks to the robot. Today, many training mission scenarios are simply omitted. A virtual trainer is thought to be a good solution for preliminary and additional support and with the cost not exceeding 10 percent of the price of the robot; its use would be warmly welcomed. According to this research, there is a need for using a virtual trainer and improving and expanding training methods.

16.4   Teleoperated robot users' needs

Survey results[8] include information about the user's preferences related to trainers. There is a need for a trainer based on additional control devices, which should be exactly the same as real ones. In the opinion of interviewees, trainees will become familiar with the steering faster and make fewer mistakes in their future work with the robot. The organization of tasks using virtual accounts is desired in order to view the training history of each trainee and estimate progress. In this way, an individual path of training for each trainee can be planned. A strong emphasis is put on the robot accessories, which are very important during missions. The most desired additions are the commonly used X-ray system, glass breaker and firearms.

Robot users carry out various practices by themselves but they cannot become fully acquainted with the devices as operators because they are not able to perform the more difficult real-life tasks. Introducing virtual practices, such as opening a car with explosives inside, working in narrow spaces, and overcoming obstacles, improves the skills of trainees the most. Applying time limits makes the training even more complicated and forces the trainee to make a decision quickly.

Additional problems to cope with are those connected with communication (losing the range, controlling lags), video channel disturbances and failure or distortion of the camera's optic system. More simulation features are added by different weather conditions, such as rain or fog, and the effect on ground behavior as a result (lubricated and poached ground is found when working in less urbanized areas or destroyed open terrain). More mission complexity through a structured training path and by adding a rigorous scoring system will enable the operators to complete the difficult tasks set.

16.5 Multilevel training for teleoperated robots

Computer training is meant to be preliminary, and carried out before the trainee works on a real machine, or it consists of additional training to fill the gaps in standard training, which cannot be carried out using the robot. The efficiency of the computer training can be achieved by dividing the training into three levels, which are described below.[3] The schematic three level training is described in Fig. 16.7.

The lowest level of training is based on a PC with its basic input devices of keyboard and mouse (see Fig. 16.7, dashed line means computer). The training stand provides a multi-display set, where the virtual robot console and the robot's environment are visualized (see Fig. 16.8). With this method, the trainee can become familiar with the robot functionalities and limitations, without causing any harm to the robot or its accessories if a mistake is made.

The first level simulator moves the robot platform and its five degrees of freedom arm. There are possible views from five different cameras, one of which is controlled in two axes. Robots can take part in various missions (different placement of bombs, training missions), in varying weather and light conditions (sunshine, fog, night) and can face simple obstacles (boxes, tunnel, car). More reality is added through the use of a glass breaker, which is equivalent to a real one, and an analog camera noise effect, which occurs when the robot and control station are separated by walls. The simulator controls the center of the mass of particular parts of the robot, so that it will fall when maneuvered badly.

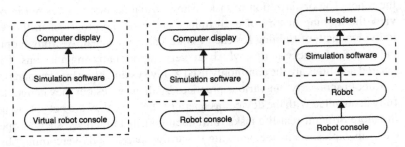

16.7 Three levels of training, from left: the lowest level, the middle level and the highest level.

16.8 Virtual robot in artificial environment.

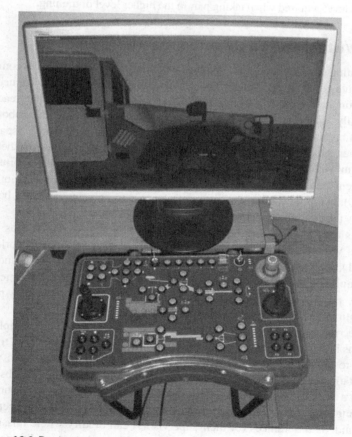

16.9 Real robot console.

The middle level of training uses a real robot control console (see Fig. 16.9), connected to the PC, which visualizes the robot movement in a selected virtual environment (see Fig. 16.7). This kind of training is a great experience for the trainee, who can become familiar with the console he or she will be working with in the future.

The highest level of training involves virtual and augmented reality to visualize either a virtual robot in a real environment or a real robot in a virtual environment. The visualization uses a 3D helmet or glasses, which enable the trainees to see a virtual model overlapping the real scene. While the operator is controlling the robot, its movements are sent to simulation software, which generates reactions to the environment (see Fig. 16.7). The proposed organization of the training can treat each trainee individually, rank him or her, and assign the trainee with a training path through the tasks, according to skill level. The computer program is able to select well-skilled trainees and apply them to more difficult tasks with higher intensity. Less-skilled trainees are obliged to redo particular tasks in order to gain better marks and progress to the skill level required when taking part in the higher level of training.

16.6   Virtual and augmented reality

Implementing virtual reality (VR) involves switching the trainees from flat computer displays into a 3D world, where depth is a very important feature. Trainees are able to estimate the distance between an object and the robot arm while reaching. Additionally, space is more visible in 3D; therefore planning the robot arm configuration and mobile platform placement during missions in narrow spaces is much easier. Also virtual environments can be 'user friendly' and mirror the same locations in which robot users work. Transforming the real world into a virtual 3D environment can be done by laser scanning and program reconstruction (see Fig. 16.10). In this way, trainees can react faster to environmental changes, because they know the environment (for example, an airport) well.[7]

Augmented reality is used to connect VR and the real world. It uses a real robot in a virtual environment or a virtual robot working in the real world (see Fig. 16.11). The first approach involves steering the robot via its console and then, using simulation software, the surroundings react to the robot's movements of grabbing, pushing, etc. The exact positioning of the robot is key to perfect manipulation possibilities, provided by simulation.

The second approach creates an opportunity to carry out some complicated operations in strictly defined conditions. By using visual printed markers, trainees are able to combine real and virtual interactions by creating their own virtual representation of a real one. In this way, driving upstairs, crossing real obstacles, etc. with a virtual robot is as simple as adding new items to the real world and introducing them into the simulation. The above solutions create an exceptional training aid, as the trainees (or their supervisors) can modify whole missions in order to train for as many crisis situations as possible in chosen conditions.

16.10 Corridor reconstruction using laser scanner.

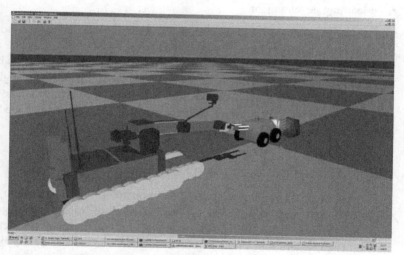

16.11 AR–virtual robot in real environment (Inspector robot is virtual and follows by smaller real and autonomous ATRVJr).

16.7   A platform for mobile robot operator training based on computer simulation using PhysX engine

This section describes the efforts undertaken in the pursuit of developing an improved training simulator of multipurpose mobile robots (including robots for de-mining). As stated previously, this software is desired by both end users and the manufacturers of robots. Work has been carried out by the Industrial Research Institute for Automation and Measurements in Poland.

16.7.1 General concept of the platform

A proposed framework for designing training for mobile robot operators aims to provide an easy and flexible way of developing the necessary elements for any simulation training. The novelty of our approach is based on the concept of a versatile system for designing a robot trainer. It is a fundamental assumption that the process itself will be minimized to use specified tools with the minimum computer programming skill requirements. The end product of the platform will be e-training, comprising a set of e-tasks.

In general, the mobile robot operator trainer is a set of hardware and software (called the e-training). The platform is designed to support the following steps of development:

- virtual robot development
- virtual control panel development
- virtual robot environment development
- specific task and events for each mission development
- e-learning development.

The platform supplies a set of tools called generators to design virtual models in a format of the platform for each step listed above.

The authors chose to base their system on a modern physics engine, as is often the case with game simulators. After reconnaissance of the available engines on the market, the Ageia PhysX engine was chosen. The main advantage of the chosen engine is that it is freeware, well documented and supported by the biggest graphics chipset manufacturer Nvidia.

16.7.2 Process of designing new training

The development of training for mobile robot operators was divided into several steps. The first step was to choose which kind of robot is going to be used, the drive model, and the sensor equipment of the robot and actuator's system. Then, where the action will be arranged. Is it indoors or maybe in an outdoor scenario? Subsequently it is necessary to plan what can happen during the mission, all events, the initial position of the robot and time needed to perform the mission. Afterwards when a ready set of indoor/outdoor scenarios has been compiled, it is possible to design the graph of a training session. All steps are supported by dedicated tools of the framework, the development of which is in progress.

Robot designing

The virtual model of the robot is the basic element of a simulation. Figure 16.12 presents the main elements of the robot model.

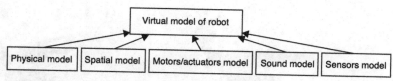

16.12 Scheme of components of virtual model of robot.

The virtual model of the robot is composed of the following models:

- physical (represents mechanical properties of a designed robot)
- spatial (represents what is visualized)
- motors/actuators (represents all driven parts of a robot)
- sound (represents a set of sounds associated with the model of motors/actuators)
- sensors (represents a set of sensors mounted onto a robot).

Each listed model is described in a single file; therefore the full virtual model of the robot requires five files and the information about their interaction. The platform offers a tool called Robot Generator for designing a robot. The window of the application is presented in Fig. 16.13.

Despite the early stage of the work on the tool, it allows different kinds of mobile robot platforms with different drive systems to be designed, even those with the manipulator mounted onto it. Figure 16.14 presents two mobile robots; a real robot equipped with a caterpillar drive and a virtual one. The tools also support the process of designing four-wheel robots driven by a differential mechanism.

The designed robot can be equipped with mostly common sensors, such as cameras, laser range finders and sonars. At this stage, it is only possible to use one end-effector, namely a gripper, to handle the dangerous material. In the future it may be possible to add new actuators, such as a Firehouse or X-ray camera.

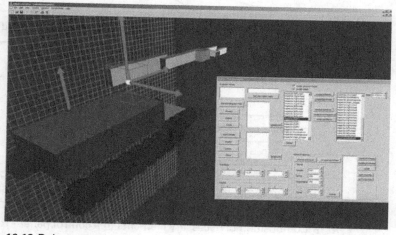

16.13 Robot generator.

16.14 Two mobile robots – Inspector, real and virtual.

Control panel design

The virtual model of the control panel is a second basic element of a simulation. Figure 16.15 presents the main elements of a control panel.

It is assumed that the control panel consists of joysticks, buttons, switches and displays. The control panel communicates with a robot via a 'communicating module' or directly, depending on whether a real or a virtual panel is in use. The control panel can be virtual or real so that the trainee can become accustomed to a real device controlling the virtual robot using a real control panel. The described technology is presented in Fig. 16.16.

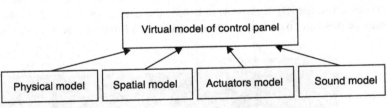

16.15 Composition of necessary elements of a control panel.

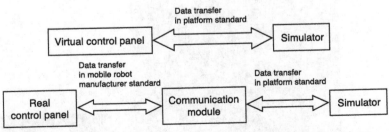

16.16 Composition of necessary elements of a control panel – robot communication.

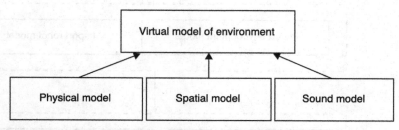

16.17 Composition of a virtual model of an environment.

Robot environment designing

The virtual model of an environment similar to the models of the control panel and robot consists of the following elements: physical model, spatial model and sound model (see Fig. 16.17). The scheme presents the main elements of an environment.

The environment in the simulator is represented by the virtual model of the environment. It is accepted that this model will be designed in a similar manner to the virtual model of the robot described previously. The platform supplies a tool that can support a designer in the process of creating the environment. It will be possible to utilize components from the components base. This base will consist of a set of sample models such as buildings, furniture (chairs, closets, tables, etc.), vehicles, trees, different kinds of surfaces (i.e. soil, road, grass). All objects from the base will have physical and spatial properties. Apart from the physical properties, there will be a spatial model attached and visualized.

E-task development

In order to design the e-task, the following models are required: robot model, control panel model and environmental model. A training designer has the ability to set the starting position of the robot and after that define all possible mission events he considers important. Next, he can define time thresholds for the mission. Such an e-task is then exported and can be used to prepare the full multilevel training. The e-task designing steps scheme is described in Fig. 16.18.

For this purpose, the framework supplies the e-tasks generator. The tool supplies the basis for the events. The e-task designer will be able to move about the scene with the robot and environment, then choose one actor or group of actors and attach an event with specific parameters. The possible events defined for the single e-task are:

- robot ride to chosen localization
- time exceeded
- move an object to chosen localization
- touch an object

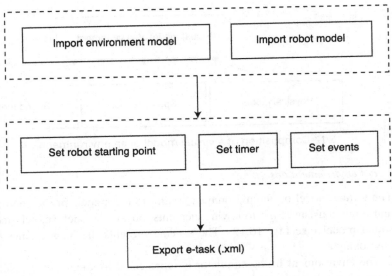

16.18 E-task designing steps.

- damage/disable robot
- neutralize/destroy an object (e.g. shoot with water gun, use explosives)
- take an X-ray.

Training composition

Once the designer has carried out the previous steps, the last task is to prepare the e-training. Using previously defined e-tasks the trainer can prepare a graph of the training. A sample graph is presented in Fig. 16.19.

The training starts with Mission 1 and continues with two or three subsequent missions. In the first case when condition C_3 is satisfied it means that Mission 2 is non-requisite, otherwise if condition C_2 is satisfied the training path will lead through to Mission 2. Under conditions C_1, C_4, C_5 and C_8, appropriate missions will be repeated. At the end of each training session, a summary is presented on the screen and saved to file.

16.7.3 Supporting tools

The platform will support specialized tools for designing physical models, spatial models and others. The standard implemented and used in the platform accepts COLLADA and PhysX files for designing physical models. For spatial models, the 3ds format is considered the best choice. Physical models can be developed using commercially available tools such as Solid Works with the Collada plug-in, or Maya with the ColladaMaya plug-in. It is also possible to use a freeware tool

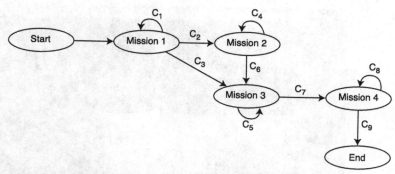

16.19 Scheme of the training graph. Each node must be one of three: start, end or mission; transition to the next mission is represented by arrow and is triggered under specific condition denoted as C_1 ... C_9.

under GNU/GPL licence, such a Blender with the Collada plug-in. Similarly for designing spatial models one can use the non-free product Autodesk 3ds Max, which has a native file standard of 3ds, or Blender, a free competitor on the market, which can handle 3ds files as well. Utilization of these tools ensures long-term support for the designed platform.

16.7.4 Running a simulation

The platform supplies a program for training new mobile robot operators. The program will be able to handle the virtual model of the robot, virtual model of the environment, virtual model of the control panel and the e-training scenario. It will support communication between the virtual model of the robot and a real control panel via a communication module as well. Additionally, it will be possible to plug in a virtual helmet and perform the simulation in a real environment with a virtual robot and artificial elements. Thus the platform will support the very modern and up-to-date technology of augmented reality. The final trainer is shown in Fig. 16.20.

16.8   Conclusions

Virtual training is needed to improve the safety of robots, users and surroundings, and to decrease the cost of the training. Preliminary and additional courses create the possibility of testing more trainees in almost any desired situation. The use of a real control console and virtual or augmented reality is indispensable in shaping training missions according to specific preferences. Virtual training introduces the standardization of procedures, a clear marking system, the establishment of an individual training path for each trainee, the ability to profile into groups in advance and it also automates particular procedures connected with task assignment or marking.

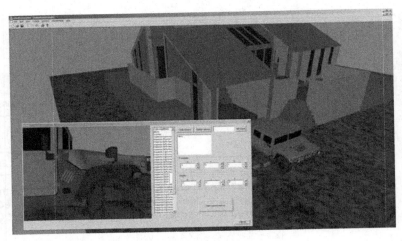

16.20 The main window of the trainer.

Modern physics engines and computer modelers are able to create a perfect copy of the real world, making the differences almost invisible. When combining realistic steering devices and an adjustable environment, formed with visual markers, trainees use a game-like tool, which is easy to adopt. This solution has proven to work well in military and non-military situations, but users of robots do not know enough about it. Benefits from the training should not be overestimated, as long as they keep people and their robots safe. After this training, operators are better skilled and able to cope with more difficult missions including finding and getting rid of explosives from buildings, vehicles and minefields.

16.9 References

1 M. A. Atya, I. El-Hemaly, A. Khozym, A. El-Emam, G. El-Qady, M. Soliman, M. Abd Alla, "Landmine detection using integration of GPR and magnetic survey", in Proceedings of the 7th IARP International Workshop on Robotics and Mechanical assistance in Humanitarian De-mining and Similar risky interventions (HUDEM'2008), Cairo, Egypt, 2008, pp. 21 and 85–89 [CD].

2 Y. Baudoin et al., "Mobile Robotic Systems Facing the Humanitarian Demining Problem State of the Art (SOTA) December 2007 ITEP 3.1.4 Task", in Proceedings of the 7th IARP International Workshop on Robotics and Mechanical assistance in Humanitarian De-mining and Similar risky interventions (HUDEM'2008), Cairo, Egypt, 2008, pp. 12–13 and 1–31 [CD].

3 J. B—dkowski, G. Kowalski, A. Masłowski, "Framework for Creation of the Simulators for Inspection Robotic Systems", in Proceedings of the 7th IARP International Workshop on Robotics and Mechanical assistance in Humanitarian De-mining and Similar risky interventions (HUDEM'2008), Cairo, Egypt, 2008, pp. 25–26 and 113–117 [CD].

4 S. Carillo, C. Santacruz, D. Botero, A. Forero, C. Parra, M. Devy, "A Complementary Multi-sensory Method for Landmine Detection", in Proceedings of the 7th IARP International Workshop on Robotics and Mechanical assistance in Humanitarian

Demining and Similar risky interventions (HUDEM'2008), Cairo, Egypt, 2008, pp. 20 and 143 [CD].

5 D. Doroftei, Y. Baudoin, "Development of a semi-autonomous De-mining vehicle", in Proceedings of the 7th IARP International Workshop on Robotics and Mechanical assistance in Humanitarian De-mining and Similar risky interventions (HUDEM'2008), Cairo, Egypt, 2008, pp. 14 and 37–40 [CD].

6 H. Itozaki, G. Ota, "Nuclear Quadrupole Resonance for Explosive Detection", in Proceedings of the 7th IARP International Workshop on Robotics and Mechanical assistance in Humanitarian De-mining and Similar risky interventions (HUDEM'2008), Cairo, Egypt, 2008, pp. 19 and 63–67 [CD].

7 G. Kowalski, J. Będkowski, A. Masłowski, "Virtual and Augmented Reality as a Mobile Robots Inspections Operators Training Aid", in Bulletin of the Transilvania University of Brasov. Proceedings of the international conference Robotics'08", vol. 15(50) series A, special issue no. 1, vol. 2, pp. 571–576, Brasov, Romania, 2008.

8 A. Maslowski, G. Kowalski, J. Będkowski, A. Kaczmarczyk, M. Kacprzak, "Elaboration of the multilevel training methodology with various stage trainers and selection of the set of the trainers enabling realization of this methodology", PIAP, Warsaw, Poland, report from realization of the stage I of the Ministry of Science and Higher Education's research project no. OR00004604, 2008.

9 Z. Zyada, T. Matsuno, T. Fukusa, "Fuzzy Template Based Automatic Landmine Detection from GPR Data", in Proceedings of the 7th IARP International Workshop on Robotics and Mechanical assistance in Humanitarian De-mining and Similar risky interventions (HUDEM'2008), Cairo, Egypt, 2008, pp. 21 and 79–84 [CD].

10 http://mecatron.rma.ac.be/index_files/Page551.htm, available 25.11.2008
11 http://www.ai.com.pl/en/products/ep09_ds.html, available 27.11.2008
12 http://www.ai.com.pl/en/products/mig21bis_ffs.html, available 27.11.2008
13 http://www.ai.com.pl/en/products/orlik_ffs.html, available 27.11.2008
14 http://www.ai.com.pl/en/products/sokol_fms.html, available 27.11.2008
15 http://www.ai.com.pl/en/products/su22m4_fms.html, available 27.11.2008
16 http://www.ai.com.pl/en/products/ts11_iskra_ffs.html, available 27.11.2008
17 http://www.ai.com.pl/en/products/twardy_ts.html, available 27.11.2008
18 http://www.amw.iq.pl/index2.php?n=1468, available 27.11.2008
19 http://www.antiterrorism.eu/airport_security.php, available 25.11.2008
20 http://www.antiterrorism.eu/combat_robot.php, available 25.11.2008
21 http://www.antiterrorism.eu/offer_inspector_p2.php, available 16.02.2009r.,
22 http://www.antiterrorism.eu/pyrotechnical_robot.php, available 27.11.2008
23 http://www.antiterrorism.eu/remote_operated_vehicle.php, available 25.11.2008
24 http://www.armedforces-int.com/categories/mine-clearing-technology/grizzly1-eodrobot.asp, available 13.02.2009r.
25 http://www.autocomp.com.pl/index.php?page=products&subpage=jaguar&lang=pol, available 27.11.2008
26 http://www.autocomp.com.pl/index.php?page=products&subpage=orlik&lang=pol, available 27.11.2008
27 http://www.autocomp.com.pl/index.php?page=products&subpage=pancernik&lang=pol, available 27.11.2008
28 http://www.autocomp.com.pl/index.php?page=products&subpage=snieznik&lang=pol, available 27.11.2008
29 http://www.autosim.no/dl/pdf/as_1300_truck_driving_simulator.pdf, available 27.11.2008
30 http://www.autosim.no/dl/pdf/autosim_car.pdf, available 27.11.2008

31 http://www.barco.com/projection_systems/downloads/AN_-_NSST.pdf,   available 27.11.2008
32 http://www.elbitsystems.com/lobmainpage.asp?id=900, available 27.11.2008
33 http://www.elbitsystems.com/lobmainpage.asp?id=904, available 27.11.2008
34 http://www.iai.csic.es/users/silo6/, available 25.11.2008
35 http://www.minewolf.com/demining-products-services/mini-minewolf-deminingmachine.html, available 25.11.2008
36 http://www.minewolf.com/humanitarian-demining.html, available 25.11.2008
37 http://www.minewolf.com/news-and-media/press.html, available 25.11.2008
38 http://www.mscsoftware.com/products/adams.cfm?Q=131&Z=396&Y=397, available 26.11.2008
39 http://www.ms.northropgrumman.com/Remotec/details/wolverine_accessories.htm, available 16.02.2009r.
40 http://www.telerob.de/telerob_online/content.php?seiten_id=36&language=english&target_medium=screen&view=1&debug=0, available 13.02.2009r.
41 http://www.telerob.de/telerob_online/content.php?seiten_id=36&language=english&target_medium=screen&view=2&debug=0, available 13.02.2009r.
42 http://www.telerob.de/telerob_online/index.html, available 16.02.2009r.
43 http://www.vortexsim.com/products/mobilecrane/, available 27.11.2008
44 http://www.vortexsim.com/products/towercrane/, available 27.11.2008
45 http://www.wcbkt.pl/beskid2mk_eng.htm, available 27.11.2008
46 http://www.wcbkt.pl/beskid2mz_eng.htm, available 27.11.2008
47 http://www.wcbkt.pl/iglica_eng.htm, available 27.11.2008
48 http://www.wcbkt.pl/iglica2_eng.htm, available 27.11.2008
49 http://www.wcbkt.pl/ortles3mk_eng.htm, available 27.11.2008
50 http://www.wmrobots.com/wmrobots/knighteodrobot.html, available 13.02.2009r.

Part IV
Robot autonomous navigation and sensors

17

A fuzzy-genetic algorithm and obstacle path generation for walking robot with manipulator

A. PAJAZITI, I. GOJANI, SH. BUZA, A. SHALA,
University of Prishtina, Kosova

Abstract: As biped robots assume more important roles in applicable areas, they are expected to perform difficult tasks like mine detecting and quickly adapt to unknown environments. Therefore, biped robots must quickly generate the appropriate gait based on information received from the visual system. In this paper a conventional PD controller and fuzzy logic controller for planning walking on flat ground for a planar five-link biped robot is presented. Both single support and double support phases are considered. The joint profiles have been determined based on constraint equations cast in terms of step length, step period, maximum step height and so on. When the ground conditions and stability constraint are satisfied, it is desirable to select a walking pattern that requires small torque and velocity of the joint actuators. Using the computed torque method gives the input–torque on the biped robot. The locomotion control structure is based on integration of kinematics and dynamics model of the biped robot. The proposed control scheme and fuzzy logic algorithm could be useful for building an autonomous non-destructive testing system based on the biped robot. The fuzzy logic–rule base is optimized using a genetic algorithm. The effectiveness of the method is demonstrated by a simulation example using Matlab software.

Key words: biped robot, conventional controller, fuzzy logic, genetic algorithm, planning walking.

17.1 Introduction

Mobile platform systems, consisting of a walking robot platform equipped with one or more manipulators, are of great importance to host applications, mainly due to their ability to reach targets that are initially outside of the manipulator's reach. Applications for such systems abound in mining, construction, forestry, planetary exploration and the military. Many of these systems employ wheeled mobile robots and walking robots which is the type of system studied in this chapter.

Walking robots continue the worldwide efforts of realizing such mechanisms with the goals of understanding walking and pushing forward technology that might be useful for several applications where wheel-driven machines are not feasible. Research on legged robots typically concentrates on the determination of a vehicle's trajectory, foothold selection and the design of a sequence of leg movements.

Motion planning for walking robots with a manipulator is concerned with obtaining open loop controls, which steer a platform from an initial state to a final

421

one, without violating the nonholonomic constraints. A comprehensive survey of developments in control of nonholonomic systems can be found in Kolmanovsky and McClamroch (1995).

Moving mobile manipulator systems present many unique problems as a result of coupling holonomic manipulators with nonholonomic bases. Seraji (1998) presented a simple on-line approach for motion control of mobile manipulators using augmented Jacobian matrices. The approach is kinematic and requires additional constraints to be met for the manipulator configuration. Perrier et al. (1998) represented the nonholonomy of the vehicle as a constrained displacement and tried to make the global feasible displacement of the system correspond to the desired one. Foulon et al. (1999) considered the problem of task execution by coordinating the displacements of a nonholonomic platform with a robotic arm using an intuitive planner, where a transformation was presented. The same authors introduced other variations of local planners, which were then combined to constitute a generalized space planner (Foulon, Fourquet, and Renaud 1998).

Papadopoulos and Poulakakis (2000) presented a planning and control methodology for mobile manipulator systems allowing them to follow desired end-effector and platform trajectories simultaneously without violating the nonholonomic constraints.

The problem of navigating a mobile manipulator among obstacles has been studied by Yamamoto and Yun (1995), who simultaneously considered the obstacle avoidance problem and the coordination problem.

To reduce the computational complexity of motion controls of mobile manipulator systems, some heuristics have also been developed by several researchers. Fuzzy logic controllers (FLCs) have been used by several researchers in the recent past (Deb et al., 1998, Pratihar et al., 2002, and Mohan and Deb, 2002) for navigation among stationary obstacles of walking robots with manipulators. The optimization problem involves finding an optimal fuzzy rule base that the walking robot should use for navigation when left in a number of scenarios of stationary obstacles. Once the optimal rule base by using the genetic algorithm (GA) is obtained off-line, the walking robot can then use it on-line to navigate in other scenarios of stationary obstacles.

In the present study, we concentrate on a navigation problem, where the objective is to find an obstacle-free path between a starting point and a destination point, requiring the minimum possible travel time.

This chapter consists of five sections: Section 17.2 describes a walking robot with manipulator model. The possible solution of the problem using FLC is proposed in Section 17.3. Section 17.3.1 discusses the fuzzy and GA approaches. The results of computer simulations are presented and discussed in Section 17.4. Some concluding remarks and scope for future work are presented in Section 17.5.

17.2   Mobile platform and manipulator system model for walking robot

The mobile platform and manipulator system model consists of two parts: a walking robot with eight degrees of freedom, and a manipulator with two degrees of freedom (see Fig. 17.1). Because nonholonomy is associated with the mobile platform, while the manipulator is holonomic, the system is studied as two connected subsystems: the holonomic manipulator and its nonholonomic platform. This allows one to find an admissible path for the mobile platform that can move it from an initial position and orientation to a final desired one. Next, using known techniques for manipulators, joint trajectories were calculated for the manipulator so that its end-effector is driven to its destination. An advantage of this approach is that it is very simple to extend the method to mobile systems with multiple manipulators on board.

17.2.1 Nonholonomic mobile platform subsystem

The eight-legged robot was adopted as a mobile platform that is 65 cm long. The minimum height of the robot is 20 cm, the maximum height (legs outstretched) is 35 cm. It consists of a central body and eight legs with three segments each (see Fig. 17.2). The legs consist of a thoracic joint for protraction and retraction, a basal joint for elevation and depression and a distal joint for extension and flexion of the leg. With six degrees of freedom (DOF) of the central body the mobile platform has 32 DOF. Each leg has three DOF, turning around the a, b and g axes, while the manipulator has two DOF (q1 and q2), and is installed on the front side of the mobile platform.

17.1 Eight-legged walking robot Scorpion with manipulator model during mine detection on flat terrain.

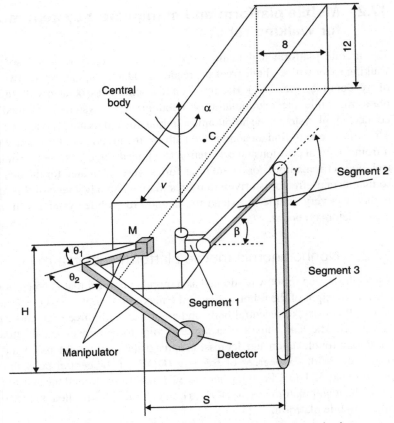

17.2 Model of the walking robot with a single leg and manipulator.

During the swing/stance phases (see Fig. 17.3) the body height H and foot distance S should remain constant:

$$H = -s_2 \sin(\beta) + s_3 \sin(\gamma - \beta) = const.$$
$$S = [s_1 + s_2 \cos(\beta) + s_3\cos(\gamma - \beta)]\cos(\alpha) = const.$$

[17.1]

17.2.2 Holonomic manipulator subsystem

The Cartesian coordinates of the center of mass C_1, center of mass C_2 and of the detector D to the mobile platform joint M are given (see Fig. 17.4). Position of C_1 – center of mass for link 1:

$$x_{C_1} = v_o \cdot t \cdot \cos(\varphi) - \frac{l_1}{2}\cos(\varphi - \theta_1)$$
$$y_{C_1} = v_o \cdot t \cdot \sin(\varphi) - \frac{l_1}{2}\sin(\varphi - \theta_1)$$

[17.2]

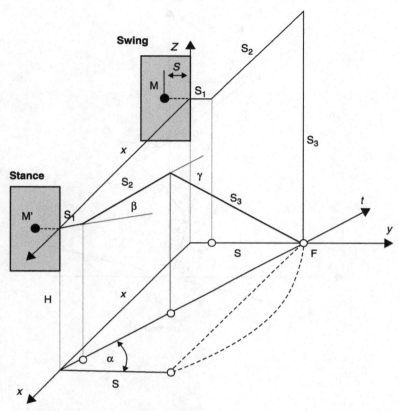

17.3 Kinematics parameters of the central body and leg L1.

Position of C_2 – center of mass for link 2:

$$x_{C_2} = v_o \cdot t \cdot \cos(\varphi) - l_1 \cos(\varphi - \theta_1) + \frac{l_2}{2} \cos(\varphi - \theta_1 + \theta_2)$$

$$y_{C_2} = v_o \cdot t \cdot \sin(\varphi) - l_1 \sin(\varphi - \theta_1) + \frac{l_2}{2} \sin(\varphi - \theta_1 + \theta_2)$$

[17.3]

or by its components:

$$x_D = v_o \cdot t \cdot \cos(\varphi) - l_1 \cos(\varphi - \theta_1) + l_2 \cos(\varphi - \theta_1 + \theta_2)$$

$$y_D = v_o \cdot t \cdot \sin(\varphi) - l_1 \sin(\varphi - \theta_1) + l_2 \sin(\varphi - \theta_1 + \theta_2)$$

[17.4]

Velocity of detector D is defined by:

$$\dot{x}_D = v_o \cos(\varphi) - v_o \cdot t \cdot \varphi \sin(\varphi) + l_1(\dot{\varphi} - \dot{\theta}_1).$$

$$\cdot \sin(\varphi - \theta_1) - l_2 (\dot{\varphi} - \dot{\theta}_1 + \dot{\theta}_2) \sin(\varphi - \theta_1 + \theta_2)$$

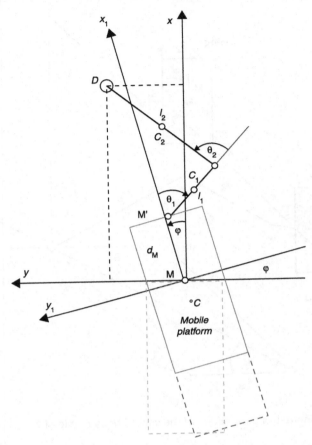

17.4 Kinematics parameters of the central body and manipulator.

$$\dot{y}_D = v_o \sin(\varphi) + v_o \cdot t \cdot \dot{\varphi} \cos(\varphi) - l_1(\dot{\varphi} - \dot{\theta}_1).$$
$$\cdot \cos(\varphi - \theta_1) + l_2(\dot{\varphi} - \dot{\theta}_1 + \dot{\theta}_2)\cos(\varphi - \theta_1 + \theta_2)$$

[17.5]

or in matrices form:

$$\dot{p}_D = \begin{bmatrix} \dot{x}_D \\ \dot{y}_D \end{bmatrix},$$

[17.6]

with

$$J_{w-r} = \begin{bmatrix} v_o \cos(\varphi) - v_o \cdot t \cdot \dot{\varphi} \cdot \sin(\varphi) \\ v_o \sin(\varphi) + v_o \cdot t \cdot \dot{\varphi} \cdot \cos(\varphi) \end{bmatrix}; J_m = \begin{bmatrix} l_1 \sin(\varphi - \theta_1) - l_2 \sin(\varphi - \theta_1 + \theta_2) \\ -l_1 \cos(\varphi - \theta_1) \ l_2 \cos(\varphi - \theta_1 + \theta_2) \end{bmatrix}; \dot{q} = \begin{bmatrix} 1 \\ \dot{\varphi} - \dot{\theta}_1 \\ \dot{\varphi} - \dot{\theta}_1 + \dot{\theta}_2 \end{bmatrix};$$

where is $J = [J_{w-r} | J_m]$ central body and manipulator Jacobian, that consists of two sub-Jacobians, J_{w-r} and J_m, where J_{w-r} multiplies the mobile platform velocities corresponding to the uncontrolled degrees of freedom, and J_m multiplies the manipulator joint velocities corresponding to the controlled degrees of freedom. $\dot{q}$ is the vector of central body and manipulator joint velocities. The matrix form of the equation (8) is given by:

$$\dot{p}_D = J \cdot \dot{q}$$

[17.7]

17.2.3 Mobile platform and manipulator system dynamics

The kinetic energy of the manipulator is given by equation [17.8]:

$$E_k = \frac{1}{2} m_D (\dot{x}_D^2 + \dot{y}_D^2) + \frac{1}{2} m_1 (\dot{x}_{C_1}^2 + \dot{y}_{C_1}^2) + \frac{1}{2} m_2 (\dot{x}_{C_2}^2 + \dot{y}_{C_2}^2) +$$
$$+ \frac{1}{2} J_{C_1} (\dot{\varphi} - \dot{\theta}_1)^2 + \frac{1}{2} J_{C_2} (\dot{\varphi} - \dot{\theta}_1 + \dot{\theta}_2)^2$$

[17.8]

where the moment of inertia for center of link 1 and 2 is:

$$J_{C_1} = \frac{1}{12} m_1 l_1^2, J_{C_2} = \frac{1}{12} m_2 l_2^2,$$

Second order Lagrange equation of motion for the mobile platform and manipulator is:

$$\frac{d}{dt} \left(\frac{\partial E_k}{\partial \dot{q}} \right) - \frac{\partial E_k}{\partial q} = \tau$$

[17.9]

where τ represents forces and torques exerted by walking robot and manipulator joint torques, E_k is the kinetic energy of system, while q and $\dot{q}$ is the vector of walking robot and manipulator joint positions and velocities.

17.2.4 Gait patterns

Figure 17.5 (left) shows the tetrapod gait, which describes the phase characteristics of the leg movement of the walking robot. The tetrapod gait, where four legs are on the ground at the same time, can be observed at low speeds of walking robots. During walking, an eight-legged robot uses its two tetrapods not unlike a biped stepping from one foot to the other – the gait is simply shifted alternately from one tetrapod to another. Since four legs are on the ground at all times, this gait is both "statically and dynamically stable". A cycle consists of a leg promotion period (swing-white) and a leg remotion period (stance-black). Phase, swing, and stance are the parameters of the robot's motion pattern (Fig. 17.5, right) together with the step size of a leg.

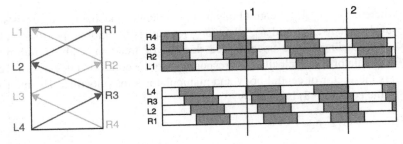

17.5 Tetrapod walking gait (left) and (right) a corresponding motion pattern beginning from the start position.

17.3 Solution of the problem by using fuzzy logic and genetic algorithm (GA) controller

The problem can be stated as follows: an eight-legged robot will have to plan its path as well as its gait simultaneously, while moving on flat terrain with occasional hurdles such as ditches and in the presence of stationary obstacles. The eight-legged robot needs to do this job by avoiding colliding with any obstacles and not falling into ditches, and all within the minimum travel time and with an optimum effort to gain ratio. Moreover, its body height (H) and foot distance (S) should always be the same to ensure static stability. To perform the above tasks, in practice, the walking robot will have to do the following sub-tasks optimally (with minimum travel time):

- Move along straight-line paths (translation only).
- Take sharp circular turns (rotation only).

In order to simplify the problem the following assumptions are made:

- The contact of the feet with flat terrain can be modeled as point contacts.
- Each obstacle is represented by its bounding circle.
- The terrain is discretized into cells and the center of each cell is considered as a candidate foothold.
- The mass of the legs is reduced into the body and the center of gravity is assumed to be at the centroid of the body.
- The detector should cover the surface with constant foot distance (S) during the forward motion of the walking robot.

17.3.1 Used controllers

In our proposed fuzzy and fuzzy genetic algorithm, two potential tools, namely fuzzy logic controller (FLC) and genetic algorithm (GA), have been merged to utilize the advantages from both. It is to be noted that the performance of an FLC depends on its rule base and membership function distributions. It is seen that optimizing the rule base of an FLC is a rough tuning process, whereas optimizing the scaling factors of the membership function distribution is a fine tuning process.

The FLC and GA controllers are used for solving the problem of combined path and gait generations simultaneously of a walking robot with manipulator. The FLC finds the obstacle-free direction based on the predicted position of the obstacles in the next time step. The inputs, namely position and velocity of the central body and manipulator, are fed to the FLC and there is a control vector to compensate for the uncertainties of the model.

Fuzzy algorithm

The fuzzy controller in our approach is based on the fuzzy control principles. Each input space is partitioned by fuzzy sets as shown in Figure 17.3. To obtain the desired response during the tracking-error convergence movement by compensating for nonlinear model dynamics, a fuzzy logic controller is designed to become a nonlinear controller. Inputs for FLC are:

Central body trajectory error:

$$d = y_d(x_d) - y(x)$$

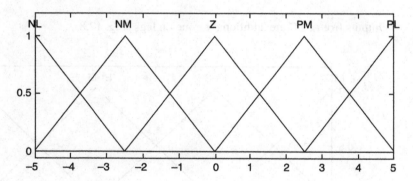

17.6 Input variable 1 – Central body trajectory error.

Where for input 1 – central body trajectory error:
NL – trajectory error is negative large.
NM – trajectory error is negative medium.
Z – trajectory error is zero.
PM – trajectory error is positive medium.
PL – trajectory error is positive large.

Central body velocity error:

$$e_v = v_0 - v$$

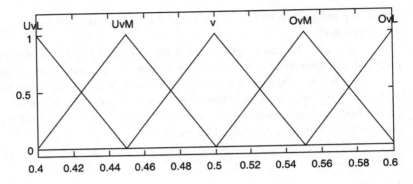

17.7 Input variable 2 – Central body velocity error.

Where for input 2 – central body velocity error:
UvL – under desired velocity large.
UvM – under desired velocity medium.
v – velocity is desired.
OvM – over desired velocity medium.
OvL – over desired velocity large.

Outputs from FLC are additional torque on legs (Fig. 17.8).

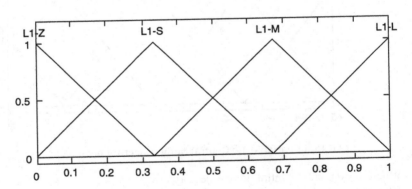

17.8 Output 1 – additional torque for first left leg L1.

Genetic algorithm

A GA is used to control the body height (H) and foot distance (S). The walking robot with manipulator will have to reach its destination starting from an initial position in minimum traveling time and its generated gaits will be optimal (with minimum number of ground-legs). This criteria is considered on

Table 17.1 Rules proposed by authors

Where for output 1 – additional torque to first left leg L1:
L1-Z – additional torque is zero.
L1-S – additional torque is small.
L1-M – additional torque is medium.
L1-L – additional torque is large.

Here, asymmetrical triangular functions that allow a fast computation, essential under real-time conditions, are utilized to describe each fuzzy set. The fuzzy sets of the two inputs d_j and e_v are calculated through 25 rules and the final output of the unit is given by a weighted average over 25 rules.

Velocity error / Trajectory error	UvL	UvM	v	OvM	OvL
NL	L1-L	L1-M	L1-S	L1-M	L1-L
	L3-L	L3-M	L3-S	L3-M	L3-L
	R2-S	R2-S	R2-S	R2-S	R2-S
	R4-S	R4-S	R4-S	R4-S	R4-S
NM	L1-M	L1-M	L1-S	L1-M	L1-M
	L3-M	L3-M	L3-S	L3-M	L3-M
	R2-S	R2-S	R2-S	R2-S	R2-S
	R4-S	R4-S	R4-S	R4-S	R4-S
Z	L1-S	L1-S	L1-Z	L1-S	L1-S
	L3-S	L3-S	L3-Z	L3-S	L3-S
	R2-Z	R2-Z	R2-Z	R2-Z	R2-Z
	R4-Z	R4-Z	R4-Z	R4-Z	R4-Z
PM	L1-S	L1-S	L1-S	L1-S	L1-S
	L3-S	L3-S	L3-S	L3-S	L3-S
	R2-M	R2-M	R2-S	R2-M	R2-M
	R4-M	R4-M	R4-S	R4-M	R4-M
PL	L1-S	L1-S	L1-S	L1-S	L1-S
	L3-S	L3-S	L3-S	L3-S	L3-S
	R2-L	R2-L	R2-S	R2-M	R2-M
	R4-L	R4-L	R4-S	R4-M	R4-M

the selection of the GA parameters, namely population size, crossover probability, mutation probability and the number of generations. Thus, the input to the GA block uses a, b and g angles through which are calculated the H and S amounts, and the output from the GA block was also the control vector that compensated the gait errors.

17.4   Simulation results

For simulation, according to the biometric approach the patterns are transferred to the model of the robot. We have use *sine function f(t)* to avoid accelerations that are too abrupt (see Fig. 17.9). The function *f*(t) for leg L4 is repeated by the other legs as *g*(t) after a partition of cycle.

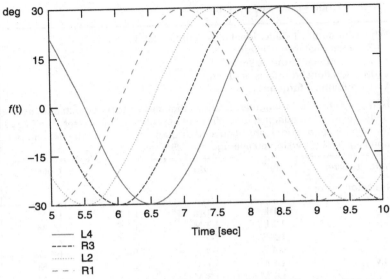

17.9 A tetrapod group of thoracic hip joints for swing/stance = 3/5 and phase φ = 0.1.

$$
f(t) = \begin{cases}
30 \cdot \sin\left(2\pi t - \dfrac{\pi}{2}\right) \\
\quad if\ 0 \le \operatorname{mod}(t, 2.5) \cdot 0.5\pi \\[2mm]
30 \cdot \sin\left(\pi \dfrac{t-0.5}{2} + \dfrac{\pi}{2}\right) \\
\quad if\ 0.5 \le \operatorname{mod}(t, 2.5) \cdot 0.5\pi + 2
\end{cases}
\qquad [17.10]
$$

The phase φ between two legs remains constant independent of the velocity.

$$
g(t) = \begin{cases}
30 \cdot \sin\left(2\pi t - \dfrac{\pi}{2} + \varphi\right) \\
\quad if\ 0 \le \operatorname{mod}(t, 2.5) \cdot 0.5\pi \\[2mm]
30 \cdot \sin\left(\pi \dfrac{t-0.5}{2} + \dfrac{\pi}{2} + \varphi\right) \\
\quad if\ 0.5 \le \operatorname{mod}(t, 2.5) \cdot 0.5\pi + 2
\end{cases}
\qquad [17.11]
$$

During the walk, all legs of the robot, when they are in swing-position, distance S = 17 [cm] (distance between central body axe and foot-end side of segment 3) is constant. Also during the walk, central body axe height is constant value H = 15 [cm]. Length of segments and manipulator links are: s1 = [5 cm].

In the control scheme shown in Fig. 17.10, through "Inverse dynamics" block, as output we have actuator torque on leg joints and manipulator joints (τ). From block "Fuzzy Logic Controller" also as output is torque (τ_{flc}). Through the Genetic Algorithm we have design as optimization algorithm for body height H and foot

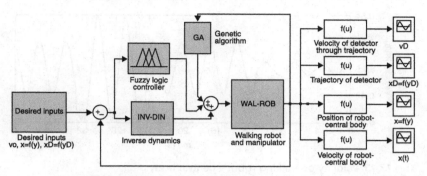

17.10 Control scheme of walking robot using fuzzy–genetic algorithm.

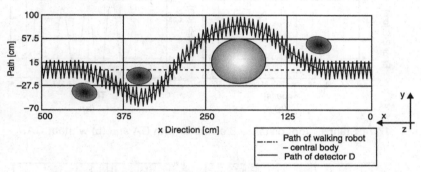

17.11 Path of the walking robot – central body and detector D.

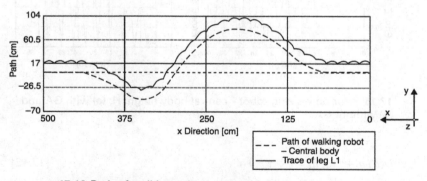

17.12 Path of walking robot – central body and leg L1.

distance S of the walking robot, which need to be constant values. Using instructions given by Kalyamoy Deb where GA implementation is using binary and real coded variables, we have developed a Genetic Algorithm Block. Inputs are estimated body height H and foot distance S. Outputs are torques (τ_{ga}). Block "Walking robot and manipulator" represents the "Direct dynamics" solution of dynamic model for walking robot and manipulator given by differential equations:

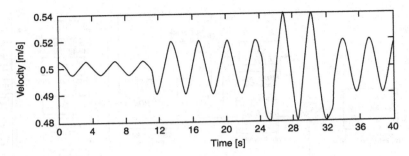

17.13 Velocity of walking robot – central body.

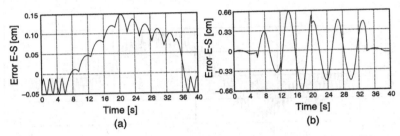

17.14 Stance – foot distance S error: (a) with GA and (b) without GA.

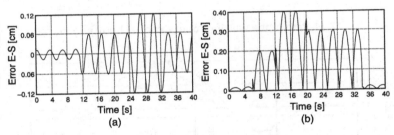

17.15 Error on walking robot – central body height H: (a) with GA and (b) without GA.

$$D(q)\cdot\ddot{q} + h(q, \dot{q}) = \tau + \tau_{flc} + \tau_{ga} \qquad\qquad [17.12]$$

where D(q) is symmetric inertia matrix, $h(q, \dot{q})$ is the vector of Coriolis, centrifugal and gravitational torques and other unmodeled disturbances, and $\ddot{q}$ is the vector of joint acceleration.

All simulated variables can be viewed: namely, the paths, velocities and errors.

In the case of keeping constant foot distance there were significant differences between the algorithms used with GA and without GA (see Fig. 17.14 (a) and (b)).

The GA algorithm has also been very influential in reducing the central body height error (see Fig. 17.15).

17.5 Conclusions

The position of the walking robot's central body and leg L1 had no significant deviation from desired values, but after 10 seconds the deviations increased considerably.

The velocity error of the walking robot's central body in comparison with desired values was not so significant.

The performance of controller GA in keeping constant values of the body height H and foot distance shows that the GA decreased four times.

This work can be extended further where the fuzzy rule will be designed by a GA that would be suitable for on-line implementation due to its computational complexity.

17.6 References

Deb, K., Pratihar, D. K., and Ghosh, A.1998. *Learning to Avoid Moving Obstacles Optimally for Mobile Robots Using a Genetic-Fuzzy Approach*. A. E. Eiben et al. (eds); *Parallel Problem Solving from Nature*, PPSN V, LNCS 1498, pp. 583–592.

Foulon, G., Fourquet, J.-Y, and Renaud, M. 1998. Planning point to point paths for nonholonomic mobile manipulators. *IEEE/RSJ Int. Conf. on Intelligent Robots and Systems (IROS)*, pp. 374–379.

Foulon, G., Fourquet, J.-Y, and Renaud, M. 1999. Coordinating mobility and manipulation using nonholonomic mobile manipulators. *Control and Engineering Practice*, vol. 7, pp. 391–399.

Gojani, I., Pajaziti, A., and Shala, A. 2002. Mobile robot navigation using cognition models and genetic algorithm-based approach. *The 13th International DAAAM Symposium*, pp. 189–190.

Klaassen, B., Linnemann R., Spenneberg D., and Kirchner, F. 2002. Biometric walking robot SCORPION, Control and Modeling, Robotic and Autonomous System, 41, pp. 69–76.

Kolmanovsky, I., and McClamroch, H. 1995. Developments in nonholonomic control problems, *IEEE Control Systems*, pp. 20–35.

Mohan, A., and Deb, K. 2002. Genetic-Fuzzy Approach in Robot Motion Planning Revisited, N.R. Pal and M. Sugeno (eds): *Advances in Soft Computing*, AFSS 2002, LNAI 2275, pp. 414–420.

Papadopoulos, E., and Poulakakis J. 2000. Planning and model-based control for mobile manipulators. *IEEE/RSJ Int. Conf. on Intelligent Robots and Systems (IROS)*, pp. 1810–1815.

Perrier, C., Dauchez, P., and Pierrot, F. 1998. A global approach for motion of nonholonomic mobile manipulators. *Proc. Of the IEEE Int. Conf. on Robotics and Automation*, pp. 2791–2796.

Pratihar, D. K., Deb, K., and Ghosh A. 2002. Optimal path and gait generations simultaneously of a six-legged robot using a GA-fuzzy approach, *Robotic and Autonomous System*, 41, pp. 1–20.

Seraji, H. 1998. A unified approach to motion control of mobile manipulators. *The International Journal of Robotics Research*, 17(2), pp. 107–118.

Yamamoto, Y. and Yun, X. 1995. Coordinated obstacle avoidance of a mobile manipulator. *Proc. of the IEEE Int. Conf. on Robotics and Automation*, pp. 2255–2260.

Synthesis of a sagittal gait for a biped robot during the single support phase

A. PAJAZITI, I. GOJANI, A. SHALA and
SH. BUZA, University of Prishtina, Kosovo and G. CAPI,
Fukuoka Institute of Technology, Japan

Abstract: This chapter presents a dynamic algorithm to control a biped robot during walking on the ground with obstacles. The proposed algorithm generates walking patterns with desired stable margin and walking speed. The hip and foot trajectories are generated based on cubic polynomial interpolation of the initial and final conditions of the biped robot's position, velocity and acceleration. This guarantees a constant velocity and a smooth transition between the single and double support phases. In order to verify the effectiveness, the proposed algorithm is implemented using the e-nuvo biped robot. The simulations and experimental results show a good performance of the proposed algorithm.

Key words: biped robot, mechanical design, gait generation, stability, dynamic walking.

18.1 Introduction

In the last few years, interest in biped robot control has grown considerably. The various works can be classified using different criteria. Based on the dynamic features of the biped walking style there are four main categories.

The first method for biped walking is the passive walking mechanism. The robots do not have actuators or control hardware, but are simple passive mechanisms that walk on a down slope. McGeer (1990) developed a passive dynamic walker to study natural passive dynamic walking. The robot can settle into a steady gait without active control or energy input when it starts on a shallow slope. The gait sequence is generated by interaction of the gravity and the mechanical system inertias.

Static walking is the simplest method for biped walking. This method is based on the basic principle of static equilibrium: a rigid body is in static equilibrium if the projection on the ground of its center of gravity (COG) is within the area of contact with the ground. Kajita et al. (1992) designed and developed a so-called "nearly ideal" 2D biped model with mass-less legs. They used four motors in the robot's body, and its legs are configured like parallel links to the mechanism. Due to simplicity, the COG of the robot is supposed to move horizontally. A control law for initiation, continuation and termination of the walking process has been developed. However, the main disadvantage of static walking is the very slow motion.

In order to increase the walking velocity, dynamic motion generation is proposed. The COG can be outside the area of contact, but the zero moment point

436

(ZMP) criteria, which is the point where the total angular moment is zero, must always be inside the sole. This criterion has been broadly used to generate biped robot gait (Vukobratovic, 1973). This method consists of planning the various movements of the robot in such a way as to assign a desired trajectory to the ZMP. Many researchers have developed control algorithms controlling the hip trajectories following the ZMP criteria. With this technique a fast and stable locomotion can be obtained, but the robot needs many degrees of freedom and high performance actuators to control hardware.

There is another category of biped robot, sometimes referred to as purely dynamic walkers. These robots do not have an actuated foot, necessary for the dynamic equilibrium. They have a simpler hardware structure, and sometimes use an external support structure for lateral balancing.

Polynomial interpolation is used on a biped's gait generation by many researchers. Shih (1999) and Huang et al. (2001) have used cubic polynomial interpolation to generate the hip and foot trajectory for uneven terrain. The work of Shih focuses only on static walking, while the work of Huang and colleagues proposes a method for dynamic walking.

This chapter describes an algorithm for planning motion patterns for the biped robot based on polynomial interpolation (Cuevas et al., 2005; Zhang and Vadakkepat, 2003). The proposed algorithm is implemented on an e-nuvo biped robot. The results show that in the algorithm produced transition is smothered in the control trajectories. The chapter is organized as follows: in 18.2, the mechanical structure of the e-nuvo robot and the part of the design considerations are introduced. In 18.3, the control algorithm is discussed. Simulation and experimental results are provided in Section 18.4. Section 18.5 is the conclusion.

18.2 Humanoid robot

ZMP Inc. has created an educational version of a humanoid robot called "e-nuvo". E-nuvo is a handy, light, desktop humanoid; with alterable system elements, a convenient Microsoft Windows-based system and a complete software development environment including "control GUI", "robot programming application programming interface (API)", "motion editor", etc. The dimensions of the e-nuvo robot are height 306.5 mm, width 193.4 mm, depth 112 mm and weight 1.2 kg. It has six degrees of freedom (DoF) in each leg and (Pitch3, Roll2, Yaw1) and one DoF in each arm including head with 1-DoF (see Fig. 18.1). Maximum walk speed is 3 m/min.

Each joint is actuated by a ZMP original DC geared motor. The body is equipped with joint angle sensors, acceleration sensors and rate gyro sensors.

The control hardware system consists of a CPU board and user interface. The central CPU is SH2, while the CPU on each joint is H8S. The user interface consists of a remote control, voice recognition and mobile phone.

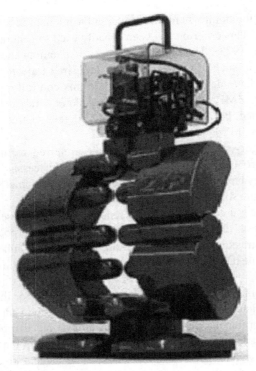

18.1 Front of the e-nuvo robot.

All joint motion commands and all sensor measurements are sent to the CPU through the CAN (controller area network). Direct control from a PC is enabled via a common serial interface. On board, there is also a small clear sound speaker and an all angle dual microphone.

18.3   Mechanical structure of a biped robot

The dynamic of a humanoid robot is closely related to its mechanical structure (Fig. 18.2). In this section, the mechanical structure for the e-nuvo robot is described. The links form a biped robot of 12 DoF as shown in Fig. 18.3.

18.4   Control algorithm of a biped robot

The walking motion of a humanoid robot is determined by the hip trajectory and the swing foot trajectory. Only the sagittal motion is discussed in this chapter. Stability can be achieved by applying the ZMP criteria. Cubic polynomial is used on sagittal motion.

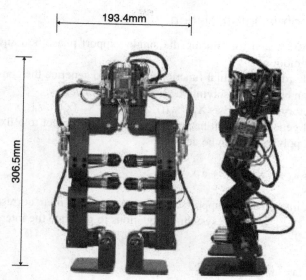

18.2 The e-nuvo robot.

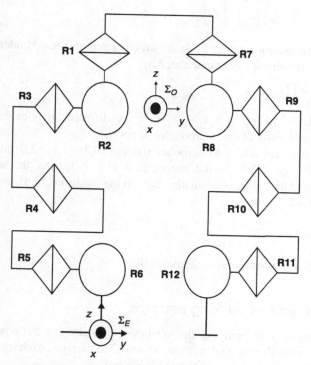

18.3 Kinematics arrangements of the e-nuvo robot.

18.4.1 Polynomial interpolation

The trajectory of hip and foot during the single support phase is a significant factor in bipedal motion.

We used third order polynomial functions in order to generate the constrained equations that can be used for solving the hip and foot trajectory.

The hip trajectory is denoted as X_h: $(x_h(t), z_h(t))$ and X_f: $(x_f(t), z_f(t))$.

A polynomial generic function has n+1 coefficients in respect to a fixed base $\{1, x, x2, x3\}$. A polynomial can be defined as:

$$P_n(x) = a_0 + a_1 x + a_2 x^2 + a_3 x^3 \qquad [18.1]$$

Considering an interpolation support of x_0, x_1, x_2, x_3 and their corresponding function values z_0, z_1, z_2, z_3 the system of equations to generate the interpolation conditions is:

$$\begin{bmatrix} 1 & x_0 & x_0^2 & x_0^3 \\ 1 & x_1 & x_1^2 & x_1^3 \\ 1 & x_2 & x_2^2 & x_2^3 \\ 1 & x_3 & x_3^2 & x_3^3 \end{bmatrix} \begin{bmatrix} a_0 \\ a_1 \\ a_2 \\ a_3 \end{bmatrix} = \begin{bmatrix} z_0 \\ z_1 \\ z_2 \\ z_3 \end{bmatrix} \qquad [18.2]$$

The matrix of this system acquires a special structure denominated Vandermonde Matrix whose determinant can be calculated by:

$$\det(B) = \prod_{i>j} (x_i - x_j) \qquad [18.3]$$

As the interpolation nodes $x_0 < x_1 < x_2 < x_3$ are different, it is evident that $\det(B) \neq 0$, independently of the interpolation support.

For example if we consider the interpolation points (0,2), (1,1), (2,0) and (3,5), the nodes support is $\{1,2,3\}$, which corresponds to n = 3, then, the searched polynomial has the third degree or smaller. By solving equation [18.3], we obtain the polynomial as follows:

$$P_3(x) = 2 + x - 3x^2 + x^3 \qquad [18.4]$$

See Fig. 18.4 showing polynomial interpolation.

18.4.2 Synthesis of walking patterns

Due to the simplicity of applying the synthesis of walking patterns we have approximated the numbers of DoF of e-nuvo from 12 to four, i.e., each leg has two DoF and the robot has a mass point located in the hip.

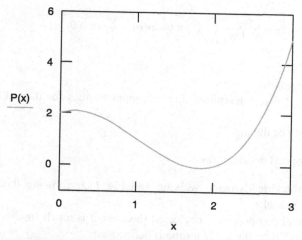

18.4 Polynomial interpolation.

One cycle of human walking can be divided into the double support phase and the single support phase.

In the single support phase, one foot supports the robot's weight while the other foot is moving forward. At the same time, the hip moves along a trajectory. In the double support phase, both feet are in contact with the ground. The robot will be statically stable if at any time the projection of the center of gravity is on the ground within the support area.

Biped dynamic walking allows the center of gravity to be outside the support region for limited amounts of time.

To achieve dynamic walking, the change between single support phase and double support phase should be smooth. At the beginning of double support phase, when the foot first contacts the ground this will affect the stability of the walking. This is a reason to use the force feedback control.

18.4.3 ZMP calculation

The position of the ZMP is computed by finding the point (X,Y,Z) where the total torque is zero. Since we are interested in the sagittal plane we assume that $Y = 0$.

Suppose that the robot has a mass point in the hip and the support knee in a constant position. Here, the inverted pendulum model can be used to model the dynamic movements of the robot. Therefore, the ankle torque can be determined:

$$\tau_{xa}(\theta) = m\left(g + \left(\frac{\tau_{xa}(\theta)}{Lm}\right) - g\sin(\theta)\sin(\theta) - \frac{v^2}{L}\cos(\theta)x_a\right)$$ [18.5]

Where:

L is the leg longitude;

$\tau_{xa}(\theta)$ and $\tau_{xb}(\theta)$ are maximum and minimum torques for the ankle in the sagittal plane;

v is the velocity of the hip;

$\frac{v^2}{L}$ is the centripetal acceleration.

See Fig. 18.5 showing dynamic walking and Fig 18.6 showing the inverted pendulum robot model.

If the centripetal component is neglected (because it is much smaller than the other components), then the ankle torque is defined as:

$$\tau_{xa} = \frac{mgx_a\,(1 - \sin^2(\theta))}{1 - \frac{\sin(\theta)}{L}x_a}$$ [18.6]

$$\tau_{xb} = \frac{-mgx_b\,(1 - \sin^2(\theta))}{1 - \frac{\sin(\theta)}{L}x_b}$$ [18.7]

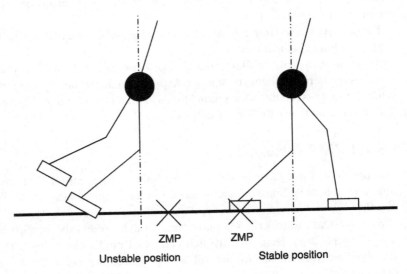

ZMP   ZMP

Unstable position   Stable position

18.5 Dynamic walking.

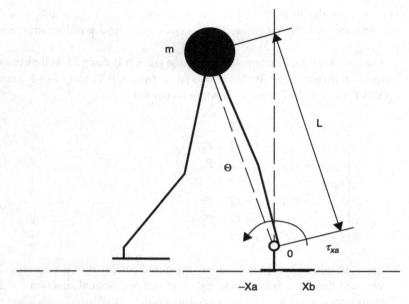

18.6 Inverted pendulum robot model.

18.4.4 Hip trajectory for single support phase

Figure 18.7 shows the hip trajectory generated by cubic polynomial when the initial and final state are known. The initial state is defined as $[x_{hs}, z_{hs}]$ and

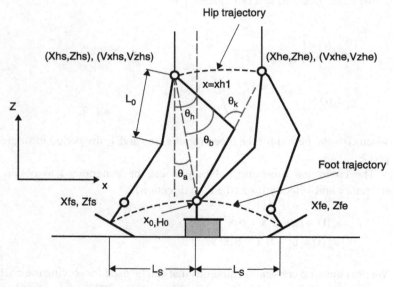

18.7 Hip and foot trajectory.

$[x_{he}, z_{he}]$ for the final state. The initial velocity $[v_{xhs}, v_{zhs}]$ is also specified in the trajectory model. When the robot arrives at its final position, the velocity is $[v_{xhe}, v_{zhe}]$.

The initial and final state positions for the cubic trajectory as well as velocity in x (the $x_h(t)$ direction) is divided in two parts: from $x_h(kT)$ to $x_h(kT+T_1)$ and from $x_h(kT+T_1)$ to $x_h(kT+T_p)$, as is shown in equation [18.8]:

$$\begin{cases} x_h(t) = x_{hs} & t = kT \\ x_h(t) = x_{h1} & t = kT + T_1 \\ x_h(t) = x_{he} & t = kT + T_p \\ \dot{x}_h(t) = v_{xhs} & t = kT \\ \dot{x}_h(t^-) = x_h(t^+) & t = kT + T_1 \\ \dot{x}_h(t) = v_{xhe} & t = kT + T_p \\ \ddot{x}_h(t) = a_0 & t = kT \end{cases}$$ [18.8]

where a_0 should be specified to satisfy the initial condition of acceleration.

We used the cubic polynomial and quadratic polynomial function for the hip trajectory and velocity for x (the $x_h(t)$ direction):

$$x_h(t) = a_0 + a_1x + a_2x^2 + a_3x^3$$
$$v_{xh}(t) = a_1 + a_2x + a_3x^2$$ [18.9]

The initial and final state positions as well as velocity for the cubic trajectory for z (the $z_h(t)$ direction) can be expressed as:

$$z_h(t) = \begin{cases} z_{hs} & t = kT \\ z_{he} & t = kT + T_s \end{cases}$$ [18.10]

$$\dot{z}_h(t) = \begin{cases} v_{zhs} & t = kT \\ v_{zhe} & t = kT + T_s \end{cases}$$ [18.11]

where T is the period for the biped robot's step and T_s the period in single support phase.

The cubic and quadratic polynomial can be generalized to obtain the hip trajectory and velocity for z (the $z_h(t)$ direction):

$$z_h(t) = b_0 + b_1x + b_2x^2 + b_3x^3$$
$$v_{zh}(t) = b_1 + b_2x + b_3x^2$$ [18.12]

We next develop constraint equations that can be used for solving the coefficients, a_i and b_j (i = 0, ..., 3 and j = 0, ..., 3) (Mu and Wu, 2003).

18.4.5 Swing and foot trajectory for single support phase

Cubic interpolation is used to generate the foot trajectory in the single support phase. The initial and final foot positions that represent the satisfied states and velocities are:

$$x_f(t) = \begin{cases} x_f(t) = x_{fs} & t = kT \\ x_f(t) = x_{fe} & t = kT + T_s \\ \dot{x}_f(t) = 0 & t = kT \\ \dot{x}_f(t) = 0 & t = kT + T_s \end{cases} \qquad [18.13]$$

$$z_f(t) = \begin{cases} z_f(t) = z_{fs} & t = kT \\ z_f(t) = z_{fe} & t = kT + T_s \\ \dot{z}_f(t) = 0 & t = kT \\ \dot{z}_f(t) = 0 & t = kT + T_s \end{cases} \qquad [18.14]$$

A smooth trajectory and velocity in x and z directions can be generated by using polynomials:

$$\begin{aligned} x_f(t) &= c_0 + c_1 x + c_2 x^2 + c_3 x^3 \\ v_{xf}(t) &= c_1 + c_2 x + c_3 x^2 \\ z_f(t) &= d_0 + d_1 x + d_2 x^2 + d_3 x^3 \\ v_{zf}(t) &= d_1 + d_2 x + d_3 x^2 \end{aligned} \qquad [18.15]$$

We next develop constraint equations that can be used for solving the coefficients, c_i and d_j ($i = 0, ..., 3$ and $j = 0, ..., 3$).

The joint position of hip and knee of the swing leg can be derived by inverse kinematics. If the hip and swing foot position in the sagittal plane at time t are $h(t) = [x_h(t), z_h(t)]$ and $f(t) = [x_f(t), z_f(t)]$, then the inverse kinematics of the swing leg can be derived as follows:

$$q = \begin{bmatrix} \theta_a \\ \theta_b \end{bmatrix} = \begin{bmatrix} \sin^{-1} \dfrac{(x_f(t) - x_h(t))}{\sqrt{(x_f(t) - x_h(t))^2 + (z_h(t) - z_f(t))^2}} \\ \cos^{-1} \dfrac{\sqrt{(x_f(t) - x_h(t))^2 + (z_h(t) - z_f(t))^2}}{2L_0} \end{bmatrix} \qquad [18.16]$$

In Fig. 18.7, the hip angle and knee angle are equal to: $\theta_h = \theta_a + \theta_b$ and $\theta_k = 2\theta_b$. L_0 is the length of calf and thigh.

18.5 Biped robot simulations

In this section, a joint profile for a four-link biped walking on the ground with obstacles during the single support phase is determined using the method discussed in Section 18.4. The values of parameters are listed as follows: $x_{hs} = -5$, $x_{he} = 0$, $z_{he} = 20$, $z_{hs} = 20$, $x_{h1} = -2.5$, $v_{xhs} = 0.1$, $v_{xhe} = 0.18$, $v_{zhs} = 0.01$, $v_{zhe} = -0.03$, $T_s = 2$, $T_1 = 1.14$, $T_p = 4$, $a_0 = -0.02$, $v_{xh1} = 4.56$, $z_{fm} = 3$ and $T_m = 1$.

Two cases have been considered as detailed below.

18.5.1 Case 1: walking when avoiding the obstacle

Figure 18.8 shows the horizontal displacement of the foot during the single support phase. The trajectories are smooth; i.e. all the velocities are continuous. Figures 18.9 and 18.10 show the profiles of the joint angles and angular velocities of leg "a" during the single support phase. It can be seen that both the joint angles and their angular velocities are repeatable and the velocities are continuous at the instant of impact.

Figure 18.11 shows the stick diagrams of a four-link bipedal presented leg "a", during the single support phase. From the diagram, one can observe the motion when the stance limb propels the upper body. The initial and the end position of each step are equal, which indicates that the repeatability condition is satisfied.

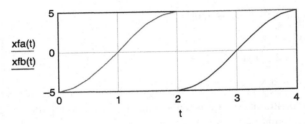

18.8 Designed trajectory of feet (legs 'a' and 'b').

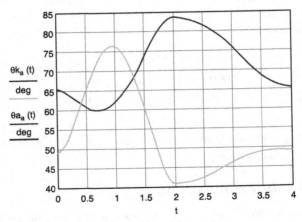

18.9 Knee and ankle joint angles of leg 'a'.

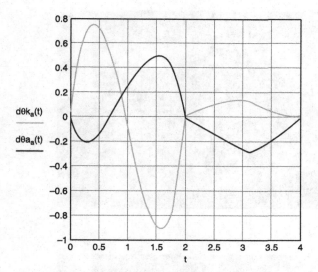

18.10 Knee and ankle joint velocities of leg 'a'.

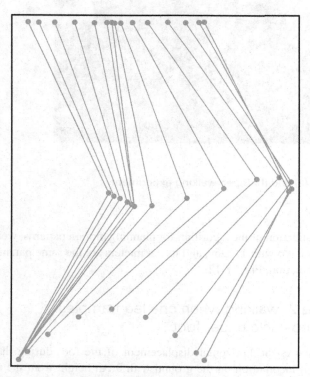

18.11 Stick diagram of single support phase of leg 'a'.

(a)

(b)

18.12 (a) and (b) Biped walking experiment.

To test the validation of the algorithm for planning motion patterns, we applied it to the robot e-nuvo with 12 DoF, and its parameters are the same parameters as those in the simulation, Fig. 18.12.

18.5.2 Case 2: walking when one leg tramples the obstacle under foot

Figure 18.13 shows the horizontal displacement of the foot during the single support phase in locomotion on the ground with 3 cm height from the nominal level. Figures 18.14 and 18.15 show the profiles of the joint angles and angular

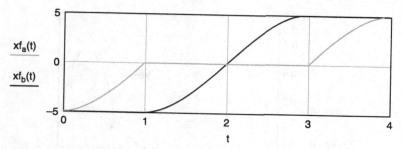

18.13 Designed trajectory of feet (legs 'a' and 'b').

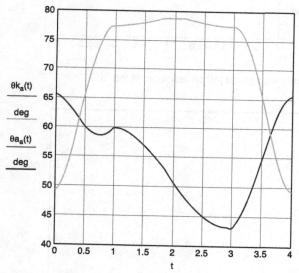

18.14 Knee and ankle joint angles of leg 'a'.

velocities of leg 'a' during the single support phase. It can be seen that both the joint angles and their angular velocities are repeatable. In addition, the velocities are continuous at the instant of impact.

Similar to Case 1, Fig. 18.16 shows the stick diagrams of a four-link bipedal presented leg 'a', during the single support phase. From the diagram, one can observe the motion when the stance limb propels the upper body. The initial and the end position of each step are equal, which indicates that the repeatability condition is satisfied.

Figure 18.17 shows the execution of the single support phase implemented on the robot software platform.

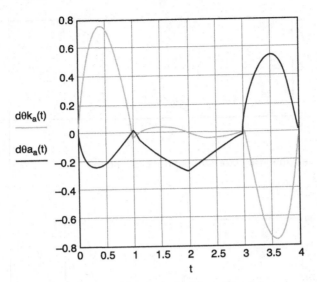

18.15 Knee and ankle joint velocities of leg 'a'.

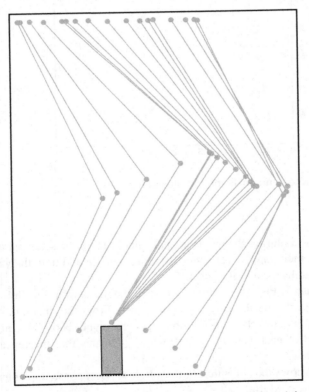

18.16 Stick diagram of single support phase of leg 'a'.

(a)

(b)

18.17 (a) and (b) Biped walking experiment when the foot touches the obstacle.

18.6   Conclusions

In this chapter, we described the algorithm that generates the walking patterns for a biped robot.

From the results obtained, it is possible to observe the values of velocity in the beginning and at the end of the single support phase (the velocities reached at the end of this phase are zero or very near to zero).

This algorithm allows the biped robot to gain stability in violent impact that is produced by the links' velocity during the single support phase.

18.7   References

Cuevas E. V., Zaldívar D., and Rojas R. (2005). Bipedal robot description. *Technical Report B-04-19*, Freie Universität Berlin, Fachbereich Mathematic und Informatik, Berlin, Germany.

Huang Q., Yokoi K., Kajita Sh., Kanaeko K., Arai H., Koyachi N., and Tanie K. (2001). Planning walking patterns for a biped robot. *IEEE Trans. Robot. Automat.*, Volume 17, No. 3.

Kajita Sh., Yamaura T., and Kobayashi A. (1992). Dynamic walking control of a biped robot along a potential energy conserving orbit. *IEEE Transaction on Robotics and Automation*, Volume 8, No. 4.

McGeer T. (1990). Passive dynamic walking. *International Journal on Robotic Research*, Volume 9, No. 2.

Mu X., and Wu Q. (2003). Synthesis of a complete sagittal gait cycle for a five-link biped robot. *Robotica*, Volume 21, pp. 581–587.

Shih Ch.-L. (1999). Ascending and descending stairs for a biped robot. *IEEE Trans. Syst. Man. Cybern.* Volume 29, No. 3.

Vukobratovic M. (1973). How to control artificial anthropomorphic system. *IEEE Trans. Syst. Man. Cyb.*, Volume SMC-3, No. 5.

Zhang R., Vadakkepat P. (2003). Motion planning of biped robot climbing stairs. In: *Proceedings of the FIRA Congress*, from mchlab.ee.nus.edu.sg.

19

Fuzzy logic control in support of autonomous navigation of humanitarian de-mining robots

A. ABBAS, British University in Egypt, Egypt

Abstract: This chapter presents a review of the various control aspects of autonomous navigation of humanitarian de-mining robots with particular emphasis on intelligent decision-making capabilities. A review of intelligent navigational functionality of autonomous robots, which examines the use of artificial intelligence techniques in the architecture of the control system, is conducted. The chapter also describes the design of a control algorithm based on a fuzzy logic controller. The viability of the control algorithm is demonstrated through a simulated experiment.

Key words: intelligent autonomous navigation, fuzzy logic control, de-mining robots.

19.1 Introduction

The aim of this chapter is to review the various control aspects of autonomous navigation of humanitarian de-mining robots with particular emphasis on intelligent decision-making capabilities. An autonomous robot is a robot that makes self-regulating and acting navigational decisions based on control algorithms mimicking human decision-making skills (Cho and Chung 1991; Aoki et al. 1994; Li 1994). This aim is achieved through the adoption of the following objectives:

- Conduct a review of intelligent navigational functionality of autonomous robots.
- Examine the use of various artificial intelligence techniques in order to realise intelligence in the architecture of the control system.
- Describe the design of a control algorithm based on fuzzy logic, which exhibits intelligent decision-making abilities.
- Demonstrate the viability of the control algorithm through a simulated experiment.

19.2 Intelligent navigational functionality of autonomous robots

To achieve navigational and performance goals, a control system which continuously attempts to seek and minimise the displacement between the current position of the robot and the target position is sought. Robot control system

453

architectures may be based on one of two modes: model-based and reactive-based. In the model-based architecture, a symbolic representation of the robot environment is perceived through the capture of sensory information and coded into the control system. This information is subsequently used to localise the position of the robot within the operating environment and to plan the path of the robot through the environment. The inference mechanism follows a sequential path through the planning level to the execution level of motor control based on this representation. This model-based mode retains several problems by nature. The main problem is its inherent inaccuracy due to ambiguity and unreliability of information in real-time operation. It is also an intensive computation task, which requires a great deal of power in a real-time setting. The sequential nature of the inference mechanism is critical in that a failure of one layer of execution will fail all subsequent layers leading to a complete failure of the control strategy. This is dissimilar to humans who mainly reason in a non-sequential manner. The use of a model-based architecture to generate a representative map of the terrain in a humanitarian de-mining operation may not be possible due to the completely unstructured environment riddled with many uncertainties (see Fig. 19.1).

On the other hand, a reactive-based control architecture is built as a set of independent but interrelated layer 'behaviours' which are reciprocally connected (see Fig. 19.2). Based on sensory data, an instantaneous map of the robot vicinity is built into the control system, which directs the mobile robot towards the reaction against perceived environmental situations. The inference mechanism is non-sequential in that each behaviour has access to sensory data and influences actions

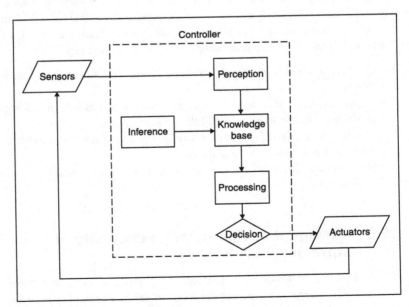

19.1 Model-based control architecture.

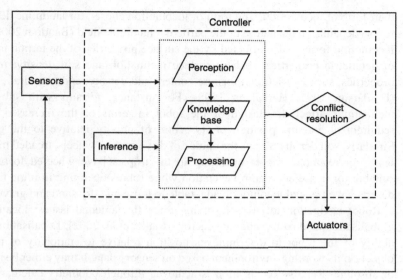

19.2 Reactive-based control architecture.

taken by the control system. The approach is similar to how humans reason, act and react. Outputs from each behaviour are blended together through a weighting factor to obtain the output behavioural action. Although the behaviour-based mode does not suffer from the problems of the model-based mode, careful attention must be paid to the issue of information overload which may be contradictory in nature. The adoption of a conflict resolution module addresses this issue through the use of weighting methods.

For a humanitarian de-mining robot, the procedure is to divide the landmine field to be inspected into narrow longitudinal segments. Therefore, the navigational task is for the robot to seek the pre-assigned target position within the mine field while avoiding the obstacles representing the detected mines. The best control approach is to utilise sensory data in order to plan the robot path in real-time in a reactive manner to any situation encountered. Hence, the behaviour-based architecture will be best suited for this application due to its inherent localised point of view (Khatib 1986; Borenstein and Koren 1989; Borenstein and Koren 1991).

The mobility and form of locomotion of autonomous robots influences the design of the control system in terms of the form of action decisions relating to the actuators. If the robot is to work in material handling in a planned factory floor layout with smooth well-structured paths, articulated wheels are commonly used. On the other end of the scale, autonomous robots operating in ill-defined harsh environments, such as underwater robots, airborne robots or humanitarian de-mining robots have a specific unique form of locomotion adapted to the operating environment. For example, underwater robots possess submarine-like propulsion, while airborne robots have plane-like or helicopter-like propulsion as their form of locomotion. The form of locomotion chosen for landmine detection robots depends on the nature of the particular landmine field (Baudoin 2008). The locomotion form is also affected by the surface properties of the terrain in terms of geometric properties, such as surface inclination and soil density; material properties, such as friction, moisture content and density; and temporal changes (Hardarson 1997; Hardarson 1998). For instance, terrains with substantial inclinations affect the design of the robot in terms of the increased torque requirement and the position of its centre of gravity relative to the ground. Similarly, soil density affects the choice of drive such as wheels, treaded tracks or legs. For example, robots operating in the tropics have wheeled locomotion suitable for dense and damp vegetation while robots operating in mine fields in deserts have tracked locomotion adaptable to loose and dry sand and gravel.

Robot navigation involves resolving three fundamental issues: localisation, globalisation, and path planning (Lacroix, Mallet et al. 2002). Localisation is the ability of the robot to determine its position relative to stationary or moving objects in its working environment based on sensory data. It may either be relative or absolute. Relative localisation is achieved through odometry calculations or inertial properties of the robot e.g. velocity and acceleration, while absolute localisation is based on measuring distance and orientation from a fixed reference point. Globalisation is the ability of the robot to relate its current position to derived destination points based on absolute or referenced terms. Path planning and trajectory generation refers to the process by which the mobile robot determines a suitable course between the current position and target position. It involves being aware of direction and distance to travel, and of obstacle avoidance techniques. Path planning also entails not only obtaining a collision-free path but also the optimisation of the generated trajectory to improve efficiency of motion in terms of travel time and computational time.

Many control strategies have been developed to resolve these fundamental navigational issues. These control strategies may be classified into four types: strategies based on geometric numerical analysis formulation, strategies based on soft computing techniques, strategies which simulate a natural process, and strategies which are biologically inspired. Geometric formulation strategies are based on using numerical analysis techniques in order to generate an optimal trajectory from the current robot position to the target position avoiding obstacles by utilising parametric curves. The practicality and flexibility of the application of these strategies in mobile robotics is self-evident. The main drawbacks are the high computational demand for the control system and its inherent lack of support for uncertainty. An example of such an approach in optimising path planning for mobile robots is given by Liang, Liu et al. (2005). The control system uses a cubic spiral based method to minimise path length while incorporating mechanisms for accommodating special constraints such as obstacles. In order to reduce the computational demands, the generated path is limited to five segments at most with three line segments at any given time. In a more recent study, a Bezier curve algorithm utilises robot target points to define the control points of the generated curve, which acts as the robot trajectory (Jolly, Sreerama Kumar et al. 2009). The generated path is optimised with respect to the initial and final velocities of the robot. As soon as collision is detected, the path-planning algorithm switches to a new suitable path encircling the obstacle and utilising the same target points as the control points of the new curve.

Soft computing techniques are based on using artificial intelligent (AI) methods to impart some form of intelligent decision-making ability onto the robot controller. Modern soft computing systems provide a generic and reliable means for implementing intelligence while maintaining computational and energy efficiency. Recent examples of the use of AI methods include the three-level reasoning approach (Velagic, Lacevic et al. 2006). Obtaining a collision-free path from the robot to the target position entails using a proportional integral multivariable angular velocity controller at the initial level to control the velocities of the wheels. At the second level, generic algorithms are used to control the position of the robot so that it converges to zero error. Subsequently, fuzzy logic is used to generate the desired path. Another technique involves utilising an instant goal algorithm to converge the robot path towards the target avoiding obstacles. When the instant goal algorithm fails to produce a feasible path temporarily, a fuzzy logic controller is consulted with the purpose of continuing the search for convergence. Additionally, a kinematic and dynamic study of the generated robot trajectory is conducted in order to reduce the abrupt changes in its curvature (Ge, Lai et al. 2007). A completely different approach is the use of the Partially Observable Markov Decision Process technique, which is a generalisation of the Markov Decision Process, which models decision-making capabilities in situations requiring optimisation of uncertainties (Foka and Trahanias 2007). In this approach, probability is used to plan the trajectory, avoiding obstacles by making observations of robot states, determining all possible actions in real-time and consequently determining the most probable trajectory.

Many of the interesting autonomous control strategies are those based on simulating a natural process. The most common of these in the field of robot path planning is simulated annealing. Simulated annealing is a method for approximating the global minimum of a generated path geometry function over a large search space possessing non-linearity and discontinuity (Martínez-Alfaro and Gómez-García 1998). The drawback of simulated annealing is its high consumption of computational needs and time. Three cases of path planning functions are considered: polylines, Bezier curves, and spline curves as a means of reducing the time to reach an optimal path (de Sales Guerra Tsuzuki, de Castro Martins et al. 2006).

On the other hand, biologically inspired control strategies depend on algorithms obtained by studying the motion of biological creatures, and insects in particular, which tend to follow optimum planning techniques. Simulating these techniques results in more efficient and superior algorithms in terms of time and computational requirements compared with simulated annealing. An example of such a technique is the two-step approach by Tan, which uses a Dijkstra graph search algorithm to generate a collision-free shortest path followed by using an ant colony system algorithm to optimise the location of the path in the environment (Tan, He et al. 2007). Another area of autonomous robot navigation inspired by creature behaviour is the development of artificial pheromone algorithms (Herianto, Sakakibara et al. 2007). Using these algorithms, navigational systems are based on ant colony behaviour where intelligent data carriers mimic ant pheromones as the communication medium.

19.3   Methodologies for controlling autonomous robots

Generally, there are three levels of control for autonomous vehicles: program control, supervised control, and total control. However, there is no clear-cut distinction between these three levels. In program control autonomy, the human operator supplies functions to the vehicle by programming motion instructions into the controller. These instructions are blindly repeated by the vehicle during its operation. Since the vehicle moves directly under its own initiative, this type of control is still considered autonomous albeit of a simple and rigid nature. This contrasts with tele-operated vehicles in which motion is achieved by remote control operation with no programming. This type of motion is not considered autonomous. In supervised control autonomy, the human operator enters the path planning techniques to be followed by the vehicle into the controller. Usually this level of autonomy possesses simple obstacle avoidance tactics. Although this level of autonomy provides some flexibility, it still requires some form of human supervision (Acar, Zhang et al. 2001). In contrast to the first two levels of autonomy, the most advanced type of autonomy of mobile vehicles is totally independent, possessing intelligent decision-making capabilities with no human interaction with the vehicle during its operation. This level provides the most flexibility of motion albeit at the expense of complexity of control. It also possesses excellent navigational and obstacle avoidance skills.

Although the control of autonomous robots employs algorithms of soft computing, robot learning is not synonymous with machine learning. The distinction being that robot learning involves interactions with the operating environment (Bekey 2005). Intelligence is expressed as the ability of the vehicle to extract information about the surrounding environment through the use of sensors and to use this knowledge to make informed decisions about actions to be taken. Artificial intelligence techniques are used to produce knowledge-based systems that encapsulate the reasoning and decision-making rationale. These knowledge-based systems provide an excellent means of extracting heuristic knowledge that is difficult to elicit by conventional algorithmic means. Knowledge-based systems also provide a means of machine learning through the accumulation of knowledge with the initially acquired knowledge.

19.3.1 Neural networks for autonomous robots

The human brain has outstanding abilities in processing information and making instantaneous decisions, even within very complex situations and in uncertain environments. The brain uses a massive network of parallel and distributed computational elements called neurons. This network provides humans with a powerful capability of learning, recognition of patterns, and reasoning within unknown environments. The artificial neural network (ANN), often simply called neural network (NN), is a processing model that is based on the behaviour of biological neurons. Neural networks are often used for classification problems or decision-making problems that do not have a simple or straightforward algorithmic solution. The advantage of a neural network is its ability to learn input to output mapping from a set of training cases without explicit programming, and subsequently being able to generalise this mapping to cases not seen previously (Haykin 1994). The tasks which neural networks can be applied to can be grouped into five classes: pattern reproduction, optimisation, pattern association or classification, feature discovery or clustering, and reward maximisation (Bekey and Goldberg 1993). See Fig. 19.3 for the layout of neural networks.

The main advantages of neural networks are that they can perform tasks that a linear program cannot. Also when an element of the neural network fails, it can continue without any problem due to its parallel nature. A neural network learns and does not need to be reprogrammed, and it can be implemented in any application without any problems. However, the neural network must be trained to operate. Neural networks have been extensively used in the control of mobile robots in order to provide autonomous navigation abilities. The complexity of finding an exact geometric location increases exponentially with an increase in

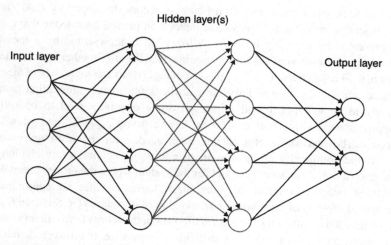

19.3 Layout of neural networks.

the number of degrees of freedom. The artificial neural networks paradigm appears to be promising for accomplishing real-time collision-free path planning (Alnajjar and Murase 2006; Wang and Liu 2008; Wang, Hou et al. 2008).

19.3.2 Genetic algorithms

Evolutionary algorithms are a family of search and optimisation methods that incorporate principles from natural evolution. They include a class called genetic algorithms, which are heuristic search algorithms that can be used to search for good solutions to difficult problems. These algorithms operate by iteratively evolving a solution to get a better one, and are manipulated by biologically inspired operations. The benefit of genetic algorithms is that they can be applied to problems without a deterministic algorithmic solution.

Genetic algorithms are fairly easy to understand and implement, and they also are applicable to a wide range of problems. They can handle huge search spaces with the ability to be easily combined with other methods, and provide many alternative solutions. However, there are several disadvantages of genetic algorithms as choosing a good representation and constructing a good fitness function can be difficult and greatly depend on the problem at hand. Although there are many operators to choose from and several parameters to set, which provide several solutions, the best choice or solution is often unclear. Lastly, no guarantee for optimal solution within a finite time, in other words, no convergence is guaranteed (Fernandez-Leon, Acosta et al. in press; Goldberg 1989; Ross and Corne 1995; Arambula Cosío and Padilla Castañeda 2004; Smith III and Nguyen 2007).

19.3.3 Fuzzy logic

Fuzzy logic is a method based on fuzzy set theory developed by L.A. Zadeh for representing approximate knowledge rather than precise knowledge that cannot be represented by the crisp methods. A fuzzy set is represented by a membership function whose element value can range between 0 and 1, in other words; a degree of truth, unlike the conventional crisp bivalent logic (Boolean logic) where the element value can only be True or False. In real life applications, information gathered is likely to be incomplete and imprecise. Some quantities tend to be subjective, approximate, and qualitative, such as 'large', 'warm', 'weak', and 'slow'. Fuzzy logic can be used to make efficient inferences and decisions based on such uncertainty.

As the complexity of the system increases, it becomes more challenging to develop a precise analytical model for the system where the model-based decision making becomes unpractical, thus fuzzy logic provides an approximate yet practical method of representing knowledge of a system (De Silva 1997). In the problem of path planning, the information about the robot environment is uncertain or incomplete, therefore it is difficult to optimise the travel distance with conventional algorithms, hence by using fuzzy logic, optimisation can be made by dealing with the various aspects of uncertainty, such as vagueness or imprecision (Ramirez-Serrano and Boumedine 1996; Al-Khatib and Saade 2003; Smith III and Nguyen 2007; Pradhan, Parhi et al. 2009).

19.4   Design of a fuzzy logic controller for autonomous navigation

Autonomous robots that operate in de-mining environments face many difficulties. First, the robot has no prior knowledge of the environment in almost all mine fields owing to the imprecise location of the mines. This situation results from the lack of, or unreliability of, maps and the shifting of mines in the ground with the passing of time. To address this issue, the robot must possess some path planning behaviour after the detection of the existence of mines. Also, the need to address the existence of obstacles and how to avoid a collision with them is a main concern. Finally, the existence of uncertainties in inputs from sensors and in control actions, in addition to descriptive linguistic uncertainties associated with control reason, constitutes a major difficulty for autonomous motion.

There are four types of uncertainties associated with autonomous navigation of mobile robots: vagueness, ambiguity, accuracy and reliability. A knowledge statement may be vague when the information conveyed is not clearly defined or is indistinguishable. Ambiguity occurs when a condition may or may not be met. On the other hand, inaccuracy occurs when the information provided is imprecise, inexact or ill-defined. Uncertainty through reliability occurs due to the dynamic nature of real-world unstructured environments. It is expressed as the dependability and consistency of information during the repeated operation of the mobile robot. Examples of these four types of uncertainty are illustrated below in Fig. 19.4.

19.4.1 Rationale for adopting fuzzy logic

Unfortunately, traditional control theory does not provide effective or efficient methods for dealing with these uncertainties due to the high demand on the control system and increased costs. Fuzzy logic control as a more suitable alternative control

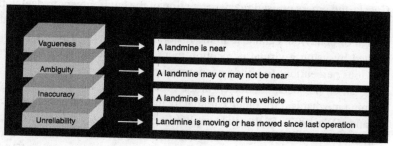

19.4 Types of uncertainty.

method for vehicles has been proposed for over 20 years (Takeuchi, Nagai et al. 1988; Palm 1989; Wakileh and Gill 1990; Ulyanov, Yamafuji et al. 1995). The classic article by Saffiotti (1997) presents a review of techniques and trends in the use of fuzzy logic control in autonomous vehicle navigation. The core of such techniques is the design of reactive behaviour-based systems, which attempt to encapsulate the tasks and actions of the vehicle. It reacts towards sensed perceptions from the environment and typically performs one or more of three tasks: goal seeking of a target, edge following, and obstacle avoidance in the direction of robot travel.

In order to take motion actions, a single decision based on multiple behaviours of often-conflicting influences must be achieved. This process is described as the fusion of behaviours into a discrete motion-action such as speed (move or stop) and orientation. The most common fusion technique is based on the weighting of each individual behaviour (Parasuraman, Ganapathy et al. 2003). The controller contains a rule base for the fuzzification of the motion parameters i.e. speed, orientation and obstacle distance. Navigation is achieved by combining results from the defuzzification of the weighted behaviours. Another technique is based on the fusion of 'allowed' and 'disallowed' directions (Yang, Moallem et al. 2003). The allowed direction is based upon the fuzzification of goal-seeking behaviour while the disallowed direction is based on the fuzzification of obstacle avoidance behaviour. An eliminator factor is used to denote the importance of obstacle avoidance behaviour when the target goal is very near to the robot.

A recent development in the area of behavioural fusion is the use of genetic algorithms for the operation of a real-time search-and-fix procedure for obstacle avoidance fuzzy rules (Hagras, Callaghan et al. 2004). This enables the controllers to fuse behaviours by partial and concurrent activation of individual behaviours. An additional technique used in the design of the rule base is fuzzy perception of the environment. The NASA Jet Propulsion Laboratory project to study autonomous robot navigation in planetary terrains uses fuzzy logic to mimic human perception of terrain visibility (Howard, Tunstel et al. 2001). This is achieved through the fuzzification of the amount and concentration of rocks to produce a fuzzy set of terrain descriptors. The controller also fuzzifies the inclination of the terrain surface. A fuzzy rule base is used to determine the transversibility of the terrain based on the fuzzified data. A different approach by Cuesta and others utilised fuzzy perception of the environment to deal with planar orientation (Cuesta, Ollero et al. 2003). Fuzzy perception is used both in the design of behaviours and also in the fusion of behaviours. This method is further refined by using a virtual perception memory in order to take into account the previous perceptions to build new perceptions.

Research has been conducted in the area of path optimisation by providing a learning mechanism for range parameters in order to minimise total vehicle travel distance (Wu 1992). Rather than explicitly elicit the fuzzy rule base directly from human experts or by using learning mechanisms, the fuzzy controller is designed to automatically extract the fuzzy rules and membership functions. This process may be preceded by the initial training of the robot trajectory by an operator followed by the extraction of the rule base from neuro-fuzzy networks (Marichal, Acosta et al. 2001). A fuzzy logic controller that requires no human training at all is described as a type-2 fuzzy logic controller (Hagras 2004). In conventional type-1 fuzzy logic controllers, uncertainty is described by precise and crisp membership functions that the developer assumes to capture uncertainty. On the other hand, type-2 fuzzy logic controllers are designed based on a fuzzy membership function rather than on crisp values. The approach taken by Novakovic and others is not to synthesise the fuzzy rule base at all (Novakovic, Scap et al. 2000). The approach taken is to analytically synthesise the fuzzy control action by determining the positions of the centres of the output fuzzy sets. Another approach is to use evolutionary self-learning algorithms to elicit the fuzzy rule base. This single rule base is used for both goal seeking and obstacle avoidance (Abdessemed, Benmahammed et al. 2004).

19.4.2 Design of the fuzzy controller

The typical design of a fuzzy logic controller (FLC) involves four steps: fuzzification of the input variables, definition of a sound fuzzy rule base, development of a suitable inference mechanism, and defuzzification of the controller outputs.

Fuzzification of variables

The inputs selected to define variables for navigation of autonomous robots are the target orientation (TN) and the obstacle or landmine distance parameters (DR, DC, DL). The TN variable indicates a measure of the angle between the robot heading and the line connecting the robot to the target direction. It is assigned one of five values representing overlapping regions measured with respect to the robot bearing: rear left (RL), front left (FL), front centre (FC), front right (FR), and rear right (RR). These overlapping regions and their associated degree of fuzziness are illustrated below in Fig. 19.5.

The landmine distance functions: distance left (DL), distance centre (DC) and distance right (DR), indicate a measure of the closeness of an obstacle or landmine to the current position of the robot. For practicality, detection is confined to the space directly ahead of the robot heading. Each one of those three functions is assigned either of two values: near or far with a degree of fuzziness associated with them. See Fig. 19.6 for detected landmine fuzzy distance.

The above-mentioned linguistic descriptors are utilised to encapsulate human perception of these inputs. Each input has a fuzzy value assigned to it according to the readings of the sensors. Fuzzification of these inputs represents a transformation of the crisp measurements obtained from sensors to the corresponding fuzzy membership functions. Discrete triangular membership functions are used to quantify the linguistic values. For the target orientation

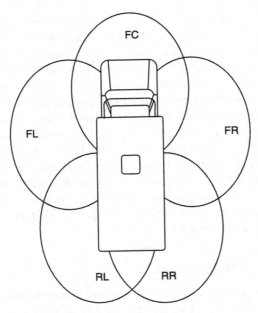

19.5 Target orientation fuzzy zones.

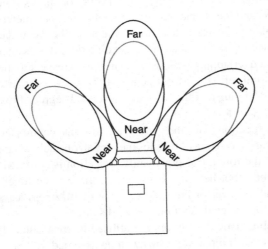

19.6 Detected landmine fuzzy distance.

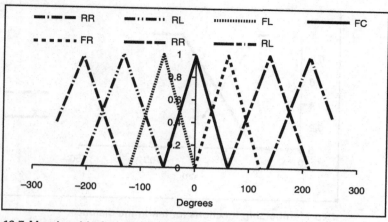

19.7 Membership functions for target orientation.

variables, the fuzzification process quantifies the certainty that the target direction is within each of the five regions. An example of the membership function for the RR variable is given in equation [19.1].

$$\mu_{RR}(x) = \frac{x - 60}{75} \quad 60° \le x \le 135°$$

$$= \frac{10 - x}{75} \quad 135° \le x \le 210°$$

$$= 0 \qquad \text{otherwise} \qquad\qquad [19.1]$$

For the landmine distance variables, the fuzzification process quantifies the certainty of whether the landmine is far, equation [19.2], or near, equation [19.3].

$$\mu_{far}(x) = \frac{x - 30}{50} \quad 30 \quad \le x \le 80 \text{ cm}$$

$$= 0 \qquad x \quad < 30 \text{ cm}$$

$$= 1 \qquad x \quad > 80 \text{ cm} \qquad\qquad [19.2]$$

$$\mu_{near}(x) = \frac{80 - x}{50} \quad 30 \quad \le x \le 80 \text{ cm}$$

$$= 0 \qquad x > 80 \text{ cm}$$

$$= 1 \qquad x < 30 \text{ cm} \qquad\qquad [19.3]$$

The output variable of the FLC is the angle that the robot is to turn relative to its current bearing. The turning angle (TA) is assigned one of five values representing overlapping regions measured with respect to the robot bearing: left big (LB), left small (LS), left right (LR), right small (RS), and right big (RB). These

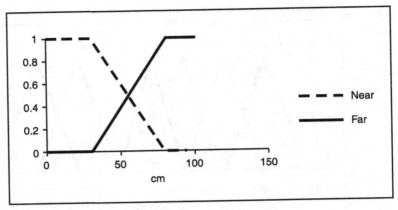

19.8 Membership functions for landmine distance.

overlapping regions and their associated degree of fuzziness are illustrated below in Fig. 19.9.

This angle is quantified using membership functions similar to those of the TN variable. An example of the membership function for the RS variable is given in equation [19.4].

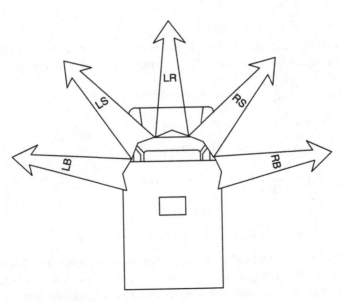

19.9 Turning angle zones.

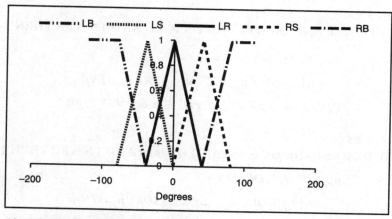

19.10 Memberships for turning angle.

$$\mu_{RS}(x) = \frac{x}{40} \qquad 0 \le x \le 40°$$

$$= \frac{80 - x}{40} \quad 40° \le x \le 80°$$

$$= 0 \qquad \text{otherwise} \qquad\qquad [19.4]$$

Fuzzy rule base

The knowledge base of the FLC contains a set of rules, which encapsulate the heuristic and intrinsic knowledge of how humans make decisions about the domain. These rules are represented by implication relationships that are expressed as fuzzy IF-THEN rules, which utilise the same linguistic descriptors in order to allow the inference mechanism to match the rules against the inputs. Each rule has one or more input variables as its antecedent and produces a fuzzy linguistic output variable as its consequent. Fuzzy implications are interpreted by several methods, which quantify the fuzzy output variable. The most widely used implication method used in FLC systems is that of Mamdani due to its simplicity, and is adopted here (Karray and De Silva 2004). The Mamdani implication is defined using the 'min' operator as:

$$\mu_{R_i}(a, b, c, d) = \min\,[\mu_A(a),\, \mu_B(b),\, \mu_C(c),\, \mu_D(d)]$$

$$\forall a \in A,\ \forall b \in B,\ \forall c \in C,\ \forall d \in D \qquad\qquad [19.5]$$

Since each of the landmine distance variables has two possible values and the target orientation variable has five possible variables, using the product counting rule, the total number of required rules in the rule base is 40 rules. Two examples of such rules and their corresponding implication functions are shown below.

Rule R9:
IF DL is far AND DC is far AND DR is near AND TN is FR THEN TA is LR

$$\mu_{R_9}(DL, DC, DR, TN) =$$
$$\min[\mu_{far}(DL), \mu_{far}(DC), \mu_{near}(DR), \mu_{FR}(TN)] \qquad [19.6]$$
$$\forall DL \in far, \ \forall DC \in far, \ \forall DR \in near, \ \forall TN \in FR$$

Rule R33:
IF DL is near AND DC is near AND DR is far AND TN is FC THEN TA is RS

$$\mu_{R_{33}}(DL, DC, DR, TN) =$$
$$\min[\mu_{near}(DL), \mu_{near}(DC), \mu_{far}(DR), \mu_{FC}(TN)] \qquad [19.7]$$
$$\forall DL \in near, \ \forall DC \in near, \ \forall DR \in far, \ \forall TN \in FC$$

Inference mechanism

The inference mechanism of the FLC has to perform two sequential steps: rule matching and rule firing. Rule matching involves the controller comparing each antecedent of all rules against the input variables. Matching is achieved against all the rules that have the membership function of all of its antecedents greater than zero. For each such rule, the implication membership function is determined by equation [19.5]. The next step is to generate the membership function for the output variables for each matched rule using modus ponens compositional rule of inference. As an example of the inference mechanism, assume that the two above rules are matched for a measured value of input

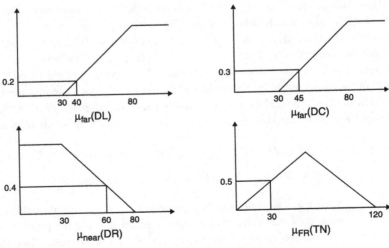

19.11 Fuzzification of rule 9 inputs.

variables from the robot sensors to be: DL = 40 cm, DC = 45 cm, DR = 60 cm and TN = 30°. For the rule R9:

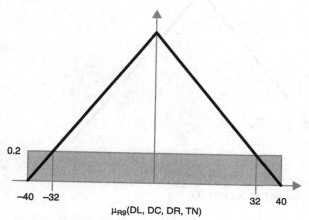

μ_{R9}(DL, DC, DR, TN)

19.12 Implication of rule 9.

Applying the Mamdani implication:

$$\mu_{R_9} (DL, DC, DR, TN) = \min [0.2, 0.3, 0.4, 0.5] = 0.2 \qquad [19.8]$$

For rule R33:

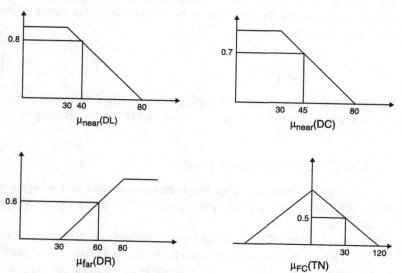

19.13 Fuzzification of rule 33 inputs.

Applying the Mamdani implication:

$$\mu_{R_{33}} (DL, DC, DR, TN) = \min [0.8, 0.7, 0.6, 0.5] = 0.5 \qquad [19.9]$$

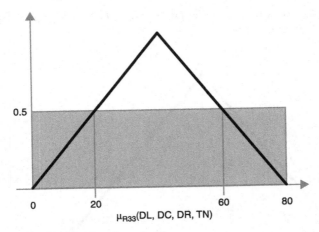

19.14 Implication for rule 33.

Defuzzification of output

Defuzzification is the process of combining the successful fuzzy output sets produced by the inference mechanism. The purpose is to produce the most certain low-level controller action. Several methods exist in the literature to perform defuzzification, the most popular of which is the centre of gravity (CoG) method. For discrete triangular linear functions the CoG method is obtained by moments of area as defined by:

$$\hat{c} = \frac{\sum_{c_i} c_i \mu_c (c_i)}{\sum_{c_i} \mu_c (c_i)}$$

[19.10]

For the matched rules R9 and R33, the application of the CoG methods (Fig. 19.15) results in a crisp control action of turn angle 28.178° with a certainty of 0.197.

19.4.3 Simulation experiment

The developed control algorithm has been implemented as a simulation in the MATLAB software using the fuzzy logic toolbox using the above-described variables (see Fig. 19.16).

The readings from the positional sensors are input to the defined fuzzy logic controller where they are operated upon to produce a discrete output value for the robot turning angle. The value of this angle is operated upon by a suitable mathematical function which determines the translation and rotation of the robot in real-time. The motion of the robot is then fed to the virtual reality component of MATLAB which in turn generates an animated output on the screen. This algorithm has been implemented in Simulink (see Fig. 19.17).

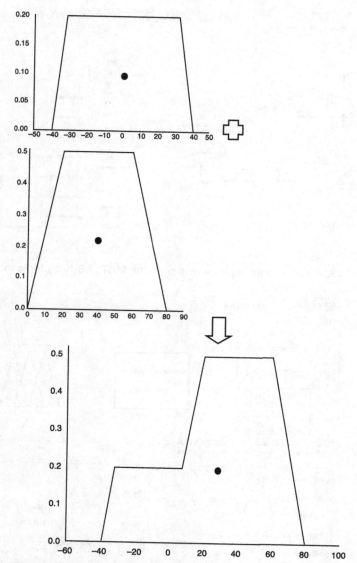

19.15 Applying the CoG method to matched rules 9 and 33.

19.5   Conclusions

In order to cater for the uncertainty associated with the autonomous navigation of landmine detection robots, a review of the various control aspects has been conducted with particular emphasis upon intelligent decision-making capabilities. These capabilities include path planning techniques, decision-making strategies and action taking algorithms. The various conventional and soft computing

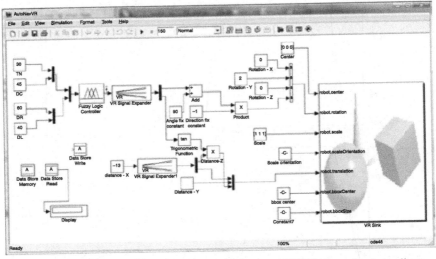

19.16 Fuzzy logic implementation in the MATLAB fuzzy logic toolbox.

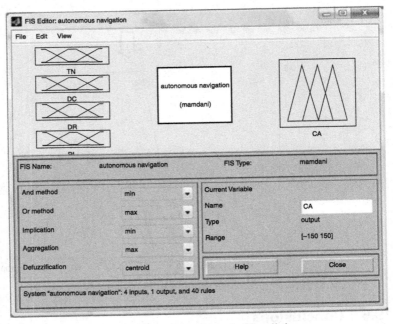

19.17 Autonomous navigation algorithm in Simulink.

control methodologies have been discussed in the context of autonomous robot navigation and the advantages and disadvantages of each are presented.

The rationale for the selection of fuzzy control is presented together with a discussion of how the fuzzy logic controller addresses uncertainty issues associated with navigation in landmine fields. An account of the design of an intelligent controller based on fuzzy logic theory for support of autonomous navigation of de-mining robots has been described. The sensed target direction and landmine distance are both fuzzified using overlapping membership functions. Heuristic knowledge of navigational decisions is encapsulated in a fuzzy rule base. A compositional rule of inference is applied on the rule base to construct a set of applicable fuzzy navigational decisions. The centre of gravity method is used to combine the results of the applicable fuzzy rules and produce a crisp control action. The feasibility of the technique is demonstrated through the simulated experiment.

19.6   References

Abdessemed, F., K. Benmahammed, et al. (2004). "A fuzzy-based reactive controller for a non-holonomic mobile robot." *Robotics and Autonomous Systems* 47(1): 31–46.

Acar, E., Y. Zhang, et al. (2001). Path Planning for Robotic Demining and Development of a Test Platform. *Proceedings of the 2001 International Conference on Field and Service Robotics*: 161–168.

Al-Khatib, M. and J. J. Saade (2003). "An efficient data-driven fuzzy approach to the motion planning problem of a mobile robot." *Fuzzy Sets and Systems* 134(1): 65–82.

Alnajjar, F. and K. Murase (2006). "Self-organization of spiking neural network that generates autonomous behavior in a real mobile robot." *International Journal of Neural Systems* 16(4): 229–239.

Aoki, T., Matsuno, M, Suzuki, T and Okuma, S (1994). Motion Planning for Multiple Obstacles Avoidance of Autonomous Mobile Robot Using Hierarchial Fuzzy Rules. *IEEE International Conference on Multisensor Fusion and Integration for Intelligent Systems*.

Arambula Cosío, F. and M. A. Padilla Castañeda (2004). "Autonomous robot navigation using adaptive potential fields." *Mathematical and Computer Modelling* 40(9–10): 1141–1156.

Baudoin, Y. (2008). Mobile Robotic Systems Facing the Humanitarian Demining Problem: State of the Art. *7th IARP WS HUDEM'2008*. American University in Cairo (AUC), Egypt.

Bekey, G. (2005). *Autonomous Robots: From Biological Inspiration to Implementation and Control*, MIT Press.

Bekey, G. A., and Goldberg, Kenneth Y., Ed. (1993). *Neural Networks in Robotics*, Kluwer Academic Publishers.

Borenstein, J. and Koren, Yoram (1991). "The Vector Field Histogram-fast Obstacle Avoidance For Mobile Robots." *IEEE Transactions on Robotics and Automation* 7(3): 278–288.

Borenstein, J. and Koren, Yorem (1989). "Real-time Obstacle Avoidance For Fast Mobile Robots." *IEEE Transactions on Systems, Man and Cybernetics* 19(5): 1179–1187.

Cho, D. K. and Chung, M. J. (1991). Intelligent Motion Control Strategy for a Mobile Robot in the Presence of Moving Obstacles. *IROS'91 IEEE/RSJ International Workshop on Intelligent Robots and Systems*.

Cuesta, F., A. Ollero, et al. (2003). "Intelligent control of nonholonomic mobile robots with fuzzy perception." *Fuzzy Sets and Systems* 134(1): 47–64.

de Sales Guerra Tsuzuki, M., T. de Castro Martins, et al. (2006). Robot path planning using simulated annealing. *Information Control Problems in Manufacturing 2006*. Oxford, Elsevier Science Ltd: 173–178.

De Silva, C. W. (1997). "Intelligent Control of Robotic Systems." *International Journal of Robotics and Autonomous Systems* 21: 221–237.

Fernandez-Leon, J. A., G. G. Acosta, et al. "Behavioral control through evolutionary neurocontrollers for autonomous mobile robot navigation." *Robotics and Autonomous Systems*, in press, corrected proof.

Foka, A. and P. Trahanias (2007). "Real-time hierarchical POMDPs for autonomous robot navigation." *Robotics and Autonomous Systems* 55(7): 561–571.

Ge, S. S., X.-C. Lai, et al. (2007). "Sensor-based path planning for nonholonomic mobile robots subject to dynamic constraints." *Robotics and Autonomous Systems* 55(7): 513–526.

Goldberg, D. E. (1989). *Genetic Algorithms in Search, Optimization, and Machine Learning*, Addison-Wesley.

Hagras, H., V. Callaghan, et al. (2004). "Learning and adaptation of an intelligent mobile robot navigator operating in unstructured environment based on a novel online Fuzzy-Genetic system." *Fuzzy Sets and Systems* 141(1): 107–160.

Hagras, H. A. (2004). "A Hierarchical Type-2 Fuzzy Logic Control Architecture for Autonomous Mobile Robots." *IEEE Transactions on Fuzzy Systems* 12(4): 524–539.

Hardarson, F. (1997). *Locomotion for Difficult Terrain*. Stockholm, Dept. of Machine Design, Royal Institute of Technology, Sweden.

Hardarson, F. (1998). *Locomotion for Difficult Terrain*. Stockholm, Dept. of Machine Design, Royal Institute of Technology, Sweden.

Haykin, S. (1994). *Neural Networks, A Comprehensive Foundation*. Englewood Cliffs, NJ, MacMillan Publishing.

Herianto, T. Sakakibara, et al. (2007). "Artificial Pheromone System Using RFID for Navigation of Autonomous Robots." *Journal of Bionic Engineering* 4(4): 245–253.

Howard, A., E. Tunstel, et al. (2001). Enhancing Fuzzy Robot Navigation Systems by Mimicking Human Visual Perception of Natural Terrain Traversability. *IFSA World Congress and 20th NAFIPS International Conference, 2001. Joint 9th*. Vancouver, BC, IEEE. 1: 7–12.

Jolly, K. G., R. Sreerama Kumar, et al. (2009). "A Bezier curve based path planning in a multi-agent robot soccer system without violating the acceleration limits." *Robotics and Autonomous Systems* 57(1): 23–33.

Karray, F. O. and C. De Silva (2004). *Soft Computing and Intelligent Systems Design*, Addison Wesley.

Khatib, O. (1986). "Real-Time Obstacle Avoidance For Manipulators And Mobile Robots." *International Journal of Robotics Research* 5(1): 90–98.

Lacroix, S., A. Mallet, et al. (2002). "Autonomous Rover Navigation on Unknown Terrains: Functions and Integration." *The International Journal of Robotics Research* 21(10–11): 917–942.

Li, W. (1994). Fuzzy Logic-based "Perception-Action" behavior Control of a Mobile Robot in Uncertain Environments. *IROS' 94 IEEE/RSJ/GI International Conference on Intelligent Robots and Systems*.

Liang, T.-C., J.-S. Liu, et al. (2005). "Practical and flexible path planning for car-like mobile robot using maximal-curvature cubic spiral." *Robotics and Autonomous Systems* 52(4): 312–335.

Marichal, G. N., L. Acosta, et al. (2001). "Obstacle avoidance for a mobile robot: A neuro-fuzzy approach." *Fuzzy Sets and Systems* 124(2): 171–179.

Martínez-Alfaro, H. and S. Gómez-García (1998). "Mobile robot path planning and tracking using simulated annealing and fuzzy logic control." *Expert Systems with Applications* 15(3–4): 421–429.

Novakovic, B., D. Scap, et al. (2000). "An analytic approach to fuzzy robot control synthesis." *Engineering Applications of Artificial Intelligence* 13(1): 71–83.

Palm, R. (1989). "Fuzzy controller for a sensor guided robot manipulator." *Fuzzy Sets and Systems* 31(2): 133–149.

Parasuraman, S., Ganapathy, V. et al. (2003). Fuzzy Decision Mechanism Combined with Neuro-Fuzzy Controller for Behavior Based Robot Navigation. *IECON '03. The 29th Annual Conference of the IEEE Industrial Electronics Society*, 2003. 3: 2410–2416.

Pradhan, S. K., D. R. Parhi, et al. (2009). "Fuzzy logic techniques for navigation of several mobile robots." *Applied Soft Computing* 9(1): 290–304.

Ramirez-Serrano, A., Boumedine, M (1996). Real-time Navigation in Unknown Environments Using Fuzzy Logic and Ultrasonic Sensing. *1996 IEEE International Symposium on Intelligent Control*: 26–33.

Ross, P., Corne, D. (1995). *Application of Genetic Algorithms*. Englewood Cliffs, NJ, Prentice Hall

Saffiotti, A. (1997). "The Uses of Fuzzy Logic in Autonomous Robot Navigation." *Soft Computing* 1(4): 180–197.

Smith III, J. F. and T. V. H. Nguyen (2007). "Autonomous and cooperative robotic behavior based on fuzzy logic and genetic programming." *Integrated Computer-Aided Engineering* 14(2): 141–159.

Takeuchi, T., Y. Nagai, et al. (1988). "Fuzzy control of a mobile robot for obstacle avoidance." *Information Sciences* 45(2): 231–248.

Tan, G.-Z., H. He, et al. (2007). "Ant Colony System Algorithm for Real-Time Globally Optimal Path Planning of Mobile Robots." *Acta Automatica Sinica* 33(3): 279–285.

Ulyanov, S. V., K. Yamafuji, et al. (1995). *Intelligent Fuzzy Motion Control of Mobile Robot for Service Use*. IEEE/RSJ International Conference on Intelligent Robots and Systems 95. "Human Robot Interaction and Cooperative Robots", Pittsburgh, PA.

Velagic, J., B. Lacevic, et al. (2006). "A 3-level autonomous mobile robot navigation system designed by using reasoning/search approaches." *Robotics and Autonomous Systems* 54(12): 989–1004.

Wakileh, B. A. M. and K. F. Gill (1990). "Robot control using self-organising fuzzy logic." *Computers in Industry* 15(3): 175–186.

Wang, M. and J. N. K. Liu (2008). "Fuzzy logic-based real-time robot navigation in unknown environment with dead ends." Robotics and Autonomous Systems 56(7): 625–643.

Wang, X., Z.-G. Hou, et al. (2008). "A behavior controller based on spiking neural networks for mobile robots." *Neurocomputing* 71(4–6): 655–666.

Wu, C.-J. (1992). Fuzzy Robot Navigation In Unknown Environments. *IEEE International Workshop on Emerging Technologies and Factory Automation, 1992*, IEEE: 624–628.

Yang, X., M. Moallem, et al. (2003). An improved fuzzy logic based navigation system for mobile robots. *Proceedings. 2003 IEEE/RSJ International Conference on Intelligent Robots and Systems, 2003*. (IROS 2003). 2: 1709–1714.

20

Human victim detection and stereo-based terrain traversability analysis for behaviour-based robot navigation

G. DE CUBBER and D. DOROFTEI, Royal Military
Academy, Belgium

Abstract: This chapter presents three main aspects of the development of a crisis management robot. First, we present an approach for robust victim detection in difficult outdoor conditions. Second, we present an approach where a classification of the terrain in the classes 'traversable' and 'obstacle' is performed using only stereo vision as input data. Lastly, we present behavior-based control architecture, enabling a robot to search for human victims on an incident site, while navigating semi-autonomously, using stereo vision as the main source of sensor information.

Key words: crisis management, visually guided robots, human victim detection, terrain traversability estimation, behavior-based robot control.

20.1   Introduction

Crisis management teams (e.g. fire and rescue services, anti-terrorist units) are often confronted with dramatic situations where critical decisions have to be made within hard time constraints. In these circumstances, a complete overview of the crisis site is necessary to make the correct decisions. However, obtaining such a complete overview of a complex site is not possible in real-life situations when the crisis management teams are confronted with large and complex unknown incident sites. In these situations, the crisis management teams typically concentrate their effort on a primary incident location (e.g. a building on fire, a crashed airplane, etc) and only after some time (depending on the manpower and the severity of the incident), do they turn their attention towards the larger surroundings, e.g. searching for victims scattered around the incident site. A mobile robotic agent could aid in these circumstances, gaining valuable time by monitoring the area around the primary incident site while the crisis management teams perform their work. However, as these teams are in general already overloaded with work and information in any medium or large-scale crisis situation, it is essential that for such a robotic agent to be useful, it should not require extensive human control (hence it should be semi-autonomous) and it should only report critical information back to the crisis management control center.

In the framework of the View-Finder project, an outdoor mobile robotic platform was developed. This semi-autonomous agent, shown in Fig. 20.1, is

20.1 The RobuDem crisis management robot used as a testbed for the presented algorithms.

equipped with a differential GPS system for accurate geo-registered positioning, and a stereo vision system. The design requirements for such a robotic crisis management system give rise to three main problems, which need to be solved for the successful development and deployment of the mobile robot:

1. How can the robot automatically detect human victims, even in difficult outdoor illumination conditions?
2. How can the robot, which needs to navigate autonomously in a totally unstructured environment, auto-determine the suitability of the surrounding terrain for traversal?
3. How can all robot capabilities be combined in a comprehensive and modular framework, so that the robot can handle a high-level task (searching for human victims) with minimal input from human operators, when navigating in a complex dynamic and environment, while avoiding potentially hazardous obstacles?

In response to the first question, we present an approach to achieve robust victim detection in difficult outdoor conditions by using the Viola–Jones algorithm for Haar-features-based template recognition and adapting it to recognize victims. Victims are assumed to be human body shapes lying on the ground. The algorithm tries to classify visual camera input images into human body shapes and background items. This approach is further explained in section 20.3.

The second problem, which is stated above, is that of the traversability estimation. This is a challenging problem, as the traversability is a complex function of both the terrain characteristics, such as slopes, vegetation, rocks, etc and the robot mobility characteristics, i.e. locomotion method, wheel properties, etc. In section 20.4 of this chapter, we present an approach where a classification of the terrain in the classes 'traversable' and 'obstacle' is performed using only stereo vision as input data.

The third question, which was raised above, relates to the robot control mechanism and the control architecture. For this application, the behavior-based control paradigm was chosen as a control mechanism due to its flexible nature, allowing the design of complex robot behavior through the integration of multiple relatively simple sensor–actuator relations. Through this control architecture, the robot is able to search for human victims on an incident site, while navigating semi-autonomously, using stereo vision as the main source of sensor information. The behavior-based control mechanism is further detailed in section 20.5 of this chapter, while section 20.2 gives an overview of the global robot control architecture.

20.2   Robot control architecture

The control architecture describes the strategy to combine the three main capabilities of an intelligent mobile agent: sensing, reasoning and actuation. These three capabilities have to be integrated in a coherent framework in order for the mobile agent to perform a certain task adequately. To combine the advantages of purely reactive and planner-based approaches, a behavior-based controller was implemented for autonomous navigation.

Figure 20.2 illustrates the general robot control architecture, set up as a test bed for the algorithms discussed in this chapter. The RobuDem robot used in this setup features two on-board processing stations, one for low-level motor control (Syndex Robot Controller), and another for all the high-level functions. A remote robot control PC is used to control the robot and to visualize the robot measurements (color images, victim data) from a safe distance. All data transfer between modules occurs via transmission control protocol (TCP) and user datagram protocol (UDP)-based connections, relying on the CORBA (Slama Garbis and Russell, 1999) and CoRoBa (Colon, Sahli and Baudoin, 2006) protocols. The wireless link from the on-board high-level PC to the remote robot control PC is set up via a wi-fi connection.

A behavior-based navigational architecture is used for semi-autonomous intelligent robot control. Behavior-based techniques have gained wide popularity in the robotics community (Jones, 2004), due to the flexible and modular nature of behavior-based controllers, facilitating the design process. Following the behavior-based formalism, a complex control task is subdivided into a number of more simple modules, called behaviors, which each describe one aspect of the sensing, reasoning and actuation robot control chain. Each behavior

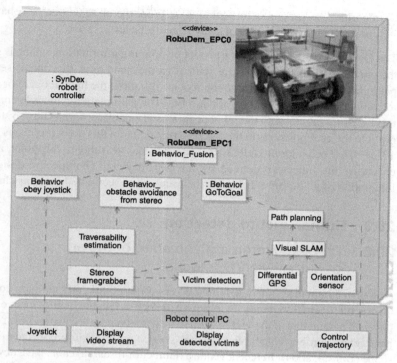

20.2 Global robot control architecture.

outputs an objective function, $o_1(x)$, ..., $o_n(x)$. These objective functions are multi-dimensional normalized functions of the output parameters, where $x = (x_1, ..., x_n) \in \mathbb{R}^n$ is an n-dimensional decision variable vector. The degree of attainment of a particular alternative x, with respect to the kth objective is given by $o_k(x)$.

Recall that the RobuDem robot is equipped with two main sensing abilities: a stereo vision system and a GPS system. The information from the stereo vision system is used threefold:

- The color images are sent over the wireless link, so that at all times the human operator receives a visual cue of the environment. This is absolutely necessary when the robot is operating under tele-operation mode.
- The (left) color image is sent to a victim detection module. This module incorporates a vision-based human victim detector, presented in section 20.3. The victim detection module will report any detected human victims back to the human operator at the remote control station.
- The calculated stereo disparity image is sent to a terrain traversability estimation module. This module incorporates a stereo-based terrain traversability analysis

algorithm, as presented in section 20.4. From the obstacle map, a behavior is constructed to steer the robot away from obstacles.

The GPS system delivers accurate robot positioning information, which is sent to the operator at the remote control station. At the same time, this data is sent to a path-planning module. From the robot control station, the human operator is able to compile a list of waypoints for the robot. The path-planning module compares this list of waypoints with the robot position and calculates a trajectory to steer the robot to the goal positions in the list. The first point on this trajectory list is sent to a *GoToGoal* behavior module, which aims to steer the robot to this point, as such executing the trajectory defined by the path planner. The specific design and integration of all different behaviors is discussed in section 20.5 of this chapter.

20.3   Human victim detection

20.3.1 Problem statement and state of the art of human victim detection

Automated human victim detection is a very difficult task, especially in complex, unstructured environments. In order to detect a human victim, one or more physical parameters of the victim need to be perceived by a sensor. These physical parameters can be: voice, temperature, scent, motion, skin color, face or body shape (Burion, 2004). In recent years, a number of research teams have developed human victim detection algorithms based on the detection of these physical parameters. In this section we give an overview of these activities and approaches used for human victim detection.

Voice

Kleiner and Kuemmerle (2007) performed audio-based victim detection by positioning two microphones at a known distance. Given an audio source left, right or between both microphones, the time difference, i.e. phase shift, is measured between both signals. This is carried out by the crosspower spectrum phase (CSP) approach, which allows the calculation of the phase shift of both signals based on the Fourier transformation. Using this approach, the bearing of the sound source can be successfully determined, if there is not too much background noise.

Temperature (body heat)

Infrared cameras give a complete picture of the environment heat, which is very useful in human detection. Although infrared cameras are very expensive, they seem the best solution to make the discrimination between human and non-human presence and, as such, are essential for a robust and efficient solution to human finding. Infrared cameras are used by Kleiner and Kuemmerle (2007), Nourbakhsh et al. (2005) and Birk et al. (2006).

A new development in this field is the hyperspectral IR imaging approach of Trierscheid et al. (2008). Hyperspectral IR images contain a contiguous spectrum in the bandwidth of IR in a spatial scanline in the scene and allow the technique of combining spectroscopy and imaging, which makes it very well suited for human victim detection.

Pyroelectric sensors are another type of heat-based human detector. These sensors are designed specifically for human detection. The sensor is made of a crystalline material that generates a surface electric charge when exposed to heat in the form of infrared radiation. It is calibrated to be sensitive to human heat wavelength (8–14 μm). These sensors are very sensitive, cheap and robust. They are composed of two infrared sensors, so they detect humans only if the human or the sensor is moving. Pyroelectric sensors have been used by Pissokas and Malcolm (2002) and Nourbakhsh et al. (2005)

Scent

CO_2 *sensors* detect carbon dioxide emissions, and even the breathing cycle of a victim. It is thus possible to determine if a person is still alive. These sensors have been used by a number of participants of the RoboCupRescue, but disadvantages are that the response time of a CO_2 sensor is very slow and the sensor has to be very close to the victim to provide useful data because it is very directional and depends greatly on air conditions such as humidity, temperature, wind, and dust. This makes it difficult to use in a disaster area.

Motion

Motion can be detected by a variety of sensors (sonar, laser, visual & IR camera) and can serve as an indication that somebody alive is present. However, motion analysis alone can never determine whether the cause of this motion field is a human being. Therefore, it is only used in combination with other characteristics (Kleiner and Kuemmerle, 2007; Nourbakhsh et al., 2005).

Skin color

Skin color is a popular parameter in the computer vision community for detecting humans. Visser et al. (2007) use skin color for human victim detection. They construct a 3D color histogram in which discrete probability distributions are learned. Given skin and non-skin histograms based on training sets, the probability that a given color value belongs to the skin and non-skin classes can then be discovered. The problem with these approaches is twofold: 1) in unstructured outdoor environments, there is no a priori data on the colors present in the environment (which could lead to a large number of false positives), and 2) the field of view of typical outdoor robot cameras is quite large, which means that a person's face only consists of a limited number of pixels (which would reduce the detection rate).

Face and body shape detection

Another popular approach in computer vision for detecting people is to perform face detection. Other detectors are specifically trained at detecting the upper body. Together, these detectors provide powerful cues for reasoning about a person's presence. The problem with these methods is that detecting victims lying on the ground using standard camera images is very different from standard person detection. These standard person detection algorithms, relying on face or upper body detection, assume that the person's face is clearly visible in the camera image and that the person is standing upright, such that the upper body can be easily detected. Victims, however, do not tend to stand up or look straight into the camera. Therefore, special techniques have to be applied, as proposed for example in Bahadori and Iocchi (2003). The approach presented in this article aims to classify the body shape of lying human victims and thus falls into this category of victim detectors.

Combined approaches

With such a multitude of detection approaches, each having their advantages and disadvantages, it is evident that the integration of multiple cues can provide better results. Therefore, several teams have investigated hybrid approaches, mixing for example motion, sound and heat (Nourbakhsh et al., 2005) or motion, sound and faces. Others (Kleiner and Kuemmerle, 2007) have focused specifically on determining the best way to integrate the information from all cues, leading to Markov random field (MRF)-based approaches.

20.3.2 The Viola–Jones based victim detector

Here, we present an approach for achieving robust victim detection in difficult outdoor conditions. The basis for this work is a learning-based object detection method, proposed by Viola and Jones (2001). Viola and Jones (2004) originally applied this technique in the domain of face detection. Their system yields face detection performance comparable to the best previous systems and is considered the fastest and most accurate pattern recognition method for faces in monocular grey-level images.

The method operates on so-called *integral images*: each image element contains the sum of all pixel values to its upper left allowing for constant-time summation of arbitrary rectangular areas.

During training, weak classifiers are selected with AdaBoost, each of them a pixel sum comparison between rectangular areas.

The object detection system classifies the images using simple features. The features are able to encode ad-hoc domain knowledge and the features-based system operates much faster than a pixel-based one. The Haar-wavelets are single-wavelength square waves, which are composed of adjacent rectangles. The algorithm does not use true Haar-wavelets, but better suited rectangle combinations. This is why they are called Haar-features instead of Haar-wavelets. The Haar-features detection procedure works by subtracting the average dark-region pixel value from the average light-region pixel value. If the difference is above a threshold, which is set during the training, the feature is present.

Hundreds of these classifiers are arranged in a multi-stage cascade. Lazy successive cascade evaluation and the constant-time property allow the detector to run fast enough to achieve an overall low latency.

In a first attempt at victim detection, we used the standard Viola–Jones detector for face and upper body detection. The first tests were executed on indoor and good quality images. These tests were very successful, 90 percent of the faces and 80 percent of the upper bodies were detected. All together the hit rate reached 95 percent while the false alarm rate stayed under 25 percent. However, the target hardware, the RobuDem, will operate in an outdoor environment where the background varies and the illumination is unpredictable. So, outdoor experiments were strongly suggested. Although the results were better than expected, the false alarm was increased dramatically while the hit rate was decreased to 70 percent for the upper body and to 30 percent for the face detection.

The conclusion from these tests is that in an outdoor environment the face detection based person detection is not viable. Frequently it only consumed the computation time without giving any results or any correct results. If the detection was more detailed, the system became too slow with a minor success rate. If the detection was tuned to be faster, the hit rate decreased below 10 percent.

The upper body detection is more robust, as it adapts itself to different illumination conditions much better. However, it gives many more false alarms.

Our first idea was to fuse the face and upper body detector to create a more robust system. Unfortunately, the face detector does not really improve the performance of the upper body detector. Also, these detectors work only for the detection of people who are standing, sitting or crouching. In the case of a person lying on the ground, all of the detectors based on the existing Haar-cascade classifiers fail. Therefore, we decided to fuse the upper body detector with a new detector, which has to be trained to detect victims.

First, we adapted the Viola–Jones technique, by training the algorithm with bodies, lying on the ground. To deal with the huge number of degrees of freedom of the human body and the camera viewpoint, the configuration space selected for human victims was reduced to victims lying face down and more or less horizontally in front of the camera. This case has been chosen because in real

20.3 Sample images for victim detection.

disasters this pose has the highest occurrence. People try to protect their head and their ventral body parts, which are the most vulnerable. Another reason is that in this position, the possible positions of the limbs form a relatively small pool compared to other cases. Also the orientation of the body must be considered because the legs have a different shape than the upper body and the head. To handle this, the sample images were taken with both body orientations (left-to-right and right-to left). To enlarge the data set, the images were then later flipped horizontally and re-used during the Haar-training.

Figure 20.3 shows some examples of the sample images. They were taken in an outdoor environment, featuring human victims in several orientations and under varying illumination. These images were taken with an on-board stereo camera system. In total, 800 positive scenes were recorded, and in each case the color rectified images of the left and right camera were recorded.

Furthermore 500 pairs of negative images were recorded outside and 100 pairs inside. These images contain no humans but the variety of the background is high in order to make the AdaBoost learning method set up good thresholds.

20.3.3 Human victim detection results

Theoretically, the cascaded Haar-classifier for victim detection has a 100 percent detection rate and less than a 10^{-6} percent false alarm rate. Of course, this is only true in the case of the positive and negative sample images. With new test images, which were taken in similar illumination conditions as the sample images but mainly in different positions, the correct detection rate was approximately 65 percent.

Figure 20.4 shows the results of the victim detection algorithm. The dark rectangles are the detector hits for the victim whose head is on the left, the light ones for the victim whose head is on the right. In the first image of Fig. 20.4, the victim was correctly found, along with a lot of false positives. These false alarms are eliminated by merging the adjacent rectangles of correct posture.

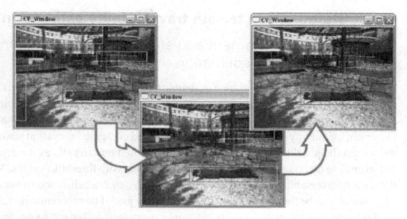

20.4 Victim detection results.

20.5 Victim detection results.

In the case of Fig. 20.5, it is more difficult to decide whether a correct classification is performed. In the first picture a small and a large rectangle cover the victim. The smaller rectangle is a true positive, but the bigger rectangle is a false alarm, which may have some typical features of a victim. As shown in the second picture, these rectangles are considered neighbors and they were merged together. The merging is done by computing an average rectangle; this is why the marked area is bigger than the actual victim.

The processing time for running the victim detector is between 60 and 80 milliseconds, which means 13–16 frames per second. This is a very good result, as it allows near real-time reactions in the robot control scheme and it also allows the integration of results of multiple detection runs over time by means of a tracking scheme, to enhance the detection rate and reduce the false positive rate. As such, the victim detection quality can be further improved, making this methodology a valuable aid for human crisis managers, as it enables a robot to scan a designated area for human survivors semi-automatically.

20.4   Stereo-based terrain traversability estimation

20.4.1 Problem statement and state of the art of terrain traversability estimation

Terrain traversability analysis is a research topic that has been the focus of the mobile robotics community for the past decade, inspired by the development of autonomous planetary rovers and, more recently, the DARPA Grand Challenge. However, already by 1994, Langer et al. (1994) had computed elevation statistics of the terrain (height difference and slope) and classified terrain cells as traversable or untraversable by comparing these elevation statistics with threshold values. Most of the terrain traversability analysis algorithms employ such a cell-based traversability map, which can be thought of as a 2.5D occupancy grid. The problem with Langer's method was that the traversability was only expressed in binary forms and soon other researchers (Singh et al., 2000; Gennery, 1999) presented solutions to lift this limitation. Seraji (2003) proposed a fuzzy-logic traversability measure, called the traversability index, which represents the degree of ease with which the regional terrain could be navigated. This degree was calculated on the basis of the terrain roughness, the slope and the discontinuity, as measured by a stereo vision system.

Schäfer et al. (2005) presented a similar stereo-discontinuities based approach without explicit calculation of a traversability map. Other researchers (Shneier et al., 2006; Kim et al., 2006; Happold et al., 2006) have embedded the stereo-based terrain traversability analysis in an on-line learning approach. The results of these methods depend greatly on the quality of the training set.

Ulrich and Nourbakhsch (2000) presented a solution for appearance-based obstacle detection using a single color camera. Their approach makes the assumption that the ground is flat and that the region in front of the robot is ground. Kim et al. (2007) present another single-camera traversability estimation method based upon self-supervised learning of superpixel regions.

Besides monocular and stereo vision, laser range finders are a useful sensor for terrain traversability estimation. Andersen et al. (2006) present a method for terrain classification using single 2D scans. The Stanford Racing Team (Thrun et al., 2006) utilized a traversability map based on data from six laser scanners registered with pose from an unscented Kalman Filter to classify grids as undrivable, drivable, or unknown. A Markov model was used to probabilistically test for the presence of an obstacle leading to an improved traversability map.

20.4.2 V-disparity based terrain traversability estimation

Detecting obstacles from stereo vision images may seem simple, as the stereo vision system can provide rich depth information. However, from the depth image, the distinction between the traversable and the non-traversable terrain is not evident, especially in outdoor conditions, where the terrain roughness and the robot mobility parameters must be taken into account. Our approach is based on the construction and subsequent processing of the *v-disparity image* (Labayrade et al., 2002), which provides a robust representation of the geometric content of road scenes. The v-disparity image is constructed by calculating a horizontal histogram of the disparity stereo image.

Consider two stereo frames, as shown in Fig. 20.6 and 20.7, and the computed disparity image ID, as shown in Fig. 20.8. The v-disparity image I_V can be constructed by accumulating the points with the same disparity that occur on a horizontal line in the image. Figure 20.9 displays the v-disparity image I_V for the given input images.

The classification of the terrain in traversable and non-traversable areas comes from the assumption that the majority of the image pixels are related to traversable terrain of the ground plane. The projection of this ground plane in the v-disparity image is a straight line, from the top left to the bottom right of the v-disparity image. Any deviations from this projection of the ground plane are likely obstacles or other non-traversable terrain items.

As such, the processing of the v-disparity image comes down to estimating the equation of the line segment in the v-disparity image, corresponding to the ground plane. This is done by performing a Hough transform on the v-disparity image and searching for the longest line segment. Then, one must choose a single parameter that accounts for the maximum terrain roughness. As this parameter depends only on the robot characteristics, it only needs to be set once. The parameter sets the maximum offset in v-disparity space to be considered part of the ground plane. Any outliers are regarded as obstacles, which enables us to compile an obstacle image I_O.

20.6 Left stereo input image.

20.7 Right stereo input image.

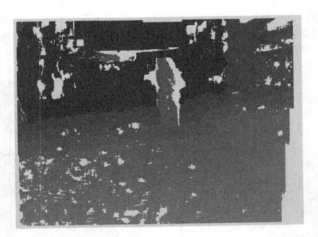

20.8 Stereo disparity image.

20.4.3 Terrain traversability estimation results

Figure 20.10 shows the v-disparity image after Hough transform. The white line indicates the largest line segment, corresponding to the ground plane. The two gray lines indicate the region in v-disparity space where pixels are considered part of a traversable region. The terrain corresponding to pixels in v-disparity space in-between the two diagonal, tinted lines is considered traversable, otherwise it is considered as an obstacle.

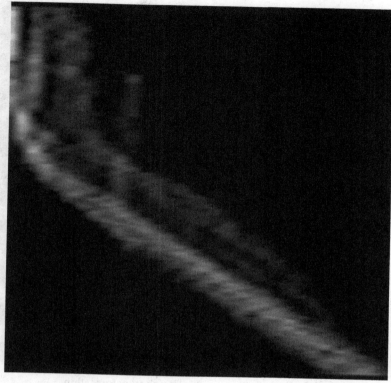

20.9 V-disparity image.

The result of this operation can be judged from Fig. 20.11, showing the obstacle image. This is a version of the color input image, where false color data corresponding to the disparity is superposed for pixels classified as belonging to non-traversable terrain.

It may be noted that the lower part of the legs of the person standing in front of the robot were not detected as obstacles. This is due to the choice of the threshold parameter for the ground plane, as discussed above. After tests in multiple environments, we used a threshold parameter of 50, which offers a good compromise between a good detection rate and a low false positive detection rate.

20.5   Behavior-based control

20.5.1 Problem statement and state of the art behavior-based control

Following the robot control architecture depicted by Fig. 20.2, the Robudem robot employs a behavior-based control architecture, which allows it to navigate to a

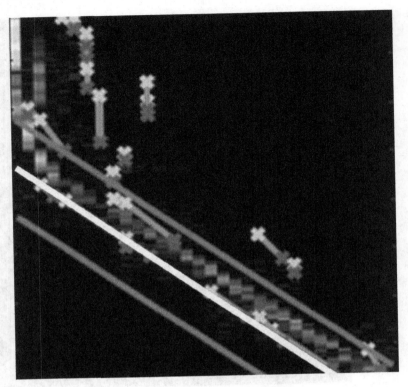

20.10 V-disparity image after Hough transformation.

number of goal positions while avoiding obstacles and detecting human victims. As can be noted from Fig. 20.2, a number of behaviors are foreseen for this application. In order for the robot to accomplish its task, two main questions need to be solved:

1. How can we design the individual behaviors, such that the robot is capable of avoiding obstacles and of navigating semi-autonomously?
2. How can these individual behaviors be combined in an optimal way, leading to a rational and coherent global robot behavior?

Indeed, the performance of the behavior-based controller depends on the implementation of the individual behaviors as well as on the method chosen to solve the behavior fusion or action selection problem. However, as in behavior-based control, the idea is to split a complex task into multiple simple tasks or behaviors, and the design of the individual behaviors in general does not pose major problems. However, the behavior fusion or action selection problem does require deeper investigation.

20.11 Obstacle image.

In general, the action selection problem can be formulated as a multiple-objective decision-making problem, as proposed by Pirjanian (1999). Mathematically, a multi objective decision problem can be represented as finding the solution to:

$$\arg\max_{\mathbf{x}} [o_1(\mathbf{x}), \ldots, o_n(\mathbf{x})]$$ [20.1]

A common method for solving a multiple-objective decision-making problem is the weighting method. This method is based on scalar vector optimization and is formulated in the following way:

$$\mathbf{x}^* = \arg\max_{\mathbf{x} \in X} \sum_{i=1}^{n} w_i o_i(\mathbf{x})$$ [20.2]

where w_i is a set of weights, normalized such that:

$$\sum_{i=1}^{n} w_i = 1$$ [20.3]

An important question is how to vary the weights w_i to generate the whole or a representative subset of the solutions. Most approaches described in the literature handle the problem in the same manner. Each weight w_i is considered to have a reasonable number of discrete values between 0 and 1. Then the equation is solved for each combination of values. The obvious limitation of this approach is the large number of computations to be performed, as the computational complexity grows exponentially with the number of objective functions.

The interval criterion weights method (Steuer, 1974) reduces the number of computations by incorporating some subjective knowledge in the process of generating optimal solutions. Basically, rather than letting each w_i assume values in the interval [0; 1], a narrower interval is used, which reflects the importance of each objective function. Apart from reducing the computational burden, the interval criterion weights method, produces the 'best' part of X*. Here 'best' reflects the decision maker's preferences, which are expressed as weight values. A further advantage of the interval criterion weights method is the fact that in practice a decision maker seems to be more comfortable defining the range within which a weight value should be, rather than estimating the single weight values.

Goal programming methods (Nakayama and Yun, 2006) define a class of techniques where a human decision maker gives his/her preferences in terms of weights, priorities, goals, and ideals. The concept of best alternative is then defined in terms of how much the actual achievement of each objective deviates from the desired goals or ideals. Further, the concept of best compromise alternative is defined to have the minimum combined deviation from the desired goals or ideals. Goal programming methods thus choose an alternative having the minimum combined deviation from the goal $\hat{o}_1$, ..., $\hat{o}_n$, given the weights or priorities of the objective functions. According to the goal programming theorem, the solution to the multi-objective decision-making problem can be expressed as:

$$\arg\max_{\mathbf{x} \in X} \sum_{i=1}^{n} w_i |o_i(\mathbf{x}) - o_i^*|^p \qquad [20.4]$$

where $1 \le p \le \infty$, $\mathbf{o}^*$ is the ideal point, w_j is the weight or priority given to the jth objective.

In section 20.5.3 of this chapter, we will present a novel way of combining the inputs of multiple behaviors, but first we will investigate the design of each individual behavior for the chosen application.

20.5.2 Behavior design

As mentioned before, a behavior is a function that relates sensor measurements to actions in the form of an objective function. In the case of robot control, the objective function of each behavior can be regarded as two-dimensional normalized function of robot steering velocity v and direction α. For this setup, three behaviors are defined, which relate the abstract sensor information into robot actions.

These three behaviors are shown below.

Obey joystick commands

If desired, the human operator can control the robot by means of a joystick. The joystick commands are directly related to the robot steering angle and direction, so the transformation of the joystick control command into an objective function can be performed straightforwardly by calculating a two-dimensional Gaussian from the joystick input ($v_{Joystick}$, $\alpha_{Joystick}$):

$$o_{Joystick}(v, \alpha) = \frac{1}{\sqrt{(2\pi)^2\,\sigma^4}} \exp\left(-\left(\frac{(v - v_{Joystick})^2}{2\sigma^2} + \frac{(\alpha - \alpha_{Joystick})^2}{2\sigma^2}\right)\right) \quad [20.5]$$

Obstacle avoidance using stereo

To drive the robot away from obstacles detected by the terrain traversability analysis algorithm, the obstacle image is analyzed. The depth values of pixels corresponding to obstacles are accumulated per vertical line in the image and the resulting function is inverted and normalized. This allows the deduction of function f of the viewing angle α as shown in Fig. 20.12. This function can be regarded as a one-dimensional objective function for obstacle avoidance from stereo input, considering only the viewing/steering angle. This one-dimensional objective function can then be extended for velocity as well, using the following formulation:

$$o_{stereo}(v, \alpha) = \frac{f(\alpha)}{1 + |vf(\alpha)/c|} \quad [20.6]$$

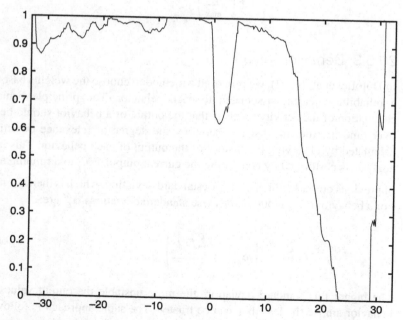

20.12 One-dimensional objective function for obstacle avoidance from stereo.

Go to goals

The goal seeking behavior is assigned two tasks. First, it points the robot to the goal position and then it varies the velocity respective to the distance to the goal. This means the development of the objective function can be split up as:

$$o_{GoToGoal}(v, \alpha) = o^{\alpha}{}_{GoToGoal}(\alpha) \cdot o^{v}{}_{GoToGoal}(v) \qquad [20.7]$$

To calculate these objective functions, the (Euclidian) distance to the goal d_{Goal} and heading to this goal θ are calculated from the current robot position given by the GPS system and the current waypoint given by the global path planner. The goal seeking behavior aims to minimize the difference between the robot heading α and the goal heading θ, which can be formulated as:

$$o^{\alpha}{}_{GoToGoal} = \frac{1}{1 + \left(\dfrac{\alpha - \theta}{\beta}\right)^2} \qquad [20.8]$$

with β the window size that is considered. $o^{v}{}_{GoToGoal}(v)$ is set up such that the velocity is always high, with the exception that when the robot approaches a goal position, the speed should be reduced. This is expressed as:

$$o^{v}{}_{GoToGoal}(v) = \begin{cases} \left(\dfrac{v}{v_{max}}\right)^2 & \text{if } d_{Goal} > d_{Threshold} \\[4mm] \dfrac{1}{1 + \left(\dfrac{v}{v_{max}}\right)^2} & \text{if } d_{Goal} < d_{Threshold} \end{cases} \qquad [20.9]$$

20.5.3 Behavior fusion

In Doroftei et al. (2007), we proposed a method to choose the weights based upon a reliability measure associated to each behavior. The principle behind the calculation of the activity levels is that the output of a behavior should be stable over time in order to trust it. Therefore, the degree of relevance or activity is calculated by observing the history of the output of each behavior. This history-analysis is performed by comparing the current output $\varpi^{b_i}_{\varpi_{j,k}}$ to a running average of previous outputs, which leads to a standard deviation, which is then normalized. For a behavior b_i with outputs ϖ_j these standard deviations $\sigma^{b_i}_{\varpi_j}$ are:

$$\sigma^{b_i}_{\varpi_j} = c_{\varpi_j} \sum_{k=i-h}^{i} \left(\varpi^{b_i}_{\varpi_{j,k}} - \frac{\sum_{l=1}^{N} \varpi^{b_i}_{\varpi_{j,l}}}{N} \right)^2 \qquad [20.10]$$

The bigger this standard deviation, the more unstable the output values of the behavior are, so the less they can be trusted. The same approach is followed for all behaviors. This leads to an estimate for the activity levels or weights for each behavior:

$$w_{b_i} = \sum_{j=1}^{\text{number of outputs}} \left(1 - \sigma_{\overline{\omega}_j}^{b_i}\right) \qquad [20.11]$$

For stability reasons, the activity level is initialized at a certain value (in general 0.5) and this estimate is then iteratively improved.

The approaches towards solving the multiple-objective decision-making problem for action selection that we have reviewed up to this point all have their advantages and disadvantages. Solving the multiple-objective decision-making problem using reliability analysis has the big advantage of incorporating direct information from the system under control into the control process. On the other hand, this architecture does not offer a human decision maker the ability to interact with the decision process. As autonomous agents increasingly have to interact with humans on the field, exchanging knowledge and learning from each other, this is a serious shortcoming. Common techniques for solving the multiple-objective decision-making problem while taking into account a decision maker's preferences involve considering these issues by offering a human operator the ability to input some objectives or ideal points. These approaches, however, suffer from the disadvantage that no reliability data from the sensing and other processes is taken into account while performing action selection. One could thus argue that while reliability analysis-based approaches are too robot centric, this second set of approaches is too human-centric. Here, we propose an approach to integrate the advantages of both theorems. This can be achieved by minimizing the goal programming and reliability analysis constraints in an integrated way, following:

$$\arg\min_{\substack{x \in X \\ w_i \in w}} \left[\lambda \left(\sum_{i=1}^{n} w_i |o_i(\mathbf{x}) - o_i^*|^p \right) + (1-\lambda) \sum_{i=1}^{n} \left| w_i - \sum_{j=1}^{\text{number of outputs}} \left(1 - \sigma_{\overline{\omega}_j}^{b_i}\right) \right|^p \right] \qquad [20.12]$$

with λ a parameter describing the relative influence of both constraints. This parameter indirectly influences the control behavior of the robot. Large values of λ will lead to a human-centered control strategy, whereas lower values will lead to a robot-centered control strategy. The value of λ would therefore depend on the expertise or the availability of human experts interacting with the robot.

It is obvious that this method increases the numerical complexity of finding a solution, but this does not necessarily lead to an increased processing time, as the search interval can be further reduced by incorporating constraints from both data sources.

20.6   Results and conclusions

In this chapter, we have discussed three main aspects of the development of a crisis management robot.

First, we have presented an approach for the automatic detection of human victims. This method is based upon the Viola–Jones (face) detector, which was adapted so that human victims lying on the ground can be detected. The results of this approach are very encouraging, although further research is required to increase the detection rate and reduce the number of false positives. This will be done by integrating the human victim detector in a tracking scheme.

Next, we presented a stereo-based terrain traversability estimation algorithm. This method is based upon the analysis of the stereo disparity, more specifically the vertical histogram of the disparity image, called the v-disparity. This approach makes it possible to robustly classify the terrain of outdoor scenes in traversable and non-traversable regions quickly and reliably, using only a single parameter to describe the robot mobility characteristics.

Ultimately, we have shown how all different robot capabilities can be integrated into a robot control architecture. A behavior-based control scheme is employed due to its flexible design. In this context, a novel approach for solving the behavior fusion problem was presented. This method combines the advantages of the traditional weighting methods and the more recent reliability analysis based method.

As a final result, the Robudem robot has become a semi-autonomous agent: it can handle a high-level task (searching for human victims) with minimal input from human operators, by navigating a complex dynamic and environment, while avoiding potentially hazardous obstacles. If required, a remote human operator can still take control of the robot via the joystick, but in normal operation mode, the robot navigates autonomously to a list of waypoints, while avoiding obstacles (thanks to the stereo-based terrain traversability estimation) and while searching for human victims.

20.7   References

Andersen J C, Blas M R, Ravn O, Andersen N A and Blanke M (2006), 'Traversable terrain classification for outdoor autonomous robots using single 2D laser scans'. *Journal of Integrated Computer-Aided Engineering*, 13(3), 223–232.

Bahadori S and Iocchi L (2003), 'Human body detection in the RoboCup rescue scenario', *First International Workshop on Synthetic Simulation and Robotics to Mitigate Earthquake Disaster*.

Birk A, Markov S, Delchev I and Pathak K (2006), 'Autonomous rescue operations on the IUB Rugbot', *International Workshop on Safety, Security, and Rescue Robotics (SSRR)*, IEEE Press.

Burion S (2004), *Human Detection for Robotic Urban Search and Rescue*, Diploma Work, EPFL.

Colon E, Sahli H and Baudoin Y (2006), 'CoRoBa, a multi mobile robot control and simulation framework', *International Journal of Advanced Robotic Systems*, 3(1), 73–78.

Doroftei D, Colon E, De Cubber G (2007), 'A behaviour-based control and software architecture for the visually guided Robudem outdoor mobile robot', *ISMCR2007*, Warsaw, Poland.

Gennery D B (1999), 'Traversability analysis and path planning for a planetary rover', *Autonomous Robots*, 1999, 6(2), 131–146.

Happold M, Ollis M and Johnson N (2006), 'Enhancing supervised terrain classification with predictive unsupervised learning', *Robotics: Science and Systems*.

Jones J L (2004), *Robot Programming: A practical guide to behavior-based robotics*. McGraw-Hill.

Kim D, Oh S and Rehg J M (2007), 'Traversability classification for UGV navigation: a comparison of patch and superpixel representations' *IEEE/RSJ International Conference on Intelligent Robots and Systems (IROS 07)*, San Diego, CA.

Kim D, Sun J, Oh S M, Rehg J M and Bobick A (2006), 'Traversability classification using unsupervised on-line visual learning for outdoor robot navigation', *IEEE Intl. Conf. on Robotics and Automation (ICRA 06)*, Orlando, FL.

Kleiner A and Kuemmerle R (2007), 'Genetic MRF model optimization for real-time victim detection in search and rescue', *International Conference on Intelligent Robots and Systems (IROS 2007)*, San Diego, California.

Labayrade R, Aubert D, Tarel J P (2002), 'Real time obstacle detection on non flat road geometry through v-disparity representation', *IEEE Intelligent Vehicles Symposium*, Versailles, France.

Langer D, Rosenblatt J and Herbert M (1994), 'A behavior-based system for off-road navigation', *IEEE Transactions on Robotics and Automation*, 10(6) 776–783.

Nakayama H, Yun Y (2006), 'Generating support vector machines using multi-objective optimization and goal programming'. *Multi-Objective Machine Learning*, 173–198.

Nourbakhsh I R, Sycara K, Koes M, Yong M, Lewis M and Burion S (2005), 'Human-robot teaming for search and rescue', *IEEE Pervasive Computing*, 4(1), 72–78.

Pirjanian P (1999), Behavior coordination mechanisms – state-of-the-art, Institute of Robotics and Intelligent Systems, School of Engineering, Univ. of South California.

Pissokas J and Malcolm C (2002), 'Experiments with sensors for urban search and rescue robots', *International Symposium on Robotics and Automation*.

Schäfer H, Proetzsch M, Berns K (2005), 'Stereo-vision-based obstacle avoidance in rough outdoor terrain', *International Symposium on Motor Control and Robotics*.

Seraji H (2003), 'New traversability indices and traversability grid for integrated sensor/map-based navigation', *Journal of Robotic Systems*, 20(3) 121–134.

Shneier M O, Shackleford W P, Hong T H, Chang T Y (2006), 'Performance evaluation of a terrain traversability learning algorithm in the DARPA LAGR program', *PerMIS, Performance Metrics for Intelligent Systems (PerMIS) Workshop*, Gaithersburg, MD, USA.

Singh S et al (2000), 'Recent progress in local and global traversability for planetary rovers', *IEEE International Conference on Robotics and Automation*.

Slama D, Garbis J and Russell P (1999), *Enterprise CORBA*, Prentice Hall.

Steuer R E (1974), 'Interval criterion weights programming: a portfolio selection example, gradient cone modification, and computational experience', *Tenth Southeastern Institute of Management Sciences Meeting*, Clemson University Press, 246–255.

Thrun S et al (2006), 'Stanley: the robot that won the Darpa grand challenge: research articles', *Journal of Robotic Systems*, 23(9) 661–692.

Trierscheid M, Pellenz J, Paulus D and Balthasar D (2008), 'Hyperspectral imaging for victim detection with rescue robots', *IEEE Int. Workshop on Safety, Security and Rescue Robotics*.

Ulrich and Nourbakhsh I (2000), 'Appearance-based obstacle detection with monocular color vision', *Proceedings of AAAI*.

Viola P and Jones M (2001), 'Robust real-time object detection', *Intl. Workshop on Statistical and Computational Theories of Vision*.

Viola P and Jones M (2004), 'Robust real-time face detection', *International Journal of Computer Vision*, 57(2), 137–154.

Visser A, Slamet B, Schmits T, González Jaime L A and Ethembabaoglu A (2007), 'Design decisions of the UvA Rescue 2007 team on the challenges of the virtual robot competition', *Fourth International Workshop on Synthetic Simulation and Robotics to Mitigate Earthquake Disaster*, Atlanta, GA.

21

Simulation using a mobile multilink robot with a virtual reality vision system

V. G. GRADETSKY, V. B. VESHNIKOV and
V. G. CHASHCHUKIN, Ishlinksky's Institute for Problems in
Mechanics of the Russian Academy of Sciences (IPMech RAS), Russia

Abstract: This chapter discusses simulation in a mobile multilink robot with a virtual reality vision system to develop distributed control for robot motions and actions in real undetermined environments such as mine fields. The chapter includes models of environment representation, a virtual vision system application for the analysis and control of multilink robot motion, and the results of the simulation for using an effective robot crawling motion over various surfaces according to specified algorithms. Taking into consideration restrictions of the motion, it is possible to identify specifics of the surfaces and the obstacles inside the space in which the robots are moving.

Key words: simulation, distributed control, mobile robot, vision system, virtual reality, undetermined environment.

21.1 Introduction

The multilink mobile robot structure is the basis for various autonomous vehicles, such as those moving over horizontal surfaces at real scale and at micro scale (Tanaka et al., 1997; Panfilov et al., 2000; Arena et al., 1999; Veshnikov & Blohnin, 1999; Gradetsky et al., 1999a), wall climbing machines (Gradetsky, 1998), walking autonomous robots (Muscato, 1998; Gradetsky, et al., 1999b; Virk et al., 1998a; Marcusek &Vitko, 1999; Waterman, 1999; Virk et al., 1998b; Muscato & Trovato, 1998), robots for motion inside tubes (Veshnikov & Blohnin, 1999; Gradetsky et al., 1996) and crawling robots (Chernousko, 2000).

The motion of the multilink mobile robot is carried out as a rule in an undetermined environment for the fulfilment of technological tasks, such as inspection or technical diagnostics, for example. The undetermined environments can be dynamically variable if obstacles or objects change their orientation and position in the space.

It is not possible to know absolutely correctly all variations of environments during every moment of a movement. So virtual reality is the instrumentation technique used to simulate various non-predicted situations for the creation of robot control algorithms.

The problem statement for this chapter is as follows: the development of a mobile multilink robot control system using the information processing of the motion data, with some virtual reality restriction for the realisation of robot

499

motion in real environments. As a rule, real environments are undetermined in full scale in advance, but using simulation in virtual reality, it is possible to predict the main peculiarities of a real situation. The restrictions can be the quality of the walls of a tube or obstacles inside the space in which the robot is moving. It is supposed that the robot receives environmental information through the vision system.

The fuzzy logic approach is used in computer simulation for the model of the environment and the robot representation. It is clear that soft computing (Panfilov et al., 2000; Muscato, 1998) of the virtual reality simulation is useful to solve the above-mentioned problem with the application of optimisation criteria for time, precision or energy. The computer language C++ was adequate for solving problems. Some examples of the robot's motion in virtual reality are listed to illustrate the obtained results.

21.2   Model of environment representation in virtual reality

It is assumed that an autonomous multilink robot is a vehicle with a concentrated mass in definite point A_1. The robot and an obstacle are moving with permanent velocity along their own trajectories, crossing at point B (Fig. 21.1), and the sensory system of the robot receives the necessary information for navigation.

The following can be seen in Fig. 21.1: V – a vehicle, moving along a trajectory 1; O – an obstacle, moving along a trajectory 2; A_1, X_1, Y_1, Z_1 – a co-ordinate system, connected with the robot, and A, X, Y, Z – a co-ordinate system, connected with an obstacle O. The robot's on-board sensory system and a distance measuring laser system work out the distance between the robot and an obstacle in every moment of time t_k in the points $k-1, k-2, k-3, \ldots k-n$. Then the robot's on-board computer calculates the trajectory of the obstacle using an interpolation method

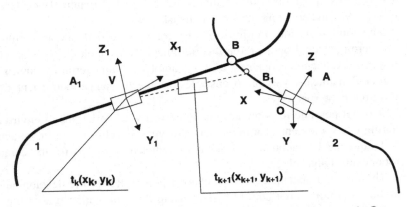

21.1 Possible trajectories of the robot V and of a moving obstacle O.

and after this is satisfied the predicted obstacle's trajectory is prolonged, using the extrapolation method.

A trajectory point extrapolation in the point $k+1$ is produced by formulas:

$$\begin{cases} Y_{k+1} = Y_k + V_{OY} \cdot \Delta t \\ X_{k+1} = X_k + V_{OX} \cdot \Delta t \cdot \gamma \end{cases}$$
[21.1]

Here γ is the coefficient of the curvature of the obstacle motion in horizontal plane in the point k of the time moment t_k.

To produce the extrapolation it is necessary to know distance l_{AB}, the velocity vector and the deviation in the horizontal plane received by means of the sensory system. Then the virtual reality computer system calculates the predicted trajectory points of the moving obstacle. If the relative velocity is known, it is possible to find the obstacle's absolute velocity under motion along the predicted trajectory.

The virtual system calculates and estimates the time of the robot and obstacle's possible meeting at point B:

$$l_{AQB(k+1)} = V_V \cdot t_k + V_V \cdot \Delta t = V_V \cdot t_k$$
$$l_{A_1B}(k+1) = V_O \cdot t_k + V_O \cdot \Delta t = V_O \cdot t_k$$
[21.2]

The meeting time in the predicted point B could be calculated as:

$$t_B = \frac{l_{AB}}{V_V}, \text{ for the robot and } t_{B_1} = \frac{l_{A_1B}}{V_O}, \text{ for the obstacle.}$$

The distance l_{AB} and l_{A_1B} in the every time-moment k are chosen on the basis of the extrapolation method from solving the triangular task (Fig. 21.2).

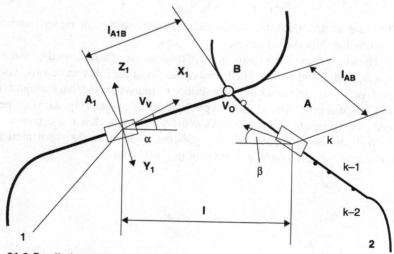

21.2 Prediction of a contrary point.

The criterion of a meeting at point B of the predicted trajectory is the expression $t_B = t_B$. The meet time is compared in every moment of time under the motion of the obstacle along the predicted trajectory in case of deviation of the obstacle motion from the extrapolatory trajectory.

The safe region calculations and the location of point B are produced using the min–max approach:

$$R_{B(safe)} = \max(\min, V_V \cdot \Delta t_B), t_B = t_s$$
$$R_{B(safe)} = \min(\max, V_V \cdot \Delta t_B), t_B = t_S$$

Expressions $V_V \cdot \Delta t_B$; $V_V \cdot t_B$ and $V_O \cdot \Delta t_B$; $V_O \cdot t_B$ are calculated at every step k.

The estimation of the mutual position of the robot and the obstacle is produced using conditions in the form of the following:

$$V_V \cdot t_B > V_O \cdot t_B \text{ or } V_V \cdot t_B < V_O \cdot t_B \text{ and then}$$

$$\begin{cases} V_V \cdot t_B - V_O \cdot t_B \geq R_{B(safe)} \\ V_O \cdot t_B - V_V \cdot t_B \geq R_{B(safe)} \end{cases} \quad\quad [21.3]$$

If these expressions are valid, it is possible to prolong the motion for the robot along the programming trajectory.

In the case of condition: $V_V \cdot t_B = V_O \cdot t_B$, is valid, so it is necessary to stop and avoid the obstacle.

In the case of condition: $V_V \cdot t_B - V_O \cdot t_B < R_{B(safe)}$, it is necessary to produce a local manoeuvre and avoid the obstacle from the left side.

In the case of condition: $V_O \cdot t_B - V_O \cdot t_B < R_{B(safe)}$, it is necessary to produce a local manoeuvre and avoid the obstacle from the right side. And so on. The acceleration on the local manoeuvre situation can be calculated as follows:

$$\dot{V} = a = \min[k_1(V_V - V_O); k_2 (l_{AB} - R_{B(safe)})].$$

The rules of actions of the various tactics depend on the safe region interaction for the robot and for the obstacles.

The kinematics of the centre mass of the robot are characterised by three angles: roll α, pitch β and yaw ψ. The lateral dynamics of the robot and control law design are based on a single-track model described by transverse and longitudinal dynamics. In the first approximation, roll and pitch angles of motion (Fig. 21.3) are neglected and it comprises front and rear links to the one fictitious link respectively.

With the above restriction, it is possible to describe the robot motion using three relevant differential equations of the first order:

$$\dot{\psi} = \frac{1}{I}[F_V l_V - F_h l_h],$$

$$\dot{\beta} = \dot{\psi} - \frac{1}{mv} [(F_i - R) \cdot \beta + F_V + F_h] \quad\quad [21.4]$$

$$\dot{\psi} = \frac{d\psi}{dt}.$$

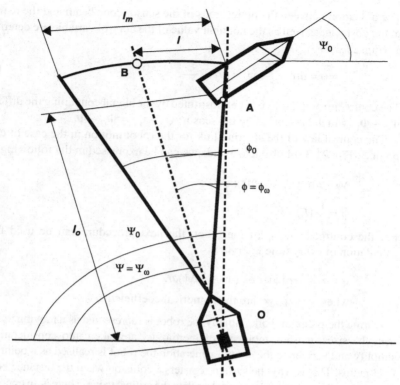

21.3 Grasping of the obstacle by robot's sensor cone beam and determination of the transverse control law.

Here $\psi, \dot{\psi}$ – yaw angle and yaw velocity, β – side slip angle, m, v – mass and velocity of a robot, I – moment of inertia, F_v, F_h – front and rear side forces, R – air resistance force.

The front side force F_v depends on the robot velocity v, the steering angled, side slip angle β and distance l_O (Mayr, 1995; Iarvis, 1992):

$$F_V = c_1 \left[\beta - \frac{l_O}{v} \dot{\psi} - \delta \right].$$

The rear side force F_h depends on the robot velocity v, the side slip angle β, distance I and ψ angle:

$$F_h = c_2 \left[\beta + \frac{I}{v} \dot{\psi} \right].$$

Coefficients c_1 and c_2 are constant. The aerodynamic resistance coefficient, atmospheric density and geometry of the robot can be estimated by the parameters c_3, ρ and S in the following way:

$$R = c_3 \frac{\rho \cdot S}{2} v^2.$$

The deviation between the optical axis of the sensor cone beam and the rear edge of the obstacle, as well as the nominal value of the control variable are determined as (Fig. 21.3):

$$\phi_v = \phi = arctg \, \frac{l}{l_0}; \, \psi_o = arctg \, \frac{l_m}{l_0}. \tag{21.5}$$

The control error $\psi_O - \psi_V$ can be substituted by the signal containing the difference $\Phi_O - \Phi_V$, which is satisfied by the sensor: $\psi_O - \psi_V = \Phi_O - \Phi_V$.

The control law of the steering link for the robot motion in the case of critical point A (Fig. 21.3) of obstacle avoidance can be expressed in the following way:

$$\Delta\phi = arctg \, \frac{l_m}{l_0} - arctg \, \frac{l}{l_0};$$
$$\dot{\psi} = \frac{1}{I}[F_V \cdot l_V - F_h \cdot l_h] \tag{21.6}$$

For the control angle ψ of the robot, the next procedure can be used for the calculation of every time interval $t_1 - t_0$:

$$\psi = \alpha_1 + \alpha_2(\phi) - \alpha_3(\dot{\psi}) - \int_{t_0}^{t}(\Delta\phi)dt,$$

where α_1, α_2, α_3 are the numerical coefficients.

During the program work, initially the robot is represented as a moving point and then the sizes, masses, moments of inertia etc. are taken into consideration. To simplify and accelerate the control generation the robot is realised as a point with a final radius. This is why the system registers a collision when the distance between the robot and any obstacle becomes less than the critical value. There is an opportunity for variation of this critical distance to imitate different sizes of robot and obstacles.

At every moment of time the robot is characterised by a set of four independent parameters: X, Y, f, V. The robot's co-ordinates are X, Y and f, V – direction and absolute value of its velocity.

The robot is to be controlled by setting two parameters: (dV/dt), (df/ds), where t is time, $s = V \cdot dt$ – the transference of the robot during the time dt.

The quantities (dV/dt), (df/ds) show the status of the control mechanisms of the real robot: (dV/dt) is an increasing function of the brake/acceleration unit, and (df/ds) is proportional to the helm divergence.

It is assumed that the robot's motion is carried out in exterior medium conditions that are uncertain beforehand. The analysis of the surrounding circumstances is fulfilled with the help of systems of different sensors. On a base of information obtained from the sensors, the system for creating the virtual reality returns a picture of the surrounding circumstances. This is represented on a monitor screen in a real-time mode to supervise the process of motion, for comparison with real circumstances and decision making by an operator to control the robot. Modification of a robot trajectory is represented in Fig. 21.4 together with the detour round a moving obstacle. Thus at the top of the figure the images show the robot motion and avoidance of the moving obstacle.

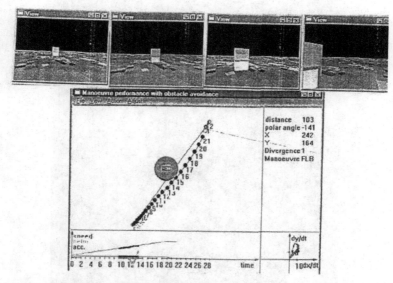

21.4 Virtual picture and trajectory of an obstacle avoidance by the robot.

In the next section the robot will be presented as a multilink system and the results of virtual reality simulation will illustrate the multilink robot behaviour in different environments, including motion inside tubes.

The initiation of the required program is indicated by the request form in the command line. The form is permitted to switch on necessary files that are the source or recipient of information. This permits the organisation of mutual information translation, information exchange and control.

21.3   Application of virtual vision system for analysis and control of multilink robot

21.3.1 Simulation system of distributed robot's control

The robot with five links was developed at the Institute of Problems in Mechanics of the Russian Academy of Sciences and is shown in Fig. 21.5. A distinctive feature of the given robot is the possibility of motion over sloping and vertical surfaces and the passage from one surface to another, if the surfaces are located at angles to each other or nearby. The robot can also overcome obstacles of a defined height and width, drive inside complicated constructions (for example, pipelines), and carry out creeping movements on a surface or inside a narrow corridor.

The robot stays on the surface of transition with the help of grippers, which are usually vacuum caps, magnets (electromagnets) or mechanical fingers, and are

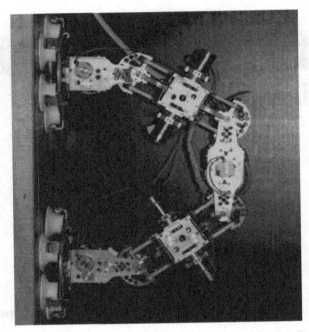

21.5 Photo of mobile multilink robot on a vertical wall.

located on the extreme links (supports). The robot is designed to carry out various technological operations in complicated mediums and conditions, which are dangerous to the health of humans. It can work in supervisory and autonomous modes on the ground, under water and in space.

The driving of the robot is fulfilled at the expense of the rotation of its separate links in hinges, which have two degrees of mobility with the help of electro motors with reducers. For example, a finite element motion can be realised with eight degrees of mobility.

This high degree of mobility allows the robot to move on surfaces with a rather complicated configuration (for example, on power installations, power reactors, the surfaces of satellites) and to enter the inside of pipelines, powering installations etc.

The construction of the robot means it is completely autonomous. Control separation of the support on a vertical surface or the robot overturning on a horizontal surface requires the creation of minimum retaining moment control. This is one of the principal conditions for robot motion at safe working capacity by means of minimal control moment generation that is necessary for the robot to stay on a surface. The minimum retaining moment has to be created using contact points on the surface of a non-moving support.

The methods of computer simulation and the virtual reality system give the possibility of investigating various modes of 'transitions' on different surfaces (horizontally, vertically, on a ceiling). The appropriate algorithms are developed for this purpose and to work out optimum conditions of loads on DC motors and on the grippers.

As was mentioned earlier, the robot can move on a surface or travel in space, with a mobile support upon which a technological tool is located, at the expense of the separate links rotation. The realisation of these rotations in a necessary sequence, and also with a defined combination or distribution, allows the realisation of rectilinear (along axes X, Y, Z or under an angle to them) motions for complicated space trajectories.

Let us consider simulation of the multilink robot motions, which was fulfilled with the use of the personal computer and programming language C++ for the following cases:

- simulation of distributed control of motion;
- simulation of distributed control of the robot moving on a vertical surface with minimum of a possible overturn moment;
- simulation of distributed control of the robot moving in a pipe;
- simulation of creeping movements on a surface inside a corridor.

21.3.2 Simulation of distributed control of multilink robot motion

Let the robot occupy an initial position I in a plane, as shown in Fig. 21.6a.

The simplest solution to transport a support in a plane from p.t_1 to p. t_5 (i.e. for realisation of one step) consists of a sequential turn of a link l_4 on an angle θ link l_3 – on an angle Y link l_2 – on an angle β and link l_1 – on an angle α. The sequence of turn angles can be changed or they can be divided into separate parts. Besides this, the turns of all links can be executed simultaneously for the reduction of the mobile support transporting time, and also distributed, if required.

Associated with this, the driving trajectories of a free support will be various. We will not detail the mathematical programs and formulas used in the robot's configuration. We will present and discuss only the outcomes of evaluations, which are usually displayed on a monitor screen for virtual representation in graphic form. The suggested method allows the creation of appropriate distributed control of drive complicated motion to satisfy set-up demands. For example, for the first simple case ('acrobat' gait), when the step is fulfilled by a sequential turn at first in the link l_4, then links l_3, l_2 and l_1, the simulation shows that it will be a curve $t_1 - t_2 - t_3 - t_4 - t_5$. For the second case, when each angle of a turn is divided into two parts, it will be a curve $s_1 - s_2 - s_3 - s_4 - s_5 - s_6 - s_7 - s_8 - s_9$ (Fig. 21.6b).

For a comparison in Fig. 21.7b the diagram of the robot motion is shown, when link 1, then 2, 3, 4 and 5 are rotated at first and after that they are turned in an inverse order. In Fig. 21.7d the simultaneous turn of all links with an identical velocity is carried out.

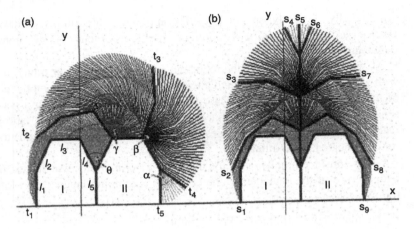

21.6 The multilink robot model.

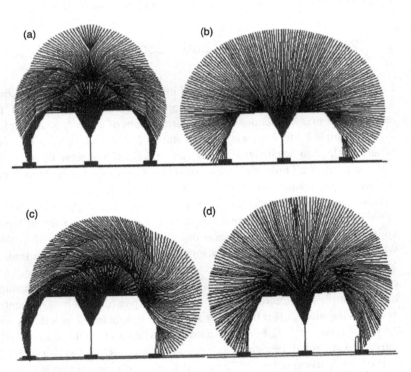

21.7 Results of the robot's motion simulation with an 'acrobat' gait.

However, from an operator's point of view, the robot's control by a sequential or distributed turns of links at the expense of switching on and off motors in supervisory mode is not an effective task. It is more effective to control when the working organ (a gripper of a support) is moved in a plane in the kernel system of co-ordinates, i.e. to the left, to the right, up, downwards or moves around a geometric centre. This motion, according to the developed algorithms, should be fulfilled automatically on command of an operator under control of the program, which is stored in the memory of the computer. An operator can supervise all changes in the robot's configuration using a virtual system at this time.

In the beginning we presented outcomes of simulation of distributed control with realisation of various rectilinear movements. It is possible to calculate a modification of a five-link robot kinematics structure for every moment of time and, accordingly, angles of links rotation, on which they are required to be turned, when the robot is fixed on a surface with the help of one support, and the other support moves in space.

The virtual vision system gives the possibility of carrying out the analysis of a series of algorithms for realisation of these movements and their practical usefulness and shortages.

For simplification we will consider the results of simulation for the robot's motion in one plane. The driving under an angle can be realised as an outcome of addition of two simple movements in appropriate planes located under an angle. Let us remember that each hinge has two degrees of mobility.

According to the first algorithm, transition of a free support (working organ) is fulfilled with the help of a minimum number of drives.

Such a transition is represented in Fig. 21.8, where a) driving along X axes; b) driving along Y axes; and c) the turn of a working organ in p. *O*. The simulation carried out shows that for transition of support with constant velocity, it is necessary to turn links with various velocities.

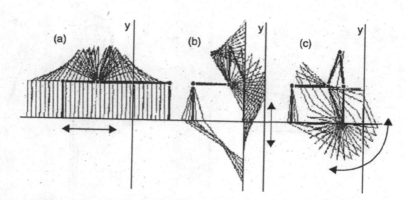

21.8 Simulation of the robot's motion with minimal number of motors.

There are defined positions or points, in which for transition of a support on the given discrete pitch, it is required to sharply change directions of a turn of separate links. As a whole, the algorithm allows the involvement of a minimum quantity of drives that reduce the energy of the supply unit. More work on this algorithm is needed to approach to check the position of outside obstacles and prevent collisions with them.

In Fig. 21.9 the results of simulation of the multilink robot's kinematics are represented for similar movements with the help of the second algorithm.

According to this algorithm the configuration of the robot is always in a maximum compress condition, i.e. the distance from the end of a free support up to a motionless support is a minimum value. This algorithm enables us to keep a value of a turn over moment in a fixing point, arising from the robot weight force, at a minimum possible. Just this moment creates a separation of a support from the surface and as a result the robot falls over. It is possible to see that the driving of a mobile support is fulfilled thus rather smoothly.

In Fig. 21.10 the results of a kinematics structure simulation are represented, when the configuration of the robot is supported always convexly in relation to a mobile and motionless supports. In comparison with the previous algorithms this mode of transition ensures the smoothest driving of a working organ. All drives work simultaneously during the robot's motion as with the previous algorithm. This mode is convenient for control in supervisory mode, when it is necessary to set various kinds of robot gaits or to fulfil driving on the given trajectories over the surface:

- Anthropoid – this is similar to the gait of a person. The action of this gait when moving is like that of a compass (thus the maximum step is achieved).
- Step by step motion – this is similarly to the movement of a caterpillar insect. It is the most simple and fastest mode of driving and involves sliding on a plane.

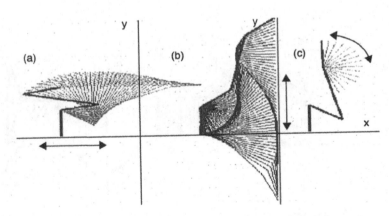

21.9 Simulation of the robot's motion with compressed configuration.

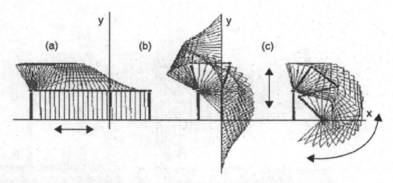

21.10 Simulation of the robot's motion with convex structure.

- 'Acrobat' mode – when the transposition of a mobile support is fulfilled in a plane above a motionless support. This mode allows the robot to overcome maximum obstacles, and also to ensure it has the minimum possible value moment of separation from a surface.

As an example, the results of simulation of the robot motion in 'step by step' mode is represented in Fig. 21.11, when the mobile support travels on a sine wave trajectory and the third algorithm of drives inclusion is used.

In Fig. 21.12 the results of simulation of the robot motion are represented, when the mobile support approaches a surface on a direct line, realises a swing in A point and moves to a surface perpendicularly and reliably fixes on it. The algorithm of motors work is the second of the above-stated.

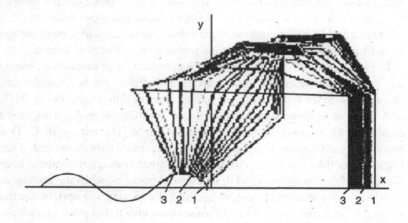

21.11 Robot's support motion on a sine trajectory.

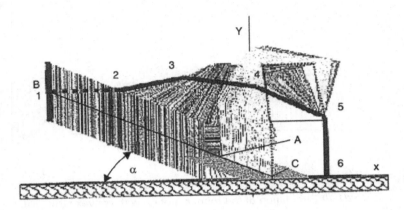

21.12 Free support goes to the surface perpendicularly.

21.3.3 Simulation of distributed control by motion of the multilink robot on a vertical surface with a minimum of possible overturn moment

We specified that with transition of the robot on a surface, it is very important to take into account forces causing a separation of it from the surface.

For a positive control, using the virtual system we investigated magnitudes of moments causing the robot to overturn or separation, when it is on a horizontal plane, vertical surface and ceiling. It is necessary to note that separation of the robot is also influenced by forces of inertia with fast turns of links. When the robot works on the ground and under water, inertia forces do not have such a large influence on its separation in comparison with weight. However, it is necessary to take into account these forces when working in the space environment.

Let us consider the simulation of the robot's movements in view of the moment from forces of weight and without the registration of forces of inertia.

In Fig. 21.13 a, b, c, d the results of a modification of overturn moment value are represented, when the robot carries out steps on a horizontal surface in 'acrobat' mode for four distributed control regimes represented in Fig. 21.7.

A moment variation is shown by a dot line (the curve A). This moment is created at the expense of the robot's full weight. The curves B, C, D and E represent modifications of moments, which are created with the weight of separate links. Under the graphs it is visible that the greatest overturn moment is created for cases of transposition b and d. In a central position, when the mobile support is strictly above motionless (a, b, d) and all links are standing upright, the moment is equal to zero (the force of weight passes through a fixing point of a support and presses the robot to the surface). For case c) the moment is equal to zero, when the links are under a defined angle from each other.

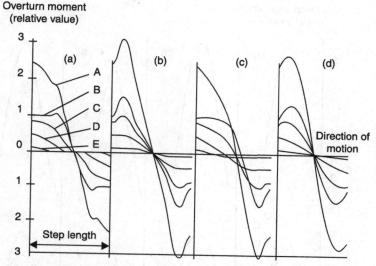

21.13 Overturn moment variation when the robot goes on a horizontal surface and a ceiling.

Reviewing this motion for a situation when the robot is on a ceiling shows that the character of a modification of an overturn moment value is similar to the above case, and the force of weight for a central disposition of the robot does not press down the robot; on the contrary, it creates a separation.

In Fig. 21.14 the graphs of a modification of an overturn moment for a case when the robot is on a vertical wall are represented. In this case an overturn moment is more likely (approximately twice).

Displaying the results of simulation of the robot's motion on various surfaces allows us to make a conclusion that the moment value depends on the configuration of the robot.

It is possible to make a preliminary conclusion that this moment decreases when the structure of the robot is in a compressed condition. With this in mind, the configuration of the robot in 'acrobat' mode, when the overturn moment is minimal, was studied.

In Fig. 21.15 the results of the robot's kinematics simulation for separate phases are represented by the realisation of a step on a vertical wall with a minimum value of overturn moment.

Under the above-mentioned conditions, the centre of the robot's mass moves up with constant velocity. The position of the robot's mass centre changes when the structure of the robot is in a compress condition. After swinging the mobile support the motion is carried out at a minimum height. Then a sequential straightening of links produced about equilibrium position occurs and the robot goes out on the final position by means of a mobile support.

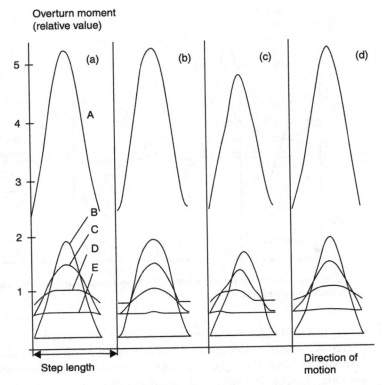

21.14 Overturn moment variation when the robot goes on a vertical surface.

21.3.4 Simulation of distributed control for multilink robot driving in a pipeline and crawling motion

The visual system can be used for purposes of investigation and control of a multilink robot during transit inside pipelines and a crawling motion over surfaces.

The model of such a robot is similar to that already considered above and represents an extended structure, which consists of several links, between which the hinges are located. The rotation of each link around the hinge can be realised with the help of an electromotor with a reduction drive.

A robot of this construction can be used in engineering purposes, for example, for diagnostics of interior surfaces of walls, deleting of outside inclusions, for delivery of equipment inside pipelines of various diameters or other purposes.

The system of virtual representation and simulation allows an operator to change values of the length of links of the robot, alter angles of a turn of each link,

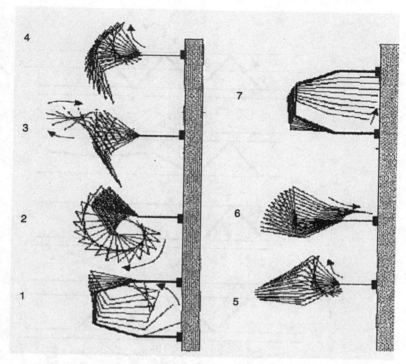

21.15 Different phases of the robot's motion on a vertical surface.

to set a configuration of the pipeline and conditions of driving (velocity of a turn in hinges), to display on a monitor screen a visual evaluation of the working zone of the robot and to accept appropriate solutions with a correction of driving by a modification of appropriate input data.

For an example of work of the given system we will consider a model of the robot consisting of five links, between which the four hinges are located. Each of the hinges has two degrees of mobility. The total number of electromotors is eight (two motors for the rotation of each link in two planes). The scheme of the mobile robot inside the pipeline is shown in Fig. 21.16a.

The movement of the robot inside the pipeline is realised by a turn of appropriate links around the hinges. The minimum number of links for realisation of longitudinal driving on the pipeline is equal to five.

One of the probable modes of driving the robot inside a pipeline involves advancing the first conducting links forwards in the direction of travel after installing the robot in a starting position. Then they are fixed at the expense of friction forces at points of contact of the hinges with walls of the pipeline. The remaining links are pulled up, with the consequent that their fixing is fulfilled.

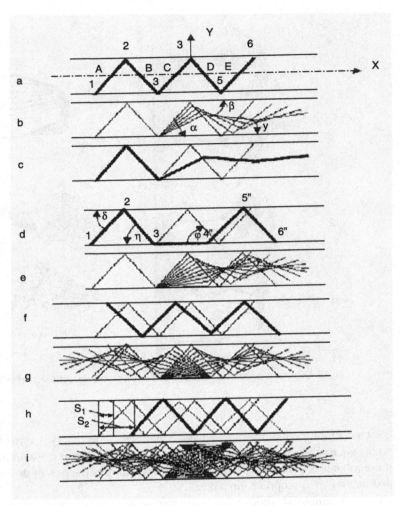

21.16 Model of multilink robot inside a pipeline and crawling motion.

We shall suppose that the robot occupies an initial position, as shown in Fig. 21.16a. The stability of its position is ensured at the expense of forces of friction, which arise in points of contact of the hinges with walls of the pipeline (points 1–6).

Let us consider the algorithm of driving the robot along axis X. For simplification of the analysis we shall consider that the robot moves by turning the links in one plane.

It is necessary for it simultaneously (or sequentially) to include those motors, which realise a turn of the fifth, fourth and third links in the direction of angles αβγδφ, as represented for calculated intermediate discrete positions in Fig. 21.16b. On this figure the results of simulation of the robot's configuration modification for discrete values of angles αβγδφ equal to five grad are represented. The system allows the analysis of driving for various ranges and absolute values of angles of a turn of each link. In Fig. 21.16c the final position of the robot (1-2-3-4ʹ-5ʹ-6ʹ) in comparison with the initial (1-2-3-4-5-6) after a turn of the third, fourth and fifth links on angles 25 grad is shown.

For a real model the moment of drive stop can be defined or calculated if the parameters of the pipeline and sizes of the robot are known, or with the help of transmitters of feed-back, which measure critical current loads in motors windings with contact of the hinges to an interior surface (wall of the pipeline).

After the realisation of full turns of the third, fourth and fifth links, the robot occupies the position represented in Fig. 21.16e. The outcomes of the mathematical simulation of the intermediate positions of links are represented in Fig. 21.16f.

After fixing the robot in this position (at the expense of contact in points 6", 5" and 4") the turn of links 1-2-3 in the direction of angles begins. The αβγδφ intermediate positions of the robot when angles are varied in a range from 0 grad αβγδφ up to 25 grad are represented in Fig. 21.16h and the configuration of the robot is compared with Fig. 21.16g. After termination of a turn the robot occupies a position in the opposite phase rather than the initial one and moves on a magnitude S1, which is determined by parameters of the robot and the diameter of the pipeline.

For completion of a full step of driving the turns of links 3-4-5, and then 3-2-1, but already in the opposite direction, are fulfilled at first. In an outcome the robot occupies the position represented in Fig. 21.16i and the length of a step thus is equal to S2. The full model of all movements of the robot for one step is shown in Fig. 21.16j.

The outcomes of simulation show that the offered model of the robot with five links and four hinges with the use of represented algorithm can move inside the rectilinear pipeline. The same character of motion can be realised for a multilink crawling robot when it moves over a plane or inclined surface. In this case the motion is limited in corridor boundaries. The working zone of the robot moving inside a pipeline or crawling over a surface is shown with the realisation of ten steps in Fig. 21.17.

From the above it is possible to make the following conclusion.

The virtual vision system allows the possibility of changing the multilink robot configuration during different modes of motion on different surfaces and to make decisions for control in optimal regimes. With the presence of obstacles on a path of motion a correction in distributed control is brought as operating single-error corrections or the appropriate restrictions. They also can be sending in memory from the processor or controller beforehand. Thus it is possible to draw the conclusion that the given system can be used effectively for display and realisation of movement inside pipelines or a crawling motion on the surface according to the proposed algorithms.

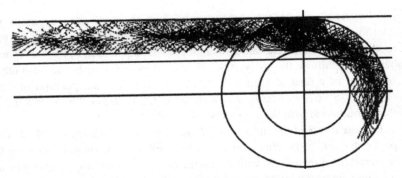

21.17 Simulation of the robot crawling motion and transition inside a pipeline.

The developed system enables the analysis of situations with transition and to accept appropriate solutions for a realisation of control.

This work was fulfilled using financial support from the Russian Foundation for Fundamental Research N 01-01-00913.

21.4 References

Arena P, Fortuna L, Branciforte M, 'Realization of a reaction-diffusion CNN algorithm for locomotion control in an Hexapode robot', *Journal of VLSI Signal Proceedings Systems*, 23, 267–280, 1999.

Chernousko F, 'Multilink robot's wave motions on horizontal surface', *Supplied Mathematics and Mechanics Journal*, 4, 518–530, 2000.

Gradetsky V, 'Wall climbing robot: evolution to intelligent autonomous vehicle', 1st International Symposium CLAWAR'98, Brussels, Belgium, 1998.

Gradetsky V, Veshnikov V, Zakharov V, Rizzotto G, Pagni A. 'Fuzzy system based on weight associated rule processor', *Proceedings of 5th IEEE International Symposium on Fuzzy Systems*, FUZZ-IEEE'96, USA, 1996.

Gradetsky V, Veshnikov V, Rizzotto G, 'Simulation of micro system control motion on the base of soft-computing approach', *Proceedings of IARP International Workshop on Micro Robots, Micro Machines and Systems*, Moscow, Russia, 1999a.

Gradetsky V, Veshnikov V, Kalinichenko S, Liapunov V, Fedorov V, 'Trajectory planning system in modular design for mobile robots', 2nd International Symposium CLAWAR'99, UK, 1999b.

Iarvis R, 'Optimal Pathways for Road Vehicle Navigation', *Proceedings of IEEE Conference*, Australia, 1992.

Marcusek J, Vitko A, 'Navigation of autonomous robots using fuzzy-neural networks', 2nd International Symposium CLAWAR'99, UK, 1999.

Mayr R, 'Lateral Vehicle Control for Automated Lane Following', *Proceedings of 2nd IFAC Conference on Intelligent Autonomous Vehicle*, Finland, 1995.

Muscato G, 'Soft computing techniques for the control of walking robots', *IEE Computing and Control Engineering Journal*, 9(4), 193–200, 1998.

Muscato G, Trovato G, 'Motion control of pneumatic climbing robot by means of a fuzzy processor', 1st International Symposium CLAWAR'98, Brussels, Belgium, 1998.

Panfilov S, Ulynov S, Kurawaki I, Ulynov V, Litvintseva L, Rizzotto G, 'Soft computing simulation design of intelligent control systems in micro-nano-robotics and mechatronics', *Soft Computing Journal*, 4(3), 147–156, 2000.

Tanaka T, Ohwi J, Litvintseva L, Yamafuji K, Ulgano S, 'Soft computing algorithms for intelligent control of a mobile robot for service use', *Soft Computing Journal*, 1(2), 89–98, 1997.

Veshnikov V, Blohnin A, 'An algorithm of a taking decision for selection of an optimal motion for a mini multilink robot', *Proceedings of IARP International Workshop on Micro Robots, Micro Machines and Systems*, Moscow, Russia, 1999.

Virk G, Kadar E, Rebour Y, 'Fuzzy Logic Based Autonomous Navigation in Unstable Diffusion Fields', 1st International Symposium CLAWAR'98, Brussels, Belgium, 1998a.

Virk G, Rebour Y, Azzi D, Ki A, Akm A. 'Fuzzy navigation for mobile robots in unstructured environments', 1st International Symposium CLAWAR'98, Brussels, Belgium, 1998b.

Waterman R, 'Industrial requirements – formulate specifications', 2nd International Symposium CLAWAR'99, UK, 1999.

22

Estimation of distance by using signal strength for localization of networked mobile sensors and actuators

J. SALES, R. MARÍN, L. NOMDEDEU and
E. CERVERA, Jaume I University, Spain

Abstract: The localization of networked mobile sensors and actuators is an active research field, and one method consists of measuring signal strength from radio transmitters. Within the context of the EU GUARDIANS (Group of Unmanned Assistant Robots Deployed In Aggregative Navigation supported by Scent detection) Project one essential aspect is having a simple localization method for the fire fighter and the mobile nodes that allows the tracking of their instant position in the space related to the base station. Several networking protocols can be useful for this purpose, using as a basis the radio signals provided by the Wi-Fi card of the mobile nodes, as well as sonar sensors. Other alternatives, such as the Zigbee connection provided by MICAz network sensors, have been studied too.

Key words: distance estimation, signal strength, networked sensors, networking protocols, Zigbee, Wi-Fi, ultrasound.

22.1 Introduction

The aim of this chapter is to show the possibilities of using signal strength for estimating distances in order to localize networked mobile sensors and actuators. Within the context of the EU GUARDIANS (Group of Unmanned Assistant Robots Deployed In Aggregative Navigation supported by Scent detection) Project one essential aspect is having a simple localization method for the fire fighter and the mobile nodes that allows tracking their instant position in the space related to the base station. The authors of this work have been studying several networking protocols that can be useful for this purpose, using as basis the radio signals provided by the Wi-Fi card of the mobile nodes, as well as sonar sensors. Moreover, as an alternative, the Zigbee connection provided by MICAz network sensors has been tested too.

First of all, in this chapter we present several state of the art methods for implementing the localization of the nodes (Li et al. 2002; Bulusu et al. 2000; Capkun et al. 2001; Niculescu and Nath 2001; Niculescu and Nath 2003; Atiya and Hager 1993; Leonard and Durrant-Whyte 1991; Tins et al. 2001). Then the chapter focuses on the experimental analysis of the radio Wi-Fi connection of the ERA-Videre mobile platform, in order to estimate the distance between two nodes belonging to the same ad-hoc network by using

520

the radio signal strength. Besides this, the estimation of the distance is experimented using xBow MICAz sensors, which use a variation of the Zigbee wireless protocol.

Results show a comparative analysis of both, the Wi-Fi ERA-Videre mobile platform and the MICAz Zigbee sensors for distance estimation using the radio signal strength.

The mobility dynamics of swarm systems present additional challenges in network communications. Several network communications protocols have recently been developed for swarm-based sensor networks (directed diffusion, geographical routing protocol, flooding protocol, etc.) (Karl 2005).

Many of those ad-hoc network protocols and applications assume knowledge of the geographic location of nodes. The absolute location of each networked node is assumed by most sensor networks, which can then present the sensed information on a geographical map (Niculescu and Nath 2001).

Finding the location without the aid of GPS of each node of an ad-hoc network is important in cases where GPS is either not accessible, or not practical to use due to power, form factor or line of sight conditions. Location would also enable routing in sufficiently isotropic large networks, without the use of large routing tables.

Several methods have recently been proposed for determining the position of a mobile node by means of measuring radio signals (measuring the time of arrival (TOA), time difference of arrival (TDOA), angle of arrival (AOA) and signal strength) (Niculescu and Nath 2003).

Mobile ad-hoc networks present a dynamic topology and overcoming routing is another problem to take into account. The research community has produced several algorithmic methods for routing (Kuhn et al. 2003, Kim et al. 2005a, 2005b; Guerrero 2006). In the case of mobile nodes applied to rescue robotics, the speed of nodes and the urgency of the operations increase the need for design routing protocols that work in a robust manner (real-time, high performance, etc.). As the loss of nodes may occur in those environments, fault-tolerance has to be taken into account in those special situations (Katevas et al. 2007).

The design of special transport protocols has to be considered for these particular mobile sensor networks, to try and avoid failure due to congestion and achieve a reasonable throughput level (Padhye et al. 1999). Those protocols need to be adapted to the field of telerobotics where the use of buffers is not feasible due to avoiding excessive time delays.

In this chapter we focus our study on the location of mobile nodes by means of radio signals and design the future integration of other types of information like those provided by ultrasound, gyroscope (compass), and odometry. We have already carried out some successful experiments in the field of teleoperation control using wireless networks (i.e. BTP-Bilateral Transport Protocol) (Wirz et al. 2008).

22.2   Available hardware for estimation of distance

The set-up for our experiments is based on the following hardware devices:

- Erratic-Videre mobile platform, equipped with an embedded personal computer (PC). The PC runs an Ubuntu 6.10 distribution with 2.6.18 kernel version. The communication capabilities of the robot are provided by a PCI-integrated WLAN card with an external antenna attached to the robot's PC (see Fig. 22.1).
- Integrated WLAN cards based on the Atheros chipset (see Fig. 22.2). For the use of this card we have used the open source drivers provided by the Madwifi project. This driver let us configure the card in ad-hoc mode, the preferred one for mobile node swarming.
- Several ultrasound sensors provided by the erratic platform. These sensors provide distance measurements through a special Player/Stage driver (Player Project). Also the platform provides odometry through the same driver.
- Wireless access point with routing capabilities (LinkSys WRT54g).

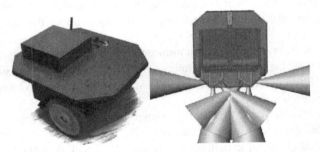

22.1 Erratic-Videre mobile platform (adapted from www.videredesign.com).

22.2 Integrated WLAN card on the robot's PC.

22.3 MICAz network sensor (CrossBow Technology) (adapted from www.xbow.com).

- MICAz network sensors provided by CrossBow Technology (see Fig. 22.3). The MICAz is a 2.4 GHz, IEEE/ZigBee 802.15.4, board used for low-power, wireless, sensor networks. This sensor can be programmed for custom sensor applications with a provided programming platform, using the TinyOS open-source operating system designed for wireless embedded sensor networks. One of the applications of the sensor nodes is to provide ZigBee capabilities to our mobile robot platforms.
- CC2431 development kit from Texas Instruments (see Fig. 22.4). The CC2431 from Texas Instruments is a SystemOnChip (SoC) with a hardware location engine targeting low-power ZigBee wireless sensor networking applications such as swarm localization and intercommunication. At the time of writing this chapter, we still have not received this kit, but we plan to test its location engine in the future.
- Hagisonic sonars (transmitters/receivers). Two different models of ultrasonic sonar (see Fig. 22.5) will be used in the future for implementing Time Difference of Arrival (TDoA) in order to derive position information. These sensors (HG-M40DAI and HG-M40DAII) are ultrasonic object detectors and range finders that offer very short- to medium-range detection. The AII model offers a special narrow directional response, 25–30 degrees in vertical direction to minimize the reflection of unwanted sound waves.

22.4 CC2431 Development kit (Texas Instruments) (adapted from www.ti.com).

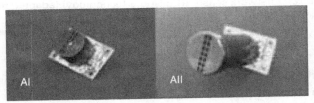

22.5 Hagisonic AniBat ultrasound tx/rx (adapted from www.hagisonic.com).

22.3   Experimental setup

A preliminary experiment has been developed in order to test the capabilities of the integrated Wi-Fi network of the robotic platform in a real indoor environment. The first experiment consisted of creating an infrastructure network involving the wireless access point and one robotic platform using its integrated Wi-Fi network card.

In Fig. 22.6 we can see the fixed location of the access point and the trajectory followed by the robot. The plan shows a 25 by 2.2 meters corridor of our university.

Measuring the received signal strength from the point of view of the robot has been done using standard wireless tools for Linux (wireless) and driver capabilities.

In Fig. 22.7 we can observe the path followed by the robot along the corridor. For each (x,y) position we can see in the bottom graph the received signal strength (RSS) for this point. As can be seen, the signal decreases when the robot gets

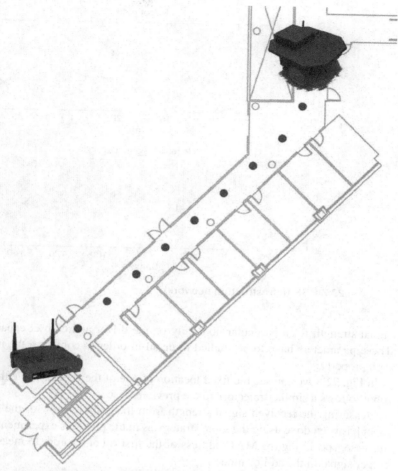

22.6 Robot trajectory (infrastructure network).

further away from the access point until the connection is lost at a distance of about 34 meters. Also a significant fluctuation of the signal can be observed along the trajectory, which can represent a drawback in being able to infer positioning of the robot.

Another similar experiment has been done to compare the results obtained in ad-hoc mode. In this case, two robotic platforms using the integrated Wi-Fi network card have been tested.

The set-up for this special mode of operation was done through the special command wlanconfig, which is a special tool provided with the driver. The possibility of changing to this mode of operation, the selection of the transmission channel, the control of the transmission power and the monitoring of the received

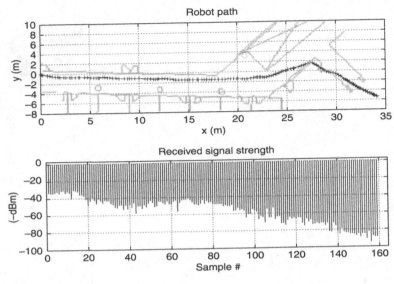

22.7 RSS (infrastructure network).

signal strength for a particular node rely on the driver and chipset capabilities. These parameters have to be studied in detail in order to configure the network card properly.

In Fig. 22.8 we can see the fixed location of one of the robots, while the other robot follows a similar trajectory to the previous case.

Measuring the received signal strength from the point of view of the moving robot has been done using the same strategy as in the previous experiment but, in this case, specifying the MAC address of the first robot in order to measure the proper signal in the ad-hoc mode.

In Fig. 22.9 we can observe the path followed by the robot along the corridor. For each (x,y) position we can see in the bottom graph the received signal strength (RSS) for this point. Like in the previous case, the signal decreases as the robot gets further away until the connection is lost at a distance of about 27 meters.

The strength of the received signal is weak in general due to this particular ad-hoc mode of operation. Like in the previous experiment, a significant fluctuation of the signal can be observed along the trajectory, which can be a problem when inferring the positioning of the robot.

22.4   Wi-fi–ZigBee experiment comparison

In order to test the feasibility of ZigBee wireless communications for our particular environment, several experiments have been done using the MICAz network sensor provided by CrossBow.

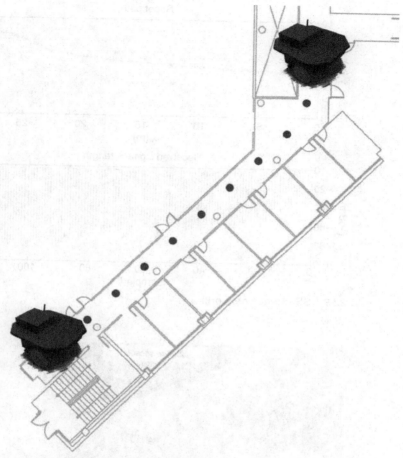

22.8 Robot trajectory (ad-hoc network).

For this purpose, two of our mobile robots have been equipped with these network sensors. For measurement purposes a monitoring software tool provided by Crossbow has been installed in a laptop mounted in the moving robot (see Fig. 22.10).

The measurements have been carried out in the same environment as in the previous experiments (see Fig. 22.8). In this case, the received signal strength has been measured both for Wi-Fi and ZigBee wireless communications. Several measures from the sensors have been taken, progressively increasing the distance from the emitter to the receiver in steps of 1 meter. For each value of distance, several values have been sampled.

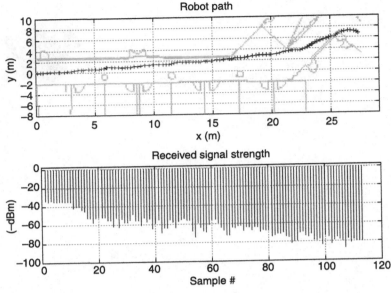

22.9 RSS (ad-hoc network).

22.10 Robot platforms equipped with MICAz ZigBee network sensors.

In order to find out if there is some influence in the received signal strength when a hidden visibility condition is included, the same measurements have been done with a human body between emitter and receiver. This scenario tries to emulate the real situation where a fire fighter is being assisted by two robots, one of them in front, and the other behind (see Fig. 22.11).

22.11 Fire fighter between robots.

22.5   Experimental results

The comparison of both Wi-Fi and ZigBee communication systems and the presence or not of a human body between robots can be observed in Fig. 22.12.

As can be seen, the Wi-Fi experiment provides higher levels of signal due to the fact that this standard allows bigger output power on transmitters. This fact makes a large distance between nodes possible (32 meters in the case of Wi-Fi versus 24 meters in the case of ZigBee).

If we take into account the effect of a hidden visibility condition (i.e. fire fighter between robots), a slight attenuation of the signal can be observed due to this fact.

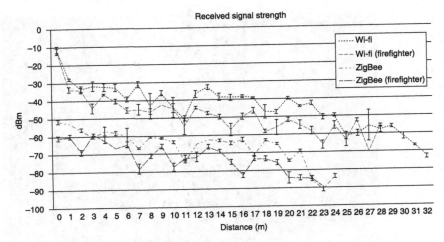

22.12 Comparative results.

The fluctuation of the signal appears to be similar in both Wi-Fi and ZigBee experiments. As stated above, this has to be taken into account when estimating positions and distances between nodes.

22.6   Conclusions

Localization of networked mobile sensors and actuators is an active research field. One traditional approach in order to localize a mobile node has been the use of laser sensors. In some special circumstances, like a smoky environment, the use of this type of optical sensor is not a good solution. The use of radio and ultrasound signals can be a feasible alternative to node localization. Several experiments using robots and several radio transmitters have been performed in order to enhance the system perception, by estimating the robot's localization using Wi-Fi/ZigBee combination. At the time of writing, we have still not carried out any experiments with the CC2431 development kit. The kit's documentation illustrates that it would be possible to get localization values with a maximum error of 3 meters, using its incorporated localization engine. In fact, we assume that a suitable solution for operating in smoke environments would be combining ultrasounds, gyroscope (compass), Wi-Fi/ ZigBee localization, and odometry prediction.

22.7   References

Atiya S. and Hager G. (1993). "Real-time vision-based robot localization." *IEEE Transactions on Robotics and Automation* 9(6): 785–800.
Bulusu N., Heidemann J. and Estrin D. (2000). "GPS-less low cost outdoor localization for very small devices." *Technical Report 00–729*, Computer Science Department, University of Southern California.

Capkun S., Hamdi M. and Hubaux J. (2001). "GPS-free positioning in mobile ad-hoc networks." *Proceedings of the 34th Annual Hawaii International Conference on Systems Science (HICSS-34)*, Volume 9, page 9008, Maui, Hawaii.

Guerrero M. (2006). "Securing and enhancing routing protocols for mobile ad hoc networks." PhD thesis, University Politecnique of Barcelona.

Karl H. and Willig A. (2005). *Protocols and Architectures for Wireless Sensor Networks.* John Wiley and Sons.

Katevas N., Pantelouka A. et al. (2007). "Test environment for VSN routing algorithms using mobile robot." *49th International Symposium ELMAR-2007*, Zadar, Croatia.

Kim Y., Govindan R. et al. (2005a). "Geographic routing made practical." *Proceedings of NSDI.*

Kim Y., Govindan R. et al. (2005b). "On the pitfalls of geographic routing." *Proceedings of the 3rd International Workshop on Discrete Algorithms and Methods for Mobile Computing and Communications (DIALM) – Principles of Mobile Computing.*

Kuhn F., Wattenhofer R. and Zollinger A. (2003). "Worst-case optimal and average-case efficient geometric ad-hoc routing." *Proceedings of 4th ACM International Symposium on Mobile Ad Hoc Networking and Computing (MobiHoc 2003).*

Leonard J. and Durrant-Whyte H. (1991). "Mobile robot localization by tracking geometric beacons." *IEEE Transactions on Robotics and Automation*, 7(3): 376–382.

Li D., Wong K. et al. (2002). "Detection, classification, and tracking of targets." *IEEE Signal Processing Magazine* 19(2): 17–30.

Madwifi Project. "Multiband atheros driver for wireless fidelity." (http://madwifi-project. org).

Niculescu D. and Nath B. (2001). "Ad hoc positioning system (APS)." *Proceedings of IEEE GLOBECOM 2001*, pp. 2926–2931, San Antonio, Texas.

Niculescu d. and Nath B. (2003). "Ad hoc positioning system using AoA." *Proceedings of INFOCOM 2003 Twenty-Second Annual Joint Conference of the IEEE Computer and Communications Societies*, pp. 1734–1743, vol. 3.

Padhye J., Kurose J. et al. (1999). "A model based TCPfriendly rate control protocol." *Proc. NOOSDAV*, pp. 137–151.

Player Project. "The player project", http://playerstage.sourceforge.net.

Tins R., Navarro-Serment L. and Paredis C. (2001). "Fault tolerant localization for teams of distribuited robots." *Proceedings of the IEEE/RSJ International Conference on Intelligent Robots and Systems*, 2: 1061–1066, Maui, Hawaii.

Wireless Tools for Linux, http://www.hpl.hp.com/personal/Jean_Tourrilhes/Linux/Tools. html.

Wirz R., Marín R. et al. (2008). "End-to-end congestion control protocols for remote programming of robots using heterogeneous networks: a comparative analysis." *Robotics and Autonomous Systems Journal* 56: 865–874.

Part V
Multi robotics systems: Navigation and cooperation

23

Experimental study on the effects of communication on cooperative search in complex environments

Ö. ÇAYIRPUNAR, V. GAZI, B. TAVLI, TOBB University of Economics and Technology, Turkey, E. CERVERA, Jaume I University, Spain, U. WITHOWSKI, University of Paderborn, Germany and J. PENDERS, Sheffield Hallam University, UK

Abstract: In this study we investigate the benefits of networked communication by experimentally evaluating the results of two search algorithms, which are spiral search and informed random search. Both simulations and real experiments are performed in order to obtain objective results. The robotic experiments were performed in an experimental area containing obstacles where the communication ranges were "simulated" with the help of an overhead camera. Each robot was allowed to keep an occupancy grid based local map of the environment also containing information about the cells it has visited and to exchange this information with the other robots within its communication range. The effect of the size of communication range on the performance of the system defined as the time of completion the search task (i.e. locating the target) was investigated.

Key words: multi-robot teams, communication network, cooperative robotic search, transmission range.

23.1 Introduction

Search and rescue operations have great importance after disaster situations like earthquakes or terrorist attacks. In such disaster relief missions, search and exploration are the initial steps of a larger operation. Traditionally these missions have been performed by human teams, however, there are intensive ongoing research efforts for developing multi-robot search teams to be deployed in such missions. Rescue robotics, or basically the use of autonomous robots in search and rescue operations, is a relatively new field of research. It is a part of the broader field of coordination of a group of mobile robots to achieve a specific objective/goal. In order to achieve cooperative behavior there is a need for effective (direct or indirect) communication methodologies. The use of a network architecture is one possible form of direct communication and will be essential in many applications that require information exchange between the robotic agents in a team and the team and human operators. In particular, in search and rescue scenarios, by combining the communication

535

network with an appropriate search algorithm, an effective search can be achieved by the robots.

Recent technological advances in control theory, electronics, electromechanical systems, and communication/networking technologies are paving the way for the development and deployment of large cooperating robot groups (swarms) (Dollarhide and Agah, 2003). The deployment of groups of relatively simple mobile robots has several advantages over a single complex (advanced) robot. These advantages include robustness to failures (the group may still be able to perform the job in case of loss/failure of one or more robots while in the case of a single robot the job will be aborted, moreover simple agents are less prone to bugs or failures compared to complex agents), flexibility (the group can re-organize/self-organize based on the situation or objective), scalability (based on the objective or task, a different number of agents can be deployed), and cost (simplicity leads to a decrease in the cost of the overall system). Moreover, autonomous robots can assist humans in risky operations during search and/or rescue missions. Furthermore, robots can have the ability to work in environments which are dangerous to humans such as collapsed or unstable buildings, in fire or gas leakages, environments with a high nuclear radiation concentration, deep under the sea, etc. Therefore, the deployment of systems of multiple cooperating robots will have great potential in search and rescue operations in the near future.

It is obvious that it is difficult or even impossible to have global information and implement centralized controllers in systems consisting of a large number of agents with limited capabilities. Therefore, recent research has concentrated on decentralized approaches. In such systems, the inter-agent communication and networking algorithms are of paramount importance. In other words, for development of effective, practical multi-robot systems, besides the need for development and verification of effective coordination and control strategies, there is a need for development and verification of robust and scalable communication and networking algorithms and protocols.

In the context of multi-robot systems the definition of communication can be made as the transfer of meaningful information between one agent and another (or the human operator). This definition is very broad and can include all kinds of communication such as information obtained from the sensors (e.g., the position of a significant object), information about the robot itself (e.g., its movements), commands or task/service request messages, etc. A more specific and narrower definition, which includes some form of intentionality, can be stated as "The intentional transfer of meaningful information between robotic agents" (Cao et al., 1997).

Communication/networking can enhance the performance of multi-robot systems from several aspects (Balch and Arkin, 1994). First of all, in the case of the group of robots, they have to fulfill a specific goal and the coordination between different agents becomes unavoidable. For example, consider a mission that includes moving a large (and possibly fragile) object by a multi-robot team.

Without communication and coordination the robots may try to push the object in different directions, which can result in undesired consequences. Second, with communication the robots can exchange valuable information and significantly improve the performance of the system. For example, in heterogeneous multi-robot teams, sensory information inquired by a robot with a specific sensor could be exchanged with other robots that do not possess this sensory set-up. Similarly, a robot not able to perform a specific task can request that service from another robot that does have that capability (a concept called service discovery). Furthermore, different tasks (or objects) can be allocated to different agents thus achieving parallel (and therefore more efficient) operation.

A group of mobile communicating robots constitutes, by its nature, a wireless ad-hoc network. In such a system there are many issues to be resolved for effective operation. First of all, since the agents will be simple, their communication capabilities (such as range, power, processing capability, etc) will also be limited. Therefore, in a case where two agents that need to communicate are out of range, they will probably need to communicate through other intermediate agents. Therefore, besides the need for the development of appropriate message structures and communication protocols, there is a need for the development of effective/ cooperative routing/networking protocols as well. A recent survey on the main issues in mobile sensor networks can be found in Akyildiz et al. (2002).

Performance of a distributed robotic system using shared communication channels is presented in Rybski et al. (2002). It is shown that for surveillance applications it is extremely important to coordinate the robots through wireless communication channels. Yet, the performance of the system is affected by the capacity of the links and the number of robots sharing the links. Rybski et al. (2004) reported that adding simple communication capabilities to robots improves the predictability of the task completion times. In Rekleitis et al. (2004) a multi-robot coverage study is presented. It is shown that by allowing robots to communicate among wireless links better algorithms for the complete coverage problem can be obtained. In Trianni et al. (2004) it can be seen through simulations that the use of direct communication (through wireless links) can be beneficial for the effectiveness of the group behavior in performing collaborative tasks.

Communication in multi-robot systems can be classified as explicit or implicit communication. Implicit communication (sometimes also called stigmergy) is communicating through the environment. In other words, if the actions taken on (or modifications made to) the environment by one agent lead to the change of behavior of the agents (the other agents and the agent itself), this is a type of implicit communication. Simply stated, in implicit communication, changes in the environment may represent some useful information. By contrast, explicit communication is the type of communication in which the robots directly pass messages to each other and/or to the human operator. Arkin (1992) has established that for certain classes of tasks, explicit communication is not a prerequisite for cooperation.

We can also divide communication in multi-robot systems into global communication and local communication. Global communication is the situation in which every agent can communicate with every other agent, whereas local communication describes the situation in which each robot can communicate only with its local neighbors. In previous studies (Yoshida et al., 1996), the efficiency of global and local communication in mobile robot systems is evaluated based on the analysis of information transmission time and probabilistic methods. However in this study the performance of a cooperative task with multiple mobile robots is studied and the effect of communication on cooperation is directly measured for different communication ranges starting from no communication to global communication.

Global communication is effective for a small number of robots in a limited area. However, when the number of robots or the size of the search space increases, this becomes difficult to achieve because of the limited communication capacity and increasing volume of communication to handle. Therefore, it is logical to choose local communication. Let us suppose that each robot has the ability to adjust its range of communication. If it is too large, the efficiency of information transmission decreases because the communication traffic becomes too congested and the robots cannot handle that traffic (Fig. 23.1a). On the other hand, the efficiency is low if the output range is too small as well (Fig. 23.1b). In addition, the selection of the communication range affects the power consumption, which is very important for a mobile robot. A higher communication range requires more power and as a result consumes the battery much faster. It is therefore essential to develop methodologies for a decision on the communication range in order to provide efficient information transmission between the agents.

One may think that louder is always better; that is, the wider a robot's communication range, the better the performance. However, this is not always the case. For example, in a simulated cooperative foraging task (Arkin, 1998) using homogenous robots, it was demonstrated that social performance can decrease substantially with increases in a robot's communication radius. The trade-off is that a call for help that is too weak prevents an agent from being heard, but too strong a call brings an entire colony together and prevents effective exploration of the environment. Loudest is indeed not best for all tasks.

A probabilistic approach to determine the optimal communication range for multi-robot teams under different conditions is presented in Yoshida et al. (1995). In that study the optimal communication area is estimated by using "information transmission probability", which represents the possibility of successful transmission. This range is determined by minimizing the communication delay time between robots, using probability of successful information transmission, assuming they are moving randomly. Equation [23.1] (which is taken from Yoshida et al., 1995) shows the relationship, where c is the information acquisition capacity, an integer representing the upper limit on the number of robots that can be received at any time without loss of information, and p is the probability of information output for each robot.

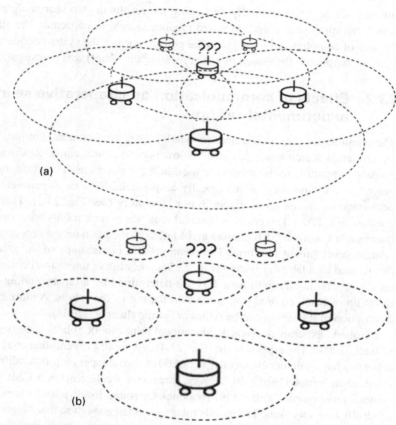

23.1 The effect of the size of communication: (a) long communication range; (b) short communication range.

$$X_{opt} = \sqrt[c]{\frac{c!}{p^c}} = \frac{\sqrt[c]{c!}}{p} \tag{23.1}$$

In this chapter, cooperative search by a team of mobile robots using communication to pass information between each other is considered. Firstly simulation based experiments are performed in a Matlab environment. Then the same experiments are repeated using real robots on an experimental set-up. The robotic experiments were performed in an experimental area containing obstacles and using e-puck robots where the communication ranges were "simulated" with the help of an overhead camera. Each robot was allowed to keep an occupancy grid-based local map of the environment also containing information about the cells it has visited and was able to exchange this information with the other robots within its communication range. Consequently, the effects of the communication range in networked communication in a multi-robot cooperative search scenario was investigated by experimentally evaluating the results of two search algorithms, which are spiral search and informed random search. In particular, the effect of the size of communication range on the performance of the system defined as the time of completion the search task (i.e, locating the target) was investigated.

23.2   Effects of communication on cooperative search: experimental set-up

The simulations are performed in Matlab. An environment similar to the experimental search space described below is created artificially. Realistic robot models compatible to the robots are used in the simulations. Therefore the robot behaviors are simulated as realistically as possible. Robotic experiments were performed in the set-up available in our laboratory (see Fig. 23.2). This set-up consists of a 120×180 cm experimental area, six e-puck robots with Bluetooth interface, a Logitech USB camera and Matlab as the main image processing and control development platform. The positions and orientations of the robots are determined by a labeling system (Fig. 23.3) consisting of three small colored dots on the robots. In addition, their IDs are determined by a binary coding system made up of black colored dots placed on the top of the robots. A more detailed description of the set-up can be found in Samiloglu et al. (2008).

In robotic experiments e-puck educational mini mobile robots have been used (e-puck, 2009). The e-puck robot (Fig. 23.4) is a small (7.0 cm diameter) mobile robot that has a microcontroller dsPIC30F6014, two stepper motors for differential drive, eight infrared proximity sensors, three axis accelerometer, a CMOS color camera, three omnidirectional microphones for sound localization, a speaker and some other sensory units. The mobile robots are small enough so that a high number of robots may be utilized simultaneously in experiments. This makes them very suitable for swarming experiments. They have Bluetooth wireless communication modules, which we have used as the medium for information exchange in our experiments. Also e-pucks can communicate with IR in small distances up to 25cm.

In order to change the transmission range, a device capable of adjusting its RF output power is necessary. The change on the RF signal output power can be matched to a proportional communication distance. The Bluetooth hardware available in the robots is a class II Bluetooth. The device has a constant RF output power and therefore cannot provide any changeable communication range. Since a communication device capable of adjusting its transmission range was not available at the time of the experiments we had to develop some other methods to simulate this feature.

To model or simulate a realistic RF wireless communication requires comprehensive work. In addition, the RF signals have some impairment such as attenuation, distortion, noise, distraction, and multipath refraction, which may vary according to the environment and can be very hard to simulate. Although it is a fact that wireless communication cannot be simply represented as an exact

23.2 Experimental set-up consisting of an arena, robots, PC and overhead camera.

distance, it is dependent on the environment and, therefore, varies relative to the changes in the environment. For example, with the obstacles around, the signal strength will be too weak to support any communication.

This chapter mainly focuses on investigating the effects of communication range on the cooperative multi robot search task. In this study we did not consider all the RF impairments and basically focused on and simulated the communication based on the disc communication model. The disc communication model assumes that the communication takes place in a circular area with a constant diameter.

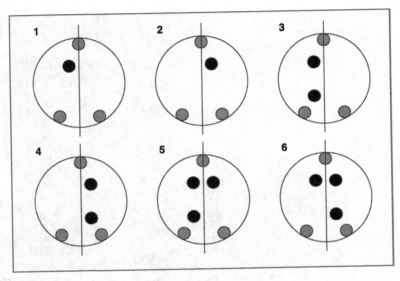

23.3 The robot labeling system for six robots.

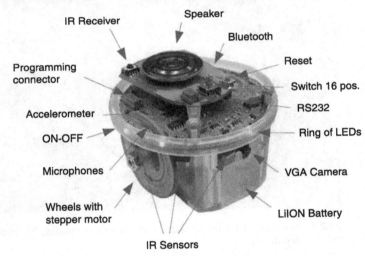

23.4 e-puck educational mini mobile robot.

Although it is a fact that in reality wireless communication cannot be simply represented as an exact distance and it is dependent on the environment. Disc communication models have been widely used in the literature since they are easy to simulate and analyze, and that is the approach we consider here as well.

The robots set their motor speeds according to the commands supplied by the computer via the Bluetooth interface. In other words, the control algorithm running on the main computer, which is based on the search strategy, decides which cell to be visited next. The robot movements are controlled by artificial potential functions. An artificial potential is binded to the target cell to be visited next and a force is applied to the negative direction to the gradient of the potential field. Then, that force is converted to the control outputs as linear and steering angle speeds to be transferred to the robots. All of the nodes of the search space are visited sequentially in that manner. Another option is to program the robots so that they receive their global position (and/or possibly the relative positions of their neighbors or all the other robots) and have their own internal decision making and control. However, we have not implemented such a strategy in this study (since conceptually it does not make much difference). Beside the higher-level control by the computer, the robots have an obstacle avoidance behavior running at low-level. In other words, the robot movements are controlled with a weighted sum of the control inputs obtained from the computer and the sensorial information collected from the environment. However, the obstacle avoidance has a higher priority to make sure that the robots do not collide with any obstacles accidentally.

A high quality USB overhead camera, which is directly connected to the computer, is used in the experiment. A resolution of 640×480 is sufficient for this set-up considering the sizes. The frame rate is not the main criteria in the selection of the camera since the image-processing unit cannot process more than 5–6 frames per second.

As was mentioned above, the frames of the arena are grabbed and processed to determine the position, orientation, and identification of the robots. This information set is supplied to the function running behavior algorithms of agents, which output the control inputs (the angular and translational speeds) to the robots. The resulting angular and translational speeds of the agents are transferred to the agents via Bluetooth communication modules. The main delay in the system occurs due to the image processing. As mentioned before, another control option could be to pass the position and orientation information to the robots and let their internal algorithm calculate the values of the control inputs which would better model more decentralized and realistic applications. The refresh rate would not be a problem since a robot can fill the gaps between the position updates with its internal odometry, thus, a continuous position estimation can be provided. This can be thought of as similar to the simultaneous use of global positioning system (GPS) and inertial measurement unit (IMU) for continuous localization of unmanned air vehicles (UAV) or unmanned ground vehicles (UGV). However, here the emphasis is to concentrate on the effect of communication ranges on the search performance of a robot team and not to deal with issues such as localization.

23.3 Effects of communication on cooperative search: problem definition

The experiment scenario is basically a search of a predefined object in a complex environment including walls and some obstacles (Fig. 23.5). The search is started individually by the robots from different locations. Robots perform the search by following a random path or a predefined path based on the environment. During the search, when the robots encounter each other (i.e., when they enter each other's communication range) they share their search information database. This concept is demonstrated in Fig. 23.5. In Fig. 23.5a robot 1 is performing the search by following the search path, which is generated by a search algorithm. The information bubble on top of robot 1 shows the explored areas in the memory of this robot, which is simply the occupancy grid map of the previously searched areas. In Fig. 23.5b a second robot joins the search from a different location. The arrows show the process of successful information sharing between robots 1 and 2. After the exchange of the search database the newly formed search maps are demonstrated in the information bubbles. Similarly, in Fig. 23.5c the cooperation of robots 2 and 3 and the resulting search maps are demonstrated. Finally, in Fig. 23.5c the communication takes place between both robots 1–2 and robots 2–3. Therefore, robots 1 and 3 communicate indirectly through robot 2 and the search maps of all of the robots are combined, which will make the continuing search more efficient (i.e., the robots will not search places which were previously searched by the other robots).

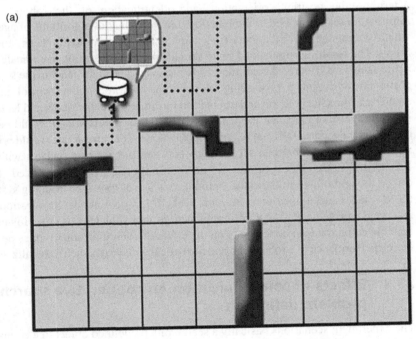

23.5 (a)–(c) Concept of cooperative search by communication.

(b)

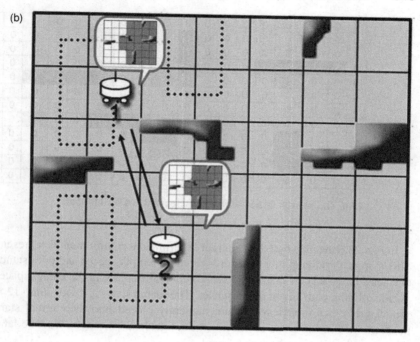

(c)

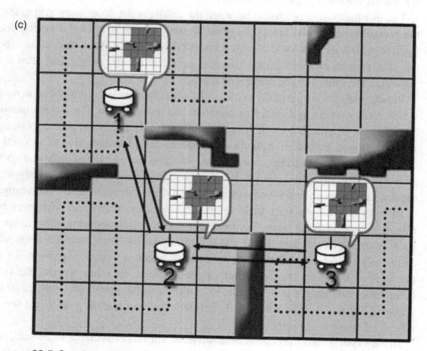

23.5 Continued

0	0	0	0	0	0	0	0	0	0	0	0	0	0	0	0	0	0
0	0	0	0	0	0	0	0	1	0	0	0	R4	0	0	0	0	0
0	0	0	0	0	0	0	0	1	0	0	0	0	0	0	0	0	0
0	1	1	1	1	0	0	0	1	0	0	0	0	0	0	0	0	0
0	0	0	R2	0	0	0	0	1	R1	0	0	1	1	1	1	0	0
0	0	0	0	0	0	0	0	0	0	0	0	0	0	0	0	0	0
0	0	0	0	0	0	0	0	0	0	0	0	0	0	0	0	0	0
0	0	0	0	0	0	T	0	0	0	1	1	1	0	0	R6	0	0
0	0	0	1	0	0	0	0	0	1	1	0	0	0	0	0	0	0
0	0	0	1	0	0	0	0	0	R3	0	0	0	0	0	0	0	0
0	0	R5	1	0	0	0	0	0	0	0	0	0	0	0	0	0	0
0	0	0	0	0	0	0	0	0	0	0	0	0	0	0	0	0	0

23.6 The search space with obstacles, robots and target.

In Fig. 23.6 the map based on our real experimental environment is represented. This map is a grid map in which zeros represent empty spaces and ones stand for obstacles in the search space. The search only takes place in the empty spaces to avoid colliding with any of the obstacles. The search space is divided into 12×18 virtual grids. Six e-puck robots are randomly placed into their initial starting positions within the arena as shown in the figure (*R1* through *R6*). The label *T* represents the object that is to be found.

The information to be shared between the robots is the occupancy grid maps of the previously searched places. In other words, it is the map of the visited cells. Each robot has its own local map of these occupancy cells. At each step the robots use that map in order to decide the next cell to visit and to prevent or at least minimize the search of the same area multiple times.

Robots share their map of the visited cells when they are in communication range, which is the maximum distance of possible data transmission. The communication distance can be changed by filtering the data transmission between the robots. As was mentioned before, in robotic experiments an overhead camera is used to calculate the robot positions, orientations and IDs. Based on that information the inter robot distances are calculated and the communication takes place only when the distance between two robots is smaller than the maximum transmission range.

Initially the robots start their search individually. However, whenever two robots encounter each other, i.e. two robots enter the communication range of each other, they exchange their local occupancy maps. The communication sizes of the robots and whether they are within that range or not are determined from the images taken by the overhead camera system. In other words, the robot communication ranges are "simulated" through the experimental set-up. In this way one can easily experiment with different communication ranges and see the effects of communication.

In the experiments the communication sizes varied between 0 and 200 cm, with steps of 20 cm. A communication range of 0 means no communication, implying that the robots search individually without cooperation. In contrast, a communication range of 200 means global communication in which each agent can communicate with every other agent.

In the following section the search strategies used in the experiments will be described in more detail.

23.4   Effects of communication on cooperative search: search strategies

Two different types of search strategies are used in the experiments. The first one is a spiral search, which uses distance transform to calculate an exploration path, and the other is informed random search, which is a simply random search having the memory of previously searched places.

23.4.1 Spiral search

We have used an altered version of spiral search as a complete search and coverage algorithm (Zelinsky et al., 1993), which is mainly focused on the search of the nearest grids first. In that search the robot sweeps all areas of free space in an environment in a systematic and efficient manner. For this reason the map of the search space should be known previous to the experiments.

To achieve the complete coverage behavior the robot follows a path that moves away from a starting point, keeping track of the cells it has visited. In other words, the robot only moves into a grid cell, which is closer to the current cell if it has visited all the neighboring cells that lie further away from the current cell. In order to do this, the search algorithm first calculates the distance transform of all the cells with respect to the starting point then generates a path for complete coverage. In Fig. 23.7 the results of the distance transform applied to robot *R1* are shown.

9	8	7	6	6	6	6	5	4	4	4	4	4	4	5	6	7	8
9	8	7	6	5	5	5	5		3	3	3	3	4	5	6	7	8
9	8	7	6	5	4	4	4		2	2	2	3	4	5	6	7	8
9				4	3	3		1	1	2	3	4	5	6	7	8	
9	8	7	6	5	4	3	2	R1	1	2					7	8	
9	8	7	6	5	4	3	2	1	1	1	2	3	4	5	6	7	8
9	8	7	6	5	4	3	2	2	2	2	2	3	4	5	6	7	8
9	8	7	6	5	4	3	3	3	3			4	5	6	7	8	
9	8	7		5	4	4	4	4		6	5	5	5	6	7	8	
9	8	8		5	5	5	5	5	5	6	6	6	6	6	6	7	8
9	9	9		6	6	6	6	6	6	6	7	7	7	7	7	7	8
10	10	10	8	7	7	7	7	7	7	7	7	8	8	8	8	8	8

23.7 Distance transform applied to robot number 1.

23.8 The complete coverage and exploration path generated for robot number 1.

Additionally, in Fig. 23.8 the complete coverage and exploration path generated for robot *R1* is presented.

Closer observation of the above-described path of complete coverage shows that the path of complete coverage produces too many turns. This is because the coverage path follows the "spiral" of the distance transform wave front that radiated from the start point. As a result the search can take longer than expected. In certain configurations of obstacles in an environment this can produce unsatisfactory performance. Therefore, complete coverage paths of the type shown in Fig. 23.8 are somewhat difficult to execute on a mobile robot. To overcome such undesirable results in our experiments the path is checked with a secondary algorithm, which looks for dead ends and handles them by changing the path to the nearest unsearched areas.

23.4.2 Random search

The second search algorithm is a type of random search in which the robots move in the search space randomly. However, robots keep a memory of the previously searched spaces. With this information the robots randomly select their next destination cell from the unvisited cells in the near vicinity. Every grid on the search map is connected to eight other cells. Therefore, the algorithm randomly chooses the next target from those neighboring eight cells. In Fig. 23.8 the exploration path generated by this algorithm is demonstrated. In addition, while exploring, information about the previously visited cells is kept on an occupancy grid map. For the later steps the algorithm takes into account the visited cells while randomly choosing the next target. Therefore, the search becomes an informed random search. To overcome unwelcome results such as in the case where all of the nearby cells have been visited, the random algorithm looks for previous cells and tries to find unsearched areas, and then selects those empty places for new target destinations.

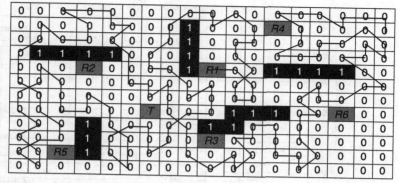

23.9 The exploration path for robot 1 generated by informed random search.

23.5 Effects of communication on cooperative search: experiment results

In all of the experiments the mission is to find a hidden object in the search space. With this objective the performance is measured as the mission's completion time (i.e., the time it takes for the robots to locate the position of the target).

Robots can only communicate when they are in communication range of each other. They share/exchange their local occupancy grids during each encounter. Then, using the information obtained from the encountered robots they update their own occupancy grid maps and modify their search path accordingly. In this manner, through intermediate robots, a robot can also obtain information about the cells searched by a robot it has never encountered. Therefore, the communication strategy has some characteristics of multi-hop communication. Due to the nature of multi-hop networking, the information can be shared between the agents although they are not in range of each other. The information can be carried over the other agents on larger distances. Therefore, it is seen that, it is not necessary to always have a wide communication range. In other words, global communication between the agents is not needed for the best performance.

The results of the experiments are given in Fig. 23.10 and 23.11. In Fig. 23.10 the simulation results are presented. These results are collected over 1000 runs. In each of the runs the robots start from random initial positions. Only the position of the target cell that is being sought and the positions of the obstacles are kept constant. Similarly Fig. 23.11 shows the results of the robotic experiments performed in the experiment set-up. These are the average results over six runs for each communication distance to be tested. More experiments could not be performed as a result of the time it takes to carry out the experiments and because of some temporary problems with the communication software-hardware.

The communication ranges in all of the experiments are distributed between zero communication and global communication. In our experiment set-up the

(a)
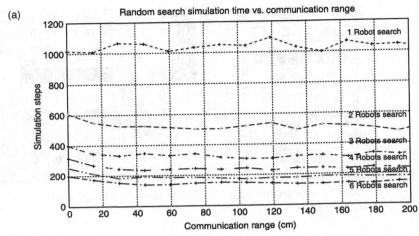

(b)
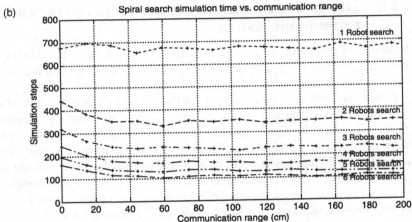

(c)
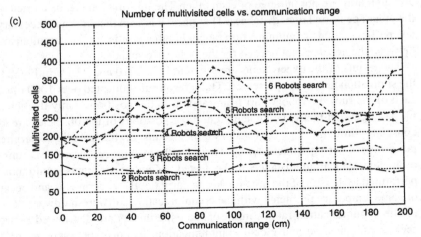

23.10 (a)–(f) Simulation results.

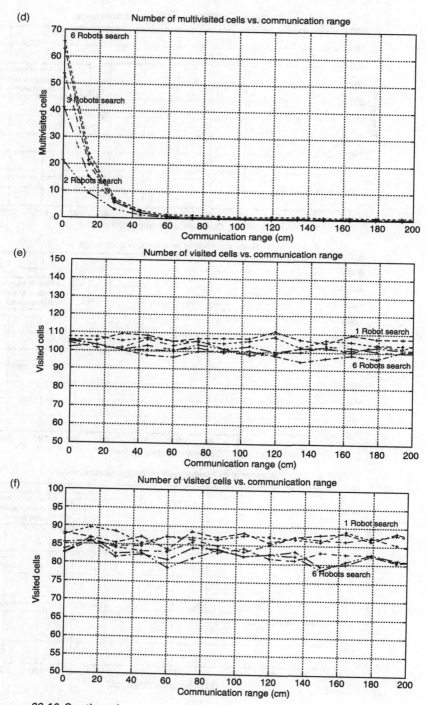

23.10 Continued

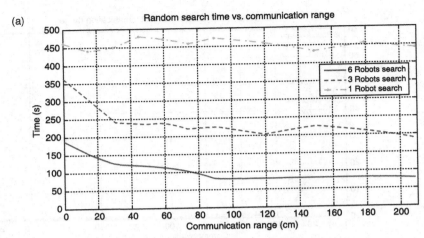

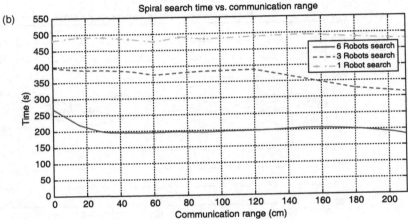

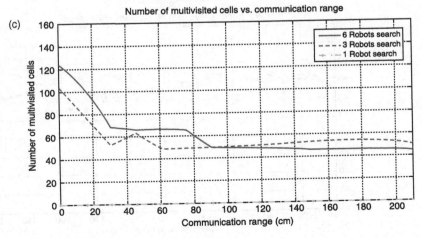

23.11 (a)–(f) Robotic experiment results.

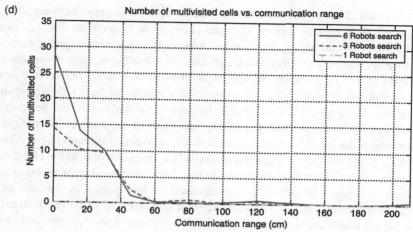

(d)

Number of multivisited cells vs. communication range

6 Robots search
3 Robots search
1 Robot search

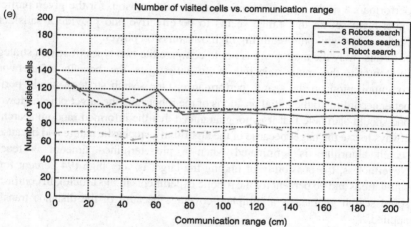

(e)

Number of visited cells vs. communication range

6 Robots search
3 Robots search
1 Robot search

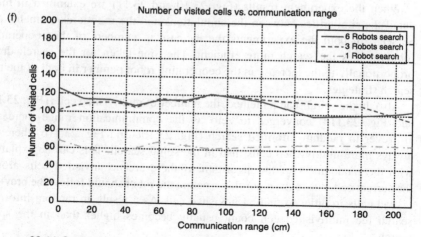

(f)

Number of visited cells vs. communication range

6 Robots search
3 Robots search
1 Robot search

23.11 Continued

maximum distance between two different robots was approximately 200 cm. Therefore, a communication distance equal to or larger than 200 cm can be described as global communication. Additionally, the search performances of different numbers of robots were collected. In simulations the experiments were repeated for cooperation of one to six robots. Similarly, the robotic search experiments are repeated for six, three and one robots and the results are presented in Fig. 23.10 and 23.11. As it can be seen from the results the performance increases proportionally to the number of robots as they are cooperating while searching. More robots means more cells to be visited at the same time. Also the effect of the number of cooperating robots is more effective than the effect of the range of communication. One can see that for a fixed number of robots initially there is some increase in the performance of the system as the size of the communication range increases. However, it settles down at around 40–60 cm and a higher communication range beyond that distance does not contribute to the performance of the system significantly. For this reason, for the given particular experiment set-up that range seems to be effective and provide a satisfactory performance while not requiring high transmission power.

In this study the main point is not to compare the two types of search strategies. In contrast, the main objective is to investigate the benefits of networked communication on the search performance. One search strategy would perform better according to the initial positions of the robots and the target object. However, one should also note that it is not guaranteed that the informed random search can always locate the target because of the algorithm's stochastic nature and because the search algorithm is terminated when a predetermined timeout is reached. Nevertheless, the possibility of finding the target is very high (94 percent for the simulations and 96 percent for the robotic experiments in this article). In contrast, the spiral search guarantees a complete coverage because of the distance transform applied.

When the experiment results (Fig. 23.10 and 23.11) are examined in more detail the similarities between the simulations and the robotic experiments can be seen. In both the experiments the increase in numbers of the cooperating robots makes the search more efficient. The time taken for the search drops proportionally to the increase in the cooperating robot number in both of the two search strategies.

Also, in the spiral search method, the number of multivisited cells (Fig. 23.10d and Fig. 23.11d) converges to zero as the communication range increases. However in random informed search (Fig. 23.10b and Fig. 23.11b) there are always more multivisited cells than in the spiral search. This can be explained with the randomness of the algorithm. If there are no unvisited cells around the current cell, due to the algorithm's nature the robot should follow the previous visited cells in order to find an unvisited cell. As a result, in random informed search the multivisited cell count usually becomes higher than in the spiral search.

Additionally the spiral search tends to be more efficient than the random informed search when the total number of visited cells is examined (Fig. 23.10e and 23.10f). In random search the mean of the total visited cells is around 102, however, in spiral search that mean is around 85 cells. This means spiral search finds the target object 17 cell before the random informed search. This effect can be seen through the simulation step counts in Fig. 23.10a and 23.10b. The spiral search shows a slightly better performance than the random informed search in the simulations. However, in the robotic experiments the random informed search shows better performance than the spiral search. The smaller number of robotic experiments can cause this result. Yet, the simulation results are more reliable because of the larger number of experiments performed.

23.6 Conclusions

In this study a cooperative search by mobile robots is investigated. The cooperation is provided by the networked communication of the agents. The results of the study are collected both by simulations and real experiments with e-puck mini mobile robots. The performance of two search strategies, which are namely a modified version of spiral search and informed random search, are measured for different communication ranges and for different numbers of robots. We observed in the experiments that the performance of the system improves with the increase of the numbers of cooperating robots. As can be seen from the experiment results (Fig. 23.10a, 23.11a) the performance of the system consisting of six robots is better than that of three robots and similarly the search performance of three robots is better than a one-robot search.

The results also show that, for the considered search scenarios, when the communication range is increased the search performance increases up to a certain point beyond which there is not much change in the performance of the system. Therefore, for this particular application set-up it is not necessary to have global communication for better performance. In other words the communication range does not need to cover all the search area. Additionally, a relatively shorter communication range means lower power consumption, therefore longer mobility of the robots.

Also it is important to point out that effective communication between the agents is highly dependent on environmental parameters such as the size of the search space and the number of the robots. Similarly the characteristic of the search algorithm is an important factor affecting the performance of the search. This is consistent with the results in related studies.

Future research can concentrate on developing algorithms for selecting the best communication range dynamically in order to minimize the power usage without significantly affecting the performance of the system. In addition, the effects of unequal communication ranges between the robots can be investigated. At the time of the experiments a communication hardware that can adjust its transmission range was not available. Therefore, we had to simulate this feature. Provided that such hardware is available, more realistic experiments can be conducted as well.

23.7   References

Akyildiz I. F., Su W., Sankarasubramniam Y., and Cayirci E. (2002) "A survey on sensor networks," *IEEE Communications Magazine* 40.8, 102–114.

Arkin R. C. (1992) "Cooperation without communication: multi-agent schema based robot navigation," *Journal of Robotic Systems* 9.3, 351–364.

Arkin R. C. (1998) *Behavior-Based Robotics*, Cambridge, Massachusetts, The MIT Press.

Balch T. and Arkin R. (1994) "Communication in reactive multi-agent robotic systems," *Autonomous Robots* 1.1, 1–15.

Cao Y. U., Fukunaga A. S., and Kahng, A. B. (1997) "Cooperative mobile robotics: antecedents and directions," *Autonomous Robots* 4.1, 7–27.

Dollarhide R. and Agah A. (2003) "Simulation and control of distributed robot search teams," *Computers and Electrical Engineering* 29.5, 625–642.

E-puck mini mobile robot from EPFL. Available from: http://www.cyberbotics.com/products/robots/e-puck.pdf [Accessed 17 July 2009].

Rekleitis I., Lee-Shue V., New A. P., and Choset H. (2004) "Limited communication, multi-robot team based coverage" *Proceedings of the IEEE international Conference on Robotics and Automation*, 4: 3462–3468.

Rybski P. E., Stoeter S. A., Gini M., Hougen D. F., and Papanikolopoulos N. P. (2002) "Performance of a distributed robotics system using shared communications channels," *IEEE Transactions on Robotics and Automation*, 18: 713–727.

Rybski P., Larson A., Veeraraghavan H., LaPoint M., and Gini M. (2004) "Communication strategies in multi-robot search and retrieval: experiences with MinDART," In *Proc. Int'l Symp. on Distributed Autonomous Robotic Systems*.

Samiloglu A. T., Cayırpunar O., Gazi V., and Koku A. B. (2008) "An Experimental Set-up For Multi-Robot Applications," *International Workshop on Standards and Common Platforms for Robotics (SCPR2008)*.

Trianni V., Labella T. H., and Dorigo M. (2004) "Evolution of direct communication for a swarm-bot performing hole avoidance," *LNCS*, 3172: 130–141.

Yoshida, E., Yamamoto, M., Arai, T., Ota, J., and Kurabayshi, D. (1995) "A design method of local communication area in multiple mobile robot system," *Proceedings of the IEEE International Conference on Robotics and Otomation*, 2567–2572.

Yoshida, E., Yamamoto, M., Arai, T., Ota, J., and Kurabayshi, D. (1996) "Evaluating the efficiency of local and global communication in distributed mobile robotic systems," *Proceedings of IROS*, 1661–1666.

Zelinsky A., Jarvis R. A., Byrne J. C. and Yuta S. (1993) "Planning paths of complete coverage of an unstructured environment by a mobile robot," *Proceedings of the International Conference on Advanced Robotics*, 533–538.

24

Mobile ad-hoc networking supporting multi-hop connections in multi-robot scenarios

U. WITKOWSKI, S. HERBRECHTSMEIER and M. EL-HABBAL, University of Paderborn, Germany

Abstract: One of the most promising applications of a multi-robot system is to assist humans in urban search and rescue (USAR) scenarios in the aftermath of natural or man-made disasters. The main disaster scenario covered by our system is a large industrial warehouse on fire, described in the GUARDIANS project funded by the European Union 6th Framework Program (project no: 045269). As described by South Yorkshire fire department (SY-Fire), which is a partner and main client of the project, fire can occur in large warehouses with dimensions over 100 by 100 square meters. In this scenario, black smoke may fill a large space of the warehouse making it very difficult for the fire fighters to orientate with the building, locate victims or find their way out and exit the building. In this chapter we describe our work and achievements regarding one main work package in the GUARDIANS project, focusing on two key arts: the ad-hoc network communication system, and communicative and non-communicative swarming behaviours. Several attempts were made to develop new topology control algorithms or modify the existing ones to best suit the required ad-hoc scenario. By investigating several techniques, we finally proposed a hierarchal cellular protocol inheriting some of its features from pro-active and hybrid hierarchical protocols like cluster-head gateway switch routing (CGSR)[1] and hierarchical state routing (HSR).[2] A special communication platform is used for implementing our communication protocol. This so called mobile communication gateway will be optimized for mobile usage and therefore will support different techniques for energy saving. Some of these techniques are dynamic frequents and voltage scaling as well as dynamic power down of non-used hardware components. Three communication standards are supported by this hardware platform: wireless local area network (WLAN), Bluetooth and Zigbee, and the first two standards are used for our project.

Key words: multi-robots, communication, swarm

24.1 Introduction

Communications and communication protocols play an important role in mobile robot systems, especially in multi-robot systems, which are optionally enhanced by humans to complement individual skills. One of the most promising applications of a multi-robot system is for assisting humans in urban search and rescue (USAR) scenarios in the aftermath of natural or man-made disasters. The main disaster scenario covered by our system is a large industrial warehouse on fire, described in the GUARDIANS project funded by the European Union 6th Framework Program (project no: 045269). In this scenario, black smoke may fill large space

557

of the warehouse making it very difficult for fire fighters to orientate with the building, find victims or find their way out and exit the building.

The main idea of the project is to send a heterogeneous team of robots inside the scenario area to assist fire fighters by detecting fire sources and hazardous gases and by providing different sensory data, localization and positioning data and maintaining communication links between fire fighters and the base station located outside the scenario area. Several teams of robots are used in the project; some are surrounding and guiding the fire fighters and providing them with sensory data, together forming a squad team, while others are autonomous teams working independently of the fire fighters for purposes like area exploration, map building and forming the communication infrastructure.

In this chapter we describe our work and achievements regarding one main work package in the GUARDIANS project, focusing on two key parts: the ad-hoc network communication system, and communicative and non-communicative swarming behaviours. In communicative mode, automatic service discovery is applied: the robots find peers to help them. The wireless network also enables the robots to support a human squad-leader operating within close range. In the case of losing network signals, the robot swarm can still be functioning using non-communicative mode and continue serving the fire fighters.

In addition to the communication capability, the ad-hoc network has to provide position data to support localization of the mobile robots and humans, which might be of great importance for guiding the humans and robots to specific targets and locations or quickly exiting the search area. In outdoor applications the GPS system is one option for position data. Finding the location of a communication node without the aid of GPS in each node of an ad hoc network is important in cases where GPS is not accessible, especially in indoor scenarios.

One important aspect that needs consideration in such scenarios is the size of the network. If many robots (maybe hundreds or more) exist in the same area, communication among them needs to be well organized and not just relying on basic sensing of medium (like carrier sense multiple access/collision detection (CSMA-CD)) or else this will result in chaotic behaviours, leading to excessive data losses, and increased error rates due to high interference, which may cause complete network failure. Additionally, in high dynamic networks, in which nodes are not static but moving fast most of the time, even the ones responsible for maintaining the communication links, special routing protocols need to be used to support this high mobility and the high probability of node failures due to the hazardous conditions inside the disaster area.

Therefore, an ad-hoc network communication system based on the mobile robots as communication nodes, with some of them static and dedicated to ensure robustness of communication coverage all over the scenario area, is deemed suitable because it can offer a robust communication infrastructure.

In this chapter we present a mobile ad-hoc network that can be used in large burning industrial warehouses. The network has to support the fire fighters and the assisting group of robots with a communication infrastructure as well as position data. One of the most important features of the network is its robustness in terms of available communication links and position data for maximizing the safety of fire fighters. Furthermore, supporting robust routing and service discovery is of further assistance to fire fighters.

The chapter is organized as follows: section 24.2 gives an overview of the system architecture and the robot teams used for our work package. A detailed description of our proposed communication system is clarified in section 24.3, regarding the communication standards used, proposed routing protocols and a short overview on the state of the art of routing protocols dedicated to similar scenarios. In section 24.4, the hardware platform on which the communication protocol is implemented is presented. Section 24.5 summarizes the experiments, demonstrations and results done so far. Finally, the chapter concludes with section 24.6.

24.2   System architecture of mobile ad-hoc networking

In the GUARDIANS project, several heterogeneous teams of robots are sent inside the scenario area to assist fire fighters by detecting fire sources and hazardous gases and by providing different sensory data, localization and positioning data and maintaining communication links between fire fighters and the base station located outside the scenario area. Figure 24.1 gives an idea of the main teams used in the project. The first team is the squad team with robots surrounding and guiding

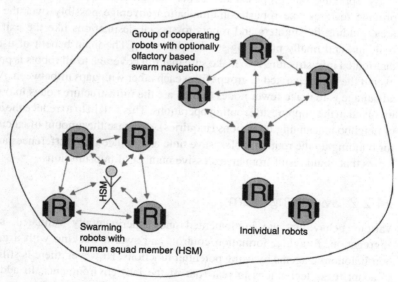

24.1 Various heterogeneous teams of robots considered and used in the GUARDIANS project.

the fire fighters and providing them with various sensory data, as shown on the left side of the figure. The second team is an autonomous team working independently of fire fighters for purposes like area exploration and map building, working in both communicative and non-communicative modes, as shown on the upper right part of the figure. A third team is optional, used for other tasks like forming the communication infrastructure in case both the second and third teams are not merged and need to act independently, as shown on the lower right part of Fig. 24.1.

A detailed description of each team and its use in the project is presented in the next subsections.

24.2.1 Communication team

Topology control techniques can be divided into two main types: infrastructure and infrastructure-less routing. After investigating various topology control algorithms, as will be discussed in detail in section 24.3.1, it was found that infrastructure-less routing techniques are not the best choice for highly dynamic networks, due to the high rate of change and update of routing tables and excessive flooding of routing information among network nodes.

Therefore, there is a need to use static nodes which take responsibility for managing the whole communication network, by monitoring the movements of robots, storing their location data and available services, assigning proper communication channels.

By applying our proposed hierarchal cell-based technique, we are able to provide features like robust communication coverage possibly over the whole area, guiding fire fighters and robots to special destinations like the exit of the building, and finally providing service discovery. The main benefit of using the distributed infrastructure is to make sure that full coverage to all robots is provided even if they are separated in groups from each other with gaps in-between. Another advantage is to make fewer nodes (which are the infrastructure nodes in our case) involved in the hopping and routing operations. This will help to reduce the amount of neighbour scanning operations (inquiries), decrease the amount of calculations for building up the routing tables, save time and reduce the interferences and data losses that could result from an excessive number of transmissions.

24.2.2 Swarming team

Various behaviours are implemented and experimentally validated, such as aggregation, foraging, formation control, and swarm tracking with a group of non-holonomic agents by using potential functions. However, there is still a need to adapt these techniques to real (out of the lab) environments. In addition, a particle swarm optimization (PSO) is developed based on high-level path planning for olfactory-based swarm navigation and a search for locating areas of

24.2 Robots spanning a communication network in a triangular mesh after driving into a room (downscaled scenario).

high odour/chemical concentrations. For this purpose a decentralized and asynchronous implementation of the PSO algorithms is used, allowing for dynamic neighbourhood of particles and time delays in information flow.

Non-communicative robot swarming

A variety of basic behaviours have been implemented like wall and path following, trajectory tracking, PSO search and Voronoi path planning. The behaviours are usually applied for a single robot. A key point for further development is to combine behaviours into meaningful complex behaviours and apply them for a group of robots. For instance, all robots in the group can apply wall following; but that does not really result in very useful behaviour. Other typical swarm behaviours have also been implemented like potential fields, which are applied for obstacle, robot and human avoidance and leader following as well as for aggregation and dispersion and formation control.

Human-robot swarm interaction

A helmet visor prototype intended for guiding a fire fighter has been built and tested with professional fire fighters. A basic finding was that they prefer very simple directions like going forward, left or right, as opposed to the more detailed device we had prototyped. In the tests it became clear that the fire fighters firstly rely on their (proven) procedure of following a wall, for a large part ignoring the instructions given by the device. It was interesting to note that the sense of orientation of the fire fighter decreased with an increased mental load.

24.3 Communication protocol in multi-robot scenarios

24.3.1 Used wireless standards

It was planned to implement a wireless communication module providing three wireless standards: wireless local area network (WLAN), Bluetooth and ZigBee. ZigBee was removed to ease design complexity and since there were no benefits for this technique over Bluetooth other than slight power consumption savings, which were not critical in our case scenario. For interconnections between infrastructure nodes, WLAN is used. WLAN is also used for special links like that between the base station and one static node inside the scenario area, and also between the base station and fire fighters via robots for hopping purposes. For intra-cell connections inside a static node cell (coverage area), Bluetooth is used. Bluetooth has some advantages over WLAN like low power consumption and low transmission latency, which is ideal for frequent low data rate transmissions like inquiry feedbacks and connection and service requests. This makes Bluetooth the best choice for usage as a communication standard between robots and each other inside a cell on one side, and between robots and the base station on the other side. Normally, the number of nodes existing inside a cell (static node's coverage range) is less than eight, but in the case of this number exceeding eight, the robots should be assigned to another static node, or additional static nodes should be used in this area to support more traffic. In the first case, scatternets could be used where a robot is chosen to act as bridge between the other piconets formed by the other robots.

24.3.2 State of the art routing protocols

Several routing methods are implemented trying to achieve high efficient routing by compromising various parameters. Depending on the application scenario, the proper technique should be chosen to achieve best results. Here is a summary of the main routing techniques used for ad-hoc networks:

- Pro-active routing (table-driven): This type of protocol maintains fresh lists of destinations and their routes by periodically distributing routing tables throughout the network. Popular examples for this technique are optimized link state routing (OLSR),[3] cluster-head gateway switch routing (CGSR) and wireless routing protocol (WRP).[4]
- For reactive routing (on-demand), this type of protocol finds a route on demand by flooding the network with route request packets. Popular examples for this technique are dynamic source routing (DSR)[5] and ad-hoc on-demand distance vector (AODV).[6]

- In flow oriented routing, this protocol finds a route on demand by following present flows. One option is to unicast consecutively when forwarding data while promoting a new link.
- The hybrid (pro-active/reactive) protocol combines the advantages of proactive and reactive routing. The routing is initially established with some proactively prospected routes and then serves the demand from additionally activated nodes through reactive flooding. The choice for one or the other method requires predetermination for typical cases.
- A similar protocol is the hierarchical routing protocol, where the choice of proactive and reactive routing depends on the hierarchic level in which a node resides. The routing is initially established with some proactively prospected routes and then serves the demand from additionally activated nodes through reactive flooding on the lower levels. The choice for one or the other method requires proper attribution for respective levels.
- Another type of protocol is the geographical routing protocol, which acknowledges the influence of physical distances and distribution of nodes to areas as significant to network performance. Popular examples for this technique are augmented tree-based routing (ATR),[7] distributed dynamic routing algorithm (DDR),[8] and hierarchical state routing (HSR).

By investigating the different methods used in the previously discussed routing techniques, one can conclude that reactive routing is not really efficient in large highly dynamic networks due to the excessive flooding that can lead to network clogging. Even though it performs well in static and low-mobility environments, the performance degrades rapidly with increasing mobility. For geographic routing, efficiency depends on balancing the geographic distribution versus occurrence of traffic. This is besides the essential need of a prior knowledge of the exact positions of nodes, which is sometimes difficult to obtain in indoors scenarios without GPS. Our idea is to combine and use the benefit of the remaining methods, by using a hierarchal, power-aware, pro-active routing in our so-called "hierarchal cell-based network". The hierarchal infrastructure has the advantage of compromising between hop and energy minimization, which uses power control and starts by using tight transmission ranges and then widening the range when required to save time and reduce hopping overhead. The static infrastructure itself has the advantage of dealing with highly dynamic changes in the network topology and involving fewer nodes for storing position information of the whole network, instead of spreading routing tables all over the network as in classic pro-active routing. The cell-based infrastructure updates the service lists with the locations of robots and available services using periodic inquiries in a pro-active manner. Our routing protocol inherits some of its features from the pro-active CGSR and HSR. CGSR uses highly dynamic destination-sequenced distance vector (DSDV)[9] as an underlying protocol. Mobile nodes are partitioned into clusters and a cluster-head is elected using a distributed algorithm. All nodes in

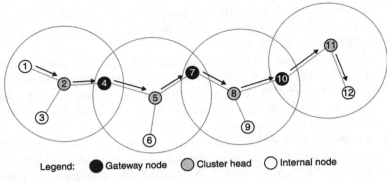

24.3 Routing in CGSR.

the communication range of the cluster-head belong to its cluster. A node that is in the communication range of two or more cluster-heads is called a gateway node. CGSR uses a least cluster change (LCC) clustering algorithm. A cluster-head change occurs only when two cluster heads come into one cluster or one of the nodes moves out of the range of all the cluster-heads, as shown in Fig. 24.3.

For the HSR protocol, it maintains a hierarchical topology, where elected cluster-heads at the lowest level become members of the next higher level. At the higher level, super-clusters are formed, and so on. Nodes that want to communicate to a node outside of their cluster ask their cluster-head to forward their packet to the next level, until a cluster-head of the other node is in the same cluster. The packet then travels down to the destination node.

24.3.3 Routing protocol

Our first idea for implementing the routing protocol is to use re-active routing, since it provides better bandwidth usage than pro-active routing, due to the limitation of unnecessary periodic updates, where a link is only searched for and established on demand. This makes re-active routing more attractive for use in large networks. The main drawbacks of re-active routing include that it is not preferred for usage in high mobility networks or in radio-harsh conditions where links tend to break more often, like in our case with warehouses full of metal and excessive reflections that lead to instability in link quality. Re-active routing is not preferred in such conditions since link failures will trigger searching and establishment of new routes, which will introduce large time delays. Consequently, this will lead to another drawback, which is the unsuitability of re-active routing for real-time continuous data transmission, which is also required in our scenario, especially for the continuous data transmission between base station and fire fighter. Some simulations were implemented for a simple but efficient reactive routing protocol that is suited for networks with harsh signal conditions, where

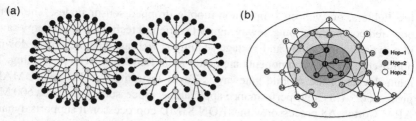

24.4 (a) Comparison between message forwarding in GSR on the left and OLSR on the right. (b) Different update rate zones in FSR.

the protocol tends to re-establish broken links very fast, which is called DSR and enhanced DSR (EDSR). The simulation results are shown in more detail in section 24.5.

According to the conditions mentioned for our scenario, pro-active routing is found to be more suitable. Most of the well-known pro-active routing protocols are originated from two families: the distance vector (DV) family and the link state (LS) family. Link state protocols are more popular nowadays since they are newer and rely on routing based on link state parameters (more stable links) rather than relying on shortest distance only. Three main LS protocols have emerged: global state routing (GSR), optimized link state routing (OLSR) and fisheye state routing (FSR). GSR is the first implemented LS protocol, which is the simplest in implementation but not preferred in large networks. OLSR tends to reduce the number of nodes used for message forwarding in the entire network by selecting special nodes called multipoint relays (MPRs). For FSR, instead of reducing the number of nodes used for forwarding messages like OLSR, the number of forwarded messages is reduced, by reducing the update rates for distanced nodes and increasing them for near-placed nodes.

OLSR is the chosen protocol for our scenario. The used software driver is called "OLSR Daemon" which is widely tested and used, and runs under Linux systems supported by the main communication module (gateway board) used in the project. Another advantage of using OLSR is that the code can be easily modified to switch to GSR or FSR. GSR is simpler and is thought to be more efficient in the case scenario since our network is mid-sized, so further tests will be done comparing GSR and OLSR and accordingly the protocol that provides better results will be chosen for our project.

24.4 Communication platform in mobile ad-hoc networking

The hardware platform used for realization of the mobile ad-hoc communication system is an optimized mobile ad-hoc communication gateway. It is optimized in terms of size and power consumption to fit on the mobile robots as well as on the fire fighters, or even on simple non-mobile dropped sensor nodes. The battery powered communication gateway supports different techniques for energy saving like dynamic frequency and voltage scaling as well as dynamic power down of non-used hardware components including radio frequency (RF) processing. The main device of the mobile gateway is Texas Instrument's (TI) new OMAP 3 processor.[10] This high-performance applications processor consists of a 600 MHz ARM Cortex-A8 processor with NEON SIMD coprocessor. It supports dynamic branch prediction and has comprehensive power and clock management, which enables high-performance, low-power operation via TI's SmartReflex adaptive voltage control. It offers more than 1200 Dhrystone MIPS with maximal power consumption from less than 2 W for the whole chip. The processor is connected to 512 mb NAND Flash and 256 mb mobile low power DDR SDRAM.

The gateway is equipped with the wireless communication standards Bluetooth and Wifi. Both have external antennas for better signal qualities and support power down to disable communication devices that are not needed. A coexistence solution ensures simultaneous operation of Bluetooth and Wifi. Bluetooth is implemented by the CSR BlueCore4-ROM, which is a commonly used Bluetooth chip. It supports Bluetooth v2.1 with enhanced data rate (EDR) for faster data transfer with 3 Mbit/s. The coupling of the Bluetooth module to the processor is done via a high-speed universal asynchronous receiver/transmitter (UART) interface. The Wifi chip 88W8686 is an ultra low-power WLAN single chip solution from Marvell. It is connected to the processor via a 4-bit SDIO interface. Both communication devices have a peak power consumption of less than 1 W during continuous transmit over Bluetooth and Wifi. Additionally, the supported wired communication standards I^2C, SPI, UART and high speed universal serial bus (USB) allow variable expansion of the gateway. Via these interfaces other communication devices like ZigBee, Sub-1 GHz-Communication or ultra-wide band (UWB) hardware or other components like sensors (e.g. chemical sensors for detection of hazardous agents), actuators, robots or computers can be easily connected to the gateway enabling optimized heterogeneous communications devices and meeting several communication demands. However, the main interface to a computer or robot is the USB device interface. An additional USB ethernet adapter offers the ability to connect the gateway to an Ethernet interface. The whole gateway is powered via a lithium battery with 3.6 V or over a USB connection with 5 V.

The software tools for the gateway are the same as for the BeBot miniature robot.[11] Everything is automatically generated via OpenRobotix.[12] OpenRobotix is an OpenEmbedded and Angstrom distribution-based open source Linux distribution for mini robots managed by the System and Circuit Technology department of the Heinz Nixdorf Institute, University of Paderborn. This software generates a complete software development environment and images for autonomous mini robot systems. The interfaces to the hardware are based on Player.[13] This allows easy integration of all important gateway information into the base station and the robot system.

The integration of computers and robots (clients) into the ad-hoc network takes place via the mobile ad-hoc communication gateway. This gateway is connected via USB to a client. Over this USB connection an Ethernet over USB protocol is implemented. This protocol is supported by the Linux Kernel USB communication device class (CDC) driver and needs no additional drivers on the client. After connecting the client to the gateway the client creates a virtual network interface and configures this interface via standard dynamic host configuration protocol (DHCP). Through this the gateway module assigns an IP address and default network gateway to the client. The IP address belongs to the gateway and allows client identification via the gateway to which it is wired and connected. The gateway module automatically publishes this client IP to every gateway in the whole network and thereby makes it available to the other clients in the network. The default network gateway configuration causes the client to route all network communication to the gateway module. Through this the complete routing of the communication is transferred to the gateway and thereby to the mobile ad-hoc communication system. Altogether this enables standard transmission control protocol/internet protocol (TCP/IP) based network communication between the gateway and client as well as the client and other clients in the network. Additionally this simplifies the integration of the mobile ad-hoc communication and separates as well as hides the network implementation from the application.

Status information and configuration of the gateway is implemented via a Player interface. This interface allows the base station to monitor the wireless connection neighbours of a gateway or gives the robot the possibility of detecting

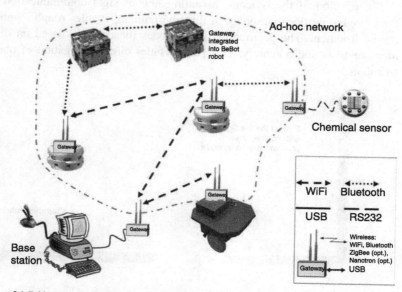

24.5 Heterogeneous mobile ad-hoc network with different robots connected by the mobile ad-hoc communication gateway.

a lost connection and automatically switching to non-communicate swarming. The configuration interfaces enable the robot to define the primary wireless connection standard. A modular software and hardware environment allows the direct equipment of the gateway with an additional sensor and simplifies the integration of this sensor into the network via a Player driver.

24.5   Experiments and results

In this section two experiments/demonstrations are presented. The first demo is to present and test the triangular distribution of the infrastructure nodes using a laser system. The second demo shows the proposed communication functionality and offers the base station the possibility of remote control and to monitor robots and sensors in this mobile ad-hoc network.

24.5.1 Laser-based triangular distribution of infrastructure nodes

This demo is to demonstrate how a group of mobile robots (beacons) detect the rough location of the fire fighter and distribute themselves with other communication nodes in triangulation forms following the fire fighter's path. This is to guarantee robust communication coverage between the fire fighter and base station, as well as helping the other swarming team with map building and area exploration. This is shown in Fig. 24.6.

The position of the beacons, assuming line of sight communication along the triangles and local scanning of the environment, enables rough mapping of the environment. The positioning of the nodes (beacons) is based on distance measurements and odometry. This part includes circumnavigation of obstacles by robots.

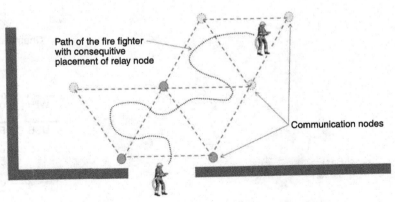

Path of the fire fighter with consequitive placement of relay node

Communication nodes

24.6 Formation of the triangles, i.e. positioning of new nodes depends dynamically on the in general arbitrary movements of the fire fighter.

The demo goes as follows: three robots are present at the beginning, close to the entrance, as marked in grey in Fig. 24.6. These robots represent the core of the triangulation distribution of communication nodes, which are equipped with laser sensors to detect the rough location of the fire fighter. The laser sensor used in this demo is the IFM single beam diffusion laser sensor. Hence, the rough location of the fire fighter will be detected according to the laser rotated angle. As shown in the fire fighter's path drawn in Fig. 24.6, the fire fighter first enters the scenario area, which means he is inside the core triangle formed by the three grey nodes in Fig. 24.6. If he moves left to the left side triangle (left sector), another (fourth) beacon will move to the dashed light grey position on the left as shown in the figure. As the fire fighter moves back to the right sector, another beacon will go to its desired position to complete the formation of the triangulation distribution. The fire fighter is equipped with laser detectors to detect the laser beam and send messages back to the scanning beacon to estimate his position. Probably each sector (triangle) will be scanned by only one robot at a time, in order to know the ID of the scanning robot to which messages (acknowledgements) are sent back from the fire fighter. The range (angle limits) of each scanning robot will be set according to its position in the scenario.

24.5.2 Simulation results of DSR and EDSR re-active routing protocols

The simulator used for network protocol implementation and evaluation is the Netlogo multi-agent simulator, which uses its own object-oriented programming language similar to C++.

As stated in section 24.3, DSR and EDSR protocols were tested on the simulator. DSR has the advantage of simplicity and cost efficiency of route establishment, which was worth testing for providing robust coverage in a network where nodes are moving rapidly, as shown in Fig. 24.7a.

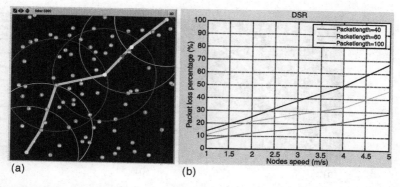

(a)   (b)

24.7 (a) Route establishment in DSR. (b) Performance of DSR with various packet lengths and speeds.

Here is a summary of the simulation parameters used: 80 nodes distributed in a 40 m × 40 m square area, moving with speed varying from 0 to 5 m/s, 115200 baud-rate, 13 m maximum coverage radius, packet sizes varying between 40 and 100 bytes, including a 20 byte overhead in each. For the tests, 4000 data bytes were to be sent from source and acknowledged by destination. Figure 24.7b shows the effect of increasing the moving speed of nodes on the connection: break down and hence packet loss. Once a link is broken, a new Route-Request is performed, in order not to waste time trying to fix an existing broken link, which might take more time compared to the fast search mechanism of DSR.

From the graph in Fig. 24.7b it can be seen that as the packet size increases, the packet loss also increases, as it takes more time for a full round trip from source to destination and acknowledged back to source, in which time the nodes would have moved further from each other and hence increased the probability of connection failure. The source and destination are steady and do not move during all simulations in order not to alter the distance between them and thus affect the simulation results.

Another test is done for the enhanced-DSR (EDSR). This is similar to the traditional DSR but with a minor enhancement to achieve a more stable performance especially in high dynamic topologies. Since DSR chooses its best route based on the minimum number of hops, two neighbouring nodes in the path might lie very close to the edge of each other's coverage, and hence the link between them becomes vulnerable to any movement of the nodes. EDSR on the other hand chooses a neighbouring node that lies in an inner-radius smaller than the maximum coverage radius to give some room for a node to move before the link breaks down, and hence provide more stability and robustness in connections. This is shown in Fig. 24.8a, where inner-radii are presented in grey. The performance of EDSR is depicted in Fig. 24.8b.

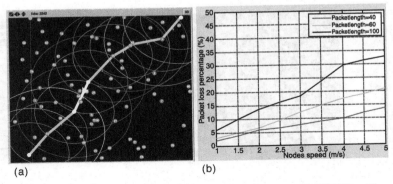

(a)   (b)

24.8 (a) Route establishment in EDSR. (b) Performance of EDSR with various packet lengths and speeds.

Tests were done with maximum coverage radius of 13m as in DSR, and inner-radius once with 11m and in another test with 9m. As the change in inner-radius affects the results very slightly, the results of only one of them are presented. Comparing the results of DSR and EDSR, one can notice that EDSR provides better performance with less packet loss for various packet lengths and node speeds.

24.5.3 Mobile ad-hoc communication

The mobile ad-hoc communication demo shows the proposed communication functionality in mobile ad-hoc networks. Based on a widely used mesh routing implementation it allows TCP/IP based communication over different communication standards. This offers functional test possibilities for the considered applications as well for the gateway design.

The build-up (Fig. 24.9) consists of three BeBot miniature robots, which can communicate with each other via Bluetooth. The first robot (1) is equipped with a Wifi adapter. It acts as a bridge between Wifi and Bluetooth and has a direct connection to the router. This router has additional Ethernet connections and acts as a bridge between the standard network and the mobile ad-hoc communication network. The base station is integrated into the ad-hoc network via the Ethernet interface of the router. The second robot (2) has a static position with a direct Bluetooth connection to the first robot. Together they build the mobile communication infrastructure and support the third robot (3) with a

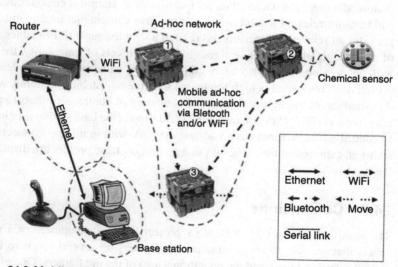

24.9 Mobile ad-hoc network with autonomous communication linked via an adapted router to a base station for mission monitoring and optionally remote control of selected robots.

network connection. The third mobile robot sends its sensor values and camera image to the base station. The base station on the other hand uses this information to remotely control the robot. While moving the robot, the base station autonomously establishes and manages Bluetooth connections to the two infrastructure nodes.

Communication network

The communication implementation is divided into different levels. At top we have network routing, which is handled by the widely used mesh routing implementation of the optimized link state routing protocol (OLSR).[14] It manages the layer 3 routing and allows TCP/IP communication in wireless ad-hoc networks. Most parts of the underlying ad-hoc network are implemented by Bluetooth. At this level Bluetooth network emulation protocol (BNEP) is used to hide the Bluetooth implementation from the routing protocol. Below this scatternet technology is used. This enables multi hop Bluetooth networks and thereby overcomes the limit of one master to maximal eight slave networks. An additional implementation for the autonomous establishing and managing of the Bluetooth connections to near gateways enables the network to support Wifi and Bluetooth in a mobile multi standard communication network.

Base station

The base station provides the means to support crew members, establish multimodal maps (obstacles, humans location, heat, chemical concentration, etc), and transmit relevant data in near real-time (live camera pictures, sound, etc). It provides all relevant information on request and at the same time enables a range of actions targeting the robot: through different levels of granularity, the robots will be possibly controlled at a high level or with much more elementary primitives. This notion is highly relevant with the adjustable autonomy, which is the paradigm of modifying the level of autonomy of humans or robots to provide effective and efficient interaction. Fig. 24.10 depicts the base station user interface to control a mobile robot via an ad-hoc network with multi hop connections. In addition, captured sensor data, i.e., images and obstacle profile, are displayed on the screen.

24.6   Conclusions

The main aim of the GUARDIANS project is the development of a team of robots that are able to support human fire fighters on several levels to increase overall safety and to extend the operational area of the fire fighters. One of the key issues is a robust communication system providing both communication between all team members and position data. We are focusing on an ad-hoc network

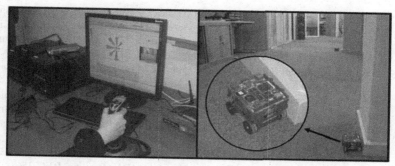

24.10 Mission monitoring and remote control of selected robots via the base station in the mobile ad-hoc network. Left: base station user interface, right: BeBot minirobot as node in the mobile ad-hoc network.

communication system based on the mobile robots as communication nodes. In our approach we use a cell-based grid with master nodes in each cell to form the basic structure of the network. Some nodes formed by special robots act as beacons to uniformly span the network. These robots will act as reference points when positioning other mobile robots or humans and at the same time form the infrastructure to support communication all over the search area. Furthermore, several routing techniques for similar ad-hoc network scenarios were investigated and accordingly the proper topology control technique was chosen to best suit our scenario. The protocol uses a similar concept as the cell-based (cluster-based) concept of CGSR, but we have used our distributed static infrastructure nodes as the cluster-heads to maintain more stability for the network. Additionally, we included the hierarchal concept and gave the cluster-heads hierarchal numbers for further reduction of the number of transmissions per routing list update, instead of massive flooding of the network and to reduce both the routing tables' sizes and packet size.

24.7   Acknowledgements

This work was supported by the Sixth Framework Program of the European Union as part of the GUARDIANS project (no. 045269, www.guardians-project.eu).

24.8   References

1 Clusterhead gateway switch routing (CGSR) – C. Chiang, Routing in Clustered Multihop, Mobile Wireless Networks with Fading, *IEEE Singapore International Conference on Networks, SICON'97*, 1997. Channel http://wiki.uni.lu/secan-lab/ClusterHead+Gateway+Switch+Routing+Protocol.html
2 Hierarchical state routing (HSR) – Scalable Routing Strategies for Ad Hoc Wireless Networks – http://wiki.uni.lu/secan-lab/Hierachical+State+Routing.html

3 Optimized link state routing protocol (OLSR) – Philippe Jacquet, Paul Muhlethaler, Amir Qayyum, Anis Laouiti, Laurent Viennot, Thomas Clausen, Optimized Link State Routing Protocol (OLSR), RFC 3626. http://www.olsr.net/, http://www.olsr.org/

4 Wireless routing protocol (WRP) – Shree Murthy, J.J. Garcia-Luna-Aveces, A Routing Protocol for Packet Radio Networks, *Proc. ACM International Conference on Mobile Computing and Networking*, pp. 86–95, November, 1995.

5 Dynamic Source Routing (DSR) – David B. Johnson, David A. Maltz, Dynamic Source Routing in Ad Hoc Wireless Networks, Mobile Computing, Thomasz Imielinski and Hank Korth (Editors), Vol. 353, Chapter 5, pp. 153–181, Kluwer Academic Publishers, 1996.

6 Ad-hoc on-demand distance vector (AODV) – C. PERKINS, E.ROYER AND S. DAS, Ad hoc On-demand Distance Vector (AODV) Routing, RFC 3561.

7 Augmented tree-based routing (ATR) – Marcello Caleffi, Giancarlo Ferraiuolo, Luigi Paura, Augmented Tree-based Routing Protocol for Scalable Ad Hoc Networks, in *Proceedings of MHWMN 2007: The Third IEEE International Workshop on Heterogeneous Multi-Hop Wireless and Mobile Networks*. http://arxiv.org/abs/0711.3099

8 Distributed dynamic routing algorithm (DDR) – NAVID NIKAEIN, HOUDA LABIOD, CHRISTIAN BONNET, Distributed Dynamic Routing Algorithm (DDR) for Mobile Ad Hoc Networks, in *Proceedings of the MobiHOC 2000: First Annual Workshop on Mobile Ad Hoc Networking and Computing*.

9 Highly dynamic destination-sequenced distance vector routing protocol (DSDV) – C. E. Perkins, P. Bhagwat, Highly Dynamic Destination-Sequenced Distance Vector (DSDV) for Mobile Computers, *Proc. of the SIGCOMM 1994 Conference on Communications Architectures, Protocols and Applications*, Aug 1994, pp 234–244.

10 OMAP™ 3 family of multimedia applications processors, Texas Instruments, http://focus.ti.com/pdfs/wtbu/ti_omap3family.pdf

11 U. Witkowski, S. Herbrechtsmeier, M. El-Habbal, U. Rückert, "Powerful Miniature Robot For Research And Education". In *IEEE Proceedings of the, 5th International Conference on Computational Intelligence, Robotics and Autonomous System* (CIRAS 2008), June 19–21, Linz, Austria, 2008.

12 OpenRobotix – OpenEmbedded based open source Linux distribution for mini robots, http://openrobotix.berlios.de/

13 The Player Project – free software tools for robot and sensor applications, http://playerstage.sourceforge.net/

14 OLSRD – an adhoc wireless mesh routing daemon, http://www.olsr.org/

25

A decentralized planning architecture for a swarm of mobile robots

M. DEFOORT, Université de Valenciennes, France, T. FLOQUET and W. PERRUQUETTI, LAGIS, FRE CNRS 3303, France, A. KOKOSY and J. PALOS, ISEN, France

Abstract: This chapter presents a decentralized planning architecture for a swarm of autonomous robots, which evolve in an unknown environment with obstacles. The swarm is composed of non-holonomic mobile robots. We assume that the robot shape is a circle or can be included into a circle. The planning algorithm takes the kinematic model of the robot and the physical limitations (maximal speeds) into account. The planning problem is described as an optimal problem with constraints. The resolution of this problem is based on the flatness property of the system and uses a receding horizon in order to guarantee a real time implementation. Experimental investigation has been conducted using a test bench made of three non-holonomic mobile robots in order to demonstrate the effectiveness of the proposed strategy.

Key words: robot swarm, decentralized intelligence, trajectory planning, obstacle avoidance, receding horizon planning, non-holonomic mobile robots.

25.1 Introduction

The coordinated control of multiple autonomous mobile robots is becoming an important robotics research field. Indeed, there are many potential advantages of such systems over a single robot, including greater flexibility, adaptability and robustness. Among all the topics of study in the field, this chapter focuses on the navigation of an autonomous mobile robots swarm evolving in environments with obstacles. Many cooperative tasks such as surveillance mapping, search, rescue or area data acquisition require the robots to have an autonomous navigation without collision (Defoort, 2007). The robots are dynamically decoupled but have common constraints that make them interact. Indeed, each robot has to avoid collision with the others. Moreover, some communication links between several vehicles must be maintained during movement. Since the available power on board the vehicle is limited, the distance between two vehicles that may exchange information will naturally be constrained. Besides maintaining communication, the feasibility of the trajectories implies the respect of the dynamic constraints, as well as avoiding obstacles and collisions. In this chapter, we will focus particularly on these dynamic and geometric aspects and ignore mobile networking factors, such as fading, cross talk, and delay, which can also affect the quality of communication between the vehicles.

Putting the interactions on a multi-robotic system to good use consists of introducing the coordination mechanisms in order to give coherence to the robot acts. By studying human behaviors, Mintzberg (1979) has identified three coordination mechanisms:

- *Coordination by adjustment*, where individuals share resources in order to achieve a common goal (nobody has control and the decision process is joint).
- *Coordination by leadership (supervision)* where a hierarchical relationship exists between individuals (some individuals have control).
- *Standardization* where some procedures are predefined in order to solve some particular cases of interaction (for example, the rules that can limit conflicts).

All the coordination mechanisms used on the robotics system are inspired by these fundamental coordination mechanisms.

Two strategies for motion planning in multi-agent systems are the centralized and decentralized (distributed) approaches. Although the centralized one has been used in different studies (see Dunbar and Murray, 2002, for example), its computation time which scales exponentially with the number of vehicles, communication requirement and lack of security make it prohibitive. To overcome these limitations, one can use a distributed strategy, which results in a formation behavior similar to the one obtained using a centralized approach.

Recently, some decentralized receding horizon planners have been proposed in Dunbar and Murray (2006), Keviczky et al. (2006b, 2006a), Kuwata et al. (2006). In Dunbar and Murray (2006), a solution is provided for unconstrained subsystems (decoupled input constraints only). Therefore, the coupling constraints between robots cannot be taken into account. In Keviczky et al. (2006b, 2006a), a distributed planner is formulated where each robot optimizes locally for itself as well as for every neighbor at each update, resulting in an increase in the computing time and decrease in decentralization. Furthermore, in order to ensure collision avoidance, some emergency strategies must be defined. In Kuwata et al. (2006), the decentralized scheme is based on a leader–follower architecture. Indeed, the vehicles update their trajectory sequentially and the feasibility is guaranteed in spite of the presence of coupling constraints between subsystems. One advantage of the leader–follower approach is that it is easy to implement. However, there is no explicit feedback from the follower to the leader. Another disadvantage is that the leader is a single point of failure. Other decentralized strategies based on potential fields are given in De Gennaro and Jadbabaie (2006) and Dimarogonas and Kyriakopoulos (2005) but they do not satisfy the non-holonomic constraint imposed by the rolling wheels.

In this chapter, a distributed implementation, without assigning any leader, of receding horizon planning is presented. Each robot optimizes only for its own trajectory at each update and exchanges information with neighboring subsystems. The outline of this chapter is as follows: Section 25.2 formally states the robot swarm navigation problem. The main results are presented in Section 25.3. In

Section 25.4, the experiment results illustrate the effectiveness of the proposed strategy and Section 25.5 states conclusions related to the analysis of this chapter.

25.2 Mobile robots: problem statement

25.2.1 Problems and assumptions

In this work, we assume that:

- The robot swarm is composed of N_a non-holonomic robots.
- The geometric shape of each robot R_i belonging to a mobile robot swarm is represented by a 2D circle of center $C_i = (x_i, y_i)$ and of radius ρ_i.
- The obstacles are static and their shape is a 2D circle or can be included in a 2D circle. The ith obstacle in the environment is denoted by $B_i = (O_i, r_i^0)$ where O_i is the obstacle $O_i = (x_i^0, y_i^0)$ and r_i^0 is its radius.
- Robots have onboard sensors, which permit them to detect surrounding obstacles and vehicles within a range. This range is described by a circle centered at C_i.
- Each robot R_i can communicate with the other robots of the swarm. The broadcasting range, $d_{i,com}$ (> 0), is limited.

In order to formulate the planning problem, it is necessary to define the set of detected obstacles $O_i(\tau_k)$, the inter-robot collision conflict set $C_{i,collision}(\tau_k)$ and the communication conflict set $C_{i,com}(\tau_k)$.

Definition 1: For each update interval $[\tau_k, \tau_k + 1]$, $\tau_k = t_0 + kT_c$, where T_c is the update period and t_0 the initial time, and for all robots R_i, $i \in \{1, \dots, N_a\}$, of the robot swarm, the detected obstacle set $O_i(\tau_k)$ is defined as the subset:

$$O_i(\tau_k) \subset \left\{ B_{1i}\left(O_{1i}, r_{1i}^0\right), B_{2i}\left(O_{2i}, r_{2i}^0\right), \dots \right\}$$

of M_i obstacles in the range of R_i robot sensors.

Note that the detected obstacle set is time dependent and evolves as long as the robot moves and discovers new obstacles (see Fig. 25.1 and 25.2).

Definition 2: The inter-robot collision conflict set $C_{i,collision}(\tau_k) \subset R$, is the subset of all

$$d(q_i(\tau_k), q_p(\tau_k)) \leq \rho_i + \rho_p + (v_{imax} + v_{pmax})(T_p + T_c), \qquad [25.1]$$

where v_{imax} and v_{pmax} are the maximal linear velocities of robot R_i and R_p respectively, $q_i = [x_i \, y_i \, \theta_i]^T$ and q_p are the state vector of robot R_i and R_p respectively, with θ_i the robot orientation, $d(q_i, q_p) = \sqrt{(x_i - x_p)^2 + (y_i - y_p)^2}$ is the Euclidian distance between robots R_i and R_p, and T_p and T_c are the planning horizon and the update period respectively.

Definition 3: The communication conflict set $C_{i,com}(\tau_k) \subset R$ is the subset of all vehicles $R_{p \neq i}$ for which the separation distance verifies:

$$d(q_i(\tau_k), q_p(\tau_k)) \geq min(d_{n,com}, d_{p,com}) - (v_{imax} + v_{pmax})(T_p + T_c) \qquad [25.2]$$

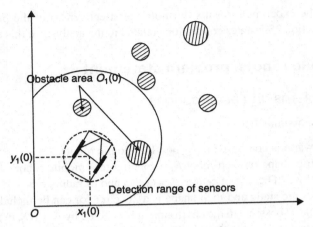

25.1 Detected obstacle set at time $\tau_k = 0$s.

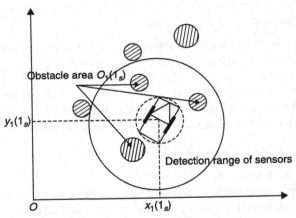

25.2 Detected obstacle set at time $\tau_k = 1$s.

25.2.2 Mobile robot modeling

Consider a multi-robot system composed of N_a wheeled mobile robots. Each mobile robot, see Fig. 25.3, is of unicycle type, with two driving wheels mounted on the same axis and independently controlled by two actuators (DC motors). The i – th robot is fully described by a three dimensional vector of generalized coordinates q_i constituted by the coordinates (x_i, y_i) of the midpoint between the two driving wheels and by the orientation angle θ_i with respect to a fixed frame:

$$q = [x_i, y_i, \theta_i]^\mathrm{T}. \tag{25.3}$$

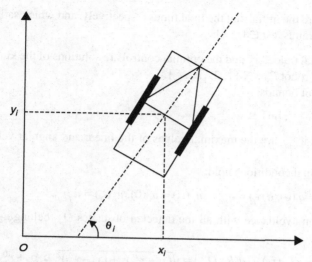

25.3 Non-holonomic differentially driven mobile robot.

Under the hypothesis of pure rolling and non-slipping condition, the vehicle satisfies the non-holonomic constraint:

$$[-\sin\theta_i \ \cos\theta_i \ 0]\,\dot{q}_i = 0$$

and the ideal kinematic equations are:

$$\dot{q}_i = \begin{bmatrix} \cos\theta_i & 0 \\ \sin\theta_i & 0 \\ 0 & 1 \end{bmatrix} u_i \qquad\qquad [25.4]$$

where $u_i = [v_i \ w_i]^T$ and v_i and w_i are the linear and angular velocities.

25.2.3 Swarm of mobile robots goal

The goal of the robots is to navigate without collision in an unknown environment with obstacles. Only the initial and final positions of each robot are known. The robots must maintain communication between themselves in order to be able to exchange information about their positions and intentions.

It is possible to express this goal as an optimal control problem with constraints.

Planning problem formulation

Find the optimal control $u_i(t)$ and the optimal trajectory $q_i(t)$ of each robot that minimizes the cost function:

$$J = \int_{t_0}^{t_f} L_i\,(q_i, u_i, t)dt \qquad\qquad [25.5]$$

where t_0 and t_f are the initial and the final times respectively, and which satisfy the following constraints $\forall\, t \in [t_0, t_f]$:

C 1 the optimal trajectory and the optimal control are solutions of the kinematic model [25.4] of robot R_i, $i \in \{1, 2, ..., N_a\}$.

C 2 the control bounds:

$$|v_i| \le v_{imax}, \ |w_i| \le w_{imax},$$

where v_{imax} and w_{imax} are the maximal values of the linear and angular velocities of robot R_i.

C 3 the terminal conditions hold:

$$q_i(t_0) = q_i(0),\ q_i(t_f) = q_{ifinal},\ u_i(t_0) = u_i(0),\ u_i(t_f) = u_{ifinal}. \qquad [25.6]$$

C 4 the collision avoidance with all the detected obstacles O_{mi} belonging to the set $O_i(\tau_k)$:

$$\forall O_{mi} \in O_{mi}(\tau_k), \quad d(R_i, O_{mi}) = \sqrt{(x_i - x_{mi}^0)^2 + (y_i - y_{mi}^0)^2} \ge \rho_i + r_{mi}^0$$

C 5 the collision avoidance with each other robot R_j belonging to the robot swarm:

$$\forall j \in \{1, ..., N_a\}, \quad i \ne j, \quad d(R_i, R_j) = \sqrt{(x_i - x_j)^2 + (y_i - y_j)^2} \ge \rho_i + \rho_j$$

C 6 the communication keeping with each other robot R_j belonging to the robot swarm:

$$\forall j \in \{1, ..., N_a\}, \ i \ne j, \ d(R_i, R_j) = \sqrt{(x_i - x_j)^2 + (y_i - y_j)^2} \ge \min(d_{i,com}, d_{j,com})$$

where $d_{i,com}$ and $d_{j,com}$ are the maximal broadcasting range of robot R_i and R_j respectively.

25.3   Mobile robots: main results

In order to implement the planning algorithm in a real robot, it is necessary to solve the problem in real time. Moreover, the robot swarm environment is unknown and each robot discovers it as it moves. So, it is impossible to calculate the robot trajectory between its initial and final position just once. Due to these constraints it is necessary to use a receding horizon planning in order to calculate in each time interval $[\tau_k, \tau_k + T_c]$, where T_c is the update period, the robot trajectory. The planning horizon is T_p, $T_p > T_c$. In order to guarantee that each robot arrives around its final position, we include in the cost function expression the terminal part: $|| q_i(\tau_k + T_c, \tau_k) - q_{ifinal}||$. On the other hand, in order to avoid the leader–follower approach and to strongly decentralize the planning algorithm, it is proposed that each robot R_i only plans its own trajectory using local information. Let $T_d \in R^+$ be the obstacle detection horizon, $T_p < T_d$. The proposed algorithm is updated each τ_k, $\tau_k = t_0 + kT_c$, and is divided into two steps:

- *Step 1.* Each robot R_i computes its trajectory by taking into account only the obstacles. Constraints C5 and C6 are not included in this first step. This trajectory is called *intuitive trajectory*. This trajectory is evaluated over the obstacles' detection horizon T_d. At the same time, robot R_i analyzes, by using its sensors, potential collisions between robots and lost communication problems, in order to build the inter-robot conflict set $C_{i,collision}(\tau_k)$ and the communication conflict set $C_{i,com}(\tau_k)$.
- *Step 2.* The robots that are in collision or lose communication with other robots of the swarm adjust their intuitive trajectories by taking into account constraints C5 and C6. This new trajectory is called *planned trajectory* and it is the trajectory that the robot must track in time interval $[\tau_k + T_c, \tau_k + 2T_c]$. This trajectory is evaluated over the planning horizon T_p.

Example: A robot formation with four robots. In the time interval $[\tau_k, \tau_{k+1}]$ the formation configuration is reported in Fig. 25.4. We can see that robot R_1 can lose communication with robot R_4 and potentially be in collision with robot R_2. These potential problems are analyzed only by using the robot sensors. Each robot analyzes potential problems at the same time as it computes its intuitive trajectory, in order to build the inter-robot conflict set $C_{i,collision}(\tau_k)$ and the communication conflict set $C_{i,com}(\tau_k)$. Robots R_1, R_2 and R_4 exchange their intuitive trajectories in order to compute the planned trajectories which guarantee that the robots can navigate without collision or lost communication. The planned trajectory of robot R_3 is its intuitive trajectory, because there is no conflict with the other robots.

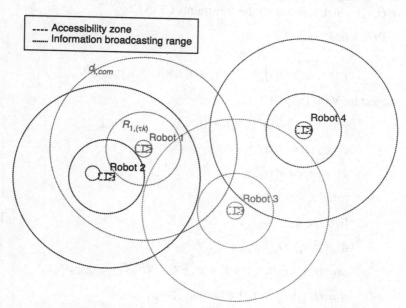

25.4 Conflict areas of robot R_1 in time interval $[\tau_k, \tau_{k+1}]$.

A more formal description follows. Over each interval $[\tau_{k-1}, \tau_k]$, let the following optimal control problem $\hat{P}_i(\tau_k)$ associated with the ith robot, which consists of determining the intuitive control input $\hat{u}_i(t, \tau_k)$ and the intuitive trajectory $\hat{q}_i(t, \tau_k)$ which only satisfies constraints C1 to C4:

Problem $\hat{P}_i(\tau_k)$:

$$\min_{\hat{q}_i(t,\tau_k), \hat{u}_i(t,\tau_k)} \int_{\tau_k}^{\tau_k+T_d} L_i(\hat{q}_i(t, \tau_k), \hat{u}_i(t, \tau_k), q_{i\,final})dt, \qquad [25.7]$$

subject to: $\forall t \in [\tau_k, \tau_k + T_d]$,

$$
\begin{cases}
\dot{\hat{q}}_i(t, \tau_k) & = f(\hat{q}_i(t, \tau_k), \hat{u}_i(t, \tau_k)), \\
\hat{q}_i(\tau_k, \tau_k) & = q_i^*(\tau_k, \tau_{k-1}), \\
\hat{u}_i(\tau_k, \tau_k) & = u_i^*(\tau_k, \tau_{k-1}), \\
|\hat{v}_i(t, \tau_k)| & \leq v_{i_{max}}, \\
|\hat{w}_i(t, \tau_k)| & \leq w_{i_{max}}, \\
d(\hat{q}_i(t, \tau_k), O_{m_i}) & \geq \rho_i + r_{m_i}^o, \quad \forall O_{m_i} \in O_i(\tau_{k-1}).
\end{cases}
\qquad [25.8]
$$

Given the conflict sets $C_{i,\,collision}(\tau_{k-1})$ and $C_{i,\,com}(\tau_{k-1})$, the intuitive trajectory $\hat{q}_i(t, \tau_k)$ and those of its neighbors, let us define the optimal control problem $P_i^*(\tau_k)$ associated with the ith robot, over each interval $[\tau_{k-1}, \tau_k]$, which consists of determining the optimal control input $u_i^*(t, \tau_k)$ and the optimal planed trajectory $q_i^*(t, \tau_k)$ which satisfies all the constraints C1 to C6:

Problem $P_i^*(\tau_k)$:

$$\min_{q_i^*(t,\tau_k), u_i^*(t,\tau_k)} \int_{\tau_k}^{\tau_k+T_p} L_i(q_i^*(t, \tau_k), u_i^*(t, \tau_k), q_{i\,final})dt, \qquad [25.9]$$

subject to: $\forall t \in [\tau_k, \tau_k + T_p]$,

$$\dot{q}_i^*(t, \tau_k) = f(q_i^*(t, \tau_k), u_i^*(t, \tau_k)), \qquad [25.10]$$

$$q_i^*(\tau_k, \tau_k) = q_i^*(\tau_k, \tau_{k-1}), \qquad [25.11]$$

$$u_i^*(\tau_k, \tau_k) = u_i^*(\tau_k, \tau_{k-1}), \qquad [25.12]$$

$$|v_i^*(t, \tau_k)| \leq v_{i_{max}}, \qquad [25.13]$$

$$|w_i^*(t, \tau_k)| \leq w_{i_{max}}, \qquad [25.14]$$

$$d(q_i^*(t, \tau_k), O_{mi}) \geq \rho_i + r_{mi}^o, \quad \forall O_{mi} \in O_i(\tau_{k-1}), \qquad [25.15]$$

$$d(q_i^*(t, \tau_k), \hat{q}_p(t, \tau_k)) \geq \rho_i + \rho_p + \xi, \quad \forall R_p \in C_{n,collision}(\tau_{k-1}), \qquad [25.16]$$

$$d(q_i^*(t, \tau_k), \hat{q}_p(t, \tau_k)) \leq \min(d_{i,com}, d_{p',com}) - \xi, \\ \forall R_{p'} \in C_{i,com}(\tau_{k-1}), \qquad [25.17]$$

$$d(q_i^*(t, \tau_k), \hat{q}_i(t, \tau_k)) \leq \xi, \qquad [25.18]$$

where $\xi \in R^+$ is a constant. Here, it is assumed that the optimal planned trajectory $q_i^*(t, \tau_k)$ is constrained to be at most at a distance ξ from the intuitive trajectory $\hat{q}_i(t, \tau_k)$ (see equation [25.18]). This constraint enforces the degree of correspondence between the planned trajectory and the intuitive trajectory known by the neighbors. Therefore, in order to guarantee collision avoidance between vehicles and preservation of communication links, deformation ξ due to the mismatch between the presumed and the optimal planned trajectories is added (see equations [25.16] and [25.17]). This process is then repeated during the robots' movement, over the interval $[\tau_{k-1}, \tau_k]$, and so on until they reach a neighborhood of their goal q_{ifinal}. As such, new information can be taken into account in the next iteration. The solution of the optimal problem is obtained by using the flatness properties of the systems, the B-spline functions and the constrained feasible sequential quadratic optimization algorithm (Lawrence and Tits, 2001).

Remark: A planning algorithm for a single robot that moves in an environment with obstacles included on a polygon shape has been proposed in Kokosy et al. (2008).

25.3.1 Receding horizon planning problems solver

An algorithm for the off-line generation of the optimal trajectory for a nonlinear system by using the flatness property of the system was proposed by Milam (2003). The optimal control problem is transformed into a parameter optimization problem by using the B-spline functions in order to define the flat outputs of the nonlinear system. However, this technique is not directly applicable on-line. It is necessary to combine it with a receding horizon strategy of planning. So, we generalize the former algorithm in order to obtain a new one that can be used on-line (Defoort et al., 2008).

Determination of the flat outputs

Definition 4: (Fliess et al., 1995) A system described by the nonlinear equation $\dot{q}_i = f(q_i, u_i)$ where $q_i \in R^n$ and $u_i \in R^m$. This system is differentially flat if a vector $z_i = [z_{1_i} \cdots z_{m_i}]^T \in R^m$ exists (dependent on the input u_i, on the input derivatives until order $l \leq n$ and on the state variable q_i) called flat output, such that the state variable q_i and the input u_i can be expressed by the flat output and its derivatives. Thus, there are three regular functions $\varphi_0 : R^n \times (R^m \times \ldots \times R^m) \to R^m$, $\varphi_1 : (R^m \times \ldots \times R^m) \to R^n$ and $\varphi_2 : R^n \times (R^m \times \ldots \times R^m) \to R^m$ such that:

$$\begin{cases} z_i = \varphi_0(x_i, u_i, \dot{u}_i \ldots u_i^{(l)}) \\ q_i = \varphi_1(z_i, \dot{z}_i, \ldots z_i^{(l-1)}) \\ u_i = \varphi_2(z_i, \dot{z}_i, \ldots z_i^{(l)}) \end{cases} \qquad [25.19]$$

For the unicycle type of robot described by equation [25.4], by choosing vector $z_i = [x_i \ y_i]^T$ as the flat output, it is possible to describe the state vector

$q_i = [x_i\ y_i\ \theta_i]^T$ and the input vector $u_i = [v_i\ w_i]^T$ by using the flat output and its derivatives:

$$
\begin{cases}
x_i = z_{1_i} \\[4pt]
y_i = z_{2_i} \\[4pt]
\theta_i = \arctan\left(\dfrac{\dot{z}_{2_i}}{\dot{z}_{1_i}}\right) \\[8pt]
u_i = \sqrt{\dot{z}_{1_i}^2 + \dot{z}_{2_i}^2} \\[8pt]
w_i = \dfrac{\dot{z}_{1_i}\ddot{z}_{2_i} - \dot{z}_{2_i}\ddot{z}_{1_i}}{\dot{z}_{1_i}^2 + \dot{z}_{2_i}^2}
\end{cases}
\qquad [25.20]
$$

Once the performance criteria (7) and (9), and constraints (8), (10)–(18) are mapped into the flat output space, the intuitive and planned trajectories are planned in this space.

B-spline function

In order to transform the problem of optimal trajectory generation into a problem of parameters optimization, a piecewise polynomial function B-spline is adopted to describe the trajectory in the flat space. Indeed, the B-spline functions have a compact support, which limits their influence on a small zone and makes them useful in an on-line implementation. Furthermore, their use allows a sufficient number of parameters to be able to satisfy the constraints of the optimal problem.

Definition 5: Let M real values t_i, called knots, with $t_0 \leq t_1 \leq \ldots \leq t_{M-1}$. A B-spline of degree d is a parametric curve $S: [t_0, t_{M-1}] \rightarrow R^2$ composed of a linear combination of basis B-splines $B_{j,d}$ of degree d

$$
S_d(t) = \sum_{j=0}^{p} C_j B_{j,d}(t)
\qquad [25.21]
$$

where C_j are the control points, and p is defined by

$$
p = n_{knot}(d-1) - \sum_{k=1}^{n_{knot}-1} S_k
$$

with

- n_{knot}: the number of intervals $[t_j, t_{j+1}]$ $(j = 1, \ldots, M-1)$ which have a non zero length,
- S_k $(k = 1, \ldots, n_{knot}-1)$: the degree of continuity of every cross point on the interval $]t_0, t_M[$. The $p+1$ basis B-spline of degree d can be defined using the recursive formula:

$$
B_{j,1}(t) = \begin{cases} 1 & \text{if } t_j \leq t < t_{j+1} \\ 0 & \text{otherwise} \end{cases}
\qquad [25.22]
$$

$$B_{j,d}(t) = \frac{t - t_j}{t_{j+d+1} - t_j} B_{j,d-1}(t) + \frac{t_{j+d} - t}{t_{j+d} - t_{j+1}} B_{j+1,d-1}(t) \qquad [25.23]$$

In our study, the three-order B-spline basis functions are used to parameterize the trajectory. For problem $\hat{P}_i(\tau_k)$ (resp. $P_i^*(\tau_k)$), the time interval $[\tau_k, \tau_k + T_d]$ (resp. $[\tau_k, \tau_k + T_p]$) is divided into n_{knot} equal segments with the corresponding union set of breakpoints:

$$[t_0 = \tau_k < t_1 < \dots < t_{n_{knot}} = \tau_k + T_d \text{ (resp. } T_p)] \qquad [25.24]$$

Figure 25.5 depicts an example of a piecewise polynomial function and the three-order B-spline basis functions associated to the breakpoints [0, 0.25, 0.5, 0.75, 1] (i.e. $n_{knot} = 4$).

The trajectories of the flat outputs are written in terms of finite dimensional B-spline curves as:

$$\begin{bmatrix} x(t, \tau_k) \\ y(t, \tau_k) \end{bmatrix} = \sum_{j=1}^{n_{knot}} C_j B_{j,3}(t) \qquad [25.25]$$

where $C_j \in R^2$ are the control points and $B_{j,3}$ is the three-order B-spline basis function (De Boor, 2001).

Nonlinear programming problem

In order to solve the optimal problem by using a computer, it is necessary to sample the time. Let:

$$\{t_0 = t_{initial} < t_1 < \dots < t_{Nech-1} = t_{final}\}$$

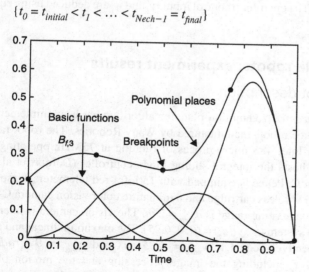

25.5 Piecewise polynomial curve expressed in terms of a linear combination of three-order B-spline basic functions.

be a uniform subdivision of the time interval $[t_{initial}, t_{final}]$ where $Nech$ is a non zero fixed integer. For each time τ_k, $\forall\ k \in \{0, ..., Nech-1\}$, the optimal criteria and the problem constraints are estimated. For example, the continuous optimization problem $\hat{P}_i(\tau_k)$ can be transformed into the following nonlinear programming one:

$$\min_{(\hat{C}_{0,i}, ..., \hat{C}_{d+n_{knot}-2,i})} J = \sum_{k=0}^{N_{ech}-1} \mu_k L_i(.)$$

where $(.) = \varphi_1(\hat{z}_i(t, \tau_k), ..., \hat{z}_i^{(l_i-1)}(t, \tau_k)), \varphi_2(\hat{z}_i(t, \tau_k), ..., \hat{z}_i^{(l_i)}(t, \tau_k)), \varphi_1(\hat{z}_i(t_{final}), ..., \hat{z}_i^{(l_i-1)}(t_{final}))$, and subject to: $\forall t \in [\tau_k, \tau_k + T_d]$,

$$\begin{cases} \dot{\varphi}_1(\hat{z}_i(t, \tau_k), ..., \hat{z}_i^{(l_i-1)}(t, \tau_k)) = f(\varphi_1(\hat{z}_i(t, \tau_k), ..., \hat{z}_i^{(l_i-1)}(t, \tau_k)), \\ \quad \varphi_2(\hat{z}_i(t, \tau_k), ..., \hat{z}_i^{(l_i)}(t, \tau_k))), \\ \varphi_1(\hat{z}_i(\tau_k, \tau_k), ..., \hat{z}_i^{(l_i-1)}(\tau_k, \tau_k)) = \varphi_1(z_i^*(\tau_k, \tau_{k-1}), ..., z_i^{*(l_i-1)}(\tau_k, \tau_{k-1})), \\ \varphi_2(\hat{z}_i(\tau_k, \tau_k), ..., \hat{z}_i^{(l_i)}(\tau_k, \tau_k)) = \varphi_2(z_i^*(\tau_k, \tau_{k-1}), ..., z_i^{*(l_i)}(\tau_k, \tau_k)), \\ \varphi_2(\hat{z}_i(\tau_k, \tau_k), ..., \hat{z}_i^{(l_i)}(\tau_k, \tau_k)) \in U_i, \\ d(\varphi_1(\hat{z}_i(t, \tau_k), ..., \hat{z}_i^{(l_i-1)}(t, \tau_k)), O_{m_i}) \geq \rho_i + r_{m_i}^o, \forall O_{m_i} \in O_i(\tau_{k-1}), \end{cases}$$

where $u_i = \{v, w_i/v_i(t) \leq v_{imax}, w_i(t) \leq w_{imax}, \forall t \in [t_{initial}, t_{final}]\}$.

Weights μ_k depend on the method of integration. The optimal control points C_j are numerically found using the constrained feasible sequential quadratic optimization algorithm (Lawrence and Tits, 2001), which allows us to calculate the flat output z_i. The open-loop control inputs v_i and w_i are deduced using equation [25.20].

25.4   Mobile robots: experiment results

25.4.1 Robot description

The proposed decentralized motion-planning algorithm has been implemented on three mobile Pekee robots manufactured by Wany Robotics. The robot radius is $\rho = 0.25\ m$. An Intel 486 micro-processor running at 75MHz operating under Linux real time hosts the integral sliding mode controller (Defoort et al. 2006; 2009), written in C. Pekee is equipped with 15 infra-red telemeters sensors, two encoders, a WiFi wireless cartridge and a miniature color vision camera C-Cam8. The maximal broadcasting range is $d_{com} = 2\ m$. The vision camera is fixed in the robot coordinate system $(x_c, y_c, z_c) = (0, 0, 0.25)$. The maximum linear and angular speeds are respectively equal to $v_{max} = 0.35\ m/s$ and $w_{max} = 0.8\ rad/s$. The computing time T_c including the image processing and the motion planning algorithm is about five minutes on the embedded 75MHz PC. In order to decrease the computing time, we used socket protocol communication and WiFi. The image data are sent to a Pentium IV *1.7GHz* PC for image processing and for generation of the time optimal trajectory. This protocol enabled us to reduce the computing time of the image to *3 s*. The localization and mapping methods reported in Defoort et al. (2007) are applied.

25.4.2 Algorithm implementation

For the experiment, we used three Pekee robots. The initial and final states and controls for the first robot are: q_1 (t_0) = *(0.5m, 0m, 0deg)*, $v_1(t_0)$ =*0m/s*, $w_1(t_0)$ = *0rad/s*, q_1 (t_f) = *(9m, 0m, 90deg)*, $v_1(t_f)$ = *0m/s*, $w_1(t_f)$ = *0rad/s*. The second robot has the following initial and final states and controls: q_2 (t_0) = *(0m, 1m, 0deg)*, v_2 (t_0) = *0m/s*, $w_2(t_0)$ = *0rad/s*, $q_2(t_f)$ = *(8m, 0m, 90deg)*, v_2 (t_f) = *0m/s*, $w_2(t_f)$ = *0rad/s*. The third robot has the following initial and final states and controls: q_3 (t_0) = *(−0.3m, −1m, 0deg)*, v_3 (t_0) = *0m/s*, $w_3(t_0)$ = *0rad/s*, q_3 (t_f) = *(7m, 0m, 0deg)*, v_3 (t_f) = *0m/s*, w_3 (t_f) = *0rad/s*. In this application, only the infra-red sensors are used for the perception of the environment. The values for the update period, the planning horizon and the obstacles detection horizon are chosen as follow: T_c = *0.5s*, T_p = *2s* and T_d = *2.5 s*. It is set the following parameters ζ = *0.25* and n_{knot} = *5*. The task is to drive these robots from the initial position to the final known position while avoiding collisions and maintaining communication constraints. Due to the existence of obstacles, robots must pass through narrow ways and constrain each other in the team. The algorithm implementation in the real robots is presented in Fig. 25.6.

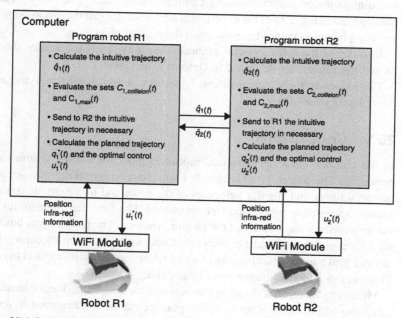

25.6 Principle of the algorithm implementation for two robots.

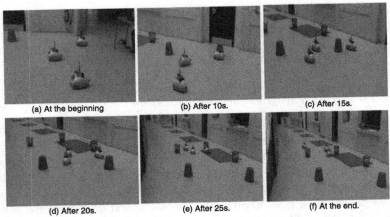

(a) At the beginning   (b) After 10s.   (c) After 15s.

(d) After 20s.   (e) After 25s.   (f) At the end.

25.7 Three Pekee mobile robots swarm from initial configurations to a desired region, avoiding collisions and maintaining communication constraints. (a) At the beginning (b) After 10s (c) After 15s (d) After 20s (e) After 25s (f) At the end.

Figure 25.7 shows six snapshots of our experiment. During the motion, each robot computes its optimal planned trajectory using local exchanged information. Figure 25.7(a) depicts the three robots at the beginning. In Fig. 25.7(a)–(f), they head toward their goal positions to form the desired geometrical shape while avoiding collision and maintaining communication links. The robot trajectories are depicted in Fig. 25.8. This experiment demonstrates that the proposed receding horizon planner manages to accomplish the desired objectives.

In order to track the optimal planned trajectory, each robot uses the robust closed-loop controller described in Defoort et al. (2006).

The videos of the experiment can be viewed at web site http://symer.ec-lille.fr/~richard/manips.htm.

25.5 Conclusions

In this chapter a new decentralized motion planner for a robot swarm has been presented. The trajectory of each robot was planned by taking into account the maximal linear and angular velocities of the robot and its kinematic model. The planner guarantees a collision-free trajectory. The fact that there is no swarm leader increases the security of the swarm missions. The planner was built with a high level of decentralization: each robot only knows its own trajectory, its own desired goal and the intuitive trajectory of robots with which conflicts may occur. This induces a low communication bandwidth.

Moreover, by using a receding horizon and the constrained feasible sequential quadratic optimization algorithm, this planner can be implemented in real time. Experimental results show the effectiveness of this approach.

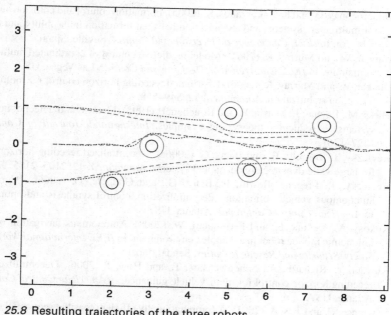

25.8 Resulting trajectories of the three robots.

25.6 Acknowledgements

This work was partially supported by the Catholic University of Lille, France, under the Entités Mobiles project.

25.7 References

De Boor, C. (2001). *A practical guide to splines*. Springer Verlag.

De Gennaro, M. and Jadbabaie, A. (2006). Formation control for a cooperative multi-agent system using decentralized navigation functions. In *American Control Conference*, Atlanta, USA.

Defoort, M. (2007). Contributions à la planification et à la commande pour les robots mobiles coopératifs. PhD thesis, Ecole Centrale de Lille.

Defoort, M., Floquet, T., Kokosy, A., and Perruquetti, W. (2006). Integral sliding mode control for trajectory tracking of a unicycle type mobile robot. Integrated Computer-Aided Engineering, 13, 277–288.

Defoort, M., Floquet, T., Perruquetti, W., and Drakunov, S. (2009). Integral sliding mode control of an extended Heisenberg system. *IET Control Theory and Applications*, to appear.

Defoort, M., Palos, J., Kokosy, A., Floquet, T., and Perruquetti, W. (2008). Performance-based reactive navigation for non-holonomic mobile robots. *Robotica*, 27, 281–290.

Defoort, M., Palos, J., Kokosy, A., Floquet, T., Perruquetti, W., and Boulinguez, D. (2007). Experimental motion planning and control for an autonomous non-holonomic mobile robot. In *IEEE International Conference on Robotics and Automation*, Roma, Italy.

Dimarogonas, D. and Kyriakopoulos, K. (2005). Formation control and collision avoidance for multi-agent systems and a connection between formation infeasibility and flocking behavior. In *IEEE Conference on Decision and Control*, Seville, Spain.

Dunbar, W. and Murray, R. (2002). Model predictive control of coordinated multi-vehicle formations. In *IEEE Conference on Decision and Control*, Las Vegas, USA.

Dunbar, W. and Murray, R. (2006). Distributed receding horizon control for multi-vehicle formation stabilization. *Automatica*, 42, 549–558.

Fliess, M., Levine, J., Martin, P., and Rouchon, P. (1995). Flatness and defect of non-linear systems: introductory theory and examples. *International Journal of Control*, 61, 1327–1361.

Keviczky, T., Borrelli, F., and Balas, G. (2006a). Decentralized receding horizon control for large scale dynamically decoupled systems. *Automatica*, 42, 2105–2115.

Keviczky, T., Fregene, Borrlli, F., Balas, G., and Godbole, D. (2006b). Coordinated autonomous vehicle formations: decentralization, control synthesis and optimization. In *American Control Conference*, Atlanta, USA.

Kokosy, A., Defaux, F., and Perruquetti, W. (2008). Autonomous navigation of a non-holonomic mobile robot in a complex environment. In *IEEE International Workshop on Safety, Security and Rescue Robotics*, Sendai, Japan.

Kuwata, Y., Richards, A., Schouwenaars, T., and How, J. (2006). Decentralized robust receding horizon control for multi-vehicle guidance. In *American Control Conference*, Atlanta, USA.

Lawrence, C. and Tits, A. (2001). A computationally efficient feasible sequential quadratic programming algorithm. *SIAM J. Optim*, 11, 1092–1118.

Milam, M. (2003). Real-time optimal trajectory generation for constrained dynamical systems. PhD thesis, California Institute of Technology.

Mintzberg, H. (1979). *The structuring of organizations*. Englewoods Cliffs, N.J, Prentice-Hall.

26

Using the NVIDIA CUDA programme to develop cognitive supervision of multi robot systems

J. BĘDKOWSKI and A. MASŁOWSKI,
Warsaw University of Technology, Poland

Abstract: This chapter describes the CUDA application in the cognitive theory-based approach to multi mobile robot control. The model of the cognitive supervisor and its main role in the robotic system is described. The new capabilities derived from the usage of GPU architecture give an opportunity for real time computation in 3D map building. The idea of real time 3D map reconstruction and analysis by autonomous navigation module is also shown. The need for supervision of the navigation module is presented. The experiments based on simulated and real environment prove the advantage of cognitive supervision. Furthermore the CUDA application shows new capabilities for robotic applications. The robot's task can be achieved quickly and the environmental map can be stored and reconstructed with high precision. The robot can be navigated autonomously in a complex and unstructured environment even from an onboard PC or remotely, therefore the approach supports the autonomous tele-operation of a remotely controlled robot by the robot assistant. Thus, any problems in communication with the base station can be partially managed.

Key words: mobile robot control, supervision.

26.1 Introduction

Cognitive supervision of the robotic system for crisis and disaster management is a new idea (Masłowski, 2006; Będkowski, 2008), which improves its application. The main goal of the approach lies in the decision selection process, the model of which is provided by the human supervisor. The new generation inspection system is remarkable for the availability of the human–robot interaction system in the structure. To achieve this interaction, the human machine interface, as a means to construe the real cognitive interaction system into modelled/simulated interaction, has to be implemented. The new idea is based on cognitive map implementation for self-reasoning using the cognitive model of the supervisor. The cognitive map has an advantage over cognitive model learning. The cognitive model is able to learn the environmental structure and establish a proper interaction between robots and environment. Therefore fully autonomous cognitive supervision solves crisis events and is inspired by human interaction with the robotic system.

The NVIDIA compute unified device architecture (CUDA) programming model for use with graphics processing units (GPUs) is used for cognitive map component development, therefore real time operation is achieved. The following algorithms are developed with CUDA: 3D map building and reconstruction, supervision of the autonomous navigation in 3D and fast classification with fuzzy ARTMAP CUDA implementation.

26.2   Cognitive model architecture

The architecture of the cognitive model of the supervisor is shown in Figure 26.1. The main component of the architecture, called the cognitive layer, solves the problem of self-reasoning for supervision and control tasks.

The cognitive model of the supervisor is composed of the cognitive layer which collaborates with the mission planning module. The cognitive layer interacts with the system of mobile agents through the behaviour layer, therefore the actual robot's behaviour can be corrected to avoid damage in hazardous environments. It is important to introduce the idea of the implementation cognitive layer as a composition of procedures solving classification, finding similarity, building geometrical representation of the environment tasks, etc. based on the data of the sensors. Figure 26.2 shows the general scheme of the cognitive layer.

The cognitive layer is equipped with mechanisms for data recording, discovering obstacles, and interpreting crisis events from all the senses. The result is a cognitive map of space, which corresponds to the knowledge of the cognitive model about its existence. Thus the reaction is based and strictly dependent on the collected knowledge on the environment and the system's condition.

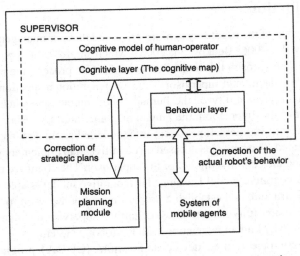

26.1 Architecture of cognitive model of the supervisor.

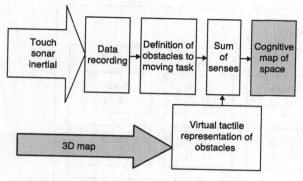

26.2 Cognitive layer.

26.3 NVIDIA compute unified device architecture (CUDA)

The compute unified device architecture (CUDA) programming model, which is based on an extended ANSI C language and a runtime environment, allows the programmer to explicitly specify parallel data computation. NVIDIA developed CUDA to open the architecture of their graphics accelerators to more general applications. The CUDA algorithm is based on the kernel execution in the grid of thread blocks (Fig. 26.3). There is a limited maximum number of threads that a block can contain. The implementation uses 256 threads (16×16). Blocks of the same dimensionality and size that execute the same kernel are batched together into a grid of blocks, therefore the number of threads that can be launched in a single kernel invocation is much larger (NVIDIA CUDA, 2007):

$$NTh = Dx\_Th * Dy\_Th * Dx\_B * Dy\_B$$

where:
NTh = number of threads
Dx_Th = number of rows in thread table
Dy_Th = number of columns in thread table
Dx_B = number of rows in grid table
Dy_B = number of columns in grid table

Each thread is identified by its thread ID, which is the thread number within the block ID_Th(x, y) = x + y*Dx_Th. Each block is identified by its block ID, which is the block number within the grid ID_B(x, y) = x+y*Dx_B. Each thread executes the kernel function for one triangle of the scene, therefore the maximum number of triangles is limited by the number of threads (NTh). Figure 26.3 shows the thread organization as a grid of thread blocks.

The thread realizes the kernel function for the input data assigned by its block ID, therefore it is suggested that the algorithm should work with a large amount

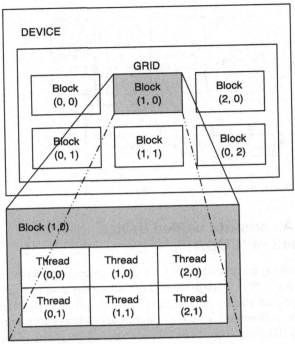

26.3 CUDA GPU thread organization as a grid of thread blocks.

of data to increase the performance of the GPU usage. The bottleneck for the CUDA is determined by the memory copy from host to device and symmetrically from device to host.

26.4   3D map building with CUDA

A geometrical map of the environment is represented by a set of triangles built from a cloud of measured points during the 3D data acquisition task. The following figures show the idea of real time map building for the mobile robot, which is equipped with 2D laser scanner mounted vertically on the chassis. Therefore the 3D cloud of points can be acquired, as the third dimension is given by robot translation. Figure 26.4 shows mobile robot ATRV Jr equipped with LRF SICK LD1000 for 3D cloud of points measurement and LMS SICK221 for SLAM purposes. The CUDA result of computing lines from the cloud of points is also shown.

The line extraction algorithm can be coded into multi parallel CUDA architecture, therefore computation time is acceptable (200 ms–400 ms for the lines extraction from single scan composed from up to 256 points). The algorithm's steps are described below:

a) Compute all lines from each pair from points (parallel).

b) For each line compute number of points with distance to line < tolerance (parallel).

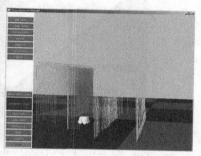

26.4 Robot ATRV Jr and the result of the CUDA line extraction algorithm.

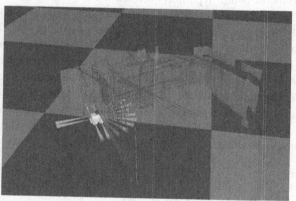

26.5 Result of the triangle computation algorithm.

c) Find the line that has the largest number of points.
d) Erase found points from set, go to c).

Then, the line extraction triangles are computed. Figure 26.5 shows the result of triangle computation.

26.5 Supervision of autonomous navigation in 3D

Autonomous navigation supervision in 3D is dependent on the geometrical map. The map should represent the environment as much as possible, therefore an increased number of calculations are required to detect potential collisions with

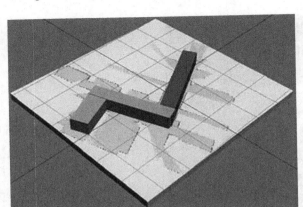

26.6 Supervision of the path planning algorithm with CUDA.

the robot on the map. The idea of robot navigation in a 3D virtual world built from 3D laser range finder (LRF) data is based on the verification of the hypothesis of motion without collision. Multi-hypothesis verification is a highly computational task; therefore CUDA capabilities are used for solving real time navigation supervision. The basic object of the algorithm is the rectangular prism. The scene is represented as a set of triangles and cloud of points. The fundamental procedure of 3D navigation is based on verification if there is an intersection of the triangle from the scene with the current rectangular prism. If intersection appears the probability of safety navigation in the current direction is low and decreases when another intersection is detected. Figure 26.6 demonstrates the supervised path, which the robot should autonomously follow without collision. The time of hypothesis computation for 102,400 triangles is less than 17 ms, therefore real time operation is achieved.

26.6   Cognitive data classification

The classifier fuzzy ARTMAP is used for building a cognitive map of space (Będkowski, 2007). The fuzzy ARTMAP system includes a pair of fuzzy ART modules (ART_a and ART_b), as shown in Fig. 26.6. During the supervised learning, ART_a receives a stream $\{a^{(p)}\}$ of input patterns and ARTb receives a stream $\{b^{(p)}\}$ of input patterns, where $b^{(p)}$ is the correct prediction given $a^{(p)}$. These modules are linked by an associative learning network and an internal controller that ensures a cognitive model of the supervisor operation in real time.

The ART_a complement coding processor transforms the M_a vector **a** into the $2M_a$-vector $A=(a,a^c)$ at the ART_a field F_0^a. A is the input vector to the ART_a field F_1^a. Similarly the input to F_1^b is the $2M_b$ vector (b, b^c). When a prediction ART_a is disconfirmed at ART_b, inhibition of map field activation induces the match tracking process. Match tracking raises the ART_a vigilance ρ_a to just above the F_1^a to F_0^a

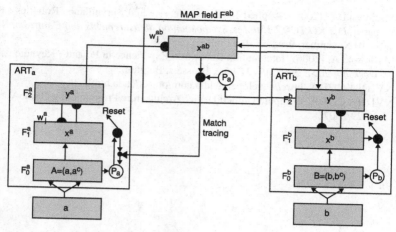

26.7 Fuzzy ARTMAP architecture.

match ratio $|x^a|/|A|$. This triggers an ART_a search which leads either to activation of an ART_a category that correctly predicts **b** or to a previous uncommitted ART_a category node. The result of supervised learning of the fuzzy ARTMAP is a set of categories of the virtual tactile representation of the 3D map, inertial, sonars and touch data, and an associated set of the predicted categories of the supervisor responds. The idea of using NVIDIA CUDA for fuzzy ARTMAP algorithm is based on using double GPU with CUDA for solving parallel ART_a and ART_b category discovery processes. The approach gives an advantage over real time supervised learning of the network of up to 102,400 categories. The time of the network response is independent from the amount of learned categories (<40 ms), therefore this solves the main problem in the previous implementations.

26.7   References

Będkowski J. (2007) 'Fast hybrid classifier DT-FAM Decision Tree Fuzzy Art Map for high dimensional problems in mobile robotics', *Proceedings of the Sixteenth International Symposium on measurement and Control in Robotics*, ISMCR, June 21–23, Warsaw, Poland.

Będkowski J. and Masłowski A. (2008) 'Cognitive model of the human supervisor for insection/intervention robotic system', *Third Conference of Young Scientists Towards the Challenges of Contemporary Technology*, 22–24 September, Warsaw, Poland.

Będkowski J., Kowalski P. and Masłowski A. (2008) 'Computational intelligence in cognitive supervision of the robotic system for crisis and disaster management', *The 4th International Conference on Information & Communication Technology and Systems*, 5 August, Surabaya, Indonesia, p. 556–561.

Masłowski A. (2004) 'Intelligent Mobile Robots supporting Security and Defense', *Proc. of the Int. Conference, IMEKO TC-17 Measurement and Control in Robotics*, NASA Space Center, Houston, Texas, USA.

Masłowski A. (2005) 'Applied Intelligent Service and Surveillance Robotics', Keynote paper, *IMEKO TC-17 15th Int. Symposium on Measurements and Control in Robotics*, 8–10 November, Brussels, Belgium.

Masłowski A. (2006) 'Intervention Robotics – Experience in Poland', Keynote paper, *Int. Workshop RISE 2006*, 19–21 June, Brussels, Belgium.

NVIDIA CUDA, http://developer.nvidia.com/object/cuda.html.

NVIDIA CUDA, Compute Unified Device Architecture Programming Guide Version 1.1 11/29/2007.

27

Laser based cooperative multi-robot map building for indoor environments

Y. ATAŞ, Ö. ÇAYIRPUNAR, S. BURAK AKAT and
V. GAZI, TOBB University of Economics and Technology, Turkey
and L. ALBOUL, Sheffield Hallam University, United Kingdom

Abstract: In this chapter a cooperative map-building scenario in a small experimental area with multiple robots is implemented. In this scenario, all the robots enter the area in a sequential manner from the same entry point, which serves as the origin for the localization information. The robots cooperatively perform division of labor such that one of the robots follows the right wall, another robot follows the left wall in order to extract the map of the outskirts of the closed environment, while the remaining robots navigate and extract the map of the interior. A base station constructs a 2D map of the environment using the data from all the robots. Artificial potential functions are used in implementation in order to control the movement of the robots and guarantee collision avoidance with obstacles or other robots.

Key words: multi-robot systems, cooperative map building, laser range finder, wireless ad-hoc network.

27.1 Introduction

Map building is an important problem in mobile robotics. The importance of the problem is due to the fact that many mobile robot systems require environment maps for their operation. Examples include the Helpmate hospital delivery robot (King and Weiman 1990), Simmons's Xavier Robot (Simmons 1996), and various museum tour-guide robots (Burgard, Cremers et al. 1999; Horswill 1994; Nourbakhsh, Boenage et al. 1999; Thrun, Beetz et al. 2000).

Over the last decade, there has been extensive work on map building for mobile robots (see e.g., Chatila and Laumond 1985; Leonard, Durrant-Whyte and Cox 1992; Rencken 1993; Thrun, Fox, and Burgard 1998). Some works seek to devise abstract, topological description of the environment (Borenstein, Everett and Feng 1996; Choset 1996; Kuipers and Byun 1996; Mataric 1994), whereas other approaches generate more detailed, metric maps (Chatila and Laumond 1985; Elfes 1989; Lu and Milios 1997; Moravec 1988). Naturally, multi-robot map building becomes increasingly robust and more efficient than single robots, since multiple robots work in coordination and jointly explore unknown environments. Increasing efficiency is one of the key reasons for using multi-robots instead of single robots. The more robots that build the map of the environment, the less time is needed for constructing the complete map. What is more, during the map

599

building somehow if one or more robots fail, the other robots will probably complete the task but with a single robot it is impossible to complete the task if the robot fails. However, compared to the single robot mapping, the extension to the multiple robots has several challenges, including coordination of robots which is a difficult problem, integration of information collected by different robots into a consistent map which needs extra calculation power and development of appropriate algorithms, and dealing with limited communication which can limit the coordination of robots.

In order to build a consistent map of the environment, the data collected by different robots has to be integrated into a single map. Furthermore, such integration should be done as early as possible to increase the reliability of the map. If the initial locations of the robots are known and we have accurate localization, map merging becomes a straightforward extension of a single robot mapping (Fenwick, Newman and Leonard 2002; Konolige, Fox et al. 2002; Thrun 2001). However, in many applications it might be difficult to integrate the information obtained from different robots due to inconsistency arising from inaccurate information about the initial postures of the robots or localization errors during operation. Therefore, direct integration of data obtained from several robots, without appropriate processing, may lead to inaccurate and possibly useless maps.

During the mapping of large-scale environments, communication between the robots and the base station might temporarily fail due to the limited communication ranges. To achieve robustness against such failures, each robot has to be able to record sensor measurements and odometry readings for mapping to be sent to the base station when communication is active again. In multi-robot systems in order to overcome this problem and increase the robustness of communication, some robots might be used as communication nodes (i.e., routers) between the other robots and the base station during the operation. So if one or more communication nodes fail, the others may keep working in the communication network and guarantee continuous information flow between the swarm of robots and the base station.

In this chapter, the problem of cooperative multi-robot online metric map building using scanning laser range finder for measuring distances and odometry information for localization during the experiment is considered. The online characteristic is important for many robot operations, such as the robot exploration problem where mapping is constantly needed for decision making as to where to move next (Burgard, Fox et al. 2000; Simmons, Apfelbaum et al. 2000). Since the mapping is online, as time progresses the map becomes more and more up-to-date and complete.

This chapter is organized as follows: in Section 27.2, we define the problem of cooperative map building with mobile robots, and in Section 27.3, the descriptions of the experimental set-up and the environment are given. In the next section, the results obtained in the experimental arena in our laboratory are presented and how the data collocated by the multiple robots can be integrated into a consistent map of the environment is briefly discussed. In the last section, the chapter is concluded with some final remarks.

27.2 Multi-robot map building: problem definition

In this chapter a system consisting of N mobile robots (agents) moving in R^2 with continuous-time non-holonomic dynamics given by:

$$\dot{x}_i(t) = \overline{v}_i(t) \cos(\theta_1(t))$$

$$\dot{y}_i(t) = \overline{v}_i(t) \sin(\theta_1(t)) \hspace{3cm} [27.1]$$

$$\dot{\theta}_i(t) = w_i(t)$$

where $x_i(t)$ and $y_i(t)$ are the Cartesian coordinates, and $\theta_i(t)$ is the steering angle of the ith agent at time t is considered. The control inputs of the ith agent are the linear speed (not velocity) $\overline{v}_i(t)$ and the angular speed $w_i(t)$.

The robots are required to build a map of an unknown enclosed area with obstacles. Robots are equipped with scanning laser sensors in order to detect obstacles at different distances or boundaries of the area. Moreover, they possess an internal indoor self-localization routine based on odometry information. The distance measurement of the laser sensor at every sample, together with the odometry information of the robots, will be used for building a map of the environment. The robot's information about the unknown area is a distance vector, which consists of the measured distance and orientation of all samples and corresponds to an array of data. For navigation purposes, this array is grouped into six clusters with 36 degrees each (see Fig. 27.1) and every cluster has 103 samples of the laser sensor data in this experiment. There is 0.351 degrees difference between two samples and the first sample of the first cluster has 108 degrees and the last sample of the last cluster has a -108 degrees steering angle in the robot's local coordinate system, which implies that the first sample corresponds to the measured distance of the nearest object at 108 degrees in the robot's local coordinate system and other samples also have corresponding angles. Let the $d^i_{pn}(t) = [d^i_{xpn}(t), d^i_{ypn}(t)]$ be the distance vector of nth sample of pth cluster for robot

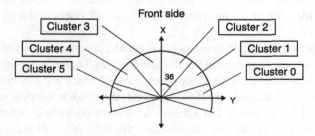

27.1 Laser cluster scheme.

i at time t. Here $d^i_{xpn}(t)$ represents the x component of the distance vector which is given by $d^i_{xpn}(t) = \|d^i_{pn}(t)\| \cos(\theta^i_{pn}(t))$, where $\|d^i_{pn}(t)\|$ represents the distance of that particular sample at time t and $\theta^i_{pn}(t)$ represents the orientation information of that particular sample at time t. Likewise $d^i_{ypn}(t)$ represents the y component of the distance vector which is $d^i_{ypn}(t) = \|d^i_{pn}(t)\| \sin(\theta^i_{pn}(t))$. Note that from the laser sensor data one can obtain both $\|d^i_{pn}(t)\|$ (the distance) and $\theta^i_{pn}(t)$ (the orientation) of the sample with high accuracy. For navigation purposes only the shortest distances in the clusters for generating the artificial potential forces (to be discussed below) are used. This simplification significantly reduces the computational complexity while the performance of the system is preserved on a satisfactory level.

Robots communicate with other robots in the system and share their information and the map-building task is performed cooperatively. The robots perform the map-building task cooperatively as one of the robots follows the wall on the left side, and another robot follows the wall on the right side in order to extract the map of the outskirts of the unknown enclosed area. The remaining robots perform an informed random walk with obstacle avoidance in the region in order to extract the map of the interior region of the area.

In order to perform the mentioned navigations of the robots, artificial potential functions are used for the low-level control algorithm. Potential functions have been successfully applied to swarm behaviors like swarm aggregation, foraging, and formation control (Gazi and Passino 2004; Gazi, Fidan et al. 2007). In particular, for the wall following behavior, a potential field which provides an attraction force is used if the distance to the wall is larger than the desired distance and a repulsion force if the distance to the wall is smaller than the desired distance and requires the robot to move along its negative gradient. Once the distance is equal to the desired distance attraction and repulsion forces are in equilibrium and there is no force acting on a given robot, the robot continues its motion without a change in its current linear speed $\bar{v}_i(t)$ and steering angle $\theta_i(t)$.

The plot of the potential function is given in Fig. 27.2. As one can see from the figure the desired distance for this particular plot is 20 cm. Only the leftmost and the rightmost clusters of the data array are considered for wall following from left side and right side, respectively. In other words, for wall following only the data from the clusters on the side of the wall contributes to the value of the potential function while the data from the other clusters is still checked for obstacle avoidance.

In random walk with obstacle avoidance, a repulsive potential force is invoked for distances which are smaller than the desired distance, i.e., only the repulsion part of the potential function is used, in order to avoid obstacles. For distances larger than the desired distance, on the other hand, the robot is moving straight with a predefined linear speed. For random walk with obstacle avoidance all clusters of the data array are considered (i.e., the data from all the clusters contribute to the value of the potential function). The gradient of the potential

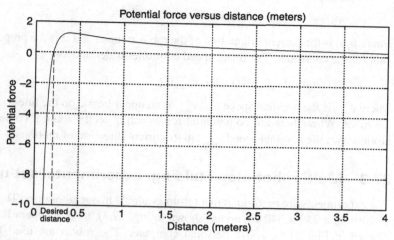

27.2 Potential force versus distance.

function for a particular cluster p is denoted as $G^i_{pn}(t)$ for robot i at time t. In the study this value is determined using the equation:

$$G^i_p(t) = d^i_p(t) \left(\frac{1}{\|d^i_p(t)\|^2} - \frac{k}{\|d^i_p(t)\|^3} \right)$$

[27.2]

where k represents the desired distance at which the robots are required to follow the walls or avoid obstacles and $\|d^i_p(t)\| = \min_{j=1,-,n} \{\|d^i_{pj}(t)\|\}$ and $d^i_p(t)$ is the vector corresponding to that sample only. In other words, as mentioned above, in calculating the potential forces from each cluster only the sample with the smallest distance value within that cluster is considered.

Using the above, the total potential field $G^i(t)$ for robot i at time t is determined as:

$$G^i(t) = \sum_{j=1}^{L} G^i_j$$

[27.3]

where L represents the number of clusters in the data array. Also the local coordinate system of the robots is used for calculating the forces acting on them.

Since the robots are non-holonomic (i.e., they cannot move in the direction of the axis connecting the two main wheels) and since their orientations may not necessarily be along the vector potential initially, the desired direction of the motion is defined as:

$$\theta_{id}(t) = atan2 \left(G_{yi}(t), G_{xi}(t) \right)$$

[27.4]

where $G_{xi}(t)$ and $G_{yi}(t)$ represent the x and y components of the potential field, respectively. Then the orientation dynamics of robot i is controlled by using a simple proportional controller:

$$w_i(t) = -\alpha(\theta_i(t) - \theta_{id}(t)) \qquad\qquad [27.5]$$

where $\theta_i(t)$ is the current orientation of the robot at time t and α is the proportional gain for the controller. The linear speed controller is defined as

$$\overline{v}_i(t) = \min\{v_{nom} + \|G_i(t)\| \quad v_{max}\} \qquad\qquad [27.6]$$

where v_{nom} is the nominal speed and v_{max} is an upper bound on the linear speed of the robot. When there are no potential forces acting on the robot it continues its motion with the nominal speed v_{nom}, in its current direction of motion.

27.3   Multi-robot map building: experimental set-up

The experiments are performed in a set-up available in our laboratory. This set-up consists of a 240 × 340 cm arena (shown in Fig. 27.3), three Khepera III robots (shown in Fig. 27.4), and a personal computer. The robots are used for map building while the computer is used as a base station. Walls and some obstacles are placed inside the arena in order to simulate a more realistic environment. The walls and obstacles are 20 cm high.

The Khepera III mini mobile robots (see Fig. 27.4) used for the experiment are produced by the K-Team Company. The robot base includes an array of nine infrared sensors for obstacle detection as well as five ultrasonic sensors for long-range object detection. The robot movements are provided with two DC servo motors with incremental encoders. Each motor is driven by its own PID controller implemented on a PIC18F4431 microcontroller and these microcontrollers are also used for measuring odometry position information. The motor control blocks act as slave devices on an I2C bus. In order to measure sensor readings a DSPIC30F5011 microcontroller working at 60 MHz is used. The communication between the microcontrollers and the main microprocessor is provided via the I2C bus.

The robots are equipped with a KoreBotLE card which is a powerful embedded platform based on a XScale PXA255 processor running at 400 MHz clock speed as the main central processing unit. KoreBot LE has 64 Mbytes memory space and 32 Mbytes of non-volatile storage space. The operating system is based on Linux kernel 2.6. The board also has a Compact Flash Type I connector on board

27.3 The experimental arena.

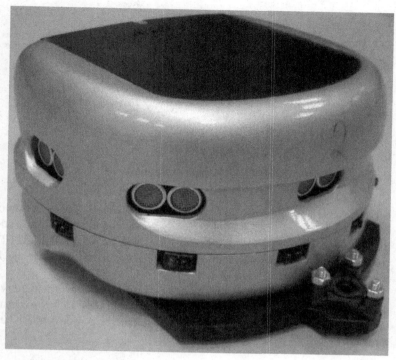

27.4 The Khepera III mini mobile robot.

and a wireless LAN IEEE 802.11b compact flash card connected to the same connector. The communication between the robots and the main computer is provided via the wireless connection interface. It is also possible to connect some extension boards available for the Khepera IIIs using the KB-250 bus, such as a USB camera, wireless camera, and other KoreBot extensions.

For the experiments the Hokuyo URG-04LX laser scanner sensor (shown in Fig. 27.5a) and an extra ball-caster have additionally been integrated with the Khepera IIIs. The URG-04LX is a compact, light weight, high performance, and low power consumption (2.5 W) laser scanner sensor for robotic applications. It has a very compact design (L: 50 mm, W: 50 mm, H: 70 mm), which enables users to attach it to a small robot such as a Khepera III. The sensor is capable of scanning a 240° circular area with 0.36° resolution and has a maximum measurement range up to 4000 mm (see Fig. 27.5b). Its power consumption, 5 V-500 mA, makes it a natural choice for battery operated vehicles.

The Hokuyo URG-04LX sensor has RS232 and USB connections onboard. RS232 serial connection is used for integrating the laser. However, it has RS232 voltage levels in contrast to the TTL 3.3 V levels of the KoreBot LE board on the Khepera III robot. Therefore, an interface card is designed for communication

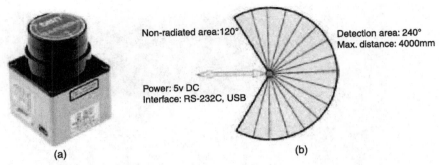

Non-radiated area:120°

Detection area: 240°
Max. distance: 4000mm

Power: 5v DC
Interface: RS-232C, USB

(a)   (b)

27.5 Hokuyo URG-04LX laser range finder.

between KoreBot LE cards and Hokuyo Laser scanners. This electronic circuit consists of a Maxim MAX3223 [22] RS232 level converter and some additional parts (shown in Fig. 27.6). MAX3223 is a +2.5V to +5.5V RS-232TTL level conversion transceiver and handles the level conversion between the Hokuyo sensor and the KoreBot LE board.

When the Hokuyo laser scanner is connected to the robots the battery life for autonomous operation drops from around 30 minutes to approximately 15 minutes. Therefore, because of the inadequate battery power of the Khepera III

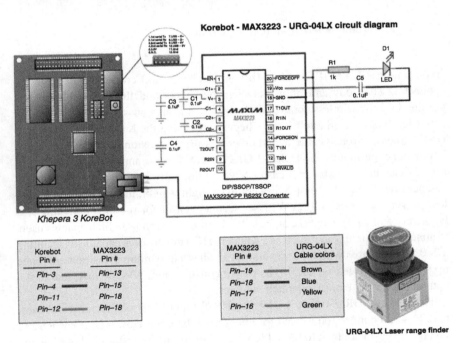

Korebot - MAX3223 - URG-04LX circuit diagram

Khepera 3 KoreBot

Korebot Pin #	MAX3223 Pin #
Pin–3	Pin–13
Pin–4	Pin–15
Pin–11	Pin–18
Pin–12	Pin–18

MAX3223 Pin #	URG-04LX Cable colors
Pin–19	Brown
Pin–18	Blue
Pin–17	Yellow
Pin–16	Green

URG-04LX Laser range finder

27.6 Connection diagram of Hokuyo URG-04LX laser range finder.

robot, a secondary lithium polymer battery is also added in order to sustain longer mobility. With the help of an external two cell lithium polymer battery at a capacity of 7.4V and 2250 mAH the robots have managed to work for about 45 minutes, which is sufficient for the experiments in this implementation. The picture of a Khepera III robot connected with a Hokuyo laser range finder and an external battery is shown in Fig. 27.7. Three Khepera III robots integrated with a Hokuyo URG-04LX laser scanner were used in the experiment.

The Khepera III robot only has two wheels for movement and a plastic piece, which is located on the bottom of the robot, at the front, and is used to sustain its balance. This plastic piece erodes over time. Eventually, the robot odometry begins to be corrupted by the plastic piece as friction between the ground and robot increases and this problem causes localization errors in the robot to occur more quickly. As a result of this problem and in order to improve the localization accuracy, the plastic has been removed and an extra ball-caster has been placed at the front of the robot with a plastic holder (see Fig. 27.7). With this replacement, localization errors are decreased.

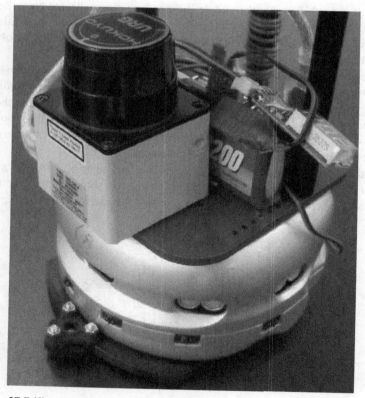

27.7 Khepera III connected with a Hokuyo URG-04LX laser range finder.

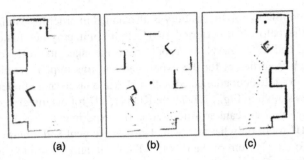

27.8 Local maps of each robot after consistent map is completed.

Finally, an external personal computer is used for map building and the final user interface. The map data is built from the individual maps collected from different robots over the wireless network (see Fig. 27.8). The figure shows three snapshots collected from the information obtained from different robots. The sub-figure on the left hand side shows the data collected from the robot following the left wall, the sub-figure in the middle is the information collected from the robot moving in between, and the sub-figure on the right hand side shows the data obtained from the robot following the right wall. This is independently obtained from the robots.

It can be seen that all these maps are incomplete on their own. Note also that for this implementation because of the particular motion control and coordination strategy considered, distances greater than 0.5 meters are not plotted on the map. The main constraint of this strategy is the error in steering angle of the robots. The small errors in steering angle of the robots cause large localization errors of objects and boundaries of area for large distances.

Matlab software is used for data acquisition and map building. In a visual window screen the growing map of the environment is shown at the same time as the robots are cooperatively moving and scanning the environment (see Fig. 27.9).

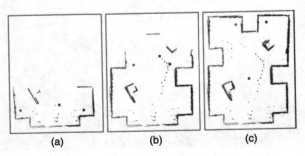

27.9 Growing map of the experimental area.

The figure shows the combined map of the environment (using the information from three robots) at different time instances. The sub-figure on the left shows the map around t = 30 s, the sub-figure in the middle shows the map around t = 120 s, while the sub-figure on the right shows the completed map of the experimental arena.

Additionally, if the distance between the base station and the robots becomes too large to sustain a reliable communication, a fourth robot might join the network and act as a communication bridge between the computer and the robots. This extends and nearly doubles the communication range. The communication bridge is not needed in this experimental set-up because of the small search area. However, if a large area is going to be mapped, such a communication bridge can easily be employed and the data flow is obtained using multi-hop communication.

The information transferred from the robots that are scanning the area to the computer is a data packet which is a combination of the robots' odometry information, the laser scan data and some other robot information like robot ID, autonomous navigation mode of the robot and communication parameters. The odometry data composes the position and orientation information relative to the starting point. The laser scan data is a 682 sample distance information which is scanned in a 240° circular area around the laser scanner with 0.36° angular resolution, 1 mm distance resolution and ±10 mm accuracy. According to the requirements a smaller sample size can be chosen.

The robots have three different autonomous navigation modes. One of the modes is informed random walk with obstacle avoidance, the flowchart of which is shown in Fig. 27.10. After the initialization, the robot is told whether to follow a wall to extract the 'outer map' or to navigate in the inner environment. If the robot is assigned the navigation of the interior it performs a random walk. At each step it reads the laser sensor data to navigate in the environment. After each navigation period, it stops for a moment and reads laser sensor data and odometry information and then sends the data to the base station and other robots. If it is needed (due to a large distance to the base station), data may be sent through an intermediate robot, which serves as a router in the communication network. Due to consistency problems with odometry and laser data information, robots suspend their navigation for a short period of time before sending their odometry and laser scan information to the base station and each other. This action makes laser data measurements for the position that will be sent to base station and other robots more reliable.

After sending data, the robots use the laser data in a potential function for navigation process. During the map-building process, the robot that is assigned to carry out a random walk receives a signal from base station when it gets near to a position that the wall-following robots have passed. Then it changes its direction to another position. With this algorithm, the robot that performs the random walk does not pass positions that the wall-following robots have passed. The other modes are left wall follow and right wall follow which operate based on the

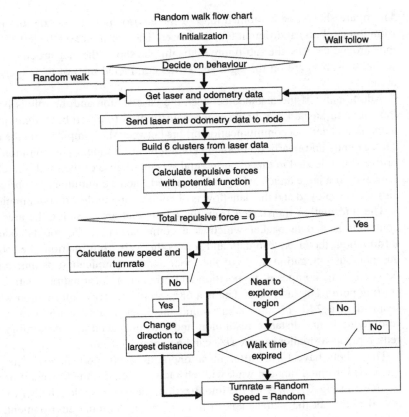

27.10 Flow chart of random walk.

second flowchart shown in Fig. 27.11. After being assigned which wall to follow, the sequence of the process is similar to the process in random walk. The robot sends data and uses laser data in a potential function for the wall-follow process. The difference is that the potential function is calculated a little differently from the random walk as discussed before.

27.4   Multi-robot map building: results

As mentioned above, the implementation of the method discussed in the earlier sections is performed by three Khepera III robots. One of the robots is assigned to follow the right wall implying that this robot is meant to start mapping on the right side of the entrance. Likewise another robot is assigned to follow the left wall implying that it is going to map the area on the left side of the entrance. The third robot performs a modified random walk with obstacle avoidance and maps the area in between the other robots. In fact all the robots in the system avoid

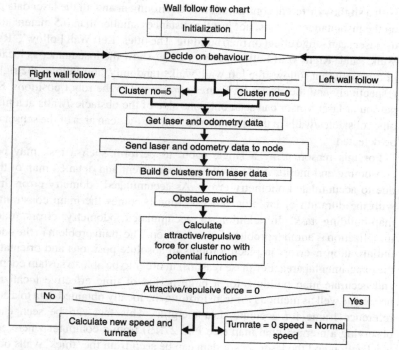

27.11 Flow chart of wall follow.

obstacles during their navigations. In addition to this process, all robots send their laser measurements and odometry information to the base computer via a wireless network, possibly through an intermediate robot. The results for an experiment were shown in Fig. 27.8 and 27.9, where the evolution of the map was also presented. At the base station the data sent by the three robots are merged in order to build a consistent map according to these data.

Figure 27.12 shows the instantaneous local maps of the individual robots according to their laser data and odometry information for another experiment

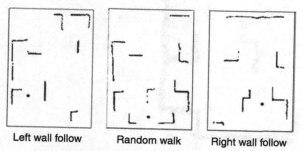

Left wall follow   Random walk   Right wall follow

27.12 Instantaneous local maps of the robots.

with a slightly different configuration of environment and all the laser data plotted on the instantaneous maps, not just the distances smaller than 0.5 meters, to show the laser scan range effect on map building. The titles 'Left Wall Follow', 'Random Walk', and 'Right Wall Follow' on Fig. 27.12 show the instantaneous local maps of the robot that follows the left wall, walks randomly, and follows the right wall respectively and the circles on figures correspond to the robot positions. Since a particular laser sensor can scan only one side of the obstacles/walls at a time, the other obstacles/walls in the environment outside the scan area of the sensor cannot be detected.

For this reason using a single robot to perform such a task may be time consuming and inefficient for building a consistent and detailed map of the area due to accumulated odometry errors. As accumulated odometry errors increase with the duration of building process, time becomes the main constraint for a map-building task. In addition to accumulated odometry errors, odometry initialization is another problem in localization. The main problem in the odometry initialization is errors in placing robots in absolute positions and orientations in the experimental area. For these reasons, in order to be able to sustain cooperative and accurate map building for a large period of time effective localization is needed as well as methods for transforming the locally obtained data to a common reference frame for accurate data merging. This fact can be seen easily by observing a merged final map (see Fig. 27.13) from a computer where errors in localization and the laser sensor data can be seen from the 'thick' walls obtained. With the proposed method, the map-building task is carried out efficiently and the task becomes less time consuming.

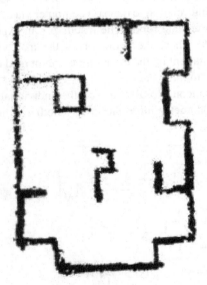

27.13 The final map of the experimental area.

27.5 Conclusions

In this chapter a cooperative map building implementation is presented for an indoor environment using Khepera III robots. Navigation of the robots is performed by using artificial potential functions during map building. According to Fox, Ko et al. (2006), cooperative map building is more efficient for building a map of the environment and less time consuming than using a single robot. Future work can focus on improving the indoor localization of the robots by using inertial measurement units (IMUs) and possibly incorporating cooperative simultaneous localization and mapping (SLAM) techniques. Moreover, in addition to the map-building task, the procedure can be integrated with effective cooperative search methods and can be used for other tasks as well.

27.6 References

Borenstein, J., B. Everett, and L. Feng (1996), *Navigating Mobile Robots: Systems and Techniques*, A. K. Peters Ltd., Wellesley.

Burgard, W., A. B. Cremers, et al. (1999), 'Experiences with an interactive museum tour-guide robot', *Artificial Intelligence*, vol. 114, no. 1–2: 3–55.

Burgard, W., D. Fox, et al. (2000), 'Collaborative multi-robot exploration', in *Proceedings of the IEEE International Conference on Robotics and Automation (ICRA)*, vol. 1, pp. 476–481, San Francisco, CA, USA.

Chatila, R. and J. P. Laumond (1985), 'Position referencing and consistent world modeling for mobile robots', in *Proceedings of the IEEE International Conference on Robotics and Automation (ICRA)*, vol. 2: 138–145.

Choset, H. (1996), 'Sensor based motion planning: The hierarchical generalized voronoi graph', PhD thesis, California Institute of Technology.

Elfes, A. (1989), 'Occupancy grids: A probabilistic framework for perception and navigation', PhD thesis, Department of Electrical and Computer Engineering, Carnegie Mellon University.

Fenwick, J. W., P. M. Newman and J. J. Leonard (2002) 'Cooperative concurrent mapping and localization', in *Proceedings of IEEE International Conference Robotics Automation (ICRA)*, vol. 2: 1810–1817.

Fox, D., J. Ko, et al. (2006), 'Distributed multi-robot exploration and mapping', in *Proceedings of IEEE*, vol. 94, no. 7.

Gazi, V., B. Fidan, et al. (2007), 'Aggregation, Foraging, and Formation Control of Swarms with Non-Holonomic Agents Using Potential Functions and Sliding Mode Techniques', *Turkish Journal of Electric*, vol. 15, no. 2: 149–168.

Gazi, V. and K. M. Passino (2004), 'A class of attractions/repulsion functions for stable swarm aggregations', International Journal of Control, vol. 77, no. 18: 1567–1579.

Horswill, I. (1994), 'Specialization of perceptual processes', *Technical Report AI TR-1511*, MIT, AI LAB, Cambridge, MA.

King, S. and C. Weiman (1990), 'Helpmate autonomous mobile robot navigation system', in *Proceedings of the SPIE Conference on Mobile Robots*, vol. 1388, pp. 190–198, Boston, MA, USA.

Konolige, K., D. Fox, et al. (2006), 'Centibots: Very large scale distributed robotic teams', *Experimental Robotics IX*, vol. 21: 131–140.

Kuipers, B. and Y. T. Byun (1991), 'A robot exploration and mapping strategy based on a semantic hierarch of spatial representations.' *Journal of Robotics and Autonomous Systems*, vol. 8: 47–63.

Leonard, J. J., H. F. Durrant-Whyte, and I. J. Cox (1992), 'Dynamic map building for an autonomous mobile robot', *International Journal of Robotic Search*, vol. 11, no. 4: 89–96.

Lu, F. and E. Milios (1997), 'Globally consistent range scan alignment for environment mapping', *Autonomous Robots*, vol. 4: 333–349.

Mataric, M. J. (1994), 'Interaction and intelligent behavior.' *Technical report AI-TR-1495*, Massachusetts Institute of Technology, Artificial Intelligence Laboratory, Cambridge, MA.

Moravec, H. P. (1988), 'Sensor fusion in certainty grids for mobile robots', *AI Magazine*: 61–74.

Nourbakhsh, I., J. Boenage, et al. (1999), 'An affective mobile robot with a full-time job', *Artificial Intelligence*, vol. 114, no. 1–2: 95–124.

Rencken, W. D. (1993), 'Concurrent localization and map building for mobile robots using ultrasonic sensors', in *Proceedings of the IEEE/RSJ International Conference on Intelligent Robots and Systems*, pp. 2129–2197, Yokohama, Japan.

Simmons, R. (1996), 'Where in the world is xavier, the robot?', *Machine Perception*, vol. 5, no. 1.

Simmons, R., D. Apfelbaum, et al. (2000), 'Coordination for multi-robot exploration and mapping', in *Proceedings of the AAAI National Conference on Artificial Intelligence*, pp. 852–858, Austin, TX, USA.

Thrun, S. (2001), 'Probabilistic online mapping algorithm for teams of mobile robots', *International Journal of Robotic Research*, vol. 20, no.5: 335–363.

Thrun, S., M. Beetz, et al. (2000), 'Probabilistic algorithms and the interactive museum tour-guide robot minerve', *International Journal of Robotics Research*, vol. 19, no. 11: 972–999.

Thrun, S., D. Fox, and W. Burgard (1998), 'A probabilistic approach to concurrent mapping and localization for mobile robots', *Machine Learning*, vol. 31, pp. 29–53. Also appeared in *Autonomous Robots*, vol. 5, no. 3–4: 253–271.

28

Heterogenous multi-agent system behaviour patterns for robotics application

L. ALBOUL, J. PENDERS and J. SAEZ-PONS, Sheffield Hallam University, UK and L. NOMDEDEU, University Jaume I, Spain

Abstract: Several behaviours required for a group of heterogeneous agents (robots and humans) in rescue and search operations are presented. The listed behaviours are especially needed in the GUARDIANS project; however they can be used in many robotics applications where a group of robots is deployed. The behaviours are linked to specific tasks to be fulfilled, and can be achieved without explicit communication and centralised control, by using only information from local sensors. Besides the theoretical framework, examples of implementations of proposed behaviours on ERA-Mobi robots are given as well. The chapter concludes with discussions on possible practical 'set-ups' for achieving the proposed behaviours in real-life environments.

Key words: multi-agent system, robot behaviours, potential functions, multi-robot formations.

28.1 Introduction

The Group of Unmanned Assistant Robots Deployed In Aggregative Navigation by Scent (GUARDIANS) project[1] is an FP6, EU-funded project, developing a mixed human–multi-robot system to assist fire fighters in search and rescue operations in an industrial warehouse in the event or hazard of fire (Penders, Alboul, Roast and Cervera 2007). The main component of the system is a team of heterogeneous autonomous robots capable of interacting with humans and able to produce certain self-organising behaviour patterns. The latter means that some techniques of swarm robotics are applicable.

The GUARDIANS multi-robot team consists of mini-robots and middle-sized robots. This means that the robots are not suitable for performing heavy physical tasks, but nevertheless, can be very useful in many fire fighting situations, as was confirmed after consulting with the South Yorkshire Fire and Rescue Service,

---

[1] GUARDIANS ran from 2007 to 2010, and involve the following partners: Sheffield Hallam University (coordinator), Robotic Intelligence Lab, Jaume-I University, Spain; Heinz Nixdorf Institute, University of Paderborn, Germany; Institute of Systems and Robotics, University of Coimbra, Portugal; Space Application Services, Belgium; K-Team Switzerland; Dept. of Electrical and Electronics Engineering, TOBB University of Economics and Technology, Turkey; Robotnik Automation, Spain; and South Yorkshire Fire and Rescue Service, UK.

615

United Kingdom, subsequently referred to as SYFIRE. SYFIRE is a member of the GUARDIANS consortium and the GUARDIANS project progresses with close collaboration with the fire fighters.

SYFIRE proposed a warehouse as the principal GUARDIANS application scenario by indicating that fires in industrial warehouses are a major concern to them. Searching for victims may be extremely dangerous due to the enormous dimensions of the warehouses and the expected low visibility when smoke develops. Large warehouses are divided into sections separated by walls, which are fire resistant for several hours. The typical section dimensions are about 100×200 m^2. In the event of a blaze, the fire may be confined to a certain area of the warehouse; however smoke might cover the whole section. In this scenario, SYFIRE pointed out that apart from the smoke, the warehouse is, in general, in a normal and neat state, and the ground is easily passable. For a robot team this implies that there are no exceptional requirements concerning the robots' motion and even wheeled mini robots are suitable.

The development of smoke represents a great challenge for fire fighters as it reduces visibility dramatically and human beings can easily become disoriented and may even get lost. Rendered without sight, fire fighters can only rely on their touch and hearing senses. However, the sense of touch is restricted by their clothing gear and the sense of hearing is reduced by the noisy breathing apparatus, with which fire fighters are equipped to provide fresh air. Besides the dimensions, low visibility and reduced perception of fire fighters, an important constraint on successful operation is the time constraint. The amount of oxygen in the breathing apparatus suffices for about 20 minutes. The average speed with which experienced fire fighters progress in the event of dense smoke (informally measured during the 'tasting exercise session' at the SYFIRE training centre), is about 12 m per minute. Given the crawling speed, fire fighters can proceed about 240 m with a full tank. Taking into account that they have approximately 20 minutes of air between getting in and getting out, the maximum advance they can make is only 120 m, which is less than the largest dimension of the modern warehouses. Another issue related to the time constraint is that a phenomenon such as *flashover* can also occur very quickly.

Note that in the UK the fire fighting procedures require a two-stage operation: first, one team lays out and fixes a guideline along a wall and then leaves, and then subsequent teams aiming for the scene of operations enter and advance following the guideline. This is done in order to avoid situations in which fire fighters can be lost, but reduces the available time even further. Besides the aforementioned problems, an important requirement is continuous and uninterrupted communication links between the crew inside and managing crews outside. In a warehouse, however, the racks form a dense lattice of metal joints, which may be packed with tins, cans or other metal based packaging. Within such a metal cave, the transmission and reception of radio signals is problematic.

In such situations, a team of GUARDIANS robots can be very useful to perform the following tasks: searching and exploring the warehouse and gathering information;

maintaining the communication link to the outside base station and supporting the fire fighters to move around and if required leading them through the site.

Mini-robots in the GUARDIANS robot team are represented mostly by Khepera III (developed by the partner K-TEAM), and middle-sized robots are ERA-Mobi robots, commonly referred to as Erratics robots (Videre Design). These robots are to be applied in a specified quantity, possibly large. The larger robot called Guardian, developed by the partner Robotnik Automation, represents an exception (see Fig. 28.1). This robot can be a member of a team, but also can perform certain tasks where a more powerful robot may be needed, such as carrying tools for fire fighters.

We focus on the basic navigation behaviours of multi-robot or human-robot teams, which have to be achieved without central and on-line control. These behaviours can be also achieved without explicit communication and therefore can be still applicable when communication links are severed. The global behaviour produced is relatively independent of the number of robots in the team. This makes the team robust to failures of individual robots and flexible with respect to its size.

28.1 Robots in the Guardians project: Khepera III, Erratic and Robot Guardian.

The basic behaviours that we cover are robot–robot and robot–fire fighter avoidance/attraction as well as robot–obstacle avoidance, and formation generation and keeping. To develop these behaviours we use the social potential field framework, which was introduced by Reif and Wang (1999). In our approach robots, humans and obstacles are considered as classes of special agents and for each pair of 'agents' different potential functions are applied. As a result, we can generate a global behaviour that leads to forming certain formation patterns of the group of robots or mixed robot–human teams. Some theoretical considerations on pattern convergence and stability are also given.

Our algorithms have been implemented using Player/Stage software (Player), which allows their direct application to real robots, and the Java Agent Development Environment (JADE) that takes care of the agent's life cycle and other agent-related issues. For testing our theoretical considerations, in particular the convergence of the behaviours, we use the programming environment NetLogo, which allows working with a large number of agents. The simulation results comply with theoretical considerations and show that the algorithms are robust and capable of dealing with teams of different sizes and the failure of individual agents, both robots and humans. We also tested our approach on the Erratic mobile platforms; and the results of the implementations confirm the results of simulation.

At the end of the chapter we briefly discuss the possible 'set-ups' of the GUARDIANS and similar robotic systems in real-life environments; in particular we indicate approaches to one of the challenging 'real-life' problems, which is robot–robot and/or robot–human detection.

The work described in this chapter is partially based on previous work by Saez-Pons, Alboul, Penders, and Veysel (2008) and Alboul, Saez-Pons, Penders and Nomdedeu (2008) and further expands the ideas presented there.

28.2   Related work

Formation control and the coordination of heterogeneous multi-agent teams (both multi-robot and mixed robot–human) has become an important and challenging research field due to the increasing use of multiple autonomous agents and swarms in various robotics applications. The research concentrates not only on techniques that achieve a desired formation given a set of parameters and limiting functions but also on the ability to modify a specific formation depending on circumstances. The team of agents should be able not only to self-configure but also to recover and reconfigure itself in the case of the loss or addition of a new member. The formation therefore should be flexible.

Various approaches have been used, however artificial potential functions (fields) methods have taken a prominent part, due to their advantage of low computational complexity, ease of implementation and combination of repulsion/ attraction, which are characteristics of all social behaviours. Artificial potential fields approaches to autonomous robot navigation were introduced in the 1980s (Krogh 1984; Khatib 1986) and since then have been widely used by the robotic scientific community. They provide a concise and effective framework for expressing various interaction patterns. Therefore, while in the early days the methods were used mostly for single robot motion planning, in due course these methods have become more and more popular for modelling the various collective behaviours and distributive control of a group of robots. Potential functions have been used, for example, in multi-robot navigation for obstacle avoidance (Krogh 1984; Penders, Alboul and Braspenning 1994), robot aggregation (Gazi and Passino 2004a, 2004b), and robot formation keeping (Song and Kumar 2002; Schneider and Wildermuth 2005).

We concentrate on two behaviours of a mixed human–multi-robot team that are crucial for the GUARDIANS project: formation generation and formation keeping. In order to achieve these behaviours, we use as the basis the social potential field (SPF) method, introduced by Reif and Wang (1999). By using force laws they were able to express such 'social' behaviours as clustering, patrolling, escorting, etc. Bruemmer et al. also utilise the SPF approach as a means to coordinate group behaviours and promote the emergence of swarm intelligence which mimics that seen in a colony of ants or swarm of bees (Bruemmer, Dudenhoeffer and McKay 2002).

The methods based on the potential field principle suffer, however, several drawbacks; the most serious one is the possibility of convergence to a local minimum. A detailed analysis of shortcomings of the potential field principle is given in Koren and Borenstein (1991). There are some attempts to avoid local minima; a recent one is presented in the paper by Fazenda and Lima (2007). However, in order to achieve the elimination of certain local minima the authors make use of strong assumptions such as the condition for a robot of having knowledge of the current configuration of the whole formation, which drastically increases the complexity of the method and makes its application to real-life situations problematic.

In our approach, a general assumption is that the robots do not have common knowledge of the environment and do not have any memory of previous positions. Nevertheless, under some not very strong conditions imposed on the robot space configurations, the robots can achieve the desired behaviours and avoid certain local minima.

We would like also to mention the works of N. Leonard and her co-authors (Ögren, Fiorelli and Leonard 2004) and Gazi and Passino (2004b), in which ideas similar to ours are expressed. However, our approach differs from the aforementioned works in several aspects. We do not assume any centralised computation and/or communication. There are no artificial virtual reference points, as in the works of Ögren et al. A human can be considered as a virtual leader in some respect, but he/she does not move with a constant velocity. The agent group is scalable and flexible, in the sense that the 'loss' or 'addition' of an agent neither influences the desired behaviour nor produces another alternative behaviour. In this respect our work has overlapping points with the work of Barnes and co-authors (2007). We do not determine in advance certain basic motions such as translation, rotation, or expansion; however they may occur as a result of the application of potential functions on a group of robots. We also do not predefine either shapes of formations, or the final robots' positions, contrary to the work by Gazi and Passino (2004b). While moving, 'our' robots still preserve formation, but the shape of the formation can be deformed in the presence of obstacles and then re-established, which is similar to the results presented in Schneider and Wildermuth (2005).

We also provide some comments on the convergence and stability of our method. In the literature stability analyses are rarely given. They are in general provided only for certain configurations, and often reduced to performance measurements (Balch and Hybinette 2000). When given, it often involves bulky computations (Song and Kumar 2002; Ögren, Fiorelli and Leonard 2004). Our analysis on stability and convergence is based on geometric considerations, following the principles exposed in several works of the second and third authors

(Penders, Alboul and Braspenning 1994; Penders, Alboul and Rodrigues 2004). While this analysis is not precise, in the sense that it may not distinguish a single solution from a class of similar solutions, it gives a good overview of possible types of configurations and their stability.

28.3   Description of heterogeneous multi-agent system

28.3.1 Agent classes and their characteristics

Depending on the circumstances we can consider several classes of agents. The classes may be the following:

- A class of robots r_i, $i = 1, 2, ..., n$
- A class of humans (fire fighters) h^j, $j = 1, 2, ..., m$
- A class of obstacles o_k, $k = 1, 2, ..., l$

28.2 Fire-fighter with assisting robots.

A class of robots, which may be *heterogeneous*, can be split into several sub-classes of homogenous robots and robots may be either holonomic or non-holonomic. A class of humans, in general, can be split into subclasses as well, but in the GUARDIANS scenario we consider only one class of human agents (consisting of fire fighters); and the number of humans m is assumed to be smaller than the number of robots. In a real-life fire-fighting situation, humans, in general, act in groups of two: one person takes the role of the leader and the second is the follower. On the whole, they are never separated, they are connected by a rope, and the follower tries to keep his/her hand on the shoulder of the leader. Therefore, without much loss of generality, we can assume that the class of humans consists of only one agent, and to a certain extent this agent plays the role of a leader and the robot team takes the role of a follower. The task of the robot team is not confined to following the human; the robots should assist him/her to navigate safely through the site. This means that the robots should avoid obstacles as well as prevent the human colliding with them. Therefore a reasonable requirement is that the robots should be able to organise themselves in a formation around the fire fighter and maintain this formation while moving. In a more general setting, a human can become a follower and be led by a robot or a team of robots. However, in this chapter we assume that a human acts independently of the robots. The classes of humans and robots can be also fused, thus forming one class of heterogeneous agents. This can be useful in such formations in which the *leader* and *follower* roles can be switched in due course.

In this chapter we assume that our system contains only one class of homogeneous robots and that the robots are *holonomic*. This is not much loss of generality, as it was shown that for a certain class of non-holonomic wheeled vehicles with differential drive, control forces can be applied on an off-wheel axis point, the kinematics of which can be made holonomic by using a suitable transformation (Lawton and Beard 2003). We also assume the robots act totally *independently* and *asynchronously* from each other and neither rely on centralised commands, nor on any common notion of time. They are capable of sensing (observing) the environment, but they are *oblivious*, meaning that they do not remember any previous observation or computation performed in preceding steps, contrary to the assumptions made in Fazenda and Lima (2007). The robots are also assumed to be *anonymous*. The latter means that they do not have any sort of identifiers that can allow them to discriminate an individual robot among other robots. However, the robots can distinguish robots from obstacles and humans. In computational simulations this is done by indicating the class of an agent, for example, by assigning a specific flag to the agents of the same class. Some considerations on how it can be done in real-life applications are given at the end of this chapter.

28.3.2 Visibility

Each robot is equipped with sensors to acquire information about the environment. How far a robot 'senses' the environment, depends on the sensing range. The sensing range of each robot may vary from zero to infinity. In what follows we will refer to the sensing range of a robot as its *visibility domain*. In the present chapter the *field-of-view* of each robot is supposed to be 360 degrees, resulting in a circular visibility domain, which in real life means that each robot is equipped with at least one omnidirectional sensor. We can determine for the visibility domain its radius R_{Visib}, which can vary from zero (a 'blind robot') to infinity. The *infinity domain* actually means that the radius takes a sufficiently large value R_{infty} that allows a robot to observe the whole surroundings in which it operates. For example, the visibility domain of the radius of n units can be considered infinite if the diameter of the minimum bounding circle which contains the site is smaller than n units.

However, the infinite visibility domain does not mean that a robot should react to any agent or object within the domain. A realistic assumption is that a robot reacts to other robots or obstacles only if they are situated at a certain distance within its visibility domain, so that the *reaction distance* d_{react} satisfies $d_{react} \leq R_{visib}$. Suppose, in the visibility domain of the robot r_i there are two obstacles O_1 and O_2, and one is hidden behind the other. In computer simulations a robot can 'sense' both obstacles, and a repulsion force can be computed by taking into consideration both of them, whereas in practice a robot will observe only one obstacle, the one closest to it, and therefore it is sufficient to apply robot–obstacle repulsion only to the first obstacle. Another reason for introducing the reaction distance is to produce more stable and realistic navigation patterns, such as avoiding unnecessary oscillations (Koren and Borenstein 1991). An example of such behaviour is navigation of a robot along an obstacle or a wall. We can also generalise the notion of visibility domain by introducing several visibility domains of a robot. A robot can posses an infinite visibility domain with respect to the human. It can sense the fire fighter in any location of the site if the fire fighter, for example, transmits a radio signal or ultrasonic signal, but it senses other robots and obstacles within a limited visibility domain.

28.3 Visibility domain of a robot.

There is no explicit communication among robots except reactions of the robots to the environment based on their observations (sometimes referred to as *implicit communication* (Parker 2000)). The robot senses other agents in their visibility domains, is capable of estimating the corresponding distances to them and acts accordingly. A human does not communicate with robots. The human has two basic behaviours: either at rest or moving.

When the human does not move, the robots gather around the human, creating a certain formation. As soon as the human is in motion, the robots start to move as well, attracted by the human, while avoiding other robots and possible obstacles.

28.3.3 Formations

Formation control of a group of agents has received a fair amount of attention in literature; however the term 'formation' is not uniquely defined. In general, one can distinguish several classes of formations, which roughly can be split into the following categories:

- A group of agents that forms a particular geometric shape (Baldasare, Nolfi and Parisi 2003; Eisenstat, 1981; Lemay et al. 2004).
- Positions of agents that satisfy a pre-defined function (Egerstedt and Hu 2001).
- The distances between the adjacent positions of agents are amid certain predefined values (Gazi 2005).

In our case, formations can be defined as groups of agents establishing and maintaining a certain configuration. The configuration does not have a predetermined shape but the agents in the group do not spread too far away from each other. Our human–robot formation has to be adapted (stretched, deformed) when obstacles are in close vicinity since the fire fighter has to be protected and escorted all times. In this respect, a formation of agents in the GUARDIANS project can be seen as a coalition: the agents act as a unit, but do not necessarily maintain a rigid geometrical shape. Considering a group of agents as a graph (network) where each agent represents a node, and agents are interconnected via their visibility domains, we can define formation as follows: over time the robots might form one or more groups (sub-graphs), so that there exists a path in each sub-graph that passes from any node to any other node, with the property that the distance d_{ri} of any individual node (robot) r_i to the agent closest to it within this path (either a robot or a human) does not exceed the certain value d_{max}. d_{max} can be defined to be smaller or equal either to the (smallest) radius of the visibility domains or reaction domains of agents. This definition is similar to the definition of the formation given in Tanner, Pappas and Kumar (2004).

From a mathematical point of view, the aforementioned condition means that each sub-graph represents a connected graph. If no robots are lost, the agents form only one sub-graph (coinciding with their graph).

Neither initial positions, nor final positions of agents are predefined. To some extent, this definition complies with the definition proposed in Lemay et al. (2004), where the group determines autonomously the most appropriate assignment of positions in the formation.

28.4   Formalism of multi-agent system

Agents of different classes perform different behaviours. Regarding three classes of agents, neither humans nor obstacles are supposed to react to robots. The motion of a human will of course be influenced by obstacles, but currently it is assumed that the human can easily find the way through the obstacles. In a more general framework robots may lead the human through obstacles, by indicating to them a direction to follow. In this case we could consider only two classes of agents: the class of mixed agents consisting of robots and humans, and the class of obstacles. The framework will not have to be changed if we swap the leading human with a leading robot. The obstacles are considered static.

The robots are influenced by all three (or two) classes of agents. We consider M autonomous mobile robots denoted by r_m, where $m = 1, 2, ..., M$, S obstacles denoted by O_s, where $s = 1, 2, ..., S$ and a human H situated in a two dimensional plane R^2. Xr_m, X_{Os}, and X_H are the positions of r_m, O_s and H, respectively.

The artificial potential functions generating the basic behaviours such as robot–robot, robot–fire fighter avoidance/attraction and robot–obstacle avoidance are described below.

28.4.1 Robot–human potential

The robots have to avoid colliding with the human and at the same time they have to be able to approach and keep the human within their sensor range. We define the potential function P_{Human} between the robot r_m and the Human H as:

$$P_{Human}(d_{rm}^H) = \frac{1}{\left(k_{hrr}\left(d_{rm}^H - w_{hrr}\right)\right)^2}$$

$$+ \frac{1}{\left(k_{hra}\left(d_{rm}^H - w_{hra}\right)\right)^2} \qquad [28.1]$$

where k_{hrr}, k_{hra}, w_{hrr} and w_{hra} are scaling parameters, and d_{rm}^H is the distance between the robot r_m and the human H.

The potential function P_{Human} between the robot r_m and the human H is composed of a repulsive term that prevents the robot from colliding with the human and an attractive term that keeps the human within its visibility domain. An example of the graph of the robot–human potential function P_{Human} is given in Fig. 28.4. In this example we have a robot r_1 and a human H in a two dimensional space R^2 and $d_{r1}{}^H$ is the distance between r_1 and H. As shown in Fig. 28.4 when r_1 is too close to H, $P_{Human}(d_{r1}^H)$ pushes r_1 away from H keeping the robot from colliding with the human. When r_1 is too far, $P_{Human}(d_{r1}^H)$ pushes r_1 closer to H maintaining the human within its sensor range.

28.4.2 Robot–robot potential

The potential function P_{Robot} between the robot r_m and the robot r_i is defined as

$$P_{Robot}(d_{rm}^{ri}) = \frac{1}{\left(k_{rr}\left(d_{rm}^{ri} - w_{rr}\right)\right)^2} \qquad [28.2]$$

where k_{rr} and w_{rr} are scaling parameters, and d_{rm}^{ri} is the distance between the robot r_m and the robot r_i. Obviously, $d_{rm}^{ri} = d_{ri}^{rm}$.

Fig. 28.5 is an example of the robot–robot potential function P_{Robot}. In this example we have two robots, r_1 and r_2 situated in a two-dimensional plane R^2 and

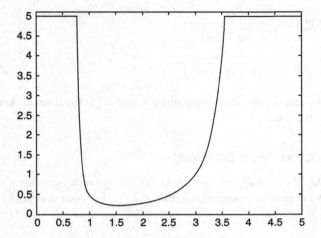

28.4 Robot–human potential function.

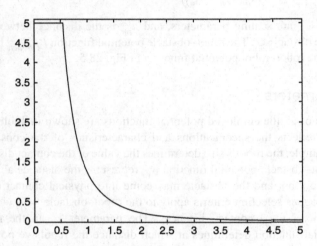

28.5 Robot–robot potential function.

d $_{r1}^{r2}$ is the distance between r_1 and r_2. In the presence of a human we assume that the robots avoid each other, by exerting on each other the repulsive force of magnitude $IR_{(m,i)}$, determined by the negative gradient of $P_{rr}(r_{mi})$:

$$\overrightarrow{IR}_{(m,i)} = -\nabla P_{Robot}(r_{mi}) \tag{28.3}$$

In the absence of a human in the visibility domain of a robot, the force acting on the robot produced by other robots in its visibility domain will become a combination of attraction and repulsion similar to the potential function between a robot and the human.

$$P_{Robot}\left(d_{rm}^{ri}\right) = \frac{1}{\left(k_{rr}\left(d_{irm}^{r} - w_{rr}\right)\right)^2}$$

$$+ \frac{1}{\left(k_{ra}\left(d_{rm}^{ri} - w_{ra}\right)\right)^2} \qquad [28.4]$$

where k_{rr}, k_{ra}, w_{rr} and w_{ra} are scaling parameters, and d_{rm}^{ri} is the distance between the robot r_m and the robot r_i.

28.4.3 Robot–obstacle potential

In a real world, there may be several static or dynamic obstacles in the environment. We define the potential function $P_{Obstacle}$ between the robot r_m and the obstacle O_s as:

$$P_{Obstacle}\left(d_{rm}^{Os}\right) = \frac{1}{\left(k_{ro}\left(d_{rm}^{Os} - w_{ro}\right)\right)^2} \qquad [28.5]$$

where k_{ro}, and w_{ro} are scaling parameters, and d_{rm}^{Os} is the distance between the robot r_m and the obstacle O_s. The robot–obstacle potential function $P_{Obstacle}$ has the same form as the robot–robot potential function in Fig. 28.5.

28.4.4 Parameters

The parameters of all the employed potential functions are shown in Table 28.1. This selection reflects the specifications and characteristics of the considered system. For example, the robot's size determines the value of the contact distance, i.e. for the robot–obstacle potential function w_{ro} represents the distance at which the edges of the robot and the obstacle may come into physical contact (w_{ro} = 0.49). The analogous selection criteria apply to the robot–obstacle (w_{rr}, w_{ra}) and robot–human (w_{hrr}, w_{hra}) contact distances. The parameter k_{ro} in the robot–obstacle potential function determines at which distance the repulsive potential

Table 28.1 Values of the parameters used in the potential functions employed for simulating required robot behaviours

Potential function	Parameter value	
Robot–Obstacle	k_{or} = 5.00,	Wor = 0.49
Robot–Robot	k_{rr} = 2.00,	W_{rr} = 0.98
	k_{ra} = 2.00,	W_{ra} = 4.00
Robot–Human	kh_{rr} = 5.00,	Wh_{rr} = 8.22
	kh_{ra} = 2.00,	Wh_{ra} = 4.00

starts pushing the robot away from the obstacle. Choosing $k_{ro} = 5$ means that the robot will not start avoiding the obstacles until approximately $d_{Os}^{Rm} = 1.5$ meters. The latter represents the reaction distance, described in section 28.4.2. Similar selection criteria have been applied for the parameters of the remaining potential functions.

28.4.5 Social potential field

The social potential function P_{Social} of r_m is defined as the sum of the aforementioned potential fields:

$$P_{Social}(X_{rm}) = P_R^O(X_{rm}) + P_{ri}^{rj}(X_{rm}) + P_r^H(X_{rm})$$

$$= \sum_{s=1}^{S} P_{Obstacle}(d_{rm}^{Os}) + \sum_{j=1, j \neq i}^{M} P_{Robot}(d_{rm}^{ri}) + P_{Human}(d_{rm}^{H}) \qquad [28.6]$$

The artificial force that 'acts' on the robot r_m is then defined accordingly:

$$\vec{F}_{Arti}(X_{rm}) = \vec{F}_{Arti\_Obstacle}(X_{rm}) + \vec{F}_{Arti\_Robot}(X_{rm}) + \vec{F}_{Arti\_Human}(X_{rm}) \qquad [28.7]$$

Note that the artificial forces are not physical forces bearing on the robots. Artificial forces are virtual forces, which are designed to control the motion of each robot. The robot 'calculates' these forces itself by measuring distances to the agents in its visibility domain, using the given parameters. The forces acting on a robot are depicted in Fig. 28.6.

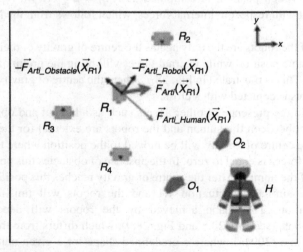

28.6 Forces acting on a robot.

28.5   Behaviour patterns and remarks on stability of multi-agent system

The application of the described potential functions generates certain formation patterns that correspond to local minima in the potential fields. We consider two set-ups: *formation generation* when the human is at rest or absent and *formation maintenance* when the human is moving. In the first set-up the agents' configurations converge to certain typical pattern forms, which are, in general, dependent on the visibility domains of the robots and the number of robots. In the second set-up these patterns will be either deformed, or the configuration can be split into several typical patterns with lesser numbers of robots. In what follows we briefly describe the typical patterns with some remarks on their stability. As an illustration tool, we use the computing software NetLogo.

28.5.1   Formation generation

First we assume that there are several robots $r_1, \ldots, r_n$, one human H at rest, and no obstacles. The robots are situated somewhere in the site.

Robots with infinite visibility domains

At present we assume that the reaction distances of the robots coincide with the radii of visibility domains. The robots are considered to be holonomic; therefore, without loss of generality, they can be represented as points in the plane. Their positions are indicated as X_{r_i}, $i = 1, \ldots, n$, where n is the number of robots. The internal forces that each pair of robots exerts on each other are equal by magnitude but opposite by direction. The centre of gravity CG of the group of robots, which position is computed as $X_{CG} = \Sigma_{i=1}^{n} Xr_i/n$, represents an invariant point with respect to internal forces, which follows from the Newton third law.

Situation 1.1 If the robots are the only agents the centre of gravity of their group will stay in the same position while internal forces will bring the group of robots into equilibrium. This is illustrated in Fig. 28.7, where the centre of gravity CG of the group of robots is depicted with a cross.

Situation 2.1 In the presence of other agents such as a human and obstacles, the forces that act between the human and the robots are external for the centre of gravity and the centre of gravity will be moved to the position where the sum of these external forces is equal to zero. In the absence of obstacles this point will be the location of the human. After the centre of gravity reaches this position, the situation will be similar to Situation 1.1 and the robots will finish in the equilibrium position. The shape achieved by the robots will depend on the number of robots (see Fig. 28.8 and Fig. 28.9), which differs from the work by Gazi and Passino (2004a), where a predefined shape for a given number of robots is considered.

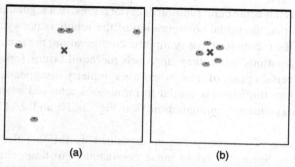

28.7 Behaviour pattern of a group of robots without a human and obstacles: (a) initial configuration, (b) final configuration.

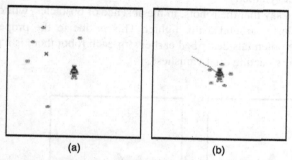

28.8 First example of a behaviour pattern of a mixed human–robot group: (a) initial configuration, (b) final configuration.

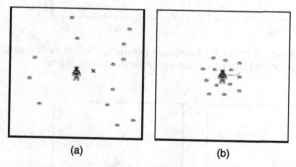

28.9 Second example of a behaviour pattern of a mixed human–robot group: (a) initial configuration, (b) final configuration.

Considering the human agent as an *Attractor*, we can pose several questions. First of all, in the absence of obstacles, how does the number of robots influence the final equilibrium configuration? As we can see, when the number of robots is sufficiently small, the robots form a regular polygon. However, when the number of robots is large, the configuration does not represent a regular polygon. A cause of this is that the initial configuration of the robots is not symmetric. However, even if the robots start in a symmetric configuration, the symmetry of the robots' configurations while they approach the equilibrium formation is preserved until a certain point of time, when the symmetry disappears, and the resulting equilibrium formation is similar to formations achieved when robots do not start in a symmetric configuration (see Fig. 28.10 and 28.11 for an illustration).

The disappearance of symmetry can be explained by the fact that time is actually discrete: an agent that might move continuously in time, changes its direction only at certain time points. As robots can move asynchronously, the virtual forces acting on them may not be always equal to the magnitude, which results in the loss of symmetry.

However, we can say that the robots, in the absence of obstacles, will organise themselves in a cluster around a fire fighter. This is due to the properties of attractive/repulsive potentials described earlier (for each robot there is a position where the virtual force acting on it vanishes).

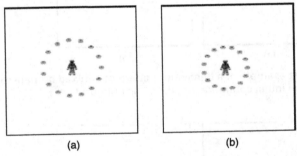

(a)   (b)

28.10 Behaviour pattern of a mixed group: (a) initial symmetric configuration, (b) intermediate symmetric configuration.

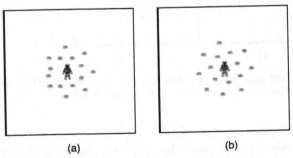

(a)   (b)

28.11 Behaviour pattern of a mixed human–robot group: (a) symmetry is lost, (b) final asymmetric configuration.

In the presence of an obstacle the robots will form a cluster that will be moving away from the obstacle. Theoretically speaking, if we assume that robots are situated in an environment without boundary (infinite environment) in which only one obstacle is present and no attractor, in the case of infinite visibility domains, the robots will move away from the obstacle infinitely in time. In the case of more obstacles the final position of the robot group may vary: they can come to rest trapped between two obstacles or move away from both obstacles depending on their initial positions.

In the presence of obstacles and a human the centre of gravity will be shifted, and at its position the vector sum of forces exerted on the robots by the obstacles and the attraction forces to the human will be equal to zero (see Fig. 28.12).

Also as robots possess infinite visibility domains a typical local minimum will occur (Fig. 28.13).

Robots with limited visibility domains

The 'set-up' with robots of limited visibility domains is more realistic. On the one hand, there is a possibility that some robots may be 'lost', but on the other hand it can reduce occurrences of local minima.

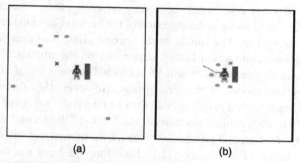

28.12 (a) and (b) Behaviour pattern of a mixed human–robot group in the presence of an obstacle.

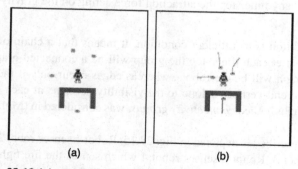

28.13 (a) and (b) Local minimum.

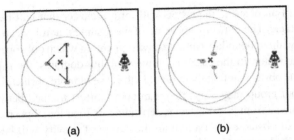

| (a) | (b) |

28.14 (a) and (b) Clustering of robots with limited visibility domains.

Situation 2.1 Robots do not 'see' the human, but sense each other. We will get a local minimum where a group will execute a clustering behaviours pattern (Fig. 28.14). The robots' behaviour is similar to the behaviour depicted in Situation 1.1.

If all robots sense the human in their visibility domains, then the robots will gather around the human, the same as in Situation 1.2.

Situation 2.2 This situation occurs when some of n robots (denoted with r_j^H, $j = 1, ..., l$) sense the human, whereas each of the other robots (denoted with r_k^r, $k = n–l$) senses at least one robot from the first group. In this case the robots of the first group will exert only repulsion forces on each other, whereas their centre of gravity CG^H will move in the direction of the human due to the attraction force exerted by the human. The robots of the second group will exert on each other either repulsion or attraction forces, depending on the mutual distances, whereas their centre of gravity CG^H will be subjected to the sum of attraction forces exerted by the robots of the first group. However, this force is not monotonically changing and it may become equal to zero if the robots of the first group disappear from the visibility domains of the robots of the second group. In this case the robots of the second group may not reach the human and will organise themselves in a cluster as in Situation 2.1. Therefore we have the following Proposition.

Proposition 6.1 The robots with limited visibility domains will gather around the human, if at any time step the attraction force acting on the gravity CG^r does not vanish.

The above proposition is a sufficient condition. It means that a chain of robots should exist so that at each time step the group will be a connected graph. The vertices of this graph will be the robots, whereas edges occur only if the robots that represent their end-vertices belong to the visibility domains of each other. A similar type of graphs, called *connectivity graphs*, was considered in (Muhammad and Egerstedt 2003).

The situation described is depicted in Fig. 28.15. Robot 1 senses robot 2, which in turn senses robot 3. Robot 3 senses robot 4 which senses the fire fighter. The robots achieve an equilibrium situation around the fire fighter.

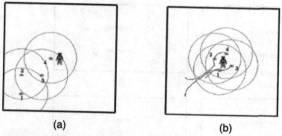

28.15 (a) and (b) A chain of robots with limited visibility domains.

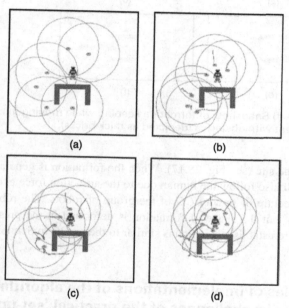

28.16 (a) – (d) Robots 'escape' the trap of the concave obstacle.

When Proposition 6.1 holds then some of the local minima can be eliminated. In Fig. 28.16 the robots escape the trap of C-concavity and gather around the fire fighter without knowing the positions of all other robots, contrary to the assumption in Fazenda and Lima (2007).

28.5.2 Formation keeping

In this subsection we show the result of one simulation when the human is moving. The robots first make a formation around the fire fighter when he/she is not close to the obstacle. Speed is important and we assume that the fire fighter moves slower than a robot. The formation is kept in the presence of obstacles while the

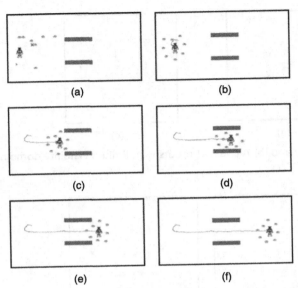

28.17 (a) – (f) Snapshots of formation keeping while moving in an environment with obstacles (depicted as thick bricks).

human moves on the site (see Fig. 28.17). When the formation is generated, the group of robots will also follow the human due to the attraction force exerted by the human on the centre of gravity CG of the group of robots. The robots still move in formation, but the shape of formation is deformed in the presence of obstacles and then re-established, which is similar to the results in Schneider and Wildermuth (2005).

28.6   Examples of implementations of the algorithms on real robots: challenges of the practical 'set-ups'

In previous sections, basic navigation behaviour patterns required from a (human-) multi-robot team in the GUARDIANS scenario have been presented, together with reasons for their feasibility. We tested our algorithms using Player/Stage software. Player is a network server for robot control, and is language and platform independent. Stage is its simulation backend. The most recent version of Stage allows the creation of a very realistic simulation environment. An example of a formation generation pattern (Situation 1.1) is given in Fig. 28.18.

The formation generation and keeping behaviours, by means of Player, were then implemented on real Erratics robots.

The algorithm for calculating corresponding forces acting on robots based on the social potential fields framework presented in section 28.4 (Formalism) has been adjusted to a 'discrete' timing. Discreteness here means that while the global

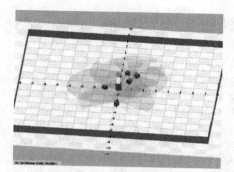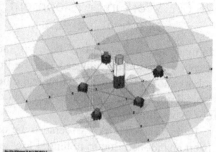

28.18 Snapshots of formation generation simulated in Stage: left – initial configuration, right – intermediate configuration.

time is, of course, continuous, the speed of the robots does not change continuously, but at certain intervals Δt_i of time. It gives more stability as robots move linearly for an (however small) interval of time. The maximum speed has been also determined in order to avoid collisions.

Robot–robot detection and consequent mutual distance estimation in a real life environment represent challenging issues, and are still under development (see below). Therefore, in order to test our algorithms the onboard computers of the robots were equipped with wi-fi, and a map of the environment was given to the robots. The map was not detailed, as obstacles inside the environment were not depicted. The map allowed robots to localise themselves by using the Adaptive Monte-Carlo localisation method. Java Agent Development Environment (JADE) was used for communicating between robots, which allows them to the estimate the positions of each other and hence, estimate the distances.

Snapshots of video of the aforementioned implementation are given in Fig. 28.19. One of the robots (with a flag) 'played' the role of fire fighter, and therefore was remotely controlled. Other robots were autonomous. The robots were also equipped with a Hokuyo laser range finder in order to avoid obstacles and other robots.

28.19 Robots follow the leading robot, 'fire fighter', while keeping the formation.

28.6.1 Work in development: approaches to practical 'set-ups'

In many real life situations the environment is unknown and no map is provided. Explicit communication can also be unavailable or severed. In such situations robots should be able to distinguish other robots from obstacles; in particular if a 'social potential fields'-based framework is applied, which requires estimation of corresponding distances.

In the GUARDIANS scenario the use of cameras on the robot's platform has been abandoned due to potential low visibility, and attention has been shifted to other sensors. The efforts have been firstly concentrated on the use of laser range finders (LRF) for robot–robot recognition and eventual distance estimation. The approach is similar to that in Howard, Matarić and Sukhatme (2002), where the laser range finder was used for distinguishing between the nodes of robots that carried a retro-reflective beacon and obstacles that did not. However the laser used had a 360-degree field-of-view (whereas the field-of-view of the Hokuyo LRF is 240 degrees) and the approach has been tested in a simulated environment. The ongoing work in GUARDIANS is on implementing a similar method on real robots.

For this purpose a special barcode is being designed, made of retro reflective material. The laser range finder will search for retro reflective targets and extract their relative range and bearing. To realise this task the laser range finder has to perform the reading of intensity values to detect and find the reflective material. A typical reflective material returns a high intensity reading value, which is clearly differentiable from the other intensity values that are reflected by the rest of the materials and colours of the environment. Equipping the robot(s) and the fire fighter(s) with a barcode made of a retro reflective material allows each robot not only to detect but also identify other robots and the fire fighter and estimate their relative position (range and bearing) within a certain range. The detection of obstacles is therefore straightforward, as obstacles will return lower intensity values than a robot or a fire fighter. This 'set-up' is under development.

However, in the case of the dense smoke, the LRF measurements become unreliable and other tools/techniques should be used. One possibility is to use a module similar to the relative positioning module developed in Pugh et al. (2009), which employs IR (infra-red) sensors for positioning as well as for communication. This approach has restrictions as by using an IR sensor-based positioning system, only very short distances can be reliably estimated. Another approach on which the consortium is currently working is the employment of ultrasound and microwave sensors.

28.7   Conclusions

We presented basic behaviours for a group of heterogeneous agents such as robot–robot and robot–fire fighter avoidance/attraction and robot–obstacle avoidance, achievable by using a social potential field-based approach. By exercising the basic behaviours, more complex behaviours emerge, namely, formation generation and formation keeping. The analysis given shows that the proposed social potential field method is a suitable technique for formation generation, obstacle avoidance, navigation, and is robust to failures of an individual robot and the team's size in complex environments. We plan to study the characteristics of formations in more complex scenarios such as dynamic environments, or environments with narrow corridors and doorways. Other shapes of visibility domains will be considered. In the present chapter the robots do not identify among themselves, but are capable of distinguishing another robot or the fire fighter from an obstacle. However it would be useful if the robots that 'sense' the fire fighter were able to 'communicate' this to other robots in their visibility domains. In this case the remaining robots will execute attraction forces only to the robots of the first group that could reduce occurrences of 'clustering' situations.

Possible practical 'set-ups' that would allow not only recognition of an agent of a certain class, but also its identification as an individual in a real-life scenario are currently under validation.

28.8 Acknowledgement

The authors would like to thank all the GUARDIANS project partners for their contribution and cooperation, especially Veysel Gazi for his valuable comments.

28.9 References

Alboul, L., J. Saez-Pons, and J. Penders, 'Mixed human-robot team navigation in the GUARDIANS project', In *Proceedings of IEEE International Workshop on Safety, Security and Rescue Robotics*, pp. 95–101, (2008).

Baldassare, G., S. Nolfi, and D. Parisi, 'Evolving mobile robots able to display collective behaviours', *Artificial Life*, Vol. 9, pp. 255–267, (2003).

Balch T. and M. Hybinette, 'Social potentials for scalable multi-robot formations', In *Proc. of the IEEE Conf. on Robotics and Automation (ICRA'00)*, Vol.1, 73–80, (2000).

Barnes, L., W. Alvis, M. Fields, K. Valavanis and W. Moreno, 'Heterogeneous swarm formation control with potential fields formed by bivariate normal functions', *International Transactions on Systems Science and Applications*, Vol. 2, No. 4, pp. 346–359, (2007).

Bruemmer, D., D. Dudenhoeffer and M. McKay, 'A robotic swarm for spill finding and perimeter formation'. In *Spectrum*, (2002).

Egerstedt, M. and X. Hu, 'Formation constrained multi-agent control', *ICRA*, 3961–3966, (2001).

Eisenstat, S. C., 'Efficient implementation of a class of preconditioned conjugate gradient methods', *SIAM Journal of Scientific and Statistical Computing*, Vol. 2, 1–4, (1981).

Fazenda, P. V.and U. Lima, 'Non-holonomic robot formations with obstacle compliant geometry', In *Proc. of IAV 2007*, Toulouse, France, (2007).

Gazi, V., 'Swarm aggregations using artificial potentials and sliding-mode control', *IEEE Transactions on Robotics*, Vol. 21, 1208–1214, (2005).

Gazi, V., and K. Passino, 'Stability analysis of swarms', *IEEE Trans. Automatic Control*, Vol. 4, 692–697, (2004a).

Gazi, V., and K. Passino, 'A class of attraction/repulsion functions for stable swarm aggregations', *International Journal Control*, Vol. 18, 1576–1579, (2004b).

Howard, M. J. Matarić, and G. S. Sukhatme, 'Mobile sensor network deployment using potential fields: a distributed, scalable solution to the area coverage problem', In *Proceedings of the 6th International Symposium on Distributed Autonomous Robotics Systems (DARS02)*, Fukuoka, Japan, June 25–27, (2002).

Jade, http://jade.tilab.com/papers-exp.htm

Khatib, O., 'Real-time obstacle avoidance for manipulators and mobile robots', *International Journal Robotics Research*, Vol. 5, 90–98, (1986).

Koren, Y., and J. Borenstein, 'Potential field methods and their inherent limitations for mobile robot navigation', In *Proc. of the IEEE Conf. on Robotics and Automation*, 1398–1404, (1991).

Krogh, B., 'A generalized potential field approach to obstacle avoidance control', In *SME conf. Proc. Robotics Research: The next five years and beyond*, 11–22, (1984).

Lawton, J. R. T. and R. W. Beard, 'A decentralized approach to formation manoeuvres', In *IEEE Transactions on Robotics and Automation*, Vol. 19(6), 933–941, (2003).

Lemay, M. et al., 'Autonomous initialization of robot formations', Vol. 3 of Robotics and Automation, 2004. *Proceedings. ICRA '04*, 3018–3023, (2004).

Muhammad, A. and M. Egerstedt, 'Topology and complexity of formations', In *Proceedings of the 2nd International Workshop on the Mathematics and Algorithms of Social Insects*, Atlanta, Georgia, USA, (2003).

NetLogo, http://ccl.northwestern.edu/netlogo/

Ögren, P., E. Fiorelli and N. E. Leonard, 'Cooperative control of mobile sensor networks: adaptive gradient climbing in a distributed environment', *IEEE Trans. on Automatic Control*, Vol. 49(8), 1292–1302, (2004).

Parker, L. E., 'Current state of the art in multi-robot teams', *Distributed Autonomous Robotic Systems*, Vol. 4, 3–12, (2000).

Penders, J., L. Alboul, and P. J. Brapenning, 'The interaction of congenial autonomous robots: Obstacle avoidance using artificial potential fields', In *Proc. ECAI-94*. John Wiley and Sons, 694–698, (1994).

Penders, J., L. Alboul, C. Roast, and E. Cervera, 'Robot swarming in the Guardians project', In *ECCS'07 Proc.*, Vol. 6, (2007).

Penders, J., L. Alboul, and M. Rodrigues, 'Modelling interaction patters and group behaviours in a three-robot team', In *Proceedings of Taros04, Colchester: Technical Report Series, CSM-415*, University of Essex, (2004).

Player, http://playerstage.cvs.sourceforge.net/viewvc/playerstage/papers/

Pugh, J., X. Raemy, C. Favre, R. Falconi and A. Martinoli, 'A fast on-board relative positioning module for multi-robot systems', *IEEE/ASME Transactions on Mechatronics, Focused Section on Mechatronics in Multi Robot Systems*, 2009

Reif, J.H. and H. Wang, 'Social potential fields: A distributed behavioural control for autonomous robots', *Robotics and Autonomous Systems*, Vol. 27, 171–195, (1999).

Saez-Pons, J., L. Alboul, J. Penders and G. Veysel, 'Non-communicative robot swarming in the Guardians project', In *Proceedings of the EURON/IARP International Workshop on Robotics for Risky Interventions and Surveillance of the Environment*. Benicàssim, Spain (2008).

Schneider, F. E. and D. Wildermuth, 'Experimental comparison of a directed and a non-directed potential field approach to formation navigation', In *Proc. 2005 IEEE Int. Symposium on Computational Intelligence in Robotics and Automation*, (2005).

Song, P., and V. Kumar, 'A potential field based approach to multi-robot manipulation', In *Proc. of the 2002 IEEE Int. Conference on Robotics and Automation*, (2002).

Tanner, H. G., G. J. Pappas and V. Kumar. 'Leader-to-formation stability', In *IEEE Transactions on Robotics and Automation*, Vol. 20(3), 443–455, (2004).

29

A light-weight communication protocol for tele-operated robots in risky emergency operations

U. DELPRATO, M. CRISTALDI, G. TUSA, Intelligence for Environment and Security – IES Solutions S.R.L, Italy

Abstract: This chapter describes a communication protocol and a software architecture designed for use in the end-to-end communication between the control station and tele-operated robots in risky emergency operations. The principles which form the basis of the design of this communication protocol are reported and discussed: first, the choice of the most suitable transport level protocol, through the explanation of the main differences between the user datagram protocol (UDP) and the transmission control protocol (TCP); then, the problems exposed by communication over the wireless medium, and the different requirements of the data types involved, which require the prioritization of data flows through the implementation of quality of service (QoS) mechanisms. The additional features provided by the proposed software architecture, which make it suitable for easy adoption in developing robotics applications, are also described.

Key words: communication protocols, user datagram protocol (UDP), transmission control protocol (TCP), quality of service (QoS), wireless.

29.1 Introduction

The use of advanced systems for handling emergency situations following natural or human-induced disaster, and for supporting human emergency workers in dangerous and hazardous environments, is one of the main challenges for the robotic research community. Such systems deal with the need to improve the speed and efficiency of emergency and rescue operations, and with extending the possibility of intervention including tasks that are considered too risky for direct human involvement, especially in the earlier stages, when there is not enough information available concerning the emergency scenario. One important task for the design of such systems is the choice of an efficient communication subsystem, especially for handling communication between the control station, where the rescue operators normally work and from where they can monitor and control the robots in the field, and the robots themselves.

This chapter presents a transport level protocol designed to deal with the problems and the requirements of such a communication subsystem, and its application to two different research platforms for tele-operated robots in risky emergency operations.

640

The design of the communication subsystem must take into account the different communication scenarios in a real work environment and, of course, the different data types involved in the communication. A first distinction is related to the inter-platform and intra-platform communication: the former mainly concerns the communication between a robot and a generic control station, where the data gathered by the robot are received and processed, while the latter relies, for example, on the communication between sensor data processing processes on board the robots, or between different processes on the control station side.

Moreover, different data types are involved in the communication between different components, such as sensor data from different sensor sources placed on board the robot, robot status messages, video streams used to monitor the emergency field, and emergency, control and navigation commands sent by the control station to the robot.

While the communication inside each main platform (control station, robot), is normally based on wired connections, which ensure enough resources for the different communication needs, the same does not apply to the wireless communication subsystem between the control station and the robot, where the available network resources, the allocation of resources for different flows, and the delays in the communication become critical due to the intrinsic characteristics of the wireless communication medium, the performances of which depend on many factors such as distance, visibility between end points, atmospheric conditions, and the number and type of data involved .

This exposes some important communication problems like higher loss probability, as well as increased and non-deterministic latency.

The Mailman protocol presented in this chapter and built on top of UDP (Information Sciences Institute University of Southern California, 1980) has been designed in order to provide the following main features:

- to take into account the issues described above, i.e. to make efficient use of wireless communication mediums, by means of its facilities for controlling the priority of different types of data involved in the communication;
- to provide the application level processes with a common layer interface, aimed at hiding the details of the lower levels processing to the user programs; this common interface can be used by the application level processes to easily set up the communication environment and address each other;
- to support both ordinary UDP transmission and reliable data transfer over UDP, upon application request.

The chapter is organized as follows. Section 29.2 concerns a discussion about the choice of the transport level protocol most suitable for the considered robotic systems, while section 29.3 and 29.4 provide a generic overview of the concepts and the functionality of the Mailman protocol and architecture. A description of the Mailman packet structure, together with an explanation of how the various fields can be used in operative scenarios, is provided in section 29.5, while section

29.6 reports on the adoption of the Mailman protocol within the framework of two different research projects for robotic applications in risky emergency scenarios. Finally, section 29.7 concludes the chapter and provides some additional ideas for future enhancement of the protocol.

29.2   Choice of the transport level protocol

The first important decision during the design of the Mailman protocol was the choice of the downstream transport level protocol to use. TCP (Information Sciences Institute University of Southern California, 1981) is a reliable, stream-based and connection-oriented protocol, while UDP is a connectionless, datagram based and unreliable transport level protocol, preferable when the speed of delivery of the data is more important than the reliability itself.

Thanks to the provided reliability, achieved by means of a complex retransmission algorithm and the maintenance of the state of the packets sent and received for each end-to-end connection, TCP has been for a long time now the Internet de-facto standard, adopted by most Internet applications (such as http and Telnet). However, a lot of TCP features make it less suitable for application in robotics systems for emergency operations (Gage, 1997).

First, TCP is a stream-oriented protocol, while the majority of data exchanged between components in robotic applications for rescue operations (e.g. robot status sent to the control station, or emergency and navigation commands sent from the control station to the robot) can be normally encoded as numbers and strings which are more easily sent and read in single message format, i.e. as datagrams.

In addition, TCP is not well suited to wireless environments in some circumstances, and was not designed for low latency and real-time traffic (Fotopoulos and Heaberline, 2002). To ensure reliability, TCP resends all data that is lost and, in order to deliver the data in the same order it is received, it needs to use a heavy buffering mechanism and to control each stream received, by adding an overhead and consuming a certain amount of resources that in many cases are unnecessary and unproductive. For streaming applications, such as video from cameras inside the robots and certain kinds of sensor data, packets that are either lost or arrive out of order should be discarded. The TCP behaviour, instead, adds further delay to new data and wastes precious bandwidth resources in the wireless environment. Moreover, being a connection-oriented protocol, it does not operate well in the face of frequent disconnections, which are common in wireless links and, by default, it is unable to recover the connection after a wireless link drop.

Conversely, it is a general consensus that UDP is the most appropriate protocol for communications requiring low latency, streaming data, connection resilience and low overhead. Being connectionless, UDP-based communication protocols are more robust to link variations or drops in the wireless environment.

Furthermore, it is also more efficient, as far as the network resources are concerned, to add reliability mechanisms at higher levels only for the kind of data that require them (Gage, 1997; Fotopoulos and Heaberline, 2002).

In addition, it is easier for the application level software to send (encode) and receive (decode) data in packet (datagram) format instead of adding a more complex encoding and decoding mechanism to obtain a clear demarcation of different messages, which are sent and received by TCP as a continuous stream of bytes.

29.3 Mailman overview

Mailman is a lightweight protocol, built on top of UDP, designed to minimize computing overhead in embedded robotic applications and for soft real-time communication. It is mainly concerned with layer 4 functionality, i.e. the transport of data received by the application layer. It aims to provide application level programs with the ability to send and receive data by means of a simple application programming interface (API) and, like some other protocols built on top of UDP (Schulzrinne et al., 2003), it provides the possibility of enhancing the traditional UDP protocol behaviour by adding additional information to the packets. This can be useful at application level for handling specific types of data. As an example, Mailman provides the capability of having reliable communication over UDP, upon the users' application request, as well as the capability to handle fragmented packets (i.e. to fragment at the sender side and reassemble at the receiver side).

Table 29.1 below illustrates the placement of Mailman with respect to the OSI and TCP/IP network reference model.

Table 29.1 OSI, TCP/IP network reference models and Mailman placement for robotic applications

Layer	OSI	TCP/IP	Robotic applications using Mailman
7	Application		Applications (e.g. sensors and other processes), additional middleware
6	Presentation	Application	
5	Session		
4	Transport	TCP and UDP	Mailman, TCP/UDP Sockets API
3	Network	IPv4, IPv6	OS IPv4 stack
2	Datalink	Device driver and	Wireless networking
1	Physical	hardware	hardware and drivers

Mailman and the Mailman library are implemented in native C language, and each process in the system can use the services provided by Mailman by simply linking the library, both in C and C++ environment. Mailman runs directly over the sockets API, using the low level UDP/IP protocol.

Further to basic and enhanced layer 4 functionality, however, Mailman offers some additional higher-level services, together with additional features suitable for an efficient handling of the network:

- Automatic setup and management of connections.
- Automatic Mailman server discovery, which is meant to ease the work of the programmers, allowing a process inside a given local area network (LAN) (e.g. robot LAN or control station LAN), to automatically establish a connection with the Mailman daemon of its LAN, being able to receive/send data from/to remote peers without having to pay attention to the IP address of the server or its peers; this also means that the network configuration can change without affecting the development and deployment of a client program.
- Grouping, which allows each process to send its data to all the interested processes in the system, and to receive the data they are interested in by different sources, by means of a group subscription mechanism, similar to the concept of multicast groups.
- Prioritization of messages. One of the main Mailman design goals has been to have a low-overhead, efficient protocol to deal with the particular nature and problems concerned with the use of the wireless communication in robotic applications for emergency operations, especially as far as the different quality of service (QoS) requirements of the different types of data are concerned. Packet scheduling is handled through priorities, and prioritization is obtained by using different priority queues, which are served accordingly by the Mailman scheduler thread. It means that critical data gets through first and that real-time data does not suffer unnecessarily long delays in network queues. Emergency messages, for example, are treated and delivered as such, with the maximum priority level. For the types of data that are sensitive to delay variations, such as video streaming, soft real-time support is provided: video frames are scheduled for delivery within an agreed maximum delay, or otherwise discarded.

29.4   Network architecture and software

From the network architecture point of view, a typical application for tele-operated robots in emergency risky operations, which uses the Mailman protocol for the communication between the robot and the control station, can be represented as depicted in Fig. 29.1. Usually, both the robot and the control station platform have a wired LAN bus attaching one or more physical computers. The two local networks are connected to each other by a point-to-point wireless link and form a wide area network (WAN).

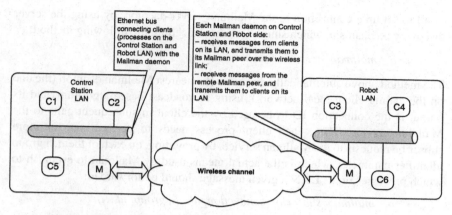

29.1 Network architecture.

From the software point of view, such a system is composed of numerous and independent software processes executing in one or more computers, and these processes are divided among subsystems on the robot and the control station sides.

From the Mailman software architecture point of view, processes on one LAN that require the Mailman services to communicate with processes on the other one, are considered as client processes.

The main components of Mailman software architecture are therefore:

- A Mailman server daemon residing on each LAN and acting as a gateway to the WAN.
- A client-to-server protocol, which allows each client to communicate with the server on its LAN.
- A server-to-server (peer-to-peer) protocol enabling two Mailman daemons to speak to each other.
- A reference implementation of the client/server protocol in the form of a program library.

Clients on one LAN can communicate with clients on the other LAN by sending messages to their nearest Mailman daemon, according to the client/server protocol. The daemon will transmit these messages across a wireless network interface to another Mailman daemon that will deliver the messages to the intended clients. The Mailman daemons act as gateways and manage the communication over the wireless channel, providing process addressing and message routing. Before a process can communicate with one or more processes it needs to find out their IP addresses and port numbers and also make its own address known. Mailman handles management of connections and addressing in a transparent fashion as far as the application level processes are concerned. In order to make use of the Mailman service (i.e. send and receive datagrams), a client application needs first of all to set up a connection to the Mailman server daemon, by using the server discovery mechanism, which simply consists of calling the following method:

> o   *mailman_connect(&connection_id)*

The method above automatically discovers the nearest Mailman daemon (the one on the calling client local network), using a broadcast based algorithm, and its output is the connection ID to be used by the client in subsequent calls to the Mailman service. Then, each client process needs to perform one or more subscribe requests to the Mailman service, by providing, on each of them, a group identifier parameter, which is the actual means used by Mailman to establish to which processes on its LAN a given message should be forwarded:

> o   *mailman_subscribe(connection_id, …, group_id …)*

Only processes subscribed to the same group ID to which a given message was sent can receive the datagram.

In order to send their data over the network, client processes need to use the following function calls:

> o   *mailman_send(…)*
> o   *mailman_recv(…)*

Data are sent and received as datagrams, with the payload being encoded as byte streams.

29.5   Mailman datagram

Figure 29.2 illustrates the structure of the Mailman datagram, composed by a header of 12 bytes, and a variable size payload.

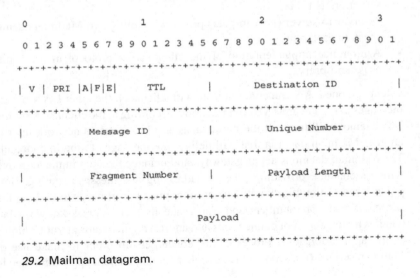

29.2 Mailman datagram.

Below is an explanation of the Mailman header fields, and how they are used for the purposes of different kind of applications:

- Version Number (V): 2 bits. This indicates the version of the Mailman protocol.
- Priority (PRI): 3 bits. Indicates the priority of the outgoing datagram. The priority values range is 0–7, with 0 representing the higher priority level.
- Ack bit (A): 1 bit. Used to enable reliable transport of data over UDP. If set, it indicates that the sending application requires the current datagram to be acknowledged. The ack message received back will indicate the sequence number of the message being acknowledged in the payload.
- Fragmentation bit (F): 1 bit. If set, indicates that the current datagram is a fragment of a bigger data packet. Used by applications that need to encode their information in large packets, such as the map plots sent by a robot and reconstructed in the control station. These data cannot be sent within a single datagram due to the limit of the maximum UDP datagram size, therefore they need to be fragmented and reassembled at the receiver side.
- End fragment bit (E): 1 bit. If set, indicates that the current datagram is the last fragment of a bigger data packet.
- Time To Live (TTL): 1 byte. Used to enable soft real-time handling of data sensitive to delay variation (e.g. video streaming). Indicates the number of delivery schedules for which a message remains valid before it expires. A value of 0 means that the message will not expire. For example, if the server schedules delivery every 10 milliseconds, then a TTL value of 50 means that the datagram should be delivered within 500 milliseconds of being received by the server. If it is not possible to make the delivery within that time frame, the datagram will be discarded. Valid values are 0–255.
- Destination ID: 2 bytes. This is the field used by the Mailman daemon to implement routing and addressing functionality. Indicates the group ID to which the datagram should be delivered, therefore only remote clients subscribing to that group ID will receive the datagram.
- Message ID: 2 bytes. This is a counter used to identify each datagram and can be used for:
  - identifying messages for the purpose of acknowledging them;
  - discarding duplicate messages.
  The protocol by itself does not acknowledge messages or discard duplicates. It is the duty of the client application to request support and provide these functions if it is needed. The purpose of this field is to make that task easier.
- Unique Number: 2 bytes. Sequence number used to identify a given sequence of fragments. It must be the same for each fragment belonging to the same fragmented packet. In general, it is different from the message ID value, which is an absolute sequence number used to identify each packet handled by Mailman.

- Fragment Number: 2 bytes. Sequence number used to identify a fragment inside a given sequence of fragments.

29.6   Mailman adoption in rescuer and ViewFinder projects

The Mailman protocol has been designed and developed within the framework of two European projects, Rescuer (http://www.rescuer-ist.net/) and ViewFinder (http://www.view-finder-project.eu/). Both ViewFinder and Rescuer are aimed at integrating mobile robots into a command and control network.

Rescuer was specifically focused on the use of an intelligent mechatronic system to support human emergency experts in tasks that are considered too risky for human involvement, such as bomb disposal and other hazardous operations, due to toxic contamination of the environment. The architecture adopted in the Rescuer project is composed of two main subsystems, the mobile control unit (MCU), where human emergency and rescue workers are expected to operate and remotely control the mobile mechatronic unit (MMU), a robot unit equipped with a certain number of sensors, a stereo head for acquiring and transmitting the video of the emergency scenario to the MCU, and two complete robot arms with dexterous grippers capable of managing objects and performing bomb disposal, driven by the operators on the MCU side.

The ViewFinder project was specifically focused on the development of robots, aimed at gathering data from hazardous and possibly contaminated environments, to establish whether the area could be entered safely by human emergency workers. The robots involved are equipped with numerous sensors, used for navigation, localization, map building and scene reconstruction, and they include chemical sensors for the detection of chemical and toxic agents. The robots are also equipped with cameras, used in parallel for collecting images to be forwarded, together with the other information collected by the sensors, to an advanced base station, where human emergency workers can examine the information and send navigation instructions to the robots.

Despite the differences in the system architecture and components used in the two projects, a lot of important common requirements and features were identified for the design of an efficient communication subsystem between the robots and the control station, which led to the design and adoption of the Mailman protocol. They are explained in the following subsections.

29.6.1   Resource consumption

Resources are very important and need to be carefully used in embedded computing systems. The choice of a simple and low-overhead protocol such as Mailman ensures an efficient use of the network, as well as low utilization of both CPU and memory resources.

29.6.2 Simplicity and ease of use during the development of other processes

The use of the services provided by Mailman for application level programs is as simple as linking a library, and calling a few methods for the set up of the addressing environment, and to send and receive data.

29.6.3 Communication requirements

On both projects the robots are connected wirelessly to the control station.

The characteristics of the wireless communication imply a lot of problems and aspects, which need to be considered in the design of the communication subsystems for such applications, as explained previously in this chapter.

Mailman takes advantage of the UDP to keep latency low and to avoid the overhead of resending lost datagrams when it is not useful. It is able to manage the QoS by using different priority queues for the received datagrams, and scheduling the outgoing datagrams according to the agreed priority. Moreover, by controlling the maximum time-to-live of flows which require it, it keeps the delay variance smaller than a maximum agreed threshold and avoids sending datagrams whose transmission would not be useful anymore (e.g., to maintain the smoothness of a video streaming, or sending status information to the control station that no longer represents the current situation, and could be incorrectly interpreted).

In addition, as far as the reliability of the communication is concerned, while a high percentage of traffic (video and most of the sensor data) do not need and should not use a reliable transport protocol, there are other types of datagrams for which the sender must be sure of their receipt. Mailman implements a reliable transport protocol over UDP, which applications can request as needed. This way, the default usage avoids all the overheads of TCP but still provides the benefit of a reliable transport protocol when needed.

Table 29.2 below shows the priority assignments and reliability requirements for different data flows in the context of the ViewFinder project.

Table 29.2 Priority assignments and reliability requirements in the ViewFinder project

Data type	Group ID	Priority level	Reliability
Emergency commands	MAILMAN_GID_EMERGENCY_TC	0	Reliable
Standard TCs sent to the robot	MAILMAN_GID_STD_TC	2	Unreliable
Standard TM from sensors (e.g. temperature)	MAILMAN_GID_LIGHT_TM	1	Unreliable
Video from camera	MAILMAN_GID_STEREO_HEAD_TM	5	Unreliable
Chemical sensors	MAILMAN_GID_CHSENSOR_TM	3	Reliable
SLAM (maps)	MAILMAN_GID_SLAM_TM	4	Reliable

Priorities are associated to group IDs, and each group ID matches a specific kind of data exchanged between processes subscribed to that group ID.

Looking at Table 29.2, it can be seen that the maximum priority level has been assigned to the emergency commands, for which a reliable communication is also enabled.

Conversely, reliability of communication is not required for what concerns standard tele commands (TCs), sent from the control station to the robot.

Reliability is required for the data acquired by the SLAM procedure to build and send the map of the scene at a given frequency to the control station, as well as for 3D map data. Maps are built and encoded in a single buffer of data, but then are sent to the control station as a sequence of smaller fragments, due to the limit imposed by the underlying UDP stack, and reassembled at the destination. Reliability in the transmission of the fragments is required for a correct reconstruction of the map at the control station side.

Standard telemetry (TM) flows have assigned the second priority level, in order to reach the base station in a timely manner, while the lower priority level has been assigned to video streaming, for which the most important requirement is smoothness, controlled by the time-to-live value placed by the sending application on each outgoing packet.

29.7   Future trends

In this chapter the design principles and functionality of Mailman, a UDP-based lightweight protocol, aimed at efficiently handling the wireless communication between remote end-points in robotic applications for risky emergency operations, has been presented. The current development of the protocol has been adopted in two different research projects (currently the ViewFinder project is still running and work on the Mailman protocol is continuing).

Possible improvements of the Mailman protocol include:

- Enhancing the peer-to-peer communication mechanism in order to allow a one-to-multi Mailman daemons communication; this would allow a control station to communicate with more than one robot unit simultaneously.
- Investigating the possibility of also using the Mailman protocol for internal communication between processes in the same platform (i.e. control station or robot). The advantage would be to simplify the design and implementation of the whole communication architecture, with the drawback of having a certain performance drop as far as the Mailman protocol speed is concerned.
- Enhancement of the priority management. The scheduling mechanism, currently based on a strict prioritization approach, could be improved by providing the opportunity to choose between different QoS scheduling mechanisms.

29.8   References

Fotopoulos V. and Heaberline C. "Reliable UDP (RDP) Transport for CORBA". OMG Embedded and Real-Time Workshop (2002).

Gage D. W., Space and Naval Warfare Systems Center San Diego. "Network Protocols for Mobile Robot Systems". Mobile robots Conference No12, Pittsburgh PA, ETATS-UNIS: vol. 3210, pp. 107–118, (1997).

Information Sciences Institute, University of Southern California. "User Datagram Protocol (UDP)", http://tools.ietf.org/html/rfc768 (1980).

Information Sciences Institute, University of Southern California. "Transmission Control Protocol (TCP)", http://tools.ietf.org/html/rfc793 (1981).

Schulzrinne H., Casner S., Frederick R. and Jacobson V. "Real Time Protocol", http://www.ietf.org/rfc/rfc3550.txt (2003).

Index

Tables, figures and illustrations are indicated in bold.

Printed in the United States
By Bookmasters